Essentials of
ECOLOGY

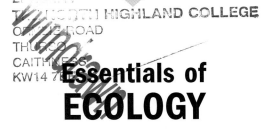

Colin R. Townsend

Department of Zoology
University of Otago
Dunedin, New Zealand

John L. Harper

Professor Emeritus in the University of Wales
Visiting Professor in the University of Exeter
Exeter, England

Michael Begon

Population Biology Research Group
School of Biological Sciences
The University of Liverpool
Liverpool, England

with assistance from Leigh Vanderklein
special assistance provided by Douglas Reed

Essentials of
ECOLOGY

Blackwell
Science

© 2000 by Blackwell Science, Inc.

Editorial Offices:

Commerce Place, 350 Main Street, Malden, Massachusetts 02148, USA
Osney Mead, Oxford OX2 0EL, England
25 John Street, London WC1N 2BL, England
23 Ainslie Place, Edinburgh EH3 6AJ, Scotland
54 University Street, Carlton, Victoria 3053, Australia

Other Editorial Offices:

Blackwell Wissenschafts-Verlag GmbH, Kurfürstendamm 57, 10707 Berlin, Germany
Blackwell Science KK, MG Kodenmacho Building, 7-10 Kodenmacho Nihombashi, Chuo-ku, Tokyo 104, Japan

Distributors:

USA
Blackwell Science, Inc.
Commerce Place
350 Main Street
Malden, Massachusetts 02148
(Telephone orders: 800-215-1000 or 781-388-8250; fax orders: 781-388-8270)

Canada
Login Brothers Book Company
324 Saulteaux Crescent
Winnipeg, Manitoba, R3J 3T2
(Telephone orders: 204-224-4068)

Australia
Blackwell Science Pty, Ltd.
54 University Street
Carlton, Victoria 3053
(Telephone orders: 03-9347-0300;
fax orders: 03-9349-3016)

Outside North America and Australia
Blackwell Science, Ltd.
c/o Marston Book Services, Ltd.
P.O. Box 269
Abingdon
Oxon OX14 4YN
England
(Telephone orders: 44-01235-465500;
fax orders: 44-01235-465555)

Acquisitions: Nancy Hill-Whilton
Development: Jill Connor
Production: Irene Herlihy
Manufacturing: Lisa Flanagan
Cover design: Meral Dabcovich, Visual Perspectives
Cover illustration: Landscape with Monkeys, ca. 1910. Rousseau, Henri "Le Douanier" (1844–1910), French—Barnes Foundation, Merion, Pennsylvania
Interior design: Janet Bollow Associates
Typeset by: Modern Graphics, Inc.
Printed and bound by: World Color

Printed in the United States of America
00 01 02 03 5 4 3 2 1

This publication includes images from Corel Stock Photos, which are protected by the copyright laws of the U.S., Canada and elsewhere. Used under license.

Library of Congress Cataloging-in-Publication Data

Townsend, Colin R.
    Essentials of ecology / by Colin R. Townsend, John L. Harper, and
  Michael Begon.
      p.   cm.
    Includes bibliographical references (p.).
    ISBN 0-632-04348-2
    1. Ecology   I. Harper, John L.   II. Begon, Michael.   III. Title.
  QH541.T66   2000
  577—dc21   99-34066
                                                                 CIP

For photo credits, see page 553

Brief Contents

Table of Contents

Part II   Conditions and Resources   77

Part III   Individuals, Populations, Communities, and Ecosystems   163

Chapter 5   Birth, Death, and Movement   165

Chapter 6   Interspecific Competition   203

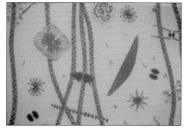

Part IV   Applied Issues in Ecology 419

Preface

By writing this book we hope to share with you some of our wonder at the complexity of nature—the Rousseau painting on the front cover portrays the beauty of the natural world. But we are also familiar with a darker side: the shadows in Rousseau's jungle. Citizens need to be ecologically literate, so that they can take part in political debate and contribute to solving the ecological problems that we carry into the new millennium. We hope our book will contribute to this objective.

The genesis of this book can be found in the more comprehensive treatment of ecology in our big book *Ecology: Individuals, Populations and Communities* (Begon, Harper & Townsend—3rd edition, 1996). This is used as an advanced university text around the world but many of our colleagues have called for a more succinct treatment of the essence of the subject. One recent reviewer of the "big book" wondered whether one would need a wheelbarrow to carry around the next edition! Thus, we have been spurred into action to produce a distinctively different book, written with clear objectives for a different audience—those taking a semester-long beginning course in the essentials of ecology. We hope that some readers will be excited enough to go on to sample the big book and the rich literature of ecology that it can lead into.

Some readers will be engaged most by the fundamental principles of how ecological systems work. Others will be impatient to focus on the ecological problems caused by human activities. We place heavy emphasis on both fundamental and applied aspects of ecology: there is no clear boundary between the two. However, we have chosen to deal first in a systematic way with the fundamental side of the subject and we have done this for a particular reason. An understanding of the scope of the problems facing us (the unsustainable use of ecological resources, pollution, extinctions, and the erosion of natural biodiversity) and the means to counter and solve these problems depend absolutely on a proper grasp of ecological fundamentals.

The book is divided into four sections. In the introduction we deal with two foundations for the subject that are often neglected in texts. Chapter 1 aims to show not only what ecology is but also how ecologists do it—how ecological understanding is achieved, what we understand (and, just as important, what we do not yet understand), and how our understanding helps us predict and manage. We then introduce "The Ecology of Evolution" and show that ecologists need a full understanding of the evolutionary biologist's discipline in order to make sense of patterns and processes in nature (Chapter 2).

What makes an environment habitable for particular species is that they can tolerate the physicochemical conditions there and find in it their essential resources. In the second section we deal with conditions and resources, both as they influence individual species (Chapter 3) and in terms of their consequences for the composition and distribution of multispecies communities, for example in deserts, rain forests, rivers, lakes, and oceans (Chapter 4).

The third section (Chapters 5–11) deals systematically with the ecology of individual organisms, populations of a single species, communities consisting of many populations, and ecosystems (where we focus on the fluxes of energy and matter between and within communities). To understand patterns and processes at each of these levels we need to know the behavior of the level below.

Finally, armed with knowledge and understanding of the fundamentals, the book turns to the applied questions of how to deal with pests and manage resources sustainably (whether wild populations of fish or agricultural monocultures) (Chapter 12), then to a diversity of pollution problems ranging from local enrichment of a lake by sewage to global climate change associated with use of fossil fuels (Chapter 13); lastly we develop an armory of approaches that may help us to save endangered species from extinction and conserve some of the biodiversity of nature for our descendants (Chapter 14).

A number of pedagogical features have been included to help you:

- Each chapter begins with a set of key concepts, which you should understand before proceeding to the next chapter.
- Marginal headings provide signposts of where you are on your journey through each chapter—these will also be useful review aids.
- Unanswered questions in ecology are highlighted—we believe passionately that the awareness of ignorance is the beginning of knowledge. Ecologists do not know all the answers.
- Each chapter concludes with a summary and a set of review questions, some of which are designated challenge questions.
- You will also find three categories of boxed text:
  1. "History boxes" emphasize some landmarks in the development of ecology.
  2. "Quantitative boxes" set aside mathematical and quantitative aspects of ecology so they do not unduly interfere with the flow of the text and you can consider them at leisure.
  3. "Topical ECOncerns" are boxes that highlight some of the applied problems in ecology, particularly those where there is a social or political dimension (as there often is). In these, you will be challenged to consider some ethical questions related to the knowledge you are gaining.

An important further feature of the book is the companion Internet website, accessed through www.blackwellscience.com. This provides an easy-to-use range of resources to aid study and enhance the content of the book. A particularly novel feature is a set of Web Research Questions that appear at the end of the four sections of the book. These explore current ecological issues in depth by exploiting the Internet as an important new research medium. Each question is repeated on the website, accompanied by a carefully chosen set of links to other relevant sites to guide the student in their own research. Other features of the website include self-assessment, multiple-choice questions for each chapter in the book and an interactive tutorial to help students understand the use of mathematical modeling in ecology.

Acknowledgments

It is a pleasure to record our gratitude to the people who helped with the planning and writing of this book. We thank Bob Campbell and Simon Rallison for getting the original enterprise off the ground and Nancy Hill-Whilton and Irene Herlihy for ably managing the project. We are also grateful to the following colleagues who provided insightful reviews of early drafts of one or more chapters: Tim Mousseau, University of South Carolina; Vickie Backus, Middlebury College; Kevin Dixon, Arizona State University, West; James Maki, Marquette University; George Middendorf, Howard University; William Ambrose, Bates College; Don Hall, Michigan State University; Clayton Penniman, Central Connecticut State University; David Tonkyn, Clemson University; Sara Lindsay, Scripps Institute of Oceanography; Saran Twombly, University of Rhode Island; Katie O'Reilly, University of Portland; Catherine Toft, UC Davis; Bruce Grant, Widener University; Mark Davis, Macalester College; Paul Mitchell, Staffordshire U. (UK); William Kirk, Keele U. (UK).

Last but not least, we are glad to thank our wives and families for supporting us, listening to us, and ignoring us precisely as was required—thanks to Laurel and Dominic, to Borgny, and to Linda, Jessica, and Robert.

The publisher and the authors gratefully acknowledge the creative and scholarly input of Douglas Reed. He was instrumental in pulling together important parts of the website package.

Introduction

Nowadays, Ecology is a subject about which almost everyone has heard and most people consider to be important—even when they are unsure about the exact meaning of the term. There can be no doubt that it *is* important; but this makes it all the more critical that we understand what it is and how to do it.

1

Ecology and How To Do It

Key Concepts

In this chapter you will

- know how to define ecology and appreciate its development as both an applied and a pure science.

- recognize that ecologists seek to describe and understand, and on the basis of their understanding to predict, manage, and control.

- appreciate that ecological phenomena occur on a variety of spatial and temporal scales, and that patterns may be evident only at particular scales.

- recognize that ecological evidence and understanding can be obtained by means of observations, field and laboratory experiments, and mathematical models.

- understand that ecology relies on scientific evidence (and the application of statistics).

1.1 Introduction

The question, What is ecology? could be translated into, How do we define ecology? and answered by examining various definitions that have been proposed, and even choosing one of them as the best (Box 1.1). But while definitions have conciseness and precision, and are good at preparing you for an examination, they are not so good at capturing the flavor, the interest, or the excitement of ecology. There is a lot to be gained by replacing that single question about definition with a series of more provocative ones: What do ecologists *do*? What are ecologists *interested* in? and Where did ecology emerge from in the first place?

the earliest ecologists   Ecology can lay claim to be the oldest science. If, as our preferred definition has it, Ecology is the scientific study of the distribution and abundance of organisms and the interactions that determine distribution and abundance (Box 1.1), then the most primitive humans must have been ecologists of sorts—driven by the need to understand where and when not only their food but also their (nonhuman) enemies were to be found—and the earliest agriculturalists needed to be even more sophisticated: having to know how to manage their living but domesticated sources of food.

Box 1.1

Definitions of Ecology

Ecology (originally in German: *Oekologie*) was first defined in 1866 by Ernst Haeckel, an enthusiastic and influential disciple of Charles Darwin—and an enthusiastic definer of biological terms. To him, ecology was "the comprehensive science of the relationship of the organism to the environment." The spirit of this definition is very clear in an early discussion of biological subdisciplines by Burdon-Sanderson (1893), in which ecology is "the science which concerns itself with the external relations of plants and animals to each other and to the past and present conditions of their existence," as contrasted with physiology (internal relations) and morphology (structure). For many, such definitions have stood the test of time. Thus, Ricklefs (1973) in his textbook defines ecology as "the study of the natural environment, particularly the interrelationships between organisms and their surroundings."

In the years after Haeckel, the fields of plant ecology and animal ecology drifted apart. Influential works defined ecology as "those relations of *plants*, with their surroundings and with one another, which depend directly upon differences of habitat among plants" (Tansley, 1904), or as the science "chiefly concerned with what may be

called the sociology and economics of *animals*, rather than with the structural and other adaptations possessed by them" (Elton, 1927). Botanists and zoologists, however, have long since agreed that they belong together and that their differences must be reconciled.

There is, nonetheless, something disturbingly vague about the many definitions of ecology that seem to suggest that it consists of all those aspects of biology that are neither physiology nor morphology—especially when nobody seriously believes this to be true. In search of more focus, therefore, Andrewartha (1961) defined ecology as "the scientific study of the distribution and abundance of organisms," and Krebs (1972), regretting that the central role of "relationships" had been lost in this interpretation, modified it to "the scientific study of the *interactions* that determine the distribution and abundance of organisms," explaining that ecology was concerned with "*where* organisms are found, *how many* occur there, and *why*." This being so, it might be better still to define ecology as "the scientific study of the distribution and abundance of organisms and the interactions that determine distribution and abundance."

These early ecologists, then, were *applied* ecologists, seeking to understand the distribution and abundance of organisms in order to apply that knowledge for their own collective benefit. They were interested in many of the sorts of things that applied ecologists are still interested in: how to maximize the rate at which food is collected from natural environments and how this can be done repeatedly over time; how domesticated plants and animals can best be planted or stocked so as to maximize rates of return; how food organisms can be protected from their own natural enemies; how to control the populations of pathogens and parasites that live on us.

In the last century or so, however, since ecologists have been self-conscious enough to give themselves a name, ecology has consistently covered not only applied but also fundamental, "pure" science. A. G. Tansley was one of the founders of ecology (Figure 1.1a). He was concerned especially to understand, for understanding's sake, the processes responsible for determining the structure and composition of different plant communities. When, in 1904, he wrote from Britain about "the problems of ecology," he was particularly worried by a tendency for too much ecology to remain at the descriptive and unsystematic stage (i.e., accumulating descriptions of communities without knowing whether they were typical, temporary, or whatever), too rarely moving on to experimental, or systematically planned, or what we might call a scientific analysis.

His worries were echoed across the Atlantic in the United States by another of ecology's founders, F. E. Clements (Figure 1.1b), who in 1905 in his *Research Methods in Ecology* complained:

> The bane of the recent development popularly known as ecology has been a widespread feeling that anyone can do ecological work, regardless of preparation. There is nothing in modern botany more erroneous than this feeling.

On the other hand, the need in applied biology *for* ecology and the contribution that applied biology can make *to* ecology were clear in the introduction to Charles Elton's (1927) *Animal Ecology* (see Figure 1.1c):

> Ecology is destined for a great future. . . . The tropical entomologist or mycologist or weed-controller will only be fulfilling his functions properly if he is first and foremost an ecologist.

In the intervening years, the coexistence of these pure and applied threads has been maintained and built upon. Many applied areas have contributed to the development of ecology and have seen their own development enhanced by ecological ideas and approaches. All aspects of food and fiber gathering, production, and protection have been involved: plant ecophysiology, soil maintenance, forestry, grassland composition and management, food storage, fisheries, control of pests and pathogens. Each of these classic areas is still at the forefront of lots of good ecology, and each has been joined by others. Biological control of pests (the use of pests' natural enemies to control them) has a history going back at least to the ancient Chinese but has seen a resurgence of ecological interest since the shortcomings of chemical pesticides began to be widely apparent in the 1950s. The ecology of pollution has been a growing concern from around the same time, and in the 1980s and 1990s focus

a pure and applied science

(a)

(b)

(c)

Figure 1.1
Three of the great founders of ecology: Arthur George
Tansley (a), Frederick E. Clements (b),
Charles Elton (c).

widened from local to global issues. The closing decades of the millennium also saw expansions both in public interest and in awareness of ecological aspects of the conservation of endangered species and the biodiversity of whole areas, of the control of disease in humans as well as many other species, and of the potential consequences of profound alterations to the global environment.

And yet, at the same time, many fundamental problems of ecology remain unanswered. To what extent does competition for food determine which species can coexist in a habitat? What role does disease play in the dynamics of populations? Why are there more species in the tropics than at the poles? What is the relationship between soil productivity and plant community structure? Why are some species more vulnerable to extinction than others? And so on. Of course, unanswered questions—if they are *focused* questions—are a symptom of the health not the weakness of any science. But ecology is not an easy science, and it has particular subtlety and complexity, in part because it has the distinction of being peculiarly confronted with uniqueness: millions of different species, countless billions of genetically distinct individuals, all living and interacting in a varied and ever-changing world. The beauty of ecology is that it challenges us to develop an understanding of very basic and apparent problems, in a way that recognizes the uniqueness and complexity of all aspects of nature but seeks to find patterns and predictions within this complexity rather than being swamped by it.

unanswered questions

Summarizing this brief historical overview, it is clear that ecologists try to do a number of different things. First and foremost ecology is a science, and ecologists therefore try to *explain* and *understand*. There are two different classes of explanation in biology: proximate and ultimate. For example, the present distribution and abundance of a particular species of bird may be "explained" in terms of the physical environment that the bird tolerates, the food that it eats, and the parasites and predators that attack it. This is a *proximate* explanation—an explanation in terms of what is going on here and now. However, we may also ask how this species of bird has come to have these properties that now appear to govern its life. This question has to be answered by an explanation in evolutionary terms: the *ultimate* explanation of the present distribution and abundance of this bird lies in the ecological experiences of its ancestors (see Chapter 2).

understanding, description, prediction, and control

In order to understand something, of course, we must first have a description of whatever it is we wish to understand. Ecologists must therefore *describe* before they explain. On the other hand, the most valuable descriptions are those whose focus is on a particular problem or need for comprehension. Undirected description, carried out merely for its own sake, is often found afterwards to have selected the wrong things to describe and has little place in ecology—or any other science. (We might, for example, describe the changing abundance of a population and the numbers of males and females, only to discover subsequently that sex ratio is irrelevant but age structure [which we ignored] crucial.)

Ecologists also often try to *predict* what will happen to a population of organisms under a particular set of circumstances and on the basis of these predictions try to *control* or exploit them. We try to minimize the effects of locust plagues by predicting when they are likely to occur and taking appropriate action. We try to exploit crops most effectively by predicting when conditions will be favorable to them and unfavorable to their enemies. We try to preserve rare species by predicting the conservation policy that will enable us to do so. Some prediction and control can be carried out without deep explanation or understanding: there is no great difficulty in predicting that if a woodland is destroyed, all the woodland birds that live there will disappear from the area—and that their disappearance can be prevented by saving the woodland. But insightful predictions, precise predictions, and predictions of what will happen in unusual circumstances can be made only when we can also explain and understand what is going on.

This book is therefore about

1. how ecological understanding is achieved.
2. what we do understand (but also what we do not understand—indeed, throughout the book, numbers of unanswered questions have been highlighted in the marginal notes).
3. how that understanding can help us predict, manage, and control.

1.2 > Scales, Diversity, and Rigor

The rest of this chapter is about the two "hows" described here: how understanding is achieved and how that understanding can help us predict, manage, and control. Later in the chapter we illustrate three fundamental points about doing ecology by examining a limited number of examples in some detail (Section 1.3). But first we elaborate on three points:

- Ecological phenomena occur at a variety of scales.
- Ecological evidence comes from a variety of different sources.
- Ecology relies on true scientific evidence and the application of statistics.

1.2.1 Questions of scale

Ecology operates at different scales: time scales, spatial scales, and biological scales. It is important to appreciate the breadth of these scales and how they relate to one another.

the biological scale

The living world is often said to comprise a biological hierarchy that begins with subcellular particles and continues through cells, tissues, and organs. Ecology then deals with the next three levels:

- Individual organisms
- Populations (consisting of individuals of the same species)
- Communities (consisting of a greater or lesser number of populations)

At the level of the *organism*, ecology deals with the way individuals are affected by (and how they affect) their environment. At the level of *population*, ecology deals with the presence or absence of particular species, with their abundance or rarity, and with the trends and fluctuations in their numbers. *Community* ecology then deals with the composition or structure of ecological communities.

We can also focus on the pathways followed by energy and matter as they move among living and nonliving elements of a fourth category of organization:

▪ Ecosystems (comprising the community, together with its physical environment)

With this level of organization in mind, Likens and Bormann (1995) would extend our preferred definition of ecology (Box 1.1) to include "the interactions between organisms and the transformation and flux of energy and matter." However, we take energy/matter transformations as being subsumed in the "interactions" of our definition.

Within the living world, there is no arena so small nor so large that it does not have an ecology. Even the popular press talks increasingly about the "global ecosystem," and, although we may have doubts about the level of understanding of some media commentators, there is no question that several ecological problems can only be examined at this very large scale. These include the relationships between ocean currents and fisheries, or between climate patterns and the distribution of deserts and tropical rain forests, or between elevated carbon dioxide levels in the atmosphere (from burning fossil fuels) and global climate change.

At the opposite extreme, an individual cell may be the stage on which two populations of pathogens compete with one another for the resources that the cell provides. At a slightly larger spatial scale, a termite's gut is the habitat for bacteria, protozoons, and other species (Figure 1.2)—a community whose diversity may

a range of spatial scales

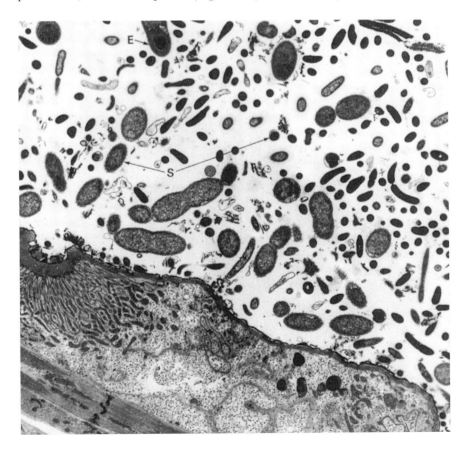

Figure 1.2
Electron micrograph of a thin section of the paunch of the termite *Reticulitermes flavipes.* Much of the flora is composed of aggregates of bacteria. Among them can be seen endospore-forming bacteria (E), spirochaetes (S), and protozoa. (After Breznak, 1975.)

reasonably be compared with that of a tropical rain forest in terms of the richness of organisms living there, the variety of interactions in which they take part, and indeed the extent to which we remain ignorant about the species identity of many of the participants. Between these extremes, different ecologists, or one ecologist at different times, may study the inhabitants of pools that form in small tree holes, the temporary watering holes of the savannahs, or the great lakes and oceans; others may examine the diversity of fleas on different species of birds, the diversity of birds in different sized patches of woodland, or the diversity of woodlands at different altitudes.

a range of time scales

To some extent related to this range of spatial scales, and to the levels in the biological hierarchy, ecologists also work on a variety of time scales. *Ecological succession*—the successive and continuous colonization of a site by certain species populations, accompanied by the extinction of others—may be studied over a period from the deposition of a lump of sheep dung to its decomposition (a matter of weeks), or from the change in climate at the end of the last ice age to the present day and beyond (around 14,000 years and still counting). Migration may be studied in butterflies over the course of days, or in the forest trees that are still (slowly) migrating into deglaciated areas following that last ice age.

the need for long-term studies

Although it is undoubtedly the case that "appropriate" time scales vary, it is also true that many ecological studies are not as long as they might be. Longer studies are more expensive and require dedication and stamina. An impatient scientific community, and the requirement for concrete evidence of activity for career progression, both put pressure on ecologists, and all scientists, to publish their work sooner rather than later. Why are long-term studies potentially of such value? The reduction over a few years in the numbers of a particular species of wildflower, or bird, or butterfly might be a cause for conservation concern—but one or more decades of study may be needed to be sure that the decline is more than just an expression of the random ups and downs of "normal" population dynamics. Similarly, a 2-year rise in the abundance of a wild rodent followed by a 2-year fall might be part of a regular cycle in abundance, crying out for an explanation. But ecologists cannot be sure until perhaps 20 years of study has allowed them to record four or five phases of such a cycle.

This does not mean that all ecological studies require 20 years—nor that every time an ecological study is extended the answer changes. But it does emphasize the great value to ecology of the small number of long-term investigations that have been carried out or are ongoing.

1.2.2 The diversity of ecological evidence

Ecological evidence has a variety of different sources. Ultimately, ecologists are interested in organisms in their natural environments (though for many organisms, the environment that is natural for them now was created by human activity). Progress would be impossible, however, if ecological studies were limited to pure "natural" environments. And even in natural habitats, unnatural acts (experimental manipulations) are often necessary in the search for sound evidence.

Many ecological studies involve careful *observation* and monitoring, in the natural environment, of the changing abundance of one or more species over time, or over space, or over both. In this way, ecologists may establish patterns, for example, that red grouse (birds shot for "sport": Figure 1.3) exhibit regular cycles in abundance that peak every 4 or 5 years, or that vegetation can be mapped into a series of zones as we move across a landscape of sand dunes. But scientists don't stop at this point—the patterns require explanation. Careful analysis of the descriptive data may suggest some plausible explanations. But establishing what causes the patterns may well require *manipulative field experiments*: ridding the red grouse of intestinal worms, suggested to underlie the cycles, and checking whether the cycles persist (they don't: Hudson et al., 1998), or treating experimental areas on the sand dunes with fertilizer to see whether the changing pattern of vegetation itself reflects a changing pattern of soil productivity.

Perhaps less obviously, ecologists also often need to turn to laboratory systems and even mathematical models. These have played a crucial role in the development of ecology, and they are certain to continue to do so. Field experiments are almost inevitably costly and difficult to carry out. Moreover, even if time and expense were not issues, natural field systems may simply be too complex to allow us to tease apart the consequences of the many different processes that may be going on. Are the intestinal worms actually capable of having an effect on reproduction or mortality of individual grouse? Which of the many species of sand dune plant are, in themselves, sensitive to changing levels of soil productivity, and which are relatively insensitive? *Controlled, laboratory experiments* often provide the best means for answering such specific questions, which may themselves be key parts of any overall explanation of the complex situation in the field.

Of course, the complexity of natural ecological communities may simply make it inappropriate for an ecologist to dive straight into them in search of understanding. We may wish to explain the structure and dynamics of a particular community of 20 animal and plant species comprising various competitors, predators, parasites,

Figure 1.3
The red grouse (*Lagopus lagopus scoticus*).

and so on (relatively speaking, a community of remarkable simplicity). But we have little hope of doing so unless we already have some basic understanding of even simpler communities of just one predator and one prey species, or two competitors, or (especially ambitious) two competitors that also share a common predator. For this, it is usually most appropriate to construct, for our own convenience, *simple laboratory systems* that can act as benchmarks or jumping-off points in our search for understanding.

and mathematical models

What is more, you have only to ask anyone who has tried to rear caterpillars from eggs, or take a cohort of shrub cuttings through to maturity, to discover that even the simplest ecological communities may not be easy to maintain and keep free from invasion by other species, be they pathogens, predators, or competitors. Nor is it necessarily possible to construct precisely the specific, simple, artificial community that interests you; nor to subject it to precisely the conditions or the perturbation of interest. In many cases, therefore, there is much to be gained from the analysis of *mathematical models* of ecological communities: constructed and manipulated according to the ecologist's design.

On the other hand, although a major aim of science is to simplify, and thereby make it easier to understand the complexity of the real world, ultimately it is the real world that we are interested in, and the worth of models and simple laboratory experiments must always be judged in terms of the light they throw on the working of more natural systems. They are a means to an end—never an end in themselves. Like all scientists, ecologists need to "seek simplicity, but distrust it" (Whitehead, 1953).

1.2.3 Statistics and scientific rigor

It is never good to overreact. For any scientist to take offense at some popular phrase or saying is to invite accusations of a lack of a sense of humor. It is nonetheless difficult to remain calm when phrases like "There are lies, damn lies, and statistics" or "You can prove anything with statistics" are used by those who know no better, in order to justify continuing to believe what they wish to believe, whatever the evidence to the contrary. There is no doubt that statistics are sometimes *misused* to derive dubious conclusions from sets of data that actually suggest either something quite different or perhaps nothing at all. But these are not grounds for mistrusting statistics in general—rather for ensuring that all those capable of doing so are educated in at least the principles of scientific evidence and its statistical analysis, so as to protect them from those who may seek to manipulate their opinions.

ecology: a search for conclusions in which we can have confidence

In fact, not only is it not true that you can prove anything with statistics, the contrary is the case: you cannot *prove* anything with statistics—that is not what statistics are for. Statistical analysis is essential, however, for attaching a level of confidence to conclusions that we may wish to draw, and ecology, like all science, is a search not for statements that have been "proved to be true" but for conclusions we can trust.

Indeed, what distinguishes science from other activities—what makes science rigorous—is that it is based not on statements that are simply assertions, but (1) on conclusions that are the results of investigations (as we have seen, of a wide variety of types) carried out with the express purpose of deriving those conclusions, and (2) even more important, on conclusions to which a level of confidence can be attached, measured on an agreed scale. These points are elaborated on, and P-values, standard errors, and confidence intervals are discussed, in Boxes 1.2 and 1.3.

(continued on page 15)

Box 1.2

Interpreting Probabilities

P-*values*

The term that is most often used, at the end of a statistical test, to measure the strength of conclusions being drawn is a P-value, or probability level. It is important to understand what these are. Imagine we are interested in establishing whether high abundances of a particular pest insect in the summer are associated with high temperatures in the previous spring, and that the data we have to address this question consist of summer insect abundances and mean spring temperatures for each of a number of years. At the outset we do not know whether there is an association, but we hope that statistical analysis of our data will allow us either to conclude, with a stated degree of confidence, that there is an association, or to conclude that there are no grounds for believing there to be an association (Figure 1.4).

Null hypotheses

To carry out a statistical test we first need a *null hypothesis*, which simply means, in this case, that we begin by proposing that there is *no* association between insect abundance and temperature (that there is no association *is* the null hypothesis). The statistical test (stated simply) then generates a probability (a P-value) of getting a data set like ours if the null hypothesis is correct.

Suppose, for example, that the probability generated by a test on our data was 0.5 (equivalently 50 percent): that is, $P = 0.5$ (Figure 1.4a). This would mean that, if the null hypothesis really were correct, if there was no association, then 50 percent of studies like ours should generate just such a data set, or one even further from the null hypothesis. Clearly, if there were no association there would be nothing very remarkable in getting data like ours. More to the point, we could have no confidence in any claim that there *was* an association.

Suppose, however, that the P-value generated by the statistical test were $P = 0.001$ (0.1 percent) (Figure 1.4b). This would mean that such a data set (or one even further from the null hypothesis) could be expected in only 0.1 percent of similar studies—one in a thousand—if there was really no association. In other words, either something very improbable has occurred, or there *was* an association between insect abundance and spring temperature. Thus, since by definition we do not expect highly improbable events to occur, we can have a high degree of confidence in the claim that there *was* an association between abundance and temperature.

Significance testing

Both 50 percent and 0.1 percent, though, make things easy for us. Where, between the two, do we draw the line? There is no objective answer to this, and so scientists and statisticians have established a convention in *significance testing*, which says that if P is less than 0.05 (5 percent), written $P < 0.05$, then results are described as statistically significant and confidence can be placed in the effect being examined (in our case, the association between abundance and temperature; Figure 1.4d), whereas if $P > 0.05$, then there is no statistical foundation for claiming the effect exists (Figure 1.4c). A further elaboration of the convention often describes results with $P < 0.01$ as highly significant.

"Insignificant" results?

Naturally, some effects are strong (for example, there is a powerful association between people's weight and their height) and others are weak (the association between

(continues)

Interpreting Probabilities, continued

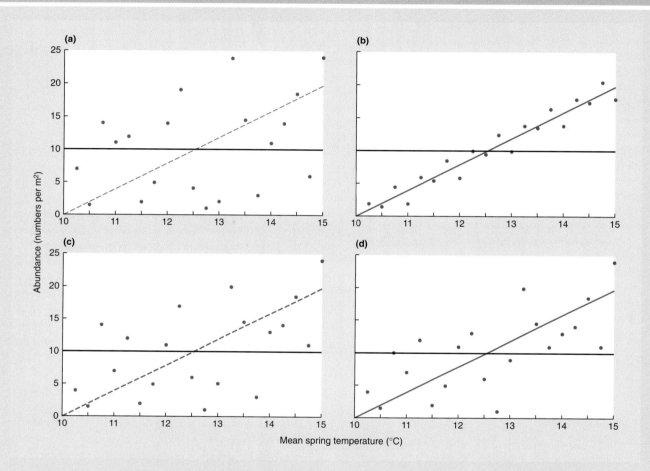

Figure 1.4

The results from four hypothetical studies of the relationship between the abundance of an insect pest in the summer and the mean temperature in the previous spring. In each case, the points are the data actually collected. Horizontal lines represent the *null hypothesis*—that there is no association between abundance and temperature, and thus the best estimate of expected insect abundance, irrespective of spring temperature, is the mean insect abundance overall. The second line is the *line of best fit* to the data, which in each case offers some suggestion that abundance rises as temperature rises. However, whether we can be confident in concluding that abundance does rise with temperature depends, as explained in the text, on statistical tests applied to the data sets: (a) The suggestion of a relationship is weak ($P = 0.5$). There are no grounds for concluding that the true relationship differs from that supposed by the null hypothesis and no grounds for concluding that abundance is related to temperature. (b) The relationship is strong ($P = 0.001$) and we can be confident in concluding that abundance increases with temperature. (c) The results are suggestive ($P = 0.1$) but it would not be safe to conclude from them that abundance rises with temperature. (d) The results are not vastly different from those in (c) but are powerful enough ($P = 0.04$, i.e., $P < 0.05$) that the conclusion that abundance rises with temperature can be considered safe.

people's weight and their risk of heart disease is real but weak, since weight is only one of many factors and is not applicable throughout its range). More data are needed to establish support for a weak effect than for a strong one. A rather obvious but very important conclusion follows from this: a P-value in excess of 0.05 (lack of statistical significance) in an ecological study may mean any one of three things:

1. There really is no effect of ecological importance.
2. The data are simply not good enough, or there are not enough of them, to support the effect even though it exists.
3. Closely related to this, the effect itself is real but weak, and extensive data are therefore needed but have not been collected.

Quoting P-values

Furthermore, applying the convention strictly and dogmatically means that when $P = 0.06$ the conclusion should be "no effect has been established," whereas when $P =$ 0.04 the conclusion is "there is a significant effect." Yet very little difference in the data is required to move a P-value from 0.04 to 0.06. It is therefore far better to quote exact P-values, especially when they exceed 0.05, and think of conclusions in terms of shades of gray rather than the black and white of "proven-effect" and "no-effect." In particular, P-values close to, but not less than, 0.05 suggest that something seems to be going on, and they indicate, more than anything else, that more data need to be collected so that our confidence in conclusions can be more clearly established.

Throughout this book, then, studies of a wide range of types are described, and their results often have P-values attached to them. Of course, as this is a textbook, the studies have been selected because their results *are* significant. Nonetheless, it is important to bear in mind that the repeated statements $P < 0.05$ and $P < 0.01$ mean that these are studies where sufficient data have been collected to establish a conclusion with which we can be confident, and where that confidence has been established by agreed means (statistical testing) and is being measured on an agreed and interpretable scale.

Statistical analyses are carried out after data have been collected and help us to interpret those data. There is no really good science, however, without forethought. Ecologists, like all scientists, must know what they are doing, and why they are doing it, *while* they are doing it. This is entirely obvious at a general level: nobody expects ecologists to be going about their work in some kind of daze. But it is perhaps not so obvious that ecologists should know how they are going to analyze their data, statistically, not only after they have collected it, nor only while they are collecting it, but even before they begin to collect it. Ecologists must plan, so as to be confident that they have collected the right kind of data, and a sufficient amount of data, to address the questions they hope to answer.

ecologists must think ahead

Ecologists typically seek to draw conclusions about groups of organisms overall: What is the birth rate of the bears in Yellowstone Park? What is the density of weeds in a wheat field? What is the rate of nitrogen uptake of tree saplings in a nursery? In doing so, we can only very rarely examine every individual in a group, or the entire sampling area, and we must therefore rely on what we hope will be *representative* samples from the group or habitat as a whole. Indeed, even if we examine a whole group (we might examine every fish in a small pond, say), we are likely to want to draw general conclusions from it: we may hope that the fish in our

ecology relies on representative samples

Box 1.3

Standard Errors

Confidence intervals

Following Box 1.2, another way in which the significance of, and confidence in, results is assessed is through reference to standard errors. Again, simply stated, statistical tests often allow standard errors to be attached either to mean values calculated from a set of observations or to slopes of lines like those in Figure 1.4. Such mean values or slopes can, at best, only ever be estimates of the "true" mean value or the true slope, because they are calculated from sets of data that themselves are only a sample of all the imaginable items of data that could be collected. The standard error, then, sets a band around the estimated mean (or slope, etc.) within which the true mean can be expected to lie with a given, stated probability. In particular, there is a 95 percent probability that the true mean lies within roughly two standard errors of the estimated mean: the *95 percent confidence interval* (for simplicity, we shall assume it is exactly two standard errors).

Hence, when we have, say, two sets of observations, each with its own mean value (for instance, the number of seeds produced by plants from two sites—Figure 1.5) the standard errors allow us to assess whether the means are statistically significantly different from one another. Specifically, roughly speaking, if each mean is more than two standard errors from the other mean, then the difference between them is statistically significant with $P < 0.05$. Thus, for the study illustrated in Figure 1.5a, it would not be safe to conclude that plants from the two sites differed in their seed production. However, for the similar study illustrated in Figure 1.5b, the means are roughly the same as they were in the first study and are roughly as far apart, but the standard errors are smaller. Hence, the difference between the means is significant ($P < 0.05$), and we can conclude with confidence that plants from the two sites differed.

When are standard errors small?

Finally, it is important to note that the large standard errors in the first study, and hence the lack of statistical significance, could have been due to data that were, for whatever reason, more variable, but are more likely to have been due to a sampling of fewer plants in the first study than the second. Standard errors are smaller, and statistical significance is easier to achieve, *both* when data are more consistent (less variable) *and* when there are more data.

---

pond can tell us something about fish of that species in ponds of that type generally. In short, ecology relies on obtaining *estimates* from representative samples. This, too, and its relationship to statistical interpretation, are elaborated on in Box 1.4.

1.3 Ecology in Practice

In previous sections we have established in a general way how ecological understanding can be achieved, and how that understanding can be used to help us predict, manage, and control ecological systems. However, the practice of ecology is easier said than done. To discover the real problems faced by ecologists and how they try to solve them, it is best to consider some real research programs in a little detail. While reading the following examples you should focus on how they illuminate three main points—ecological phenomena occur at a variety of scales; ecological evidence comes from a variety of different sources; ecology relies on scientific

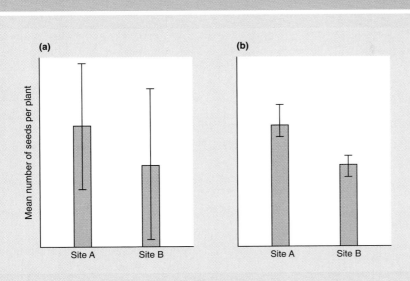

Figure 1.5
The results of two hypothetical studies in which the seed production of plants from two different sites was compared. In all cases, the heights of the bars represent the mean seed production of the sample of plants examined, and the lines crossing those means extend one standard error above and below them: (a) Although the means differ, the standard errors are relatively large and it would not be safe to conclude that seed production differed between the sites ($P = 0.4$). (b) The differences between the means are very similar to those in (a), but the standard errors are much smaller, and it can be concluded with confidence that plants from the two sites differed in their seed production ($P < 0.05$).

evidence and the application of statistics. Every other chapter in this book will contain descriptions of similar studies, but in the context of a systematic survey of the driving forces in ecology (Chapter 2–11) or of the application of this knowledge to solve applied problems (Chapters 12–14). For now, we content ourselves with seeking an appreciation of how four research teams have gone about their business.

1.3.1 The brown trout in New Zealand—effects on individuals, populations, communities, and ecosystems

It is rare for a study to encompass more than one or two of the four levels in the biological hierarchy (individuals, populations, communities, ecosystems). For most of the twentieth century, physiological and behavioral ecologists (studying individuals), population ecologists, and community and ecosystem ecologists have tended to follow separate paths, asking different questions in different ways. However, there

(*continued on page 20*)

Box 1.4

Sampling, Accuracy, and Precision

The discussion in Boxes 1.2. and 1.3 about when standard errors will be small or large, or when our confidence in conclusions will be strong or weak, has implications not only for the analysis and interpretation of data after they have been collected. It also carries a general message about planning the collection of those data.

In undertaking a sampling program, the aim is to satisfy a number of criteria (Figure 1.6):

- That the estimate should be accurate or unbiased: that is, neither systematically too high nor too low as a result of some flaw in the program
- That the estimate should have as narrow confidence limits (be as precise) as possible
- That the time, money, and human effort invested in the program should be used as effectively as possible (because these are always limited)

Random and stratified random sampling

To understand these criteria, consider another hypothetical example. Suppose that we are interested in the density of a particular weed (say wild oat) in a wheat field. To prevent bias, it is necessary to ensure that each part of the field has an equal chance of being selected for sampling. Sampling units should therefore be selected at random. We might, for example, divide the field into a measured grid, pick pairs of coordinates at random, and count the wild oat plants within a 50-cm radius of the selected grid point. This unbiased method can be contrasted with a plan to sample only weeds from between the rows of wheat plants, giving too high an estimate, or within the rows, giving too low an estimate (Figure 1.6a).

Remember, however, that random samples are not taken as an end in themselves, but because random sampling is a means to truly representative sampling. Thus, randomly chosen sampling units may end up being con-

centrated, by chance, in a particular part of the field that, unknown to us, is not representative of the field as a whole. It is often preferable, therefore, to undertake *stratified random sampling* in which, in this case, the field is divided up into a number of equal sized parts (*strata*) and a random sample taken from each. This way, the coverage of the whole field is more even, without our having introduced bias by selecting particular spots for sampling.

Separating subgroups and directing effort

Suppose now, though, that half the field is on a slope facing southeast and the other half on a slope facing southwest, and that we know that aspect (which way the slope is facing) affects weed density considerably. Random sampling (or stratified random sampling) ought still to provide an unbiased estimate of density for the field as a whole, but for a given investment in effort, the confidence interval for the estimate will be unnecessarily high. To see why, consider Figure 1.6b. The individual values from samples fall into two groups a substantial distance apart on the density scale: high from the southwest slope; low (mostly zero) from the southeast slope. The estimated mean density is close to the true mean (it is accurate), but the variation among samples leads to a very large confidence interval (it is not very precise).

If, however, we acknowledge the difference between the two slopes and treat them separately from the outset, then we obtain means for each that have much smaller confidence intervals. What is more, if we average those means and combine their confidence intervals to obtain an estimate for the field as a whole, then that interval too is much smaller than previously (Figure 1.6b).

But has our effort been directed sensibly, with equal numbers of samples from the southwest slope, where there

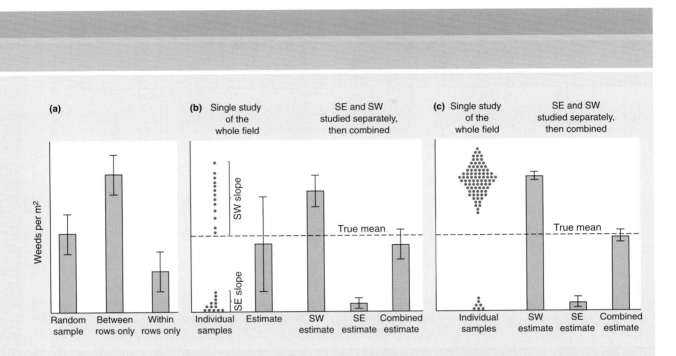

Figure 1.6
The results of hypothetical programs to estimate weed density in a wheat field: (a) The three studies have equal precision (95% confidence intervals) but only the first (from a random sample) is accurate. (b) In the first study, individual samples from different parts of the field (southeast and southwest) fall into two groups (left); thus, the estimate, although accurate, is not precise (right). In the second study, separate estimates for southeast and southwest are both accurate and precise—as is the estimate for the whole field obtained by combining them. (c) Following on from (b), most sampling effort is directed to the southwest, reducing the confidence interval there, but with little effect on the confidence interval for the southeast. The overall interval is therefore reduced: precision has been improved.

are lots of weeds, and the southeast slope, where there are virtually none? The answer is No. Remember that narrow confidence intervals arise from a combination of a large number of data points *and* little intrinsic variability (Box 1.3). Thus, if our efforts had been directed mostly at sampling the southwest slope, the increased amount of data would have noticeably decreased the confidence interval (Figure 1.6c), whereas less sampling of the southeast slope would have made very little difference to that confidence interval because of the low intrinsic variability there. Careful direction of a sampling program can clearly increase overall precision for a given investment in effort. And generally, sampling programs should, where possible, identify biologically distinct subgroups (males and females, old and young, etc.) and treat them separately, but sample at random within subgroups.

can be little doubt that, ultimately, our understanding will be enhanced considerably when the links among all these levels are made clear—a point that can be illustrated by examining the impact of the introduction of an exotic fish to streams in New Zealand.

Prized for the challenge they provide to anglers, brown trout (*Salmo trutta*) have been transported from their native Europe all around the world; they were introduced to New Zealand from 1867, and self-sustaining populations are now found in many streams, rivers, and lakes there. Until quite recently, few people cared about native New Zealand fish or invertebrates, so little information is available on changes in the ecology of native species after the introduction of trout. However, trout have colonized some streams but not others. We can therefore learn a lot by comparing the current ecology of streams containing trout with those occupied by certain native fish in the genus *Galaxias* (Figure 1.7).

the individual level—consequences for invertebrate feeding behavior

Mayfly nymphs of various species commonly graze microscopic algae growing on the beds of New Zealand streams, but there are some striking differences in their activity rhythms depending on whether they are in *Galaxias* streams or trout streams. In one experiment, nymphs collected from a trout stream and placed in small artificial laboratory channels were less active during the day than the night, whereas those collected from a *Galaxias* stream were active both day and night (Figure 1.8a). In another experiment, with another mayfly species, records were made of individuals visible in daylight on the surface of cobbles in artificial channels placed in a real stream. Three treatments were each replicated three times—no fish in the channels, trout present, and *Galaxias* present. Daytime activity was significantly reduced in the presence of either fish species—but to a greater extent when trout were present (Figure 1.8b).

These differences in activity pattern appear to reflect the fact that trout rely principally on vision to capture prey, whereas *Galaxias* rely on mechanical cues. Thus, invertebrates in a trout stream are at considerably more risk of predation during daylight hours. And these conclusions are all the more robust because they derive both from the readily controlled conditions of a laboratory experiment and from the more realistic, but more variable, circumstances of a field experiment.

Figure 1.7

A brown trout and a galaxiid fish in New Zealand streams—is the galaxiid hiding from the introduced predator?

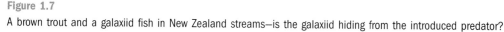

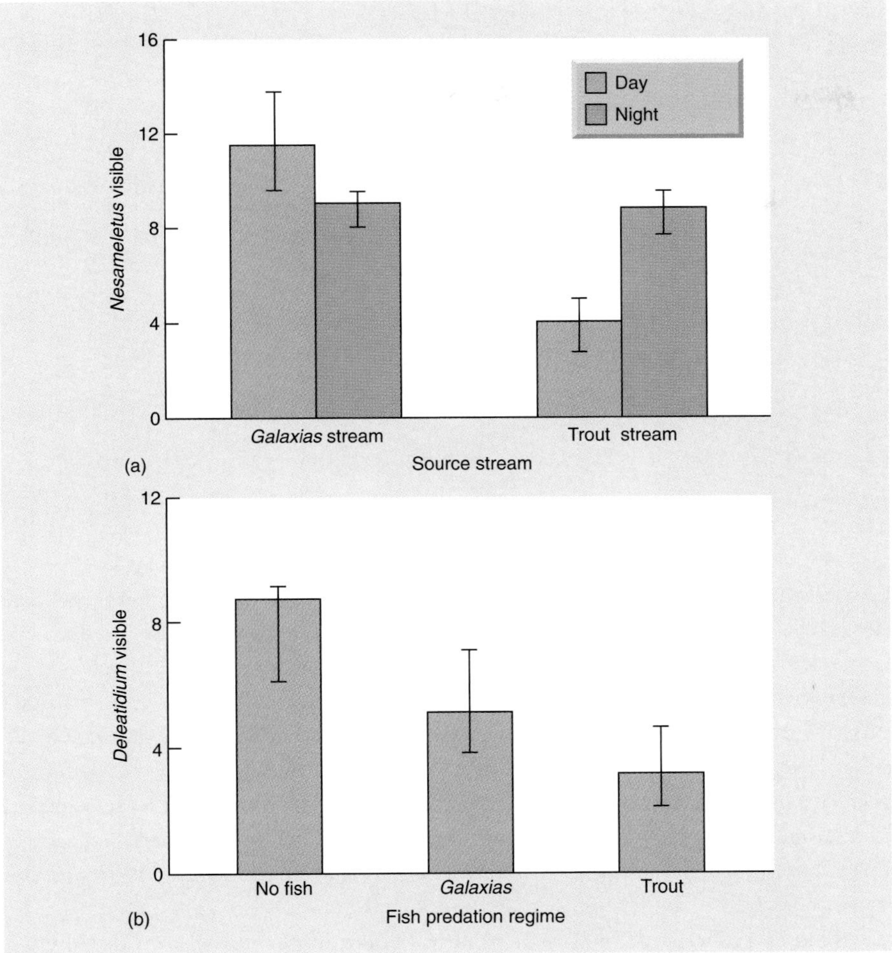

Figure 1.8
(a) Mean number (± standard error) of *Nesameletus ornatus* mayfly nymphs collected either from a trout stream or from a *Galaxias* stream that were recorded by means of video as visible on the substrate surface in laboratory stream channels during the day and night (in the absence of fish). (After McIntosh & Townsend, 1994.) (b) Mean number (± standard error) of *Deleatidium* mayfly nymphs observed on the upper surfaces of cobbles during late afternoon in channels (placed in a real stream) containing no fish, trout, or *Galaxias*. (After McIntosh & Townsend, 1996.) In all cases, the standard errors were sufficiently small for differences to be statistically significant ($P < 0.05$).

In the Taieri River in New Zealand, 198 sites were selected in a stratified manner by choosing streams of similar dimensions at random in each of three tributaries from each of eight subcatchments of the river. Care was taken not to succumb to the temptation of choosing sites with easy access (near roads or bridges), in case such a selection would bias the results. The sites were classified as containing (1) no fish, (2) *Galaxias* only, (3) trout only, or (4) both *Galaxias* and trout. At every site a variety of physical variables was measured (stream depth, flow velocity, phosphorus concentration in the stream water, percentage of the stream bed composed of gravel, etc.). A statistical procedure called *multiple discriminant functions analysis* was then used to determine which physical variables, if any, distinguished one type of site from another. Means and standard errors of these key environmental variables are presented in Table 1.1.

Trout occurred almost invariably below waterfalls that were large enough to prevent their upstream migration; they tended to occur at low elevations because sites without waterfalls downstream tended to be at lower elevation. Sites containing *Galaxias* (or no fish) were always upstream of one or several large waterfalls. The

the population level—brown trout and the distribution of native fish

Table 1.1

Means and standard errors for important discriminating variables for fish assemblage classes in 198 sites in the Taieri River

| | | VARIABLES* | | |
SITE TYPE	NUMBER OF SITES	NUMBER OF WATERFALLS DOWNSTREAM	ELEVATION (METERS ABOVE SEA LEVEL)	PERCENTAGE OF BED COMPOSED OF COBBLES
No fish	54	4.37 (0.64)	339 (31)	15.8 (2.3)
Brown trout only	71	0.42 (0.05)	324 (28)	18.9 (2.1)
Galaxias only	64	12.3 (2.05)	567 (29)	22.1 (2.8)
Trout + *Galaxias*	9	0.0 (0)	481 (53)	46.7 (8.5)

* Standard errors in parentheses.

few sites that contained both trout and *Galaxias* were below waterfalls, at intermediate elevations, and in sites with cobble beds; the unstable nature of these stream beds may have promoted the coexistence (at low densities) of the two species. This descriptive study at the population level therefore takes advantage of a "natural" experiment (streams that happen to contain trout or *Galaxias*) to determine the effect of the introduction of trout. The most probable reason for the restriction of populations of *Galaxias* to sites upstream of waterfalls, which cannot be climbed by trout, is direct predation by trout on the native fish below the waterfalls (a single small trout in a laboratory aquarium has been recorded consuming 135 *Galaxias* fry in a day).

the community—brown trout cause a cascade of effects

That an exotic predator such as trout has direct effects on *Galaxias* distribution or mayfly behavior is not surprising. However, we can ask whether these changes have community consequences that cascade through to other species. In the relatively species-poor stream communities in the south of New Zealand, the plants are mainly algae that grow on the stream bed. These are grazed by various insect larvae, which in turn are prey to predatory invertebrates and fish. As we have seen, trout have replaced *Galaxias* in many of these streams. An experiment involving artificial flow-through channels (several meters long, with mesh ends to prevent escape of fish but to allow invertebrates to colonize naturally) placed into a real stream was used to determine whether trout affect the stream food web differently than the displaced *Galaxias* do. Three treatments were established (no fish, *Galaxias* present, trout present, at naturally occurring densities) in each of several randomized blocks located in a stretch of a stream with each block separated by more than 50 m. Algae and invertebrates were allowed to colonize for 12 days before introducing the fish. After a further 12 days, invertebrates and algae were sampled (Figure 1.9).

A significant effect of trout in reducing invertebrate biomass was evident ($P = 0.026$), but the presence of *Galaxias* did not depress invertebrate biomass from the no-fish control. Algal biomass, perhaps not surprisingly then, achieved its highest values in the trout treatment ($P = 0.02$). It is clear that trout do have a more

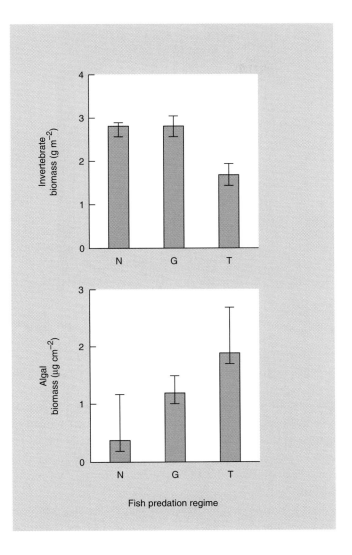

Figure 1.9
Total invertebrate biomass and algal biomass (chlorophyll a) (± standard error) for experiments performed in summer in a small New Zealand stream. N = no fish; G = *Galaxias* present; T = trout present. (After Flecker & Townsend, 1994.)

pronounced effect than *Galaxias* on invertebrate grazers and, thus, on algal biomass. The indirect effect of trout on algae occurs partly through a reduction in invertebrate density, but also through trout restriction of the grazing behavior of the invertebrates that are present (see Figure 1.8b).

A strong *trophic cascade*, with effects flowing down from one trophic level to the next and then the next, such as that produced by the introduced brown trout, may be expected to have consequences at the ecosystem level, but this has not been documented before, for two likely reasons. First, it is difficult to find two communities with contrasting predation regimes but the same physical settings; second, particularly great effort and expense are required for such studies (with the added problem that replication of treatments is usually not feasible—as in this case). However, the sequence of studies described here provided the impetus for a detailed energetics investigation of two neighboring tributaries of the Taieri River (with very similar physicochemical conditions), one occupied by just trout and the other (because of

the ecosystem—trout and energy flow

Figure 1.10
Annual estimates for "production" of biomass at one trophic level, and the "demand" for that biomass (the amount consumed) at the next trophic level, for primary producers (algae) on the left, invertebrates (which consume algae) in the center, and fish (which consume invertebrates) on the right. Estimates are for a trout stream and a *Galaxias* stream. In the former, production at all trophic levels is higher, but because the trout consume essentially all of the invertebrate production (center), the invertebrates consume only ~21% of primary production (left). In the *Galaxias* stream, these fish consume only ~18% of invertebrate production, "allowing" the invertebrates to consume ~75% of primary production. (After Huryn, 1998.)

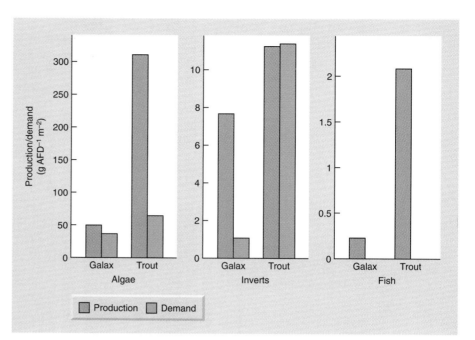

a waterfall downstream) containing only *Galaxias*. No other fish were present in either stream.

The hypothesis under examination was that the rate at which radiation energy was fixed in photosynthesis by the algae would be greater in the trout stream because the rate at which grazing invertebrates consume algae would be less in the presence of trout. Annual net *primary production* (the rate of production of plant, in this case, algal, biomass) was indeed six times greater in the trout stream than in the *Galaxias* stream (Figure 1.10).

Measurement of *secondary production* (the rate at which consumers produce new biomass per unit area per unit time) is a very time-consuming exercise that requires frequent sampling and identification of all animal taxa, analyses of separate size classes of each taxon, and knowledge of their life cycles. The results show that primary consumers (invertebrates that eat algae) in the trout stream produced new biomass at about 1.5 times the rate in the *Galaxias* stream, and trout themselves produced new biomass at roughly 9 times the rate of *Galaxias* (Figure 1.10).

Ecosystem ecologists are also interested in the efficiency with which each trophic level uses energy available from the trophic level below (Chapter 11). The results here indicate that *Galaxias* consume only about 18 percent of available prey production each year, but the grazing invertebrates consume about 75 percent of primary production in the *Galaxias* stream. In stark contrast, trout consume virtually 100 percent of annual invertebrate production in their stream while the grazing invertebrates consume only about 21 percent of primary production (Figure 1.10). That is, strong control by trout of invertebrate abundance releases algae to produce and accumulate biomass at a fast rate.

This series of studies, therefore, illustrates some of the variety of ways in which ecological investigations may be pursued, and both the range of levels in the biological hierarchy that ecology spans and the way in which studies at different levels may complement one another. Although it is necessary to be cautious when interpreting the results of an unreplicated study (only one trout and one *Galaxias* stream in the "ecosystem study"), the conclusion that a trophic cascade is responsible for the patterns observed at the ecosystem level can be drawn with some confidence because of the variety of other corroborative studies conducted at the individual, population, and community levels.

1.3.2 Successions on old fields in Minnesota—a study in time and space

Ecological succession is a concept that must be familiar to many who have simply taken a walk in open country—the idea that a newly created habitat, or one in which a disturbance has created an opening, will be inhabited, in turn, by a variety of species appearing and disappearing in some recognizably repeatable sequence. Widespread familiarity with the idea, however, does not mean that we understand fully the processes that drive or fine-tune successions; yet developing such understanding is important not just because succession is one of the fundamental forces structuring ecological communities, but also because human disturbance of natural communities has become ever more frequent and profound, and we need to know how communities may respond—and, it is hoped, recover—from such disturbance, and how we may aid that recovery.

One particular focus for the study of succession has been the old agricultural fields of the eastern United States, abandoned as farmers moved west in search of "fresh fields and pastures new." One such site is now the Cedar Creek Natural History Area, roughly 50 km north of Minneapolis, Minnesota. The area was first settled by Europeans in 1856 and was initially subject to selective and clear-cut logging. Clearing for cultivation then began about 1885, and land was first cultivated between 1900 and 1910. Now there are agricultural fields that are still under cultivation and others that have been abandoned at various times since the mid-1920s. Cultivation led to depletion of nitrogen from what was already naturally nitrogen-poor soil.

First, studies at Cedar Creek illustrate the value of natural experiments. We particularly want to know the successional sequence of plants that occurs in fields in the years following abandonment and want to be able to account for that sequence. We could plan an artificial manipulation, under our control, in which a number of fields currently under cultivation were "forcibly" abandoned and the communities in them sampled repeatedly into the future. (We would need a *number* of fields because any single field might be atypical, whereas studying several would allow us to calculate mean values for, say, "number of new species per year," and place confidence intervals around those means.) But the results of this experiment would take decades to accumulate. The natural-experiment alternative, therefore, is to use the fact that records already exist of when many of the old fields were abandoned. Thus, Figure 1.11 illustrates data from a group of 22 old fields, surveyed in 1983, having been abandoned at various times between 1927 and 1982 (i.e., between 1 and 56 years previously). Interpreted cautiously, these can be treated as 22 "snapshots" of the continuous process of succession in old fields at Cedar Creek in general, even though each field was itself only surveyed once (Inouye et al., 1987).

the use of natural experiments

in generating correlations

Figure 1.11

Twenty-two fields at different stages in an old-field succession were surveyed to generate the following trends with succession age: (a) introduced species decreased, (b) prairie species increased, (c) annual species decreased, (d) perennial species increased, (e) soil nitrogen content increased. *r* in each case is the "correlation coefficient," which can vary between 0 (no correlation) and 1 (perfect correlation). (After Inouye et al., 1987.)

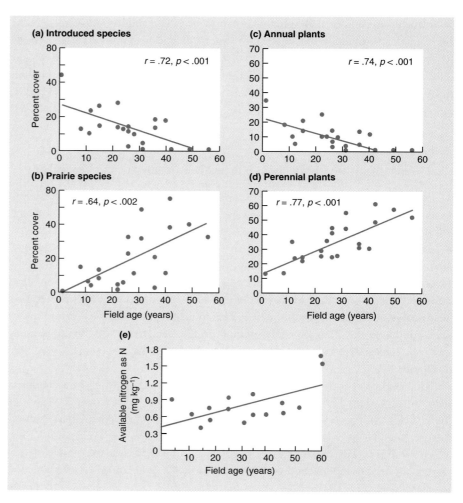

A number of the shifting balances during succession are clear from Figure 1.11 as statistically significant trends. Over the 56 years, the cover of introduced species (mostly agricultural weeds) decreased (Figure 1.11a) while the cover of species from nearby prairies increased (Figure 1.11b): the natives reclaimed their land. Of more general applicability, the cover of annual species decreased over time, while the cover of perennial species increased (Figures 1.11c, d). Annual species (those that complete a whole generation from seed to adult through to seeds again within a year) tend to be good at increasing in abundance rapidly in relatively empty habitats (the early stages of succession), whereas perennials (those that live for several or many years and may not reproduce in their early years) are slower to establish but more persistent once they do.

On the other hand, natural experiments like this, although frequently suggestive and stimulating (and too good an opportunity to miss), usually generate only *correlations*. They may therefore fail to establish what actually causes the observed patterns. In the present case, we can see the problem by noting, first, that field age is itself strongly correlated with concentration of nitrogen—perhaps the single most important plant nutrient (Figure 1.11e)—in the soil. The question therefore arises, Are the correlations in Figures 1.11a–d an effect of field age itself? Or is the causal agent nitrogen, with which age is correlated?

Furthermore, the pattern of correlation changes with the spatial and temporal scale of our observations. Given the correlations in Figure 1.11, it is no surprise that in field-to-field comparisons (which are also age-to-age comparisons) introduced species cover decreased, and prairie species cover increased, with increased soil nitrogen. The story told by *within-field comparisons*, however (that is, between patches of the same age) is quite different. Of the 12 fields where introduced species cover and soil nitrogen were significantly related, introduced species *increased* with soil nitrogen in 10, and of the 13 fields where prairie species cover and soil nitrogen were significantly related, prairie species *decreased* with soil nitrogen in 9.

patterns change with the spatial and temporal scale

One possible plausible explanation of these patterns runs as follows: the "large-scale" differences from field to field (and hence from time to time) are mostly a reflection of differences in opportunity. The perennials and prairie species take more time to arrive and establish thriving populations. They therefore increase in abundance over time, forcing out the annuals and introduced species, which therefore decrease over time. Time, in this sense, *causes* the pattern, whereas the correlations of these species' abundances with nitrogen simply occurs because soil nitrogen content also builds up over time. The contradictory correlations *within* fields, on the other hand, according to this explanation, arise because the annuals and introduced species have faster growth rates, especially on richer soils. In fields where these plants are present, therefore, they are particularly abundant in the nitrogen-rich patches.

Can manipulative field experiments help support—or refute—what so far is no more than a plausible explanation based on correlation? It seems to follow from the proposed explanation (time matters) that nitrogen itself has little role to play in driving these successions, and that manipulating nitrogen should do little to alter the species sequences that these fields have followed. It is of interest, therefore, that three of the old fields, last cultivated in 1934, 1957, and 1968, were selected, and over an 11-year period starting in 1982, six replicate 4-meter by 4-meter plots at each site were subjected to one of eight treatments: nitrogen added at rates ranging from 0 to 27.2 g per square meter per year. Two questions in particular were asked: Do patches receiving different supply rates of nitrogen become less similar in species composition over time? Do patches receiving similar supply rates of nitrogen become more similar in species composition over time (Inouye & Tilman, 1995)? Some of the results obtained are shown in Figure 1.12.

artificial experiments: the search for causation

At the start of the experiment, plots within a field were similar to one another, but 10 years later, plots receiving different amounts of nitrogen had diverged in species composition—and the greater the difference in nitrogen input, the greater the divergence (Figure 1.12a). Moreover, whereas at the start of the experiment, fields of different age tended to be very different in species composition, 10 years later plots within them that had been subjected to similar rates of nitrogen input had become remarkably similar, despite there being, in one case, 34 years' difference in their age (Figure 1.12b). Thus, this experiment tends to refute the simplicity of our proposed explanation. Time itself is not the only cause of successional changes in species composition of these old fields. Differences in available nitrogen cause

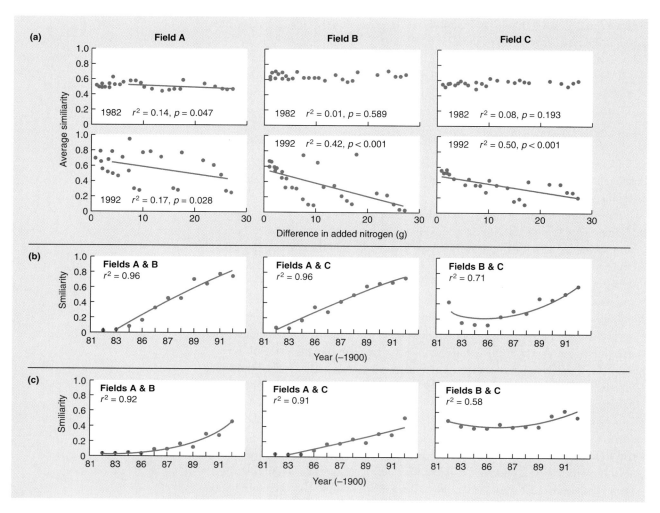

Figure 1.12
Results from an experiment in which three old fields from Figure 1.11 were given artificial nitrogen treatments starting in 1982: field A (abandoned 1968), field B (1957), and field C (1934). (a) In 1982, plots within each field were similar in species composition (though a slight trend is apparent in A): each point represents the comparison of a pair of plots. By 1992, however, plots receiving different nitrogen treatments had diverged in composition. (b) Between 1982 and 1992, plots receiving 17 g of nitrogen per square meter per year in pairs of fields became increasingly similar in composition. (c) Similar to (b) but with only 1 g of nitrogen per square meter per year and less convergence. (After Inouye & Tilman, 1995.)

insight into the effects of nitrogen pollution

successions to diverge; similarities cause them to converge much more quickly than they would otherwise. Time ("opportunity," "colonization") and nitrogen are clearly intimately intertwined, and further experiments will be required to disentangle their web of cause and effect—just one of many unanswered ecological questions.

Finally, experimental manipulations over extended periods like these may also provide important insights into the possible effects of more chronic human disturbances to natural communities. The lowest rates of nitrogen addition in the experiment were similar to those experienced in many parts of the world as a result of increased atmospheric deposition of inorganic nitrogen. Even these low levels apparently lead to convergence of previously dissimilar communities over a 10-year period (Figure 1.12c). Experiments like this are crucial in helping us to predict the effects of pollutants, a point that is taken further in the next example.

1.3.3 Hubbard Brook—a long-term commitment of large-scale significance

The Cedar Creek study took advantage of a temporal pattern (a succession that takes decades to run its course) reflected more or less accurately by a pattern in space (fields abandoned for different periods). The spatial pattern has the advantage that it could be studied within the time bite of most research projects (3–5 years). It would have been better still to follow the ecological pattern through time, but rather few researchers or institutions have risen to the challenge of designing research programs that last for decades.

A notable exception has been the work of Likens and associates at the Hubbard Brook Experimental Forest, an area of temperate deciduous forest drained by small streams in the White Mountains of New Hampshire, United States. The researchers were pioneers with no precedents to follow. They decided to think big, and their work has shown the value of large-scale studies and long-term data records. The study commenced in 1963 and continues to the present. In the second edition of their classic *Biogeochemistry of a Forested Ecosystem*, Likens and Bormann (1995) make poignant reference to three of their original collaborators who had died since the study began—long-term indeed.

The research team developed an approach called the *small watershed technique* to measure the input and output of chemicals from individual catchment areas in the landscape. Because many chemical losses from terrestrial communities are channeled through streams, a comparison of the chemical composition of stream water with that of incoming precipitation can reveal a lot about the differential uptake and cycling of chemical elements by the terrestrial biota. The same study can reveal much about the sources and concentrations of chemicals in the stream water, which in turn may influence the productivity of stream algae and the distribution and abundance of stream animals.

The watershed—the extent of terrestrial environment drained by a particular stream—was taken as the unit of study because of the role that streams play in chemical export from the land. Six small watersheds were defined and their outflows were monitored (Figure 1.13). A network of precipitation gauges recorded the incoming amounts of rain, sleet, and snow. Chemical analyses of precipitation and stream water made it possible to calculate the amounts of various chemical elements entering and leaving the system. In most cases, the output of chemicals in stream flow was greater than the input from rain, sleet, and snow (Table 1.2). The source of the excess chemicals was weathering of parent rock and soil, estimated at about 70 g per square meter per year. The exception was nitrogen: less was exported in stream water than was added to the catchment by precipitation, dry deposition, and fixation of atmospheric nitrogen by microorganisms in the soil.

the catchment area as a unit of study

Table 1.2

Annual chemical budgets for forested watersheds at Hubbard Brook (kilograms per hectare per year). Inputs are for dissolved materials in precipitation or as dry fall (gases or associated with particles falling from the atmosphere). Outputs are losses in stream water as dissolved material plus particulate organic material.

	NH_4^+	NO_3^-	SO_4^{2-}	K^+	Ca^{2+}	Mg^{2+}	Na^+
Input	2.7	16.3	38.3	1.1	2.6	0.7	1.5
Output	0.4	8.7	48.6	1.7	11.8	2.9	6.9
Net change*	+2.3	+7.6	−10.3	−0.6	−9.2	−2.2	−5.4

* Net change is positive when the catchment gains matter and negative when it loses it.
(After Likens et al., 1971.)

Figure 1.13
The Hubbard Brook experimental forest.
(Courtesy of Gene Likens.)

Likens and his associates had the brilliant idea of performing a large-scale experiment in which all the trees were felled in one of Hubbard Brook's six watersheds. In terms of experimental design, statistical purists might argue the study was flawed because it was unreplicated. However, the scale of the undertaking rather precluded replication. Adjacent watersheds were kept uncut for reference. In any case, it was the asking of a dramatically new question that made this study a classic rather than an elegant statistical design.

Within a few months of felling all the trees in the drainage basin, the consequences were evident in the stream water. The overall export of dissolved inorganic substances from the disturbed watershed rose to 13 times the normal rate (Figure 1.14). Two phenomena were responsible. First, the enormous reduction in transpiring surfaces (leaves) led to the passing of 40 percent more precipitation through the groundwater to be discharged to the streams, and this increased outflow caused greater rates of leaching of chemicals and weathering of rock and soil. Second, and more significantly, deforestation effectively broke the within-system nutrient cycling by uncoupling the decomposition process from the plant-uptake process. In the absence of nutrient uptake in spring, when the deciduous trees would have started production, the

insights from a large-scale field experiment

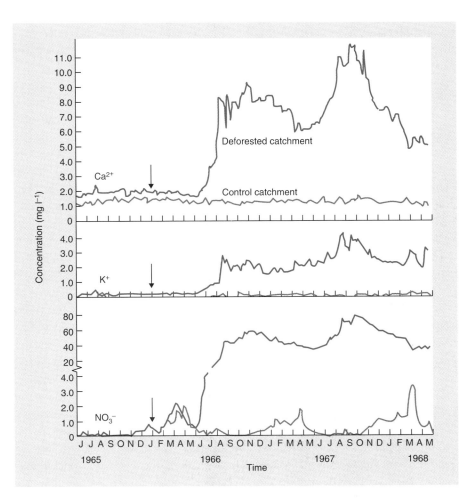

Figure 1.14
Concentrations of ions in stream water from the experimentally deforested watershed 2 and the control (un-manipulated) watershed 6 at Hubbard Brook. The timing of deforestation is indicated by arrows. Note that the nitrate axis has a break in it. (After Likens & Borman, 1975.)

inorganic nutrients released by decomposer activity were available to be leached in the drainage water. The major effect of deforestation was on nitrate, whereas increased decomposition of nitrogenous organic matter produced ammonium, which in turn was nitrified to nitrite and nitrate. Because of its mobility in the soil and the absence of uptake by trees, nitrate was lost at a high rate in stream water. These results emphasize the normally efficient cycling to which inorganic nitrogen is subject in forest ecosystems.

to allow statistically significant trends to become evident many years of data may be required

Likens knew from the beginning of the study in 1963 that the rain and snow at Hubbard Brook were quite acid, but it was some years before the widespread nature of acid rain in North America became clear. In fact, Hubbard Brook is more than 100 km from the nearest urban-industrial area, yet precipitation and stream water were both markedly acid as a result of atmospheric pollution from fossil fuels. The long-term records kept so meticulously since 1963 at Hubbard Brook have proved invaluable in monitoring progress in the war against acid rain and its long-term consequences. The value of such records of stream water concentrations can be seen for hydrogen, sulfate, and nitrate, three ions associated with acid rain (which in simple terms is a mixture of dilute nitric and sulfuric acids; sulfuric acid is the dominant acid in the eastern United States). There have been linear, statistically significant declines in average annual concentrations of H^+ and SO_4^- since 1964/1965, and also of NO_3^-, though the latter has been subject to much greater year-to-year variation (Figure 1.15). Of note, however, is the fact that the results for shorter periods suggest quite different trends. Consider the hydrogen ion graph, where three periods of 4 years are highlighted in different colors. The first suggests an increasing trend, the second no change, and the third a decreasing trend. In fact, no statistically significant long-term trend was established until nearly two decades of data had been amassed (Likens, 1989).

long data runs reveal the history of acid rain

It is thought that acid rain began in the United States in the early 1950s (before monitoring began at Hubbard Brook). After the passage of the Clean Air Act in 1970, emissions of SO_2 and particulates were reduced, and this trend has been clearly reflected in both precipitation and stream water chemical composition (Figure 1.15). Additional reductions in emissions are expected as a result of the 1990 amendments to the Clean Air Act. However, critical questions remain—will forest and aquatic ecosystems recover from the effects of acid rain, and how long will it take (Likens et al., 1996)?

UNANSWERED QUESTION:

how long to recover from acid rain?

Using long-term data from Hubbard Brook and predictions of reductions in SO_2 emissions as a result of government legislation, Likens and Bormann (1994) estimated that by the year 2000 the sulfur loading in the atmosphere would still be three times higher than values recommended for protection of sensitive forests and aquatic communities (many plants, fish, and aquatic invertebrates are intolerant of acid conditions). Moreover, declining inputs to Hubbard Brook of base cations, such as calcium, may be causing the forests and streams to become even more sensitive to acidic inputs. Likens and Bormann (1995) hypothesize that a dramatic decline in forest growth rates during recent years may be related to a decline in the level of calcium, a critical nutrient for tree growth, in the soil. Acid rain may be responsible for the calcium deficiency. An associated reduction in bird populations in the forest may even be linked to this scenario. These unanswered questions are the subject of new phases of research at Hubbard Brook.

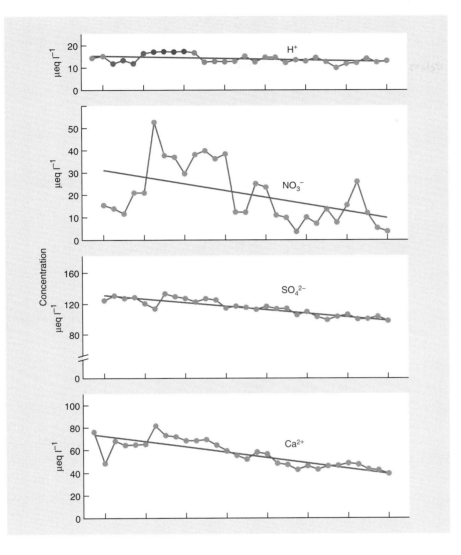

Figure 1.15
Long-term changes in concentrations (microequivalents per liter) of H$^+$, NO$_3$, SO$_4$, and Ca^{2+} in stream water from Hubbard Brook watershed 6 from 1963/1964 to 1992/1993. The regression lines for all these ions have a probability of being significantly different from zero (no change) of $P < 0.05$; in other words, there is a statistically significant pattern of decline in each. (After Likens & Bormann, 1994.)

1.3.4 Fox rabies in Europe—a model study

The value, indeed the necessity, of mathematical models in ecology can be illustrated by examining the problem of controlling the spread of rabies. Rabies is a viral disease of the central nervous system that excites considerable fear—perhaps because of its behavioral symptoms, which suggest "madness," perhaps because once the first symptoms are apparent, death is very frequently inevitable. It is *endemic*, present at a relatively unchanging prevalence, in many parts of the world, but in Europe its prevalence seems, over the centuries, to have waxed and waned. In the most recent outbreak, for example, rabies appears to have spread westward and southward since the 1940s from what is now Poland (Figure 1.16). Wherever rabies occurs, it tends to have a relatively high prevalence in species of wildlife (though the particular species vary: wild dogs in parts of Africa, for example, and skunks and raccoons in different parts of the United States), but it causes most problems when it passes from these wildlife *reservoirs* to domestic dogs, or to cattle, or to people. The current

Figure 1.16
The gradual spread of rabies across
Europe in fox populations after an
initial outbreak in Poland. (After
MacDonald, 1980.)

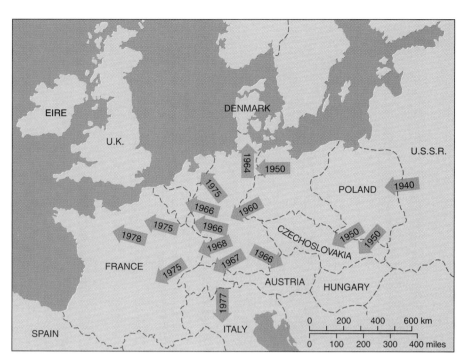

Figure 1.16
The gradual spread of rabies across Europe in fox populations after an initial outbreak in Poland. (After MacDonald, 1980.)

European outbreak is largely associated with red foxes (*Vulpes vulpes*) (Figure 1.17); while among the species of human concern, cattle have been the main victims.

keeping rabies out and planning in case it arrives

Those countries beyond the present limit of rabies's spread in Europe are, not surprisingly, keen to remain so. In particular, rabies has not spread across the Channel from mainland Europe to the British Isles. To protect itself, Britain has had a strict policy of subjecting all incoming dogs to an extended period of quarantine (though the value of this is questionable and the policy itself is under review at the time of

Figure 1.17
A red fox pouncing on a mouse.

writing), but it is also necessary to think ahead: to have plans for what *should* be done if rabies *were* to enter Britain and start to spread.

Clearly, for a government to have such advance plans is no more than behaving responsibly—but what should those plans be based on? Ideally, on an ability to predict how rabies would spread from an initial point of invasion, and what the consequences of different strategies to control its spread would be. Broadly, there are two such strategies (assuming that the risk would come, as it has elsewhere in Europe, from foxes). The first would be to cull foxes: to kill enough of them, on a regular basis, for there to be insufficient hosts to support rabies's spread. The second would be to halt its spread by attracting foxes to bait containing an oral vaccine against rabies, such that there would be insufficient *susceptible* foxes to support the disease.

should fox populations be culled or made less susceptible to the disease?

If and when rabies arrives in Britain, the authorities must have already reached a decision about which alternative is more likely to be effective. One conventional scientific route to such a decision is to carry out carefully controlled experiments. But in this case experiments would be impossible, since they would presumably involve artificially introducing rabies repeatedly, sometimes trying to control it by culling, sometimes by baiting. There is another option though: to explore the consequences of the different alternatives through a mathematical model. We might never be as confident in the results obtained as we would have been in the results of a large-scale experiment with real foxes and real rabies. But since these are unobtainable, mathematical results and predictions are certainly far better than no predictions at all.

One of the great virtues of a mathematical model is that it makes explicit the connection between what is fed into the model—in this case, what we believe we know about foxes and rabies—and the conclusions and predictions that emerge from the model. By using a precise and clear mathematical logic, the model is able to say, *If* we proceed from this particular set of assumptions, then *these* are our conclusions. But it is important to be cautious: the conclusions of a mathematical model can, at best, be no better than the assumptions from which they stem. And, as we shall see, some at least of the assumptions of a mathematical model are inevitably only approximations to the truth.

The steps in building and applying a mathematical model of fox–rabies popula-tions are as follows:

steps in the modeling process

1. Construct a general model of interacting host and pathogen populations that is appropriate for foxes and rabies.
2. Use the available data on fox and rabies biological characteristics to make the model a more specific reflection of their interaction (*parameterizing the model*).
3. Check that the model "works"—that it is good enough, at least, to generate fox–rabies population dynamics that look like those actually observed in nature.
4. Use the model, if it is good enough, to ask which of the alternative control strategies is likely to be the most effective.

It is neither necessary nor appropriate to go into great detail here about any of these stages. What is important is that both the value and the limitations of mathematical models as ecological tools should be recognized—and that modeling should be demystified.

An appropriate host–pathogen model is set out as a flow diagram in Figure 1.18. In order to be analyzed and explored, this flow diagram has to be converted into a series of equations, but neither the equations themselves nor the details of their analysis need concern us. The model contains three types of host:

1. Those that are uninfected and hence susceptible to infection: their number, denoted by S, can be added to by birth but can be reduced by both natural death and infection.
2. Those that are infected but not yet infectious, because the disease has a latent phase: their number, L, can be added to when susceptibles become infected through contact with infectious hosts but can be reduced by both natural death and by hosts' passing from the latent to the fully infectious class.
3. Those that are infectious: their number, I, can be added to when individuals pass through the latent phase, but can be reduced by both natural and disease-induced death.

Note that, unlike in many other types of disease, there is no recovered or immune class here because such recovery to immunity is very rare in fox rabies. Note, too, though, that no recovery is just one of several approximations that have been made in order to make the model manageable (actually, recovery is rare but not unknown). Among the other approximations are the assumptions that movement of foxes into and out of populations can be ignored, and that the transmission of rabies from infectious to susceptible foxes can be modeled by imagining that foxes simply bump into one another at random.

judging the success of a model

Despite these approximations, can the model capture the essence of the fox-rabies interaction? The model overall can generate various sorts of population dynamics (Figure 1.19). At low host densities the pathogen is unable to sustain itself, whereas at higher densities the pathogen is sustained and tends to generate regular fluctuations (cycles) in host abundance.

Figure 1.18
A model of a host-pathogen interaction, expressed as a flow diagram, in a population consisting of hosts that are either susceptible (S), latently infected (L), or fully infectious (I). (After Anderson et al., 1981.)

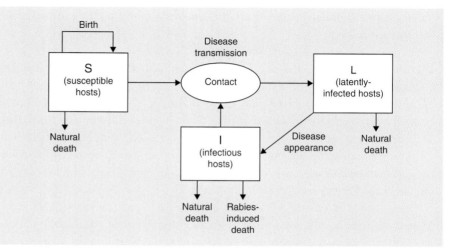

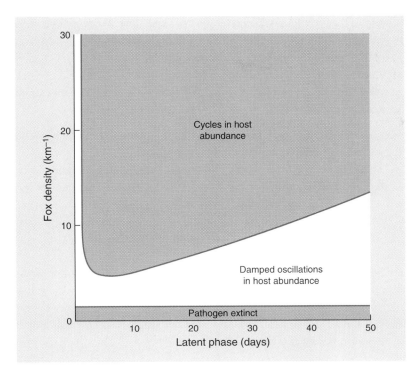

Figure 1.19
The various types of population dynamics generated by the model in Figure 1.18 for different values of the natural, disease-free density of hosts (vertical axis) and the length of the latent phase of the disease (horizontal axis). The overall pattern (pathogen not sustained at low host densities, etc.) is applicable to host–pathogen interactions of this type in general, but the particular values are those appropriate to the interaction between red foxes and rabies. (After Anderson et al., 1981.)

Which of these patterns is predicted for fox rabies? To get an answer, estimates of various parameters for these species are fed into the model. For example, foxes typically live for 2 years and give birth to a cub every year; rabies typically gives rise to a latent phase lasting 28 days; rabid foxes typically die 5 days after becoming infectious. With these and other appropriate values, the model predicts first (Figure 1.19) that rabies will only be sustained in fox populations with a density of at least one individual per square kilometer; and also that at the densities frequently attained by fox populations with endemic rabies in mainland Europe, the fox populations should exhibit cycles of abundance with peaks around every 4 years, whereas no cycles should be observed in the absence of rabies. These predictions, broadly, are borne out in practice. Hence, the model does indeed appear capable of capturing the essence of the interaction. This in turn can give us some confidence in applying the model to a comparison of alternative potential methods of rabies control.

For culling to be effective it would have to keep the fox population below the rabies threshold of one fox per square kilometer (Figure 1.19). However, given that fox densities in Britain are estimated often to be around five per square kilometer, and in urban areas may be much higher, the level of sustained fox killing that would be required to make this successful is simply not achievable. The model therefore predicts that fox culling would not work as a means of rabies control in Britain. What about baits laced with an oral vaccine? Here, too, the aim would be to reduce the density of *susceptible* (nonvaccinated) foxes to below one per square kilometer, and since these baits are capable of attracting up to 80 percent of a fox population, the strategy seems to have a good chance of success in those many areas where densities are less than five per square kilometer. Overall, therefore, the model predicts that baiting would be a much more effective and achievable strategy for controlling the spread of rabies than culling, but that in urban areas (densities much higher than five per square kilometer) even baiting may not be successful.

This example, then, has illustrated a number of important general points about mathematical models in ecology:

1. Models can be valuable for exploring scenarios and situations for which we do not have, and perhaps cannot expect to obtain, real data.
2. They can be valuable, too, for summarizing our current state of knowledge and generating predictions in which the connections among current knowledge, assumptions, and predictions are explicit and clear.
3. In order to be valuable in these ways, a model does not have to be (indeed, cannot possibly be) a full and perfect description of the real world it seeks to mimic—all models incorporate approximations.
4. Caution is therefore always necessary—all conclusions and predictions are provisional and can be no better than the knowledge and assumptions on which they are based.
5. Nonetheless, a model is inevitably applied with much more confidence once it has received support from real sets of data.

Summary

Ecology as a pure and applied science

We define *ecology* as the scientific study of the distribution and abundance of organisms and the interactions that determine distribution and abundance. From their origins in prehistory as an applied science of food gathering and enemy avoidance, the twin threads of pure and applied ecology have developed side by side, each depending on the other. This book is about how ecological understanding is achieved, what we do and do not understand, and how that understanding can help us predict, manage, and control.

Questions of scale

Ecology deals with four levels of ecological organization—individual organisms, populations (individuals of the same species), communities (a greater or lesser number of populations), and ecosystems (the community together with its physical environment). Ecology can be done at a variety of spatial scales, from the community within an individual cell to that of the whole biosphere. Ecologists also work on a variety of time scales. Ecological succession, for example, may be studied during the decomposition of animal dung (weeks) or during the period of climate change since the last ice age (millennia). The normal period of a research program (3–5 years) may often miss important patterns that occur over long time scales.

The diversity of ecological evidence

Many ecological studies involve careful observation and monitoring, in the natural environment, of the changing abundance of one or more species over time, or over space, or over both. Establishing the cause(s) of any patterns observed often requires manipulative field experiments. For complex ecological systems (and most of them are) it is often appropriate to construct simple laboratory systems that can act as jumping-off points in our search for understanding. Mathematical models of ecological communities also have an important role to play in unraveling ecological complexity. However, the worth of models and simple laboratory experiments must always be judged in terms of the light they throw on the working of natural systems.

Statistics and scientific rigor

What makes the science of ecology rigorous is that it is based not on statements that are simply assertions but on conclusions that are the results of carefully planned investigations with well-thought-out sampling regimes, and on conclusions, moreover, to which a level of statistical confidence can be attached. The term that is most often used, at the end of a statistical test, to measure the strength of conclusions being drawn is P-value or *probability level*. The statements $P < 0.05$ (significant) or $P < 0.01$ (highly significant) mean that these are studies where sufficient data have been collected to establish a conclusion with which we can be confident.

Ecology in practice

Studies of the impacts of brown trout, introduced to New Zealand last century, have spanned all four ecological levels (individuals, populations, communities, ecosystems). Trout have replaced populations of native galaxiid fish below waterfalls. Laboratory and field experiments have established that grazing invertebrates in trout streams show an individual response, spending more time hiding and less time grazing. Trout cause a *cascading community effect* because the grazers impact less on the algae. Finally, a descriptive study has revealed an ecosystem consequence—primary productivity by algae is higher in a trout than in a galaxiid stream.

In the Cedar Creek Natural History Area are agricultural fields that are still under cultivation and others that have been abandoned at various times since the mid-1920s. This natural experiment was exploited to provide a description of the species sequence associated with succession on such abandoned fields. However, the fields differed not only in age but also in soil nitrogen level. A set of field experiments, in which soil nitrogen level was augmented in a systematic way in fields

of different age, showed that time and nitrogen interacted to cause the observed successional sequences.

The Hubbard Brook Experimental Forest study has been running since 1963. A large-scale experiment, involving the felling of all the trees in a single watershed, resulted in a dramatic increase in chemical concentrations (particularly nitrate) in stream water. The loss of nitrate from the land and its increase in water can be expected to have consequences for the communities on both sides of the land–water interface. Monitoring of chemical concentrations for more than three decades in undisturbed watersheds has revealed how acid rain has been diminishing as a result of the Clean Air Act. However, neither the forest nor the stream is immune from the continuing effects of the pollution that caused acid rain.

In case rabies spreads to the British Isles, it is necessary to have a strategy planned. Broadly, there are two possibilities: to kill enough foxes (the main agent of rabies spread in Europe) or to halt the spread of rabies by making sufficient foxes immune. A fox–rabies mathematical model, based on the limited data available and a number of assumptions and approximations, predicts that baiting with an oral vaccine would be the more effective strategy, but that in urban areas even baiting may not be successful.

Review Questions

▲ = Challenge Question

1. ▲ Discuss the different ways that ecological evidence can be gained. How would you go about trying to answer one of ecology's unanswered questions, namely, Why are there more species in the tropics than at the poles?

2. ▲ The variety of microorganisms that live on your teeth have an ecology like any other community. What do you think might be the similarities in the forces determining species richness (the number of species present) in your oral community as opposed to a community of seaweeds living on boulders along the shoreline?

3. Why do some temporal patterns in ecology need long runs of data to detect them, whereas others need only short runs of data?

4. Discuss the pros and cons of descriptive studies as opposed to laboratory studies of the same ecological phenomenon.

5. What is a "natural field experiment"? Why are ecologists keen to take advantage of them?

6. Discuss the assertion, Seek simplicity, but distrust it, from the point of view of ecological research.

7. ▲ In a study of stream ecology, you need to choose 20 sites to test the hypothesis that brown trout have higher densities where the stream bed consists of cobbles. How might your results be biased if you chose all your sites to be easy to access because they are near roads or bridges?

8. How might the results of the Cedar Creek study of old-field succession have been different if a single field had been monitored for 50 years, rather than simultaneously comparing fields abandoned at different times in the past?

9. ▲ When all the trees were felled in a Hubbard Brook watershed, there were dramatic differences in the chemical characteristics of the stream water draining the watershed. How do you think stream chemical properties would change in subsequent years as plants began to grow again in the watershed?

10. What are the main factors affecting the confidence we can have in predictions of a mathematical model?

As the great Russian-American biologist Dobzhansky said, "Nothing in biology makes sense, except in the light of evolution." But equally, very little in evolution makes sense, except in the light of ecology: ecology provides the stage on which the "evolutionary play" is performed. Ecologists and evolutionary biologists need a thorough understanding of each other's disciplines to make sense of key patterns and processes.

The Ecology of Evolution

Key Concepts

In this chapter you will

- appreciate that Darwin and Wallace, who were responsible for the theory of evolution by natural selection, were both ecologists who studied the relationships between organisms and their environments.

- understand that the populations that make up a species vary from place to place on both geographic and fine local scales and some of the variation is heritable.

- realize that natural selection can act very quickly on heritable variation—we can study it in action and control it in experiments.

- understand that reciprocal transplanting of individuals of a species into each other's habitats can show a finely specialized fit between organisms and their environments.

- appreciate that the origin of species requires the reproductive isolation of populations (e.g., on islands) as well as natural selection forcing them to diverge.

- realize that the environment in which evolution occurs is continually changing and natural selection fits organisms to their past—it does not anticipate the future.

- realize that the evolutionary history of species leaves its fingerprints in their form and distribution and constrains what future selection can achieve.

- understand that natural selection may force similar form or behavior to evolve from widely different ancestral lines (convergent evolution) or the same range of ecological differences to evolve in populations that have become separated (parallel evolution).

2.1 ⟩ Introduction

It is not too difficult to imagine a planet like Earth that is inhabited by just one sort of organism. This might be limited in its distribution to just a very tiny subset of the multiplicity of available environments. It might range widely over many physical environments if it possessed broad temperature tolerance. It might range even more widely if it could tolerate periods of desiccation, and even farther if it could also function when immersed in water. But it could live only where there was liquid water, a source of energy and access to inorganic resources for growth. Even one such ideal species would give the Earth an ecology and a biogeography. Ecologists visiting from another planet would find plenty to keep themselves busy studying the relationships between this unique species and its environment!

In fact, of course, the Earth is inhabited by a multiplicity of types of organism that are distributed neither randomly nor as a homogeneous mixture over the surface of the globe. Any sampled area, even on the scale of a whole continent, contains only a subset of the variety of species present on earth. One of the greatest of all ecological generalizations is that all species are so specialized that they are always absent from almost everywhere. A great part of the science of ecology tries to explain why there are so many types of organism and why their distributions are so restricted. A proper answer to these ecological questions depends fundamentally on an understanding of the processes of evolution that have led to present-day diversity and distribution.

all species are so specialized that they are always absent from almost everywhere

Until relatively recently in the history of biology, the emphasis on diversity was to use it (e.g., for medicine, food, and fiber), to exhibit it in zoos and botanic gardens, and to catalogue it in museums (Box 2.1). In the absence of an understanding of how this diversity developed, the catalogues can be likened to stamp collecting rather than science. The enduring contribution of Charles Darwin and Alfred Russel Wallace was to provide ecologists with the scientific foundations to comprehend patterns in diversity and distribution over the face of the Earth.

2.2 ⟩ Evolution by Natural Selection

Darwin and Wallace were both ecologists

Darwin and Wallace (Figure 2.1) were both ecologists (although their seminal work was performed before the term was coined) who were exposed to the diversity of nature in the raw. Darwin sailed around the world as naturalist on the 5-year expedition of H.M.S. *Beagle* (1831–1836) recording and collecting in the enormous variety of environments that he explored on the way. He gradually developed the view that the natural diversity of nature was the result of a process of evolution in which *natural selection* favored some variants within species through a "struggle for existence." He developed this theme over the next 20 years through detailed study and an enormous correspondence with his friends as he prepared a major work for publication with all the evidence carefully marshaled. But he was in no hurry to publish.

In 1858, Wallace wrote to Darwin spelling out in all its essentials the same theory of evolution. Wallace was a passionate amateur naturalist. He had read Darwin's journal of the voyage of the *Beagle* and after a visit to the Jardin des Plantes

Box 2.1

A Brief History of the Study of Diversity

An awareness of the diversity of living organisms, and of what lives where, is part of the knowledge that the human species accumulates and hands down through the generations. Hunter–gatherer peoples needed (and still need) detailed knowledge of the natural history of their environments to obtain food successfully and at the same time escape the hazards of being poisoned or eaten. The Arawaks of the South American equatorial forest know where to find and how to catch the species of large animal around them and also the names of trees and how they can be used.

The Chinese emperor Shen Nung had compiled what was perhaps the first written herbal of useful plants before 2000 B.C. and by the first century A.D. Dioscorides had described five hundred species of medicinal plants and illustrated many of them.

Collections of living specimens in zoos and gardens also have a long history—certainly back to Greece in the seventh century B.C. The urge to collect from the diversity of nature developed in the West in the seventeenth century when some individuals made their living by finding interesting specimens for other people's collections. John Tradescant the father (died 1638) and John Tradescant the son (1608–1662) spent most of their lives collecting plants and importing live specimens for the gardens of royalty and the nobility. The father was the first botanist to visit Russia (1618), and he brought back many living plants; and his son made three visits (1637, 1642, and 1654) to the New World to collect specimens in the American colonies.

Wealthy individuals built up vast collections into personal museums and traveled or sent travelers in search of novelties from new lands as they were discovered and colonized. Naturalists and artists (often the same people) were sent to accompany the major voyages of exploration to report and take home, dead or alive, collections of the diversity of organisms and artifacts that they found. Sciences of taxonomy and systematics developed and flourished—taxonomy gave names to the various types of organism and systematics provided systems for classifying and pigeonholing them.

When big national museums were established (the British Museum in 1759 and the Smithsonian in Washington in 1846), they were largely from the gifts of personal museums and collections.

Like zoos and gardens, the museums' main role was to make a public display of the diversity of nature, especially "the new and curious." Much of the seventeenth-, eighteenth-, and nineteenth-century fascination with the diversity of natural history was like stamp collecting, acquiring diversity for its own sake, especially rarities.

There was no need to explain the diversity—the biblical theory of the seven-day creation of the world sufficed. However, the idea that the diversity of nature had "evolved" over time by progressive divergence from preexisting stocks was beginning to be discussed in the early nineteenth century. In 1844 an anonymous publication, "The Vestiges of Creation," put the cat among the pigeons with a popular account of the idea that animal species had descended from other species.

in Paris and the insect room at the British Museum he wrote in 1847, "I should like to take some one family to study thoroughly, principally with a view to the theory of the origin of species." From 1847 to 1852, with his friend H. W. Bates, he explored and collected in the river basins of the Amazon and Rio Negro, and from 1854 to 1862 he made an extensive expedition in the Malay Archipelago. He recalled lying on his bed in 1858 "in the hot fit of intermittent fever, when the idea [of natural selection] suddenly came to me. I thought it all out before the fit was over, and . . . I believe I finished the first draft the next day."

Today, competition for fame and financial support would commonly lead to fierce conflict about priority. Instead, in an outstanding example of personal greatness in science (Darwin wrote to his friend Hooker, "It is miserable in me to care at all about priority"), sketches of Darwin's and Wallace's ideas were presented together

an outstanding example of personal greatness in science

Figure 2.1
Charles Robert Darwin in 1854, 46 years old (left), and Alfred Russel Wallace in 1862, 39 years old (right), were passionate naturalists and explorers. They saw the interaction between organisms and their environments as an essential part of the process of natural selection: their theory of evolution by natural selection is an ecological theory.

in science (Darwin wrote to his friend Hooker, "It is miserable in me to care at all about priority"), sketches of Darwin's and Wallace's ideas were presented together at a meeting of the Linnean Society in London. Darwin's "On the Origin of Species" was then hastily prepared and published in 1859 as an "abstract" (in reality, a book) of what was intended eventually to become his "big book." (In fact, most of what Darwin called his big book, with all its detail, footnotes and references, was not published until 1975 [Stauffer, 1975].) *On the Origin of Species* may be considered the first major textbook of ecology, and aspiring ecologists would do well to read at least the third chapter.

Both Darwin and Wallace had read "An Essay on the Principle of Population," published by Malthus in 1798. Malthus's essay was concerned with the human population and its intrinsic rate of natural increase, which, if unchecked, he calculated, would be capable of doubling every twenty-five years and overrunning the planet. Malthus realized that limited resources slowed the growth of populations and placed absolute limits on their size and that disease, wars, and other disasters also checked population growth. Wallace commented, "The most interesting coincidence in the matter, I think, is that I, as well as Darwin, was led to the theory itself through Malthus." Malthus had written with humankind in view but, as experienced field naturalists, Darwin and Wallace realized that the Malthusian argument applied with equal force to the whole of the plant and animal kingdoms.

influence of Malthus's essay on Darwin and Wallace

The living world of nature is dominated by reproduction, overcrowding, and death. Darwin and Wallace were almost obsessed by this great truth. They appreciated that all organisms possess a potential to multiply that is impossible to realize. "Every being, which during its natural lifetime produces several eggs or seeds, must suffer destruction during some period of its life, and during some season or occasional year, otherwise, on the principle of geometrical increase, its numbers would quickly become so inordinately great that no country could support the product" (Darwin [1859] *Origin of Species*). Noting the immense fecundity of some species (e.g., a single individual of the sea slug *Doris* may produce 600,000 eggs and the parasitic roundworm *Ascaris*, 64 million), Darwin used as an example of the absurd consequences of unimpeded population growth a population of fish which in eight generations, each fish laying two thousand eggs, "would cover like a sheet the whole globe, land and water." In one of the earliest examples of population ecology Darwin counted all the seedlings that emerged from a piece of cultivated ground three feet long and two wide: "Out of 357 no less than 295 were destroyed, chiefly by slugs and insects."

the forces of reproduction, crowding, and death

There is no species in which individuals produce only one egg or seed, yet averaged over a number of generations each individual organism can leave no more than one descendant. On average all but one must die. On average only that one survivor can become an ancestor; only that one can leave descendants.

The theory of evolution by natural selection rests on a series of established truths:

fundamental truths of evolutionary theory

1. Individuals that form a population of a species are not identical.
2. Some of the variation between individuals is heritable.
3. All populations are capable of exponential growth that would greatly overwhelm the environment, but most individuals die before reproduction and most (usually all) reproduce at less than their maximal rate.
4. Different ancestors leave different numbers of descendants (N.B. descendants, *not* just offspring): they do not all contribute equally to subsequent generations.

Given these four truths, the heritable features that define a population will change over time. Evolution will happen.

The philosopher Herbert Spencer described the process of evolution by natural selection as "the survival of the fittest," but if we then ask which are the fittest we find ourselves answering "those that survive" and if we ask which organisms survive

"the survival of the fittest"

we are rather forced to answer, "those that are fittest." Clearly this is not very helpful. But the idea of *fitness* redefined remains central to evolutionary thinking. An individual will survive, reproduce, and leave descendants in some environments but not in others. It is in this sense that some environments may be described as favorable or unfavorable, and it is in this same sense that some organisms can be considered to be fit or not.

Fitness is a relative, not an absolute term. The numbers of seeds produced by a plant, or eggs produced by an insect, their fecundity, are not direct measures of their fitness; nor indeed are the *numbers* of descendants that they leave. Rather it is the proportionate contribution that an individual makes to future generations that determines its fitness: the fittest individuals in a population are those that leave the greatest number of descendants *relative to* the number of descendants left by other individuals in the population. Those individuals that leave the greatest proportion of descendants in a population have the greatest influence on the heritable characteristics of that population. If, as a result, the heritable characteristics of a population change from generation to generation, evolution *will have* occurred.

human selection and natural selection are very different processes

Darwin had been greatly influenced by the achievements of plant and animal breeders—for example, the extraordinary variety of pigeons, dogs, and farm animals that had been deliberately bred by selecting individual parents with exaggerated traits. He and Wallace saw nature doing the same thing; "selecting" those individuals that survived from their excessively multiplying populations, hence the phrase "natural selection." But the phrase gives a wrong impression. There is a great difference between human and natural selection. In particular human selection has an aim for the future—to breed a cereal with a higher yield, a more attractive pet dog, a better hunter, or a cow that will yield more milk. But nature has no aim. Evolution happens because some individuals have survived the death and destruction of the past, not because they were somehow chosen or selected by "Mother Nature" as improvements for the future!

Darwin and Wallace placed slightly different emphases on the forces that drive evolution. Wallace emphasized the killing forces of physical conditions such as frost, drought, and predators. Darwin laid more emphasis on competition for limited resources and the lethal effects of crowding that result from overpopulation (we pick up these three powerful ecological forces in Chapters 3, "Physical Conditions and the Availability of Resources"; 5, "Birth, Death, and Movement"; and 8, "Predation, Grazing, and Disease"). Both authors emphasized that most individuals die before they can reproduce and contribute nothing to future generations. Both tended to ignore the important fact that those individuals that do survive in a population may leave different numbers of descendants.

2.3 ▸ Evolution within Species

to understand the evolution of species
we need to understand evolution
within species

The natural world is not composed of a continuum of types of organism each grading into the next. We recognize boundaries between one sort of organism and another, and in one of the great achievements of biological science, Linnaeus in 1789 devised an orderly system for naming the different sorts. Part of his genius was to recognize that there were features of both plants and animals that were not easily modified by the organisms' immediate environment—conservative characteristics that were useful for classifying organisms into pigeonholes. In flowering plants the form of the flowers was particularly stable, whereas the sizes of leaves and stems were much more easily changed by heat and cold, watering and drought, and giving or withholding fertilizers. Nevertheless, within what we recognize as Linnaean species there is often considerable variation, and some of this is heritable. It is on this intraspecific variation that the plant and animal breeders worked to make the improvements that so impressed Darwin. In nature some of this intraspecific variation is clearly correlated with variations in the environment and represents local specialization.

Darwin called his classic book *On the Origin of Species by Means of Natural Selection*, but evolution by natural selection does far more than create new species. Natural selection and evolution occur *within* species, and we now know that we can study them in action and within our own lifetime. Moreover, we need to study the way that evolution occurs within species if we are to understand the origin of new species.

2.3.1 Variation within the distributional range of a species

The perennial plant *Achillea lanulosa* is widely distributed in the Sierra Nevada range of California. Where the plants grow at low altitudes they are normally tall and vigorous, but at higher altitudes they are small and depauperate, becoming quite puny at the altitudinal limits of the species. Such differences may be the direct effects of the different environments on the growth of the individual plants or of inherited specializations—these hypotheses can be tested by growing the plant elsewhere. Plants of this species are perennial and can be lifted and transplanted, so Jens Clausen and his colleagues sampled plants from various positions on a transect across the mountain range and transplanted them all into a garden at Stanford. The plants from different positions on the transect retained clear differences even when growing side by side in the same garden (Figure 2.2), strongly suggesting that the differences among the plants in their natural habitats were inherited specializations, each to its local environment.

Evolution forces the characteristics of populations to diverge from each other only if there is sufficient heritable variation on which selection can act and provided that the forces of selection favoring divergence are strong enough to counteract the hybridization which promotes mixing. Two populations will not diverge completely if their members (or, in the case of plants, their pollen) are continually migrating between them, mating and mixing their genes. The balance between the differentiating forces of selection and the mixing forces of hybridization was studied by Bradshaw and his students at a site called Abraham's Bosom on the island of Anglesey off the coast of North Wales. Here there was an intimate mosaic of very different habitats at the margin between maritime cliffs and grazed pasture. A common species, creeping bent grass (*Agrostis stolonifera*), was present in many of the habitats. Figure 2.3 shows a map of the site. Plants were sampled from the positions marked with numbers on the map. Lengths of shoot taken from this species of grass readily form roots, so that a number of independent rooted plants can be cloned from each single

the characteristics of a species may vary over its geographical range

variation over very short distances

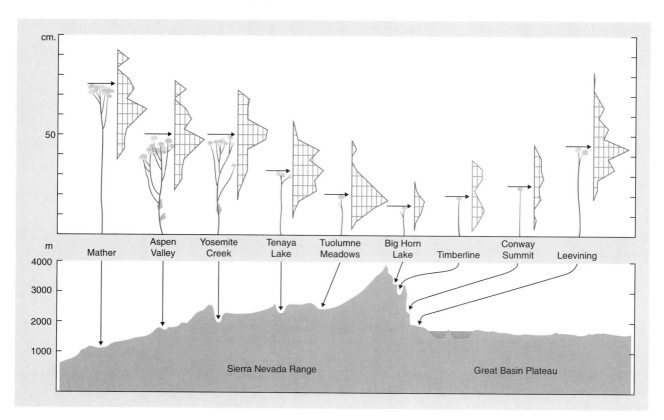

Figure 2.2

Samples of the perennial plant *Achillea lanulosa* were taken from sites along a transect across east central California at latitude approximately 38°N. The samples, each of about 60 individuals, were transplanted into a common garden at Stanford, California, and allowed to develop together in this uniform environment. The figure records the growth made by the various individuals in the Stanford garden. The arrows point to the mean height of each population; the frequency and range of heights are shown in a form appropriate for the time at which the experiment was made. (Clausen et al., 1948.)

plant in the field. These were planted in an experimental garden in a careful statistical design so that each of four plants sampled from a site in the field was represented by five rooted clonal replicates of itself. The plants spread by sending out shoots along the ground surface (stolons), and these form roots at the nodes. The growth of the plants in the experimental garden was compared by measuring the lengths of the stolons. Stolon lengths in the experimental garden were strikingly different, depending on the original habitat of the plants. Samples 6, 7, 8, 9, 10, 12 in the field represent a sharp transition from cliff to pasture over just a few meters. In the experimental garden plants from the cliffs still formed short stolons and differed sharply from plants that had come from the pasture, which retained a long stolon habit. Sites 1, 3, 5, 11, 12 represent a gradually changing habitat, and when transplanted to the experimental garden the plants formed a continuous gradation from short to long stolons as they had in the field. Here again is suggestive evidence of evolutionary divergence between plants of the same species but over much shorter distances than in the *Achillea* study and certainly within the range of pollen dispersal

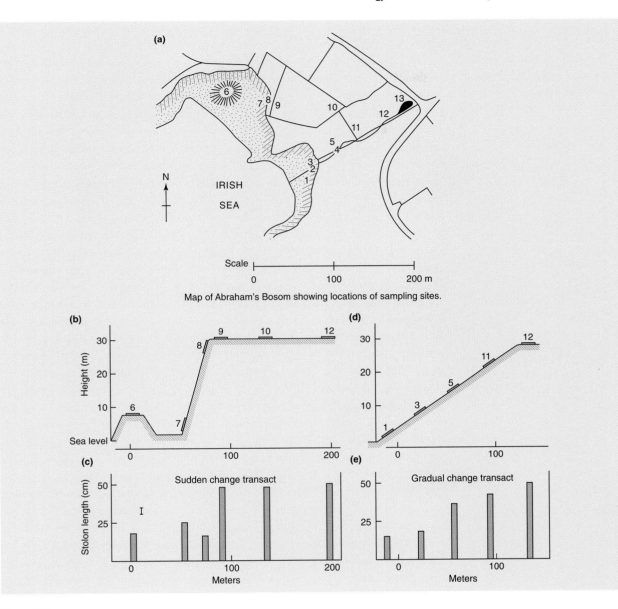

Figure 2.3

(a) Map of Abraham's Bosom, the site chosen for a study of evolution over very short distances. The unshaded area is grazed pasture; the shaded area is cliffs falling to the sea. The numbers indicate sites from which the grass *Agrostis stolonifera* was sampled. A small stream flows from site 13 down to the cliffs. Note that the whole area is only 200 m long. The prevailing winds are from the west. (b) A vertical transect across the study area showing sudden change from pasture to cliff conditions. (c) The mean length of stolons produced by plants in the experimental garden grown from samples taken from positions shown on the "sudden change" transect in (b). (d) A vertical transect across the study area showing gradual change from pasture to cliff conditions. (e) The mean length of stolons produced in the experimental garden from samples taken from positions shown on the "gradual change" transect in (d). (From Aston & Bradshaw, 1966.)

between each other. It seems again that the forces of selection outweigh the mixing forces of hybridization.

We can also test whether organisms have evolved to become specialized to life in their local environment by comparing their performance when they are grown "at home" (i.e., in their original habitat) with their performance "away" (i.e., in the habitat of others). Such *reciprocal transplant* experiments are an extension of the transplant experiments described earlier, which showed that plants sampled from different areas behaved differently when grown in a common garden. In contrast, reciprocal transplant experiments compare performance in each other's habitats.

It can be difficult to detect the local specialization of animals by transplanting them into each other's habitat: if they don't like it, most species of animal will run away. But invertebrates like corals and sea anemones are sedentary, and some can be lifted from one place and established in another. The sea anemone (*Actinia tenebrosa*) is found abundantly in pools on headlands around the coast of New South Wales, Australia. Ayre (1985) chose three colonies on headlands within 4 km of each other on which the anemone was abundant. Within each colony, he selected three transplant sites (each of 3–5 m length of shore) and at each he set aside three 1-m-wide strips for transplant experiments—two to receive anemones from the away sites and one to receive transplanted individuals from the site in question (home site). Ayre cleared the experimental sites of anemones that were present and transplanted into them anemones that had been lifted from the other sites or, for "home" transplants, from within 4 m of the home site. The anemone multiplies clonally by producing broods of asexual juveniles. The number of juveniles brooded per adult was used as a measure of the performance of the anemones in the various pools (home and away).

Eleven months later all survivors were collected and dissected to determine how many were brooding young. This was taken to be a measure of their fitness in the home and away environments. The proportion of adults that were found brooding is shown in Table 2.1. Anemones originally sampled from Green Island were rather successful in brooding young after being transplanted both home and away and did not show any specialization to their home environment. However, in all the other transplant experiments a greater proportion of anemones brooded young at home than at away sites: strong evidence of evolved local specialization. In later experiments Ayre (1995) lifted anemones from a variety of sites as before, but he then kept them for a period to acclimate at a common site before transplanting them in a reciprocal experiment. This more severe test convincingly confirmed the results in Table 2.1.

Another reciprocal transplant experiment was carried out with white clover (*Trifolium repens*), which forms clones in grazed pastures. Individual clones differ in features such as the pattern of white markings on the leaves, ability to release hydrogen cyanide when damaged or bitten, and susceptibility to various diseases. Of 50 clones sampled from a 1-hectare field of old pasture in North Wales, 43 had unique combinations of ecologically relevant heritable features. To determine whether the characteristics of individual clones matched local features of their environment, Turkington and Harper (1979) removed plants from marked positions in the field and multiplied them into clones in the common environment of a greenhouse. They then transplanted samples from each clone into the place in the sward of vegetation

Table 2.1

A reciprocal transplant experiment of the sea anemone *Actinia tenebrosa*

SITE OF ORIGIN		TRANSPLANTED TO SITES A, B, AND C AT		
		GREEN ISLAND	SALMON POINT	STRICKLAND BAY
Green Island	a	**0.42**	0.68	0.78
	b	**0.80**	0.63	0.75
	c	**0.67**	0.62	0.61
Salmon Point	a	0.11	**0.42**	0.13
	b	0.18	**0.43**	0.28
	c	0.00	**0.50**	0.40
Strickland Bay	a	0.11	0.06	**0.33**
	b	0.00	0.06	**0.27**
	c	0.04	0.20	**0.27**

a, b, and c are the three replicate sites in each colony. In each case the proportion of adults that were found brooding young is shown. Transplants back to the home sites are shown in bold print.

(Adapted from Ayre, 1985.)

from which it had originally been taken, and also to the places from where all the others had been taken. The plants were allowed to grow for a year before they were removed, dried, and weighed. The mean weight of clover plants transplanted back into their home sites was 0.89 g but at away sites it was only 0.52 g, a statistically highly significant difference.

The clover plants studied were not random samples but had been chosen from patches dominated by four different species of grass. In a second experiment, dense experimental plots of the four grasses were created from sown seed and clonal samples of the different clovers were planted into them. Again the clovers were removed after twelve months, dried, and weighed; the results are shown in Figure 2.4. The mean yield of clovers grown with their original neighbor grass was 59.4 g; the mean yield with "alien" grasses was 31.9 g, again a highly significant difference. Both parts of the clover experiment provide strong direct evidence that clover clones in the pasture were specialized and tend to perform best (make most growth) in their local environment and with their local neighbors.

2.3.2 Variation within a species where the selection pressure is known

It is evident from the previous section that variation within a species can be observed at large and small scales within its distributional range and that this is an expression of evolved specialization rather than simply being an immediate response to local environmental conditions. However, the selection pressure responsible for the differences was not known in any of the studies described. We now turn to some classic studies involving variation in populations where the selective forces are understood.

Figure 2.4
Plants of white clover (*Trifolium repens*) were sampled from a field of permanent grassland from local patches dominated by four different species of grass. The clover plants were multiplied into clones and transplanted (in all possible combinations) into plots that had been sown with seed of the four grass species. The histograms show the average weights of the transplanted clones after 12 months' growth. Clover types were sampled from patches dominated by *Agrostis tenuis (At)*, *Cynosurus cristatus (Cc)*, *Holcus lanatus (Hl)*, *Lolium perenne (Lp)*. I indicates the difference between the height of any pair of columns that is statistically significant at $P < 0.05$. (From Turkington & Harper, 1979.)

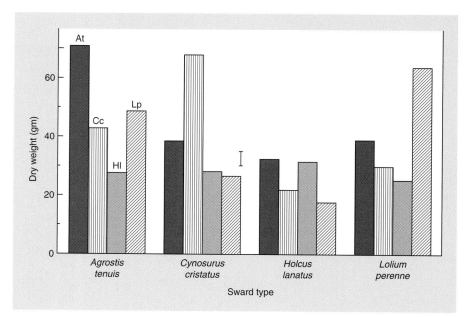

natural selection by pollution—the evolution of a melanic moth

It is not surprising that some of the most dramatic examples of natural selection in action have been driven by the ecological forces of environmental pollution—these can provide rapid change in powerful selection pressures. Pollution of the atmosphere in and after the Industrial Revolution has left evolutionary fingerprints in the most unlikely places. *Industrial melanism* is the phenomenon in which black or blackish forms of species of moths and other organisms have come to dominate populations in industrial areas. In the dark individuals, a dominant gene is responsible for producing an excess of the black pigment melanin. Industrial melanism is known in most industrialized countries, including some parts of the United States (e.g., Pittsburgh), and more than a hundred species of moth have evolved forms of industrial melanism.

The earliest recorded species to evolve in this way was the peppered moth (*Biston betularia*); the first black specimen was caught in Manchester (England) in 1848. By 1895, about 98 percent of the peppered moth population in the Manchester area was melanic. Kettlewell gathered collaborators into a large-scale survey of pale and melanic forms of the peppered moth in Britain and recorded more than twenty thousand specimens between 1952 and 1970, from which he made the map shown in Figure 2.5. The winds in Britain are predominantly westerlies, spreading industrial pollutants (especially smoke and sulfur dioxide) toward the east. Melanic forms were concentrated toward the east and were completely absent from unpolluted western parts of England and Wales, northern Scotland, and Ireland.

The moths are preyed upon by insectivorous birds that hunt by sight. Kettlewell reared large numbers of melanic and typical moths and released them in equal numbers in batches of about fifty each day in a rural and largely unpolluted area of southern England. Of the 190 moths that were captured by birds, 164 were melanic and 26 were typicals. A second study was made in an industrial area near the city

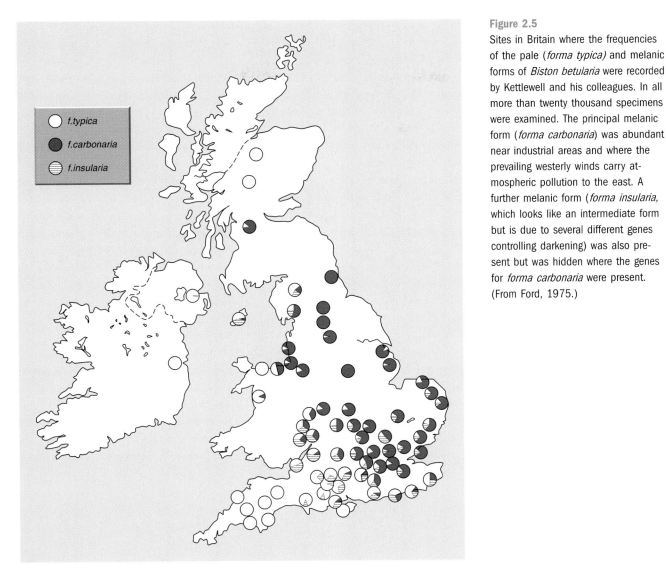

Figure 2.5
Sites in Britain where the frequencies of the pale (*forma typica)* and melanic forms of *Biston betularia* were recorded by Kettlewell and his colleagues. In all more than twenty thousand specimens were examined. The principal melanic form (*forma carbonaria*) was abundant near industrial areas and where the prevailing westerly winds carry atmospheric pollution to the east. A further melanic form (*forma insularia*, which looks like an intermediate form but is due to several different genes controlling darkening) was also present but was hidden where the genes for *forma carbonaria* were present. (From Ford, 1975.)

of Birmingham, where equal proportions of marked melanic and typical moths were released and then recaptured. In the first year, 13 percent of those recaught were typicals and 27.5 percent melanics. In the second year, 25 percent were typicals and 53 percent melanics. This showed that a significant selection pressure was exerted through bird predation and that moths of the typical form were clearly at a disadvantage in the polluted industrial environment (where their light color stood out against a sooty background), whereas the melanic forms were at a disadvantage in the pollution-free countryside (Kettlewell, 1955). Industrialized environments in Western Europe and the United States started to change in the 1960s, as oil and electricity began to replace coal and legislation was passed to impose smoke-free zones and to reduce industrial emissions of sulfur dioxide (see Chapter 13). The frequency of melanic forms then fell in what was perhaps the fastest example of evolutionary change ever recorded in any multicellular organism (Figure 2.6). The melanic form had changed from having a selective advantage in the population to being at a strong disadvantage.

The forces of selection that favored and then disfavored melanic forms have clearly been related to industrial pollution, but the idea that melanic forms were favored simply because they were camouflaged against smoke-stained backgrounds may be only part of the story. The moths rest on tree trunks during the day, and nonmelanic moths are well hidden against a background of mosses and lichens. Industrial pollution has not just blackened the moths' background; atmospheric pollution, especially SO_2, has also destroyed most of the moss and lichen on the tree trunks. Indeed the distribution of melanic forms in Figure 2.5 closely fits the areas in which tree trunks were likely to have lost lichen cover as a result of SO_2 and so ceased to provide such effective camouflage for the nonmelanic moths. Thus SO_2 pollution may have been as important as smoke in selecting melanic moths.

natural selection by pollution—evolution of heavy-metal tolerance in plants associated with mining

Forms of some grasses and other flowering plants are tolerant of another form of pollution, the presence of toxic heavy metals such as lead, zinc, and copper, which contaminate the soil after mining. Populations of plants on contaminated areas may be tolerant of heavy metals, while at the edge of these areas a transition from tolerant to intolerant forms can occur over very short distances (Figure 2.7).

In some cases it has been possible to measure the speed of evolution. Zinc-tolerant forms of two species of grass (*Agrostis canina* and *Festuca ovina*) were found to have evolved under a zinc galvanized fence and under galvanized electricity pylons within thirty years of their erection (Al-Hiyaly et al., 1993).

natural selection by predation—a controlled field experiment in fish evolution

The guppy (*Poecilia reticulata*), a small freshwater fish from northeastern South America, has been the material for a classic series of evolutionary experiments. In Trinidad many rivers flow down the northern range of mountains, and these are

Figure 2.6
Change in the frequency of the *carbonaria* form of the peppered moth *Biston betularia* in the Manchester area since 1950. Vertical lines show the standard error and the horizontal lines show the range of years included. (Cook et al., 1998.)

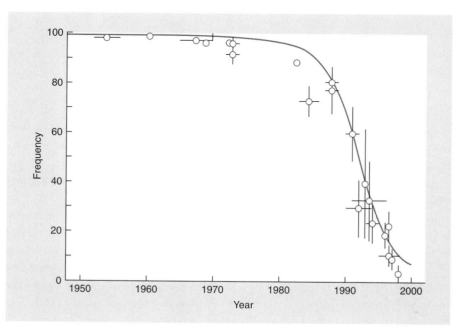

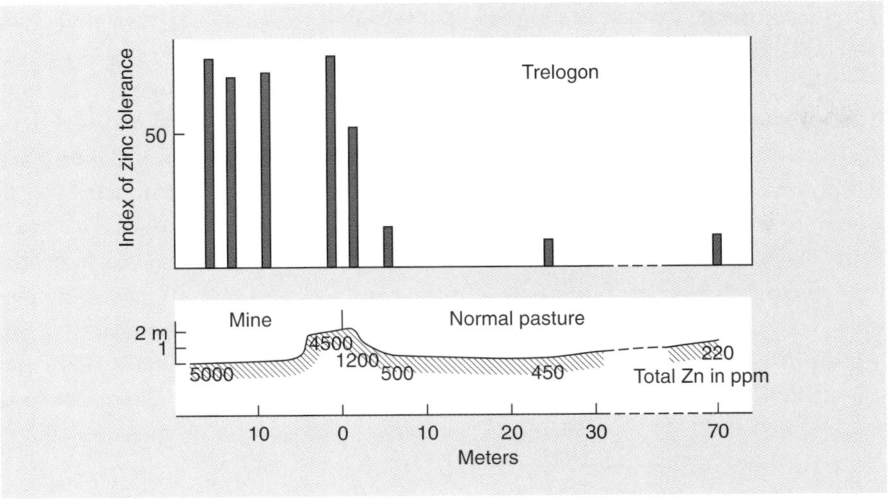

Figure 2.7
The grass *Anthoxanthum odoratum* colonizes land heavily contaminated with zinc on an old mine. This is possible because the grass has evolved zinc-tolerant forms. Samples of the grass were taken along a transect from a mine (Trelogan) into surrounding grassland (Zn concentrations in soil shown as parts per million [ppm]) and tested for zinc tolerance by measuring the length of roots that they produced when grown in a culture solution containing excess zinc. The index of zinc tolerance falls off steeply over a distance of 2–5 meters at the mine boundary. (Putwain, in Jain & Bradshaw, 1966.)

subdivided by waterfalls that isolate fish populations above the falls from those below. Guppies are present in almost all these water bodies, and in the lower waters they meet various species of predatory fish that are absent higher up the rivers. The populations of guppies in Trinidad differ from each other in almost every feature that biologists have examined. Forty-seven of these traits tend to vary in step with each other (they *covary*) and with the intensity of the risk from predators. This correlation suggests that the guppy populations have been subject to natural selection from the predators. But the fact that two phenomena are correlated does not prove that one causes the other. Only controlled experiments can establish cause and effect.

Where guppies have been free or relatively free from predators the males are brightly decorated with different numbers and sizes of colored spots (Figure 2.8). Females are dull and dowdy and (at least, to us!) inconspicuous. Whenever we study natural selection in action it becomes clear that costs and balances are involved. For every selective force that favors change there is a counteracting force that resists the change. Color in male guppies is a good example. Female guppies prefer to mate

Figure 2.8
Male and female guppies (*Poecilia reticulata*) showing two flamboyant males, courting a typical dull-colored female. (Photograph kindly provided by Anne Magurran and Iain Matthews.)

with the most gaudily decorated males—but these are more easily seen by predators, which capture more of them.

This sets the stage for some revealing experiments on the ecology of evolution. Endler established guppy populations in ponds in a greenhouse and exposed them to different intensities of predation. The number of colored spots per guppy fell sharply and rapidly when the population suffered heavy predation (Figure 2.9a). In a field experiment he moved two hundred guppies from a site far down the Aripo River where predators were common and introduced them to a site high up the river where there were neither guppies nor severe predators. The transplanted guppies thrived in their new site and changed their patterns of color with extraordinary speed. Within just two years the males had more and bigger spots of more varied color (Figure 2.9b). The females' choice of the more flamboyant males had dramatic effects

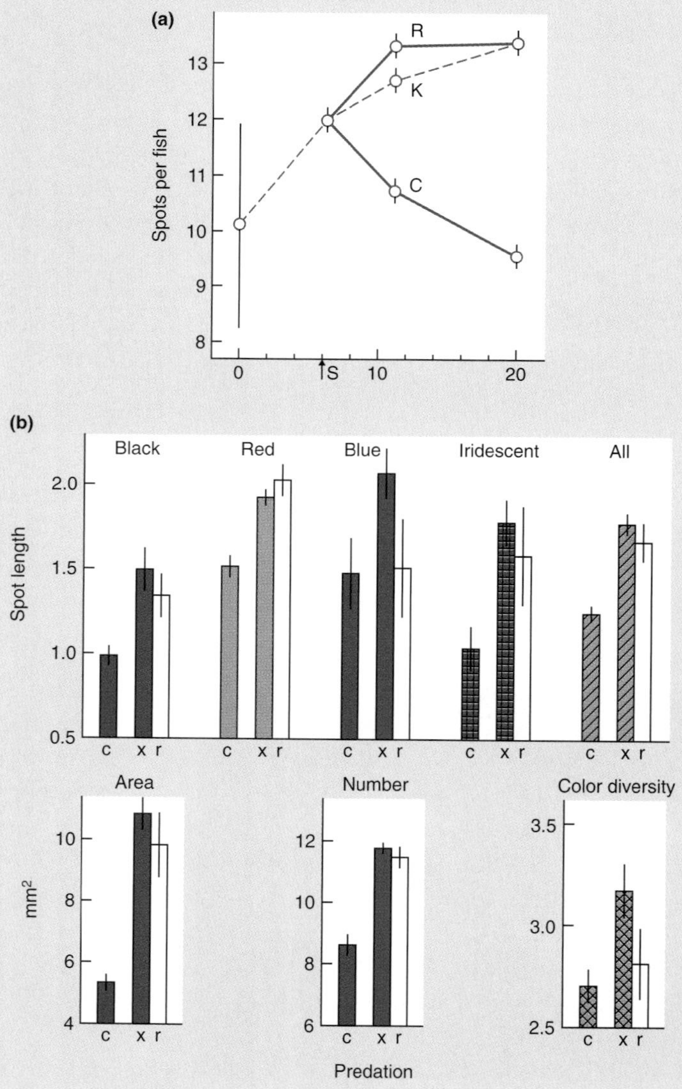

Figure 2.9
(a) An experiment showing changes in populations of guppy *Poecilia reticulata,* exposed to predators in experimental ponds. The graph shows changes in the number, size, and diversity of colored spots per fish in ponds with different populations of predatory fish. The initial population was deliberately collected from a variety of sites so as to display high variability and was introduced to the ponds at time 0. At time S weak predators (*Rivulus hartii*) were introduced to ponds R, a high intensity of predation by the dangerous predator *Crenicichila alta* was introduced into ponds C, while ponds K continued to contain no predators (the vertical lines show ± two standard errors). (b) Results of a field experiment. A population of guppies originating in a locality with dangerous predators (*c*) was transferred to a stream having only the weak predator (*Rivulus hartii*) and, until the introduction, no guppies (*x*). Another stream nearby with guppies and *R. hartii* served as a control (*r*). The bar diagrams show the lengths of the various colored spots, their area, number, and diversity on male fish collected at the three sites 2 years after the introductions. Note how *x* and *r* have converged and changed dramatically from *c*.

on the gaudiness of their descendants, but this was only because predators did not reverse the direction of selection. The significance of this remarkable study is backed up by experiments formally proving that male color pattern is genetically determined and influences female choice. This is an example of *sexual selection*, which some authors regard as an evolutionary force quite distinct from natural selection.

The speed of evolutionary change in Endler's experiment in nature was as fast as that in artificial selection experiments. There was plenty of overpopulation (as many as 14 generations of fish occurred in the 23 months during which the experiment took place) and there was considerable genetic variation in the populations upon which natural selection could act. The guppies transplanted into the nearly predator-free environment evolved in other respects too. The females were larger and older at maturity, they produced fewer but bigger offspring, and they began to lose the habit of moving together in schools, behavior that defends them against predators (Endler, 1980; Magurran, 1998).

It should be clear from Sections 2.3.1 and 2.3.2 that ecologists and evolutionary biologists need a thorough understanding of each other's disciplines to make sense of geographic patterns of intraspecific variation in the form and behavior of plants and animals.

2.3.3 Adaptive peaks and specialized abysses

Natural selection changes the character of a population by sifting out and eliminating much of its variation and leaving behind a residue for future generations with a narrower range and more restricted potential. This is commonly pictured as a force that drives populations toward peaks of *perfection*—of perfect match between organism and environment. This is an optimist's view (Figure 2.10a). An alternative picture of natural selection is that it forces populations into ever-narrowing ruts of overspecialization—ever-deepening traps (Figure 2.10b). This pessimist's view emphasizes how the effects of natural selection are to limit and constrain. The emphasis is then on how it is the result of natural selection that all species are always absent from almost everywhere and have such specialized environmental needs that they risk extinction when the environment changes.

It is easy to see that a population of plants faced with repeated drought is likely to develop tolerance of water shortage and an animal repeatedly faced with cold winters to develop habits of hibernation or a thick protective coat. But droughts do not become any less severe or winters milder as a result. Physical conditions are not heritable, they leave no descendants, they are not subject to natural selection. But the situation is quite different when two species interact, predator on prey, parasite on host, competitive neighbor on neighbor.

natural selection does not act on physical conditions

Natural selection may select from a population of parasites those forms that are more efficient at infecting their host. But this immediately sets in play forces of natural selection that favor more resistant hosts. As they evolve they put further pressure on the ability of the parasite to infect. Host and parasite are then caught in never ending reciprocating selection. A result is that both host and parasite become increasingly specialized—caught in an ever deepening rut. Eventually only a specialized form of the parasite can infect and can do so only on a highly

but parasites, predators, and competitors can all be both forces of natural selection and objects of selection

Figure 2.10
Two models of natural selection in action: (a) an optimistic and (b) a pessimistic interpretation of natural selection. In both (a) and (b) a horizontal plane is drawn to represent two dimensions of a range of environmental conditions. On both diagrams ellipses are drawn to show the range of this variation that is tolerated by four populations of organisms. In the optimistic view (a) the vertical scale is a measure of the range of fitness of the organisms in the populations and natural selection is shown as arrows driving the population to ever higher "adaptive" peaks. Population 1 is highly variable and tolerates a wide range of conditions. Natural selection is relatively weak. Population 2 is a very uniform population, and natural selection is fierce and driving the population to a very high degree of specialization and local fitness. In the pessimistic view (b) natural selection is shown as arrows driving the populations into ruts, troughs, and pits. The vertical downward axis is a measure of the intensity of specialization. The highly variable population with weak selection is rather safe if the environment changes but population 2 is at extreme risk of extinction. The two depictions should not be seen as one right and one wrong impression of natural selection in action, but rather as two views of the same truths.

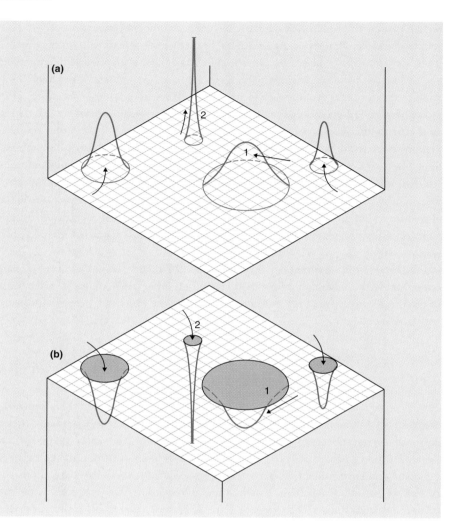

specialized form of the host. We will find examples of this extreme form of natural selection—coevolution—when we consider organisms as habitats in Chapter 7.

2.4 > The Ecology of Speciation

We have seen that natural selection can force populations of plants and animals to change their character—to evolve. But none of the examples we have considered has involved the evolution of a new species. Indeed Darwin's *On the Origin of Species* is about natural selection and evolution but is not really about the origin of species! Those who studied the evolution of melanism in the peppered moth named the black and normal forms *forma carbonaria* and *forma typica*: they classified them as forms within a species, not as different species. Likewise the different growth forms of the grasses on the cliffs and pastures of Abraham's Bosom and the dull and flamboyant races of Endler's guppies are just local genetic classes. None qualifies for the status of distinct species. But when we ask just what criteria justify naming two populations as different species we meet real problems.

2.4.1 What do we mean by a "species"?

Cynics have said, with some truth, that a species is what a competent taxonomist regards as a species! Darwin himself regarded species (like genera) as "merely artificial combinations made for convenience." In the 1930s two American biologists, Ernst Mayr and Theodosius Dobzhansky, proposed an empirical test that could be used to decide whether two populations were part of the same or of two different species. They recognized organisms as being members of a single species if they could, at least potentially, breed together in nature to produce fertile offspring. They called a species tested and defined in this way a *biospecies*. In the examples that we have used earlier in this chapter we know that melanic and normal peppered moths can mate and that the offspring are fully fertile; this is also true of colored and dull guppies and of plants from the different types of *Achillea* and *Agrostis*. They are all variations within species—not species (certainly not biospecies).

UNANSWERED QUESTION:

is there a definition of "species" that can be applied in all cases?

In practice biologists do not apply the Mayr–Dobzhansky test before they recognize every species (there is simply not enough time and resources), but it is there to resolve arguments if they arise. What is more important is that the test recognizes a crucial element in the evolutionary process. If the members of two populations are able to hybridize and their genes are combined and reassorted in their progeny, natural selection can never make them truly distinct. Although natural selection may tend to force a population to evolve into two or more distinct forms, sexual reproduction and hybridization mix them up again. The evolutionary biologist George Williams likened the process to the fate of the legendary king Sisyphus, who was condemned in hell forever to push to the top of a mountain a rock that always fell down into the valley just before it had reached the peak. Natural selection continually narrows the range of variation in a population (pushes it up the hill), but whenever sexual reproduction (hybridization) occurs, it releases more variation and undoes some of the selection.

biospecies do not exchange genes

Two parts of a population can evolve into distinct species only if some sort of barrier prevents gene flow between them; they might, for example, be isolated on different islands. While isolated from each other they may then evolve and become so different that, if they meet again, they can no longer hybridize and their populations can no longer exchange genes. They are now different biospecies. Figure 2.11 illustrates this process.

Differences that are particularly effective in keeping newly evolved species distinct are different rituals of courtship, different signals of attraction between the sexes, and different species of insect pollinator in flowering plants. It may sometimes happen that hybrids form between two evolving species but their parental chromosomes have become so different that they fail to pair at meiosis: the hybrids are then sterile (for example, the horse–donkey hybrid is the sterile mule).

The evolution of species and the balance between natural selection and hybridization are illustrated by the remarkable case of two species of sea gull—an example that seems almost to have been designed to provide a conundrum for taxonomists.

evolution in sea gulls

The role of isolation in the evolution of species from the model of Mayr (1942). A uniform species with a large range (1) differentiates (2) into local forms, varieties, or subspecies, which (3) become genetically isolated from each other, for example, separated by geographic barriers or dispersed onto different islands. After evolution in isolation they may meet again when they (4) are unable to hybridize and have become true biospecies *or* (5) retain an ability to hybridize and form hybrid zones—not true biospecies.

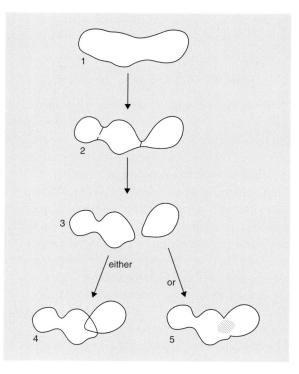

The lesser black-backed gull (*Larus fuscus*) originated in Siberia and colonized progressively to the west, forming a chain or *cline* of different forms, spreading from Siberia to Britain and Iceland (Figure 2.12). The neighboring forms along the cline are distinctive, but they hybridize readily in nature. Neighboring populations are therefore regarded as part of the same species and taxonomists give them only subspecific status (e.g., *Larus fuscus graelsii, Larus fuscus fuscus*). Populations of the gull have, however, also spread east from Siberia, again forming a cline of freely hybridizing forms. Together the populations spreading east and west encircle the northern hemisphere and they meet and overlap in northern Europe. The eastward cline and the westward cline have diverged so far that, where they meet, it is easy to tell them apart and they are recognized as two different *species*, the lesser black-backed gull (*Larus fuscus*) and the herring gull (*Larus argentatus*). Moreover, the two "species" do not hybridize and they have clearly become true biospecies.

In this remarkable example we can see how two distinct biospecies have evolved from one primal stock and the stages of their divergence remain frozen in the cline that connects them. But it is where a population becomes split into completely isolated populations, for example, dispersed onto different islands, that they most readily diverge into distinct species.

2.4.2 Islands and speciation

The most celebrated example of evolution and speciation on islands is the case of Darwin's finches in the Galapagos archipelago. Darwin had been chided for underestimating the importance of isolation in the evolution of species and responded

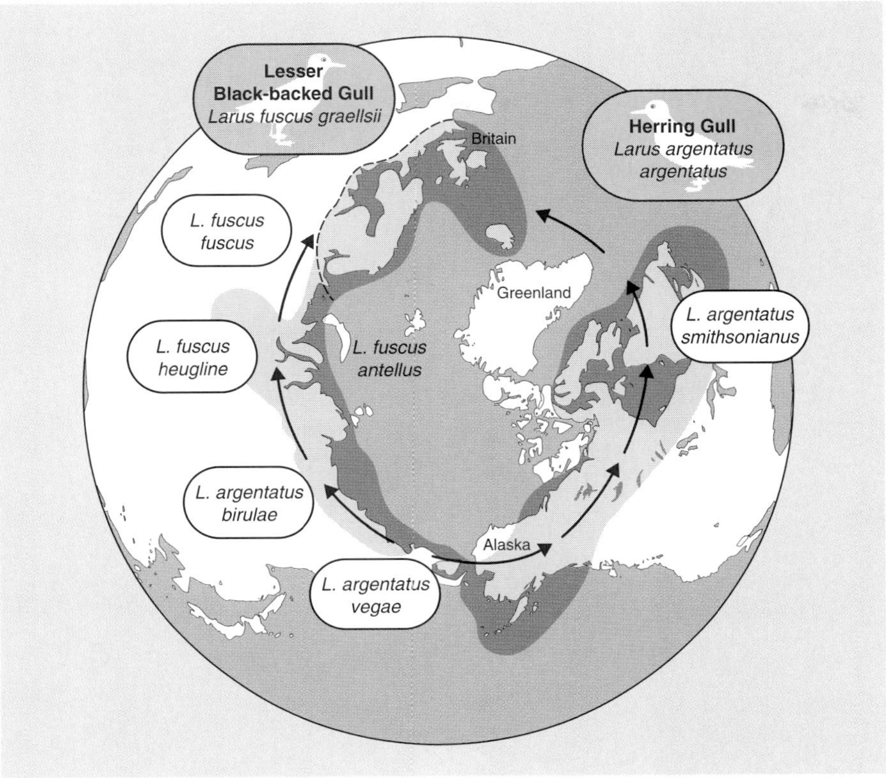

Figure 2.12
Two species of gull, the herring gull and the lesser blackbacked gull, have diverged from common ancestry as they have colonized and encircled the northern hemisphere. Where they occur together in northern Europe they fail to interbreed and are clearly recognized as two distinct species. However, they are linked along their range by a chain of freely interbreeding races or subspecies. (After Brookes, 1998.)

(in a letter, 1876), "It would have been a strange fact if I had overlooked the importance of isolation, seeing that it was such cases as that of the Galapagos archipelago which chiefly led me to study the evolution of species."

The Galapagos Islands are volcanic and are isolated in the Pacific Ocean about 1000 km west of Equador and 750 km from the island of Cocos, which is itself isolated 500 km from Central America. At more than 500 m above sea level the vegetation is open grassland. Below this is a humid zone of forest that grades into a coastal strip of desert vegetation with some endemic species of prickly pear cactus (*Opuntia*). Fourteen species of finch are found on the islands, and there is every reason to suppose that these evolved from a single ancestral species that invaded the islands from the mainland of Central America.

Darwin's finches

In their remote island isolation, the population of Galapagos finches has radiated into a variety of species in groups with contrasting ecologies (Figure 2.13). Members of one group, including *Geospiza fuliginosa* and *G. fortis*, have strong bills and hop and scratch for seeds on the ground. *Geospiza scandens* has a narrower and slightly longer bill and feeds on the flowers and pulp of the prickly pears as well as on seeds. Finches of a third group have parrot-like bills and feed on leaves, buds, flowers, and fruits, and a fourth group with a parrot-like bill (*Camarhynchus psittacula*) has become insectivorous, feeding on beetles and other insects in the canopy of trees. A so-called woodpecker finch, *Camarhynchus* (*Cactospiza*) *pallida*, extracts insects from crevices by holding a spine or a twig in its bill. Yet a further group includes

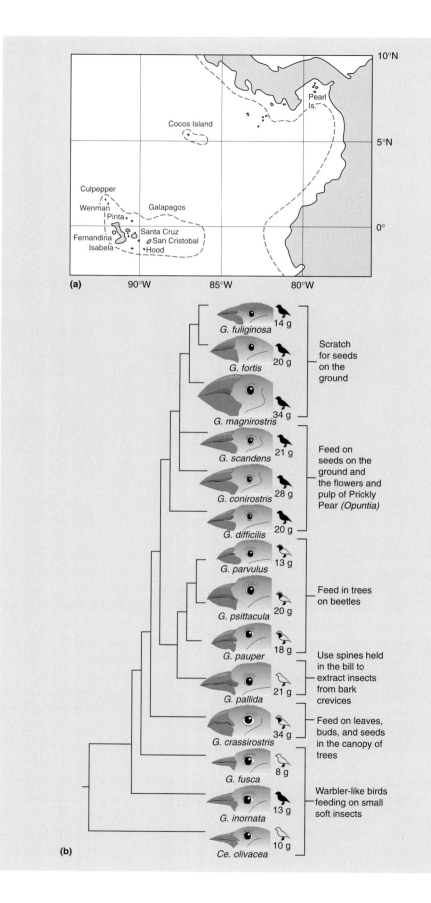

Figure 2.13

(a) Map of the Galapagos Islands showing their position relative to Central America; on the equator 5° equals approximately 560 km. (b) A reconstruction of the evolutionary history of the Galapagos finches based on variation in the length of microsatellite deoxyribonucleic acid (DNA). The feeding habits of the various species are also shown. Drawings of the birds are proportional to actual body size. The maximum amount of black coloring in male plumage and the average body mass are shown for each species. The *genetic distance* (a measure of the genetic difference) between species is shown by the length of the horizontal lines. Notice the great and early separation of the warbler finch (*Certhidea olivacea*) from the others, suggesting that it may closely resemble the founders that colonized the islands. G = *Geospiza*, C = *Camarhynchus*, Ce = *Certhidea*. (Adapted from Petren et al., 1999.)

a species (*Certhidea olivacea*) that, rather like a warbler, flits around actively and collects small insects in the forest canopy and in the air. Populations of ancestor species became reproductively isolated, most likely after chance colonization of different islands within the archipelago, and evolved separately for a time. Subsequent movements between islands may have brought nonhybridizing biospecies together, and subsequently these have evolved to fill different niches. We will see in Chapter 6 that when individuals from different species compete, natural selection may act to favor those individuals that compete least with members of the other species. An expected consequence is that among a group of closely related species, such as Darwin's finches, differences in feeding and other aspects of their ecology are likely to become enhanced with time.

The entire process of evolutionary divergence of these species appears to have happened in less than 3 million years. Very rarely hybridization occurs among the species that have similar ecologies, for example, between those related species that live and feed on the ground. However, these occasions are so rare that the species are true or emerging biospecies. We need to remember that the origin of a species is normally a process rather than an event. For the formation of a new species, like the boiling of an egg, there is some freedom to argue about when it is completed!

The evolutionary relationships among the various Galapagos finches have been traced by molecular techniques (analyzing variation in the length of microsatellite deoxyribonucleic acid [DNA]) (Petren et al., 1999) (see Figure 2.13). These accurate modern tests confirm the long held view that the family tree of the Galapagos finches radiated from a single trunk (i.e., was *monophyletic*) and also provides strong evidence that the warbler finch (*Certhidea olivacea*) was the first to split off from the founding group and is likely to be the most similar to the original colonist ancestors.

The flora and fauna of many other archipelagos show similar examples of great richness of species with many local *endemics* (i.e., species known only from one island or area). The Hawaiian Islands are home to an extraordinary diversity of picture-winged fruit flies (species of *Drosophila*) and closely related species of honey creeper that have diverged in feeding habit and bill shape remarkably like the Galapagos finches. Lizards of the genus *Anolis* have evolved a kaleidoscopic diversity of species on the islands of the Caribbean, and isolated groups of islands, such as the Canaries off the coast of North Africa, are treasure troves of endemic species.

Invaders onto marine islands may become isolated from other parts of their population and are then free to diverge under natural selection and become different species. But there are other kinds of "islands" in which colonists can also become genetically isolated from the rest of a population. Mountains isolate valleys from each other and valleys isolate mountains. A few individuals that chance to be dispersed to a habitable site in a mountain range can form the nucleus of an expanding new species. Its character will have been colored by the particular genes that were represented among the colonists—unlikely to be a perfect sample of the parent population. What natural selection can do with this *founder population* is limited by what is in its limited sample of genes (plus occasional rare mutations). Indeed much of the deviation between populations isolated on islands appears to be due to a *founder effect*—the chance composition of the pool of founder genes puts limits and constraints on what variation there is for natural selection to act upon.

The evolutionary biologist's understanding of island patterns depends on a thorough appreciation of ecological processes such as dispersal (Chapter 5) and interspecific competition (Chapter 6). Likewise, the ecologist's understanding of ecological specialization, species distributions, species diversity, and niche partitioning, among many other ecological phenomena, would be rudimentary indeed without the underpinning provided by the evolutionary processes discussed in this chapter.

2.4.3 The special case of the evolution of species by polyploidy

One of the most important ways in which populations may diverge when reproductively isolated from each other is that the chromosomes themselves evolve (they may lose, repeat, double, gain, or invert segments). A consequence is that if populations meet after evolving in isolation, they may hybridize to form fully vigorous hybrids—but the chromosomes that the two parents contributed to the hybrids are unable to pair and complete normal meiosis: they are therefore sterile. In plants an occasional error in cell division may cause the total number of chromosomes to be doubled. The process, called *polyploidy*, gives every chromosome in the polyploidized sterile hybrid an exactly similar partner. All the chromosomes can now form pairs at meiosis and the hybrids have become fully fertile. A famous example is the formation toward the end of the nineteenth century of sterile hybrids between two marine grasses, the European *Spartina maritima* and a chance introduction of *Spartina alterniflora* from North America. The hybrids proved very vigorous and spread rapidly by clonal growth. One or more polyploids formed among the hybrids and these were fully fertile. *Spartina maritima* has 56 chromosomes and *S. alterniflora* has 70. The sterile hybrid has 63 chromosomes and the fertile polyploid 126.

2.5 ⟩ The Effects of Climatic Change on the Evolution and Distribution of Species

relics of climatic change

Much of what we see in the present distribution of species and animals on Earth represents phases in a recovery from past climatic change. Changes in climate, particularly during the ice ages of the Pleistocene (the past 2–3 million years), bear a lot of the responsibility for the present patterns of distribution of plants and animals. As climates have changed, species populations have advanced and retreated; have been fragmented into isolated patches, terrestrial "islands" in a sea of other vegetation; and have then rejoined. The techniques available for analyzing and dating biological remains (particularly the analysis of buried pollen left from previous vegetation) have become very sophisticated, and the extent of climatic and biotic change in the Pleistocene is beginning to be unraveled. These methods allow us to detect just how much of the present distribution of organisms is a precise local evolved match to present environments, and how much is a fingerprint left by the hand of history.

cycles of glaciation have occurred repeatedly

For most of the past 2–3 million years the Earth has been very cold. Evidence from the distribution of oxygen isotopes in cores taken from the deep ocean floor shows that there may have been as many as sixteen glacial cycles in the Pleistocene, each lasting for up to 125,000 years (Figure 2.14). Each glacial phase may have lasted for as long as 50,000–100,000 years with brief intervals of only 10,000–20,000 years when the temperatures rose to, or above, those we experience today. If these time scales are correct, it is present floras and faunas that are unusual, because they have developed at the warm end of one of a series of unusual catastrophic warm periods!

During the 20,000 years since the peak of the last glaciation, global temperatures have risen by about 8°C. The analysis of buried pollen shows the rate at which vegetation has changed during this period. Woody species produced most of the buried pollen, and the pollen records from Connecticut (Figure 2.14) show that, as the ice retreated, various forest species advanced in turn, spruce first and chestnut most recently. Each new arrival added to the number of species present, which has

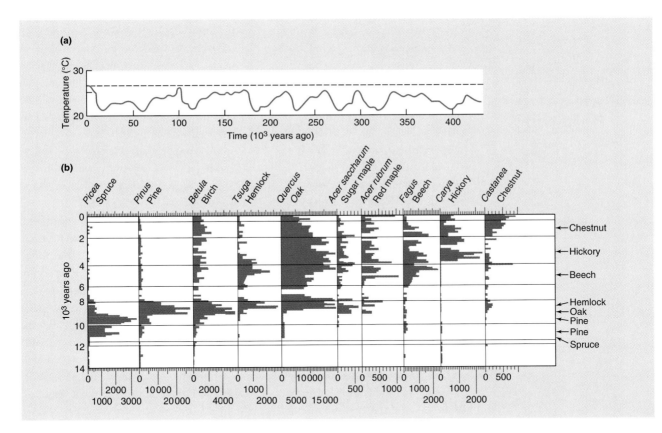

Figure 2.14

(a) An estimate of the temperature variations with time during glacial cycles over the past 400,000 years. The estimates were obtained by comparing oxygen isotope ratios in fossils taken from ocean cores in the Caribbean. The dashed line corresponds to the ratio 10,000 years ago, at the start of the present warming period. Periods as warm as the present have been rare events, and the climate during most of the past 400,000 years has been glacial. (After Emiliani, 1966; Davis, 1976.) (b) Records of past vegetation can be obtained by determining the numbers of pollen grains of different species deposited in peat or lake sediments. The figure shows changes in frequency of pollen from different species of tree accumulated in the sediments of Rogers Lake, Connecticut. The pollen counts were made on cores taken down through layers of sediment that were deposited continuously from late glacial times to the present. The estimated date of arrival of each species in Connecticut is shown by arrows at the right of the figure. (After Davis et al., 1973.)

the distribution of trees has changed
gradually since the last glaciation

UNANSWERED QUESTION:

are animals and plants still recovering
from the last glaciation?

predicted global warming by the
"greenhouse effect" is nearly a
hundred times faster than postglacial
warming

increased continually over a period of fourteen thousand years. The story was the same in Europe.

As the number of pollen records has increased, it has become possible not only to plot the history of change in the vegetation at a point in space, but to begin to map the movements of the various species as they have spread across the continents. The work of Davis and colleagues has been especially important in showing that assemblies of species that currently form distinctive communities did not move together. Different species spread at different speeds, and not always in the same direction. The species composition of vegetation has continually been changing and is almost certainly still doing so.

In the invasions that followed the retreat of the ice in eastern North America, spruce invaded first, followed by jack pine or red pine, which spread northward at a rate of 350–500 m per year for several thousands of years. White pine started its migration about one thousand years later, at the same time as oak. Hemlock was also one of the rapid invaders (200–300 m per year), and arrived at most sites about one thousand years after white pine. Chestnut moved slowly (100 m per year) but became a dominant species once it had arrived; it appeared on the Allegheny Mountains some three thousand years before it reached Connecticut.

We do not have such good records for the postglacial spread of the animals associated with the changing forests, but it is at least certain that many species could not have spread faster than the trees on which they feed. Some of the animals may still be catching up with their plants, and tree species are still returning to areas they occupied before the last ice age! It is quite wrong to imagine that our present vegetation is in some sort of equilibrium with (adapted to) the present climate. Climatic change has been too fast for change in the vegetation to keep pace. Vegetation is always out of step with its climate (Davis, 1976).

As vegetation has moved back and forth after glaciations, populations have been left behind as isolated patches—"islands" that are ripe for evolution. Reconstruction of the vegetation that has been present in the northwestern Great Basin of the United States suggests that some species scarcely moved. *Juniperus osteospermum,* for example, has been present in the fossil record of the Great Basin continuously over the last thirty thousand years (Nowak et al., 1994). Species that remained in place as the climate froze and then warmed were probably genetically highly diverse populations or even species containing many subpopulations with different tolerances and only the most tolerant survived natural selection by climatic change. Other species such as *Pinus monophylla* persisted through the coolest phases as relict "islands" in the "oceans" of surrounding vegetation in the South and what remained of the populations then served to seed migration farther north as the climate became warmer.

Evidence of changes in vegetation that followed the last retreat of the ice hint at the likely consequences of the global warming (3°C in the next hundred years) that is predicted to result from continuing increases in atmospheric carbon dioxide (see references to the "greenhouse effect" in Chapter 13). But the scales are quite different. Postglacial warming of about 8°C occurred in just twenty thousand years. Change in the vegetation failed to keep pace with this natural speed of environmental warming. The predicted pace of global warming in the "greenhouse effect" is nearly a hundred times faster and must presumably cause widespread waves of species migration and extinction. These in turn will leave behind isolated islands of populations ripe for new evolutionary divergence.

The records of climatic change in the tropics are far less complete than those for temperate regions. But a picture is emerging of vegetational change in the tropics that paralleled changes occurring in temperate regions. Patches occupied by tropical forest expanded during warmer, wetter periods, and patches of savanna expanded in the cooler and drier periods (Prance, 1987). The present distributions of both plant and animal species are largely accidents of history, evidence of the positions once occupied by these "tropical islands in a sea of savanna" (Figure 2.15).

2.6 ▶ The Effects of Continental Drift on the Ecology of Evolution

The patterns of species formation that occur on islands appear on an even larger scale in the evolution of genera and families across continents. The occasional rare events of dispersal undoubtedly account for some of the invasions and subsequent isolation across oceans. But many curious distributions of organisms between continents seem inexplicable as the result of dispersal over vast distances. Biologists, especially Wegener (1915), met outraged scorn from geologists and geographers when they argued that it must have been the continents that had moved rather than the organisms that had dispersed.

landmasses have moved

Eventually measurements of the directions of the Earth's magnetic fields required the same, apparently wildly improbable, explanation and the critics capitulated. The discovery that the tectonic plates of the Earth's crust move and carry the migrating continents with them reconciles geologist and biologist (Figure 2.16). While major

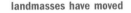

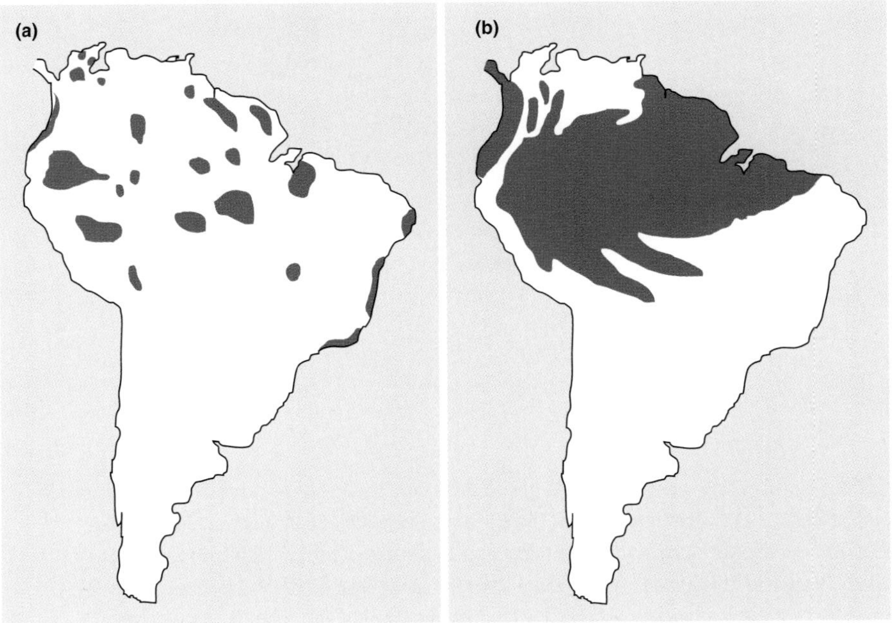

Figure 2.15
(a) The estimated distribution of tropical forest in South America at the time when the last glaciation was at its peak. Gallery forest along river sides may have prevented the patches from being quite as isolated as the map suggests. (b) Present day distribution. (After Prance, 1987.)

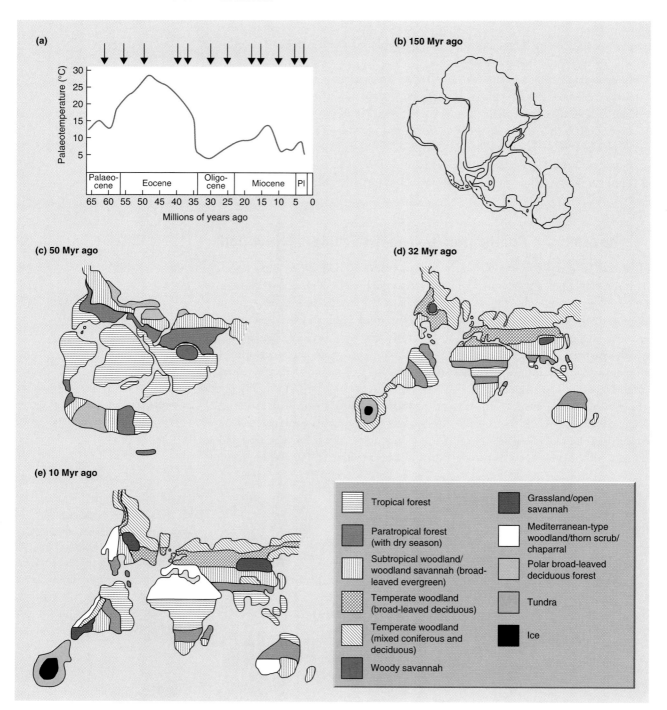

Figure 2.16

(a) Changes in temperature in the North Sea over the past 60 million years. During this period there were large changes in sea level that allowed dispersal of both plants and animals between landmasses. (b–e) Continental drift. (b) The ancient supercontinent of Gondwanaland began to break up about 150 million years ago. (c) About 50 million years ago (early Middle Eocene) recognizable bands of distinctive vegetation had developed and (d) by 32 million years ago (early Oligocene) these had become more sharply defined. (e) By 10 million years ago (early Miocene) much of the present geography of the continents had become established but with dramatically different climates and vegetation from today: the position of the Antarctic ice cap is highly schematic. (Adapted from Janis, 1993; Norton & Sclater, 1979; and other sources).

Large flightless birds are found in three major continents. (a) The ostrich (*Struthio camelus*) is African and commonly occurs together with herds of zebra and antelope in savanna or steppe grasslands. (b) The rhea (*Rhea americana*) is found in similar grasslands in South America (e.g., Brazil and Argentina) commonly together with herds of deer and guanocos. (c) The emu (*Dromaius novaehollandiae*) inhabits equivalent habitats in Australia. Many other species of these very large, mainly herbivorous birds have been sought after by humans for food and become extinct. The presence of these evolutionarily related and ecologically similar species in three widely separated continents is explained by the drifting apart of the continents from the time (150 Myr ago) when they were portions of the primitive continent of Gondwanaland (see Figure 2.17).

evolutionary developments were occurring in the plant and animal kingdoms, their populations were being split and separated, and land areas were moving across climatic zones. This was happening while changes in temperature were occurring on a vastly greater scale than the glacial cycles of the Pleistocene episode.

The established drift of the continents answers many questions in the ecology of evolution. The curious world distribution of large flightless birds is one example (Figure 2.17). The presence of the ostrich in Africa, the emu in Australia, and the very similar rhea in South America could scarcely be explained by dispersal of some common flightless ancestor.

Techniques of molecular biology make it possible to analyze the time at which the various flightless birds started their evolutionary divergence (Figure 2.17). The tinamous seem to have been the first to diverge and became evolutionarily separate from the rest, the *ratites*. Australasia became separated from the other Southern continents leaving the ancestral stock of ostriches and rheas to be separated when the Atlantic opened up between Africa and South America. The Tasman Sea opened up about 80 million years ago and ancestors of the kiwi are thought to have made their way, by island hopping, about 40 million years ago across to New Zealand, where divergence into the present species happened relatively recently. The unraveling of this particular example implies the early evolution of the property of flightlessness

and divided populations that have then evolved independently

Figure 2.17
(a) The distribution of terrestrial flightless birds. (b) The phylogenetic tree of the flightless birds and the estimated times (million years [Myr]) of their divergence. (After Diamond, 1983; from data of Sibley & Ahlquist.)

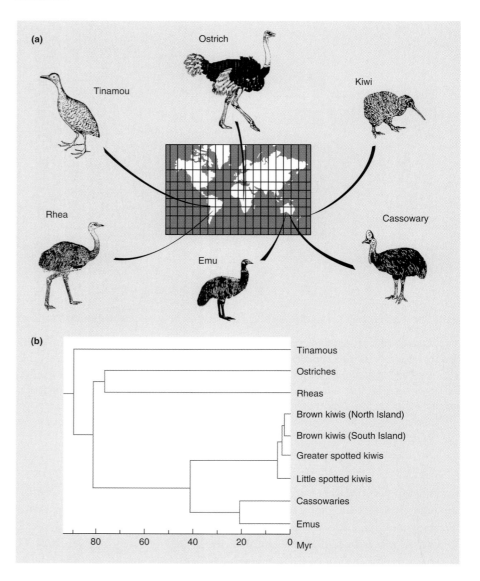

and only subsequently the isolation of the different types between the emerging continents.

2.7 Interpreting the Results of Evolution: Convergent and Parallel Evolution

Flightlessness did not evolve independently on the different continents. However, there are many examples of organisms that have evolved in isolation from each other and then converged on remarkably similar forms or behavior. Similarity in form and behavior has often developed between organisms that live in a similar environment but have quite different ancestry. This is particularly striking when similar roles are played by structures that have quite different evolutionary origins—that is, when the structures are *analogous* (similar in superficial form or function) but not *homologous* (derived from an equivalent structure in a common ancestry). When this occurs, it is termed *convergent evolution*.

Large swimming carnivores have evolved in four quite distinct groups: among fish, reptiles, birds, and mammals (Figure 2.18). There are profound differences between these groups in their internal structure and physiology, but they have all converged in form onto a hydrodynamically efficient shape.

Such convergence is direct evidence of the power of evolutionary forces to shape the same form from quite different starting material. The French geneticist Jacob said that evolution was like "tinkering." It did not create ideal forms from ideal beginnings—rather it tinkered together what it could from what was available at the time (a good tinker can make a saucepan from a bicycle or a bicycle from saucepans).

A comparable series of examples can be used to show the parallels in the evolutionary pathways of ancestrally related groups that diverged into the same pattern of diversity after they were isolated from each other. The classic example of such parallel evolution is found among the placental and marsupial mammals. Marsupials arrived on what would become the Australian continent in the Cretaceous

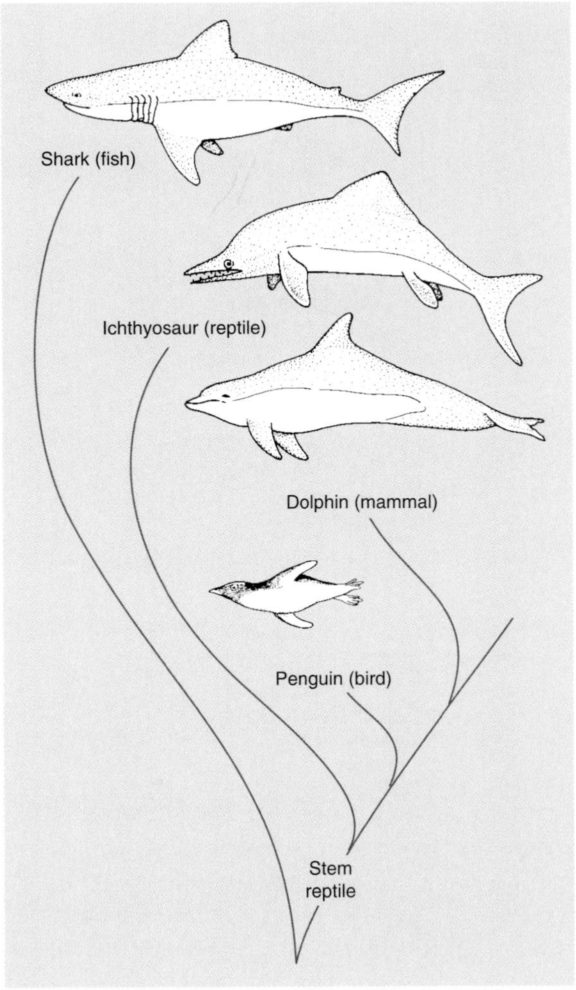

convergent evolution

parallel evolution

Figure 2.18
An example of convergent evolution of body form among large marine carnivores from different evolutionary lines. (After Hildebrand, 1974; and others.)

Shark (fish)

Ichthyosaur (reptile)

Dolphin (mammal)

Penguin (bird)

Stem
reptile

period (around 90 million years ago; see Figure 2.16), when the only other mammals present were the curious egg-laying monotremes (now represented only by the spiny anteaters and the duck-billed platypus *Ornithorhynchus anatinus*). An evolutionary process of radiation then occurred among the Australian marsupials that in many ways accurately paralleled what was occurring among the placental mammals on other continents (Figure 2.19).

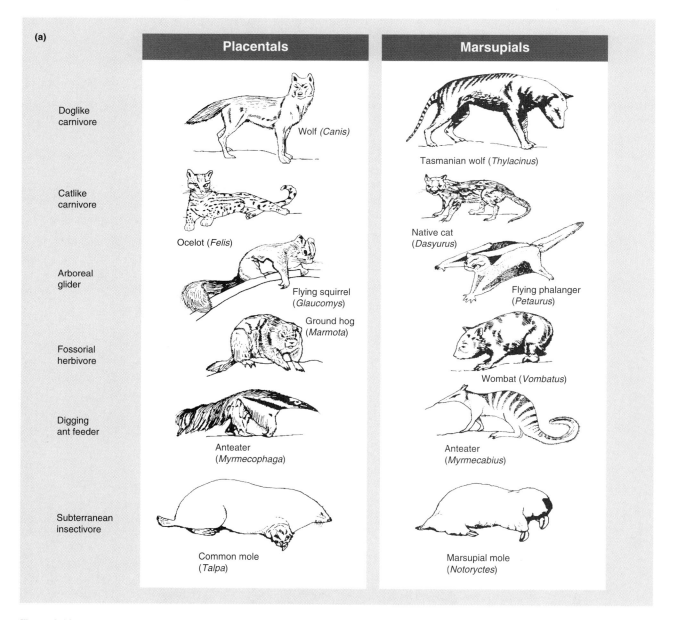

(a)

Placentals	Marsupials

Doglike carnivore — Wolf (*Canis*) / Tasmanian wolf (*Thylacinus*)

Catlike carnivore — Ocelot (*Felis*) / Native cat (*Dasyurus*)

Arboreal glider — Flying squirrel (*Glaucomys*) / Flying phalanger (*Petaurus*)

Fossorial herbivore — Ground hog (*Marmota*) / Wombat (*Vombatus*)

Digging ant feeder — Anteater (*Myrmecophaga*) / Anteater (*Myrmecabius*)

Subterranean insectivore — Common mole (*Talpa*) / Marsupial mole (*Notoryctes*)

Figure 2.19
Parallel evolution of marsupial and placental mammals. The pairs of species are similar in both appearance and habit and usually (but not always) in life-style.

The subtlety of the parallels between marsupial and placental mammals in both their form and their life-style is so striking that it is hard to escape the view that the environments of placentals and marsupials contained ecological pigeonholes (niches) into which the evolutionary process has neatly "fitted" ecological equivalents. (It is important to remember that, in contrast to convergent evolution, the marsupials and placentals started to diversify from a common ancestral line and both inherited a common set of potentials and constraints.)

interpreting the match between organisms and their environment

When we marvel at the diversity of complex specializations by which organisms match their varied environments there is a temptation to regard each case as an example of evolved perfection. But there is nothing in the process of evolution by natural selection that implies perfection. For example, no population of organisms can contain all the possible genetic variants that might exist and might influence fitness. The evolutionary process works on the genetic variation that is available. It favors only those forms that are fittest from *among the range of variety available,* and this may be a very restricted choice. The very essence of natural selection is that organisms come to match their environments by being "the fittest available" or "the fittest yet": they are not "the best imaginable."

It is particularly important to realize that past events on the Earth can have profound repercussions on the present. Our world has not been constructed by taking each organism in turn, testing it against each environment, and molding it so that every organism finds its perfect place. It is a world in which organisms live where they do for reasons that are often, at least in part, accidents of history (analogous to golf balls fitting neatly into egg cups). Moreover the ancestors of the organisms that we see around us lived in environments that were profoundly different from those of the present. Evolving organisms are not free agents—some of the features acquired by their ancestors hang like millstones around their necks, limiting and constraining where they can now live and what they might become. It is very easy to wonder and marvel at how beautifully the properties of fish fit them to live in water—but just as important to emphasize that these same properties prevent them from living on land. Indeed it was the few that chanced to escape from their ancestral watery home that started the evolutionary lines that diversified into the dinosaurs and mammals.

> Summary

The force of natural selection

Life is represented on Earth by a diversity of specialist species, each of which is absent from almost everywhere. Early interest in this diversity was from explorers, and collectors and the idea that the diversity had arisen by evolution from earlier ancestors over geological time was not seriously discussed until the first half of the nineteenth century. Charles Darwin and Alfred Russel Wallace (strongly influenced by having read Malthus's essay "The Principle of Population") independently proposed that natural selection provided a force that would drive a process of evolution. The theory of natural selection is an ecological theory. The reproductive potential of living organisms leads them inescapably to compete for limited resources. Success in this competition is measured by leaving more descendants than others to subsequent generations. When these ancestors differ in properties that are heritable the character of populations will necessarily change over time and evolution will happen.

Darwin had seen the power of human selection to change the character of domestic animals and plants and he recognized the parallel in natural selection. But there is one big difference:

humans select for what they want in the future, but natural selection is a result of events in the past—it has no intentions and no aim.

Natural selection in action

We can see natural selection in action within species in the variation within species over their geographic range and even over very short distances where we can detect powerful selective forces in action and recognize ecologically specialized races within species (*ecotypes*). Transplanting plants and animals between habitats reveals tightly specialized matches between organisms and their environments. The evolutionary responses of animals and plants to pollution demonstrate the speed of evolutionary change, as do experiments on the effects of predators on the evolution of their prey. The evolution of specialization can be interpreted as matching organisms more accurately with their environment or as forcing them into ever tighter straitjackets of form and behavior.

The origin of species

Natural selection does not normally lead to the origin of species unless it is coupled with the reproductive isolation of populations from each other—as occurs for example on islands and is illustrated by the finches of the Galapagos Islands.

Biological species are recognized when they have diverged enough to prevent them from forming fertile hybrids if and when they meet. The formation of new species normally occurs over long periods, but plants are unusual in that fully fertile new species can arise from sterile interspecific hybrids by polyploidy—involving a doubling of the number of chromosomes.

Parallel and convergent evolution

The origin of species and of the heritable variation within them reflects their past—which ancestors have left most descendants in a changing world. Properties that have been selected in the past may constrain what evolution is possible in the future (and its direction). The drift of continents over tens of millions of years and the repetitive ice ages (over tens of thousands of years) have provided the fragmentation of populations and the diversity of selective forces that favor the evolution of species. Evidence of the power of ecological forces to shape the direction of evolution comes from *parallel evolution* (in which populations long isolated from common ancestors have followed similar patterns of diversification) and from *convergent evolution* (in which populations evolving from very different ancestors have converged on very similar form and behavior).

Review Questions

⚠ = Challenge Question

1. ⚠ What do you consider to be the essential distinction between *natural selection* and *evolution?*

2. What was the contribution of Malthus to Darwin's and Wallace's ideas about evolution?

3. Why is "the survival of the fittest" an unsatisfactory description of natural selection?

4. What is the essential difference between natural selection and the selection practiced by plant and animal breeders?

5. What are *reciprocal transplants*? Why are they so useful in ecological studies?

6. Is sexual selection, as practiced by guppies, different from or just part of natural selection?

7. In what ways do the results of natural selection by parasites and predators differ from selection by physical conditions of the environment?

8. What is *polyploidy*? What are its consequences?

9. What is it about the Galapagos finches that has made them such ideal material for the study of evolution?

10. What is the difference between convergent and parallel evolution?

11. ⚠ The process of evolution can be interpreted as optimizing the fit between organisms and their environment or as narrowing and constraining what they can do. Discuss whether there is a conflict between these interpretations.

Web Research Questions

1. The words "ecology," "ecologist," "ecological," and so on have spread from the science of ecology to the press, to pressure groups, and to the general public. Survey and describe briefly the range of meanings these words are now being given and by whom. Some have suggested that the words have been hijacked and misused: do you see any evidence of that? Does it/would it matter if they were?

2. Discuss the pros and cons of long-term ecological research, using the Hubbard Brook Experimental Forest study as an example. This program is run by the Institute of Ecosystem Studies (IES). To get some idea of the scale of long-term ecosystem studies find out roughly how many scientists and related staff work at the IES. The Hubbard Brook Experimental Forest is one of a number of long-term ecological research (LTER) sites in North America. Compare and contrast the objectives of the Hubbard Brook study with three of the other sites. Should all long-term ecological programs go on indefinitely? If not, what criteria would you suggest to determine whether and when a long-term program should be terminated?

3. Like museums, the traditional role of zoos was to make a public display of the diversity of nature. Is this still the case? Or has the conservation of endangered species come to assume more importance? Visit the websites of local zoos, and others around the world, and discuss the relative roles of display and conservation in the different institutions. Describe the variety of endangered species that zoos are working with.

Conditions and Resources

For ecologists, organisms are really only worth studying where they can live. Perhaps the most fundamental features of an environment that make it habitable are that the organisms can tolerate the local conditions and that their essential resources are being provided. We cannot expect to go very far in understanding the ecology of any species without understanding its interactions with conditions and resources.

3

Physical Conditions and the Availability of Resources

Key Concepts

In this chapter you will

- understand the nature of, and contrasts between, conditions and resources.

- understand how organisms respond to the continuum of a condition like temperature, but also to "extreme" conditions and to the timing both of variations and of extremes.

- appreciate how a plant's responses to, and its consumption of, the resources of solar radiation, water, minerals, and carbon dioxide are intertwined.

- appreciate the importance of contrasting body compositions in the consumption of plants by animals, and of overcoming defenses in the consumption of animals by other animals.

- understand the effects of intraspecific competition for resources.

- appreciate how responses to conditions and resources interact to determine ecological niches.

3.1 ▷ **Introduction**

It is convenient to distinguish conditions and resources as two properties of environments that determine where organisms can live. Conditions are physicochemical features of the environment such as its temperature and humidity. Among important features are their diurnal and annual cycles and the frequency of extreme events such as coldest nights and hottest days. In aquatic environments, important conditions include osmotic pressure, pH, aeration, and the physical forces of water flow and wave action. The presence of an organism always alters the conditions in its immediate environment. Sometimes this is on a very large scale. A tree, for example, maintains a zone of higher humidity on the ground beneath its canopy. On a microscopic scale, though, the presence of an algal cell in a body of water alters the pH in the shell of water that surrounds it. It is characteristic of conditions, however, that they are not consumed or used up by the activities of organisms.

resources, unlike conditions, are consumed

Environmental resources *are* consumed by living organisms in the course of their growth and reproduction. Green plants photosynthesize and obtain both energy and materials for growth and reproduction from inorganic materials. Their resources are solar radiation, carbon dioxide, water, and mineral nutrients. Almost all nonphotosynthetic organisms use the bodies of other organisms as their food resource. (*Almost* all because many of the primitive Archaebacteria obtain "chemosynthetic" energy by oxidizing methane, ammonium ions, hydrogen sulfide, or ferrous iron; they live in environments like hot springs and deep sea vents using resources that were abundant during early phases of life on Earth). Hence, with resources, unlike conditions, what has been consumed is no longer available to another consumer. The rabbit that has been eaten by an eagle is no longer available to another eagle. The quantum of solar radiation that has been absorbed and photosynthesized by a leaf is no longer available to another leaf. This has an important consequence: organisms may compete with each other to capture a share of a limited resource.

In this chapter we consider, first, examples of the ways in which environmental conditions limit the behavior and distribution of organisms. We draw most of our examples from the effects of temperature, which serve to illustrate many general effects of environmental conditions. We consider next the resources used by photosynthetic green plants and then the ways in which organisms as resources have to be captured, grazed or even inhabited before they are handled and digested. Finally we consider the ways in which organisms of the same species may compete with each other for limited resources.

3.2 ▷ **Environmental Conditions**

3.2.1 What do we mean by "harsh," "benign," and "extreme"?

It seems quite natural to describe environmental conditions as "extreme," "pleasant," "harsh," "benign," "adverse," "favorable," "equable," or "stressful." But these characterizations describe how we, human beings, feel about them. At first thought, it may seem obvious when conditions are "extreme": the midday heat of a desert, the cold of an Antarctic winter, the salt concentration of the Dead Sea or the Great Salt Lake. What this means, however, is only that these conditions are extreme *for us,* given our particular physiological characteristics and tolerances. But to a cactus there is nothing extreme about the typical desert conditions in which cacti have evolved; nor are the icy fastnesses of Antarctica an extreme environment for penguins. Indeed, a tropical rain forest would be a harsh environment for a penguin though it might be benign for a macaw or a banana plant. A lake is a harsh environment for a cactus but benign for a trout.

There is a relativity in the ways organisms respond to conditions: what appears unpleasantly hot to one (so that it shrinks away and withdraws) may be easily tolerated or even sought by another. It is too easy and dangerous for the ecologist to assume that all other organisms sense the environment in the way we do. Rather, the ecologist should try to gain a "worm's-eye" (or "plant's-eye") view of the environment; to see the world as others see it. Emotive words like *harsh* and *benign,* even relativities such as hot and cold, are to be used by ecologists only with care.

3.2.2 Effects of conditions

Temperature, relative humidity, and other physicochemical conditions induce a range of physiological responses in organisms. These responses largely determine whether the physical environment is habitable or not. There are three basic types of response curve that describe the effect of a condition on an organism (Figure 3.1a–c). In the first (Figure 3.1a), extreme conditions are lethal, but between the two extremes there is a continuum of more favorable conditions. Organisms are typically able to survive over the whole continuum, but can grow actively only over a more restricted range and can reproduce only within an even narrower band. This is a typical response curve for the effects of temperature or pH. In the second (Figure 3.1b), the condition

Penguins do not find the Antarctic in the least bit "extreme."

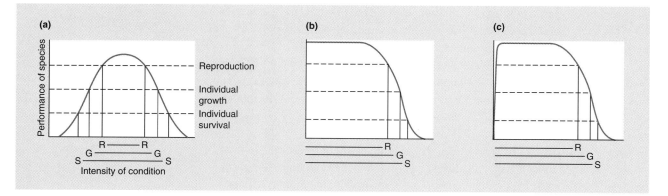

Figure 3.1

Response curves illustrating the effects of a range of environmental conditions on individual survival (S), growth (G), and reproduction (R). (a) Extreme conditions are lethal; less extreme conditions prevent growth; only optimal conditions allow reproduction. (b) The condition is lethal only at high intensities; the reproduction-growth-survival sequence still applies. (c) Similar to (b), but the condition is required by organisms, as a resource, at low concentrations.

is lethal only at high intensities. This is the case for poisons. At low or even zero concentration the organism is typically unaffected, but there is a threshold above which performance decreases rapidly: first reproduction, then growth, and finally survival. The third (Figure 3.1c), then, applies to conditions that are required by organisms at low concentrations but become toxic at high concentrations. This is the case for some minerals, such as copper and sodium chloride which are essential resources for growth when they are present in trace amounts but become toxic conditions at higher concentrations.

Of these three, the first is the most fundamental. It is accounted for, in part, by changes in metabolic effectiveness. For each 10°C rise in temperature, for example, the rate of biological processes often roughly doubles (Figure 3.2). The increase is brought about because high temperature increases the speed of molecular movement and speeds up chemical reactions. Thus, at lower temperatures (though "lower" varies from species to species, as explained earlier) performance may be impaired simply as a result of metabolic inactivity.

high and low temperatures   At extremely high temperatures, on the other hand, enzymes and other proteins become unstable and break down, and the organism dies. Difficulties may set in before these extremes are reached, however. At high environmental temperatures, terrestrial organisms are cooled by the evaporation of water (from open stomata on the surfaces of leaves or through panting in dogs), but this may lead in turn to serious, perhaps lethal, problems of dehydration. Moreover, as water supplies and reserves run low, body temperature may rise rapidly. However, even where loss of water is not a problem, among aquatic organisms, death is usually inevitable if temperatures are maintained for long above 60°C. The exceptions, *thermophiles*, are mostly specialized fungi and the primitive archaebacteria. One of these, *Pyrodictium*

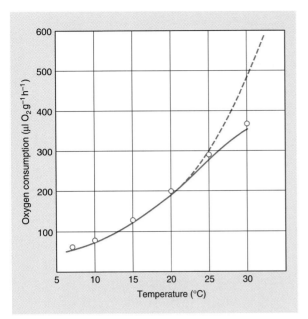

Figure 3.2
The rate of oxygen consumption of the Colorado beetle (*Leptinotarsa decemlineata*), which doubles for every 10°C rise in temperature up to 20°C but increases less fast at higher temperatures. (After Marzusch, 1952.)

occultum, can live at 105°C—something that is only possible because, under the pressure of the deep ocean, water does not boil at that temperature.

At temperatures a few degrees above zero, organisms may be forced into extended periods of inactivity and the cell membranes of sensitive species may begin to break down. This is known as *chilling injury,* which affects many tropical fruits—for example, bananas blacken at even a few degrees above 0°C. Many species of both plants and animals can tolerate temperatures well below zero provided that ice does not form. If it is not disturbed, water can supercool to temperatures as low as −40°C without forming ice, but a sudden shock allows ice to form quite suddenly within plant cells and this, rather than the low temperature itself, is then lethal. Ice formed within a cell is likely simply to disrupt and destroy it. If, however, temperatures fall slowly, ice can form between cells and draw water from within them. The effects on plants are then very much like those of drought.

The absolute temperature that an organism experiences is important. But the timing and duration of temperature extremes may be equally important. For example, unusually hot days in early spring may interfere with fish spawning or kill the fry but otherwise leave the adults unaffected. Similarly, a late spring frost might kill seedlings but leave saplings and larger trees unaffected. The duration and frequency of conditions are often critical. In many cases, a periodic drought or tropical storm has a greater effect on distribution than the average condition. To take just one example: the saguaro cactus is liable to be killed when temperatures remain below freezing for 36 hours, but if there is a daily thaw it is under no threat. In Arizona, the northern and eastern edges of the cactus's distribution correspond to a line joining places where on occasional days it fails to thaw. Thus the saguaro is absent where there are occasionally lethal conditions—an individual need only be killed once.

the timing of extremes

The saguaro cactus can survive only short periods at freezing temperatures.

3.2.3 Conditions as stimuli

Environmental conditions act primarily to modulate the rates of physiological processes. In addition, though, many conditions are important stimuli for growth and development and prepare an organism for conditions that are to come.

photoperiod is commonly used to time dormancy, flowering, or migration

The idea that animals and plants in nature can anticipate, and be used by us to predict, unusual seasons ("a big crop of berries means a harsh winter to come") is the stuff of folklore. But there are important advantages to an organism that can predict and prepare for repeated events such as the seasons. For this the organism needs an internal clock that can be used to check against an external signal. The most widely used external signal is the length of day—the photoperiod. On the approach of winter, bears, cats, and many other mammals develop a thickened fur coat, and birds such as ptarmigan put on winter plumage. Very many insects can interpose a *dormant phase* (*diapause*) within the normal activity of their life cycle, and this is very often photoperiodically timed to anticipate a period when conditions will become adverse (Figure 3.3). Other photoperiodically timed events are the seasonal onset of reproductive activity in animals, the onset of flowering, and the seasonal migration (and the stocking of bodily reserves for it) in birds.

An experience of chilling is needed by many seeds before they will break dormancy. This prevents them from germinating during the moist warm weather after ripening and then being killed by winter cold. Temperature and photoperiod

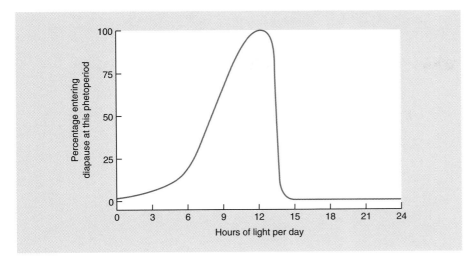

Figure 3.3
The influence of day length (photoperiod) on the start of diapause by larvae of the Oriental fruit moth (*Grapholitha molesta*). The highest percentage of larvae enter diapause at photoperiods of between 13 and 14 hours of daylight; this characteristic has the effect of preparing them to be in state to tolerate the frosts of fall and winter. (Redrawn from a variety of sources.)

interact to control the seed germination of birch (*Betula pubescens*). Seeds that have not been chilled need a specific photoperiod before they will germinate, but if the seed has been chilled, it starts growth without the light stimulus. Either way, growth should be stimulated once winter has passed. The seeds of lodgepole pine, on the other hand, remain protected in their cones until they are heated by forest fire. This stimulus is an indicator that the ground has been cleared and that new seedlings have a chance of becoming established.

Seasonal stimuli such as shortening day length in the fall may also trigger increasing tolerance to forthcoming conditions. This is the process of *acclimatization* (it is called *acclimation* when induced in the laboratory). When exposed for several days to a relatively low temperature, for example, the whole temperature response of a plant or animal may shift downward along the temperature scale (Figure 3.4). One way in which this is done is by forming chemicals that act as antifreeze compounds: they prevent ice from forming within the cells and protect their membranes if ice does form (Figure 3.5). Acclimatization in some deciduous trees (frost hardening) can increase their tolerance of low temperatures by as much as 100°C.

acclimatization

3.2.4 The effects of conditions on interactions between organisms

Although organisms respond to each condition in their environment, the effects of conditions may be determined largely by the responses of other community members. Temperature, for example, does not act on one species: it also acts on its competitors, prey, parasites and so on. Most especially, an organism will suffer if its food is another species that cannot tolerate an environmental condition. This is illustrated by the distribution of the rush moth (*Coleophora alticolella*) in England. The moth lays its eggs on the flowers of the rush (*Juncus squarrosus*) and the caterpillars feed on the developing seeds. The plant can grow at elevations above 600 m and the moths and caterpillars are little affected by the low temperatures above this elevation. However, the rush fails to ripen its seeds above 600 m. This, in turn, limits the

conditions may affect the availability of a resource

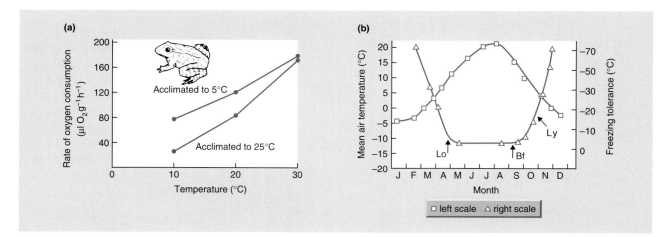

Figure 3.4

Acclimation to low temperatures. (a) The way in which the rate of oxygen consumption by frogs (*Rana pipiens*) depends on temperature is influenced by their previous experience—the temperature to which they had been *acclimated*. (After Rieck et al., 1960.) (b) The ability of the willow (*Salix sachalinensis*) to survive freezing changes through the season on Mt. Kurodake, Japan. Tolerance increases with the onset of winter but is low throughout the spring, summer, and fall. Bf, buds forming; Lo, leaves opening; Ly, leaves yellowing. (After Sakai & Otsuka, 1970.)

distribution of the moth, because caterpillars that hatch in the colder elevations will starve as a result of insufficient food (seeds).

the development of disease

The effects of conditions on disease may also be important. Conditions may favor the spread of infection (winds carrying fungal spores), or favor the growth of the parasite, or weaken the defenses of the host. For example, temperature and soil moisture interact to determine the rate of mortality from fungal diseases in corn (Figure 3.6).

Conditions can affect interactions on very local scales. During an epidemic of southern corn leaf blight (*Helminthosporium maydis*) in a corn field in Connecticut, the disease was more severe in local areas of the crop that were shaded by trees. The plants closest to the trees which were shaded for the longest periods were the most heavily diseased (Figure 3.7).

or competition

Competition between species can also be profoundly influenced by temperature. The distinguished American experimental ecologist Thomas Park (1954) cultured two species of flour beetle, *Tribolium castaneum* and *T. confusum*, at various combinations of temperature and humidity in the laboratory. When the species were grown on their own they survived under a wide range of conditions. But when they were cultured together, only *T. castaneum* survived at high temperature and humidity, and only *T. confusum* survived at low temperature and humidity.

3.2.5 Responses by sedentary organisms

Motile animals have some choice over where they live: they can show preferences. They may move into shade to escape from heat or into the sun to warm up. Such choice of environmental conditions is denied to fixed or sedentary organisms. Plants

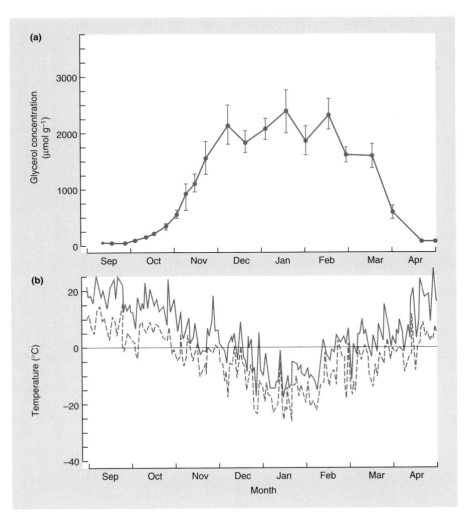

Figure 3.5

(a) Changes in the glycerol content of the larva of the goldenrod gall moth, which protects it from freezing. (b) Daily maximum and minimum temperatures. (After Rickards et al., 1987.)

are obvious examples, but so are many aquatic invertebrates such as sponges, corals, barnacles, mussels, and oysters.

In all except equatorial environments, physical conditions follow a seasonal cycle. Indeed, there has long been a fascination with organisms' responses to these (Box 3.1). Morphological and physiological characteristics can never be ideal for all phases in the cycle and the jack of all trades is master of none. One solution is for the morphological and physiological properties of the organism to keep changing with the seasons (or even anticipating them, as in acclimatization). But continually changing may be costly: a deciduous tree may have leaves ideal for life in spring and summer but faces the costs of making new ones every year. Acclimatization may involve a change in biochemical characteristics to make costly antifreeze compounds. An alternative is to economize by having long lasting leaves like those of pines, heathers, and the perennial shrubs of desert and garrigue but paying the cost of their more sluggish physiological processes.

Aquatic plants experience a quite different set of environmental conditions from those on land. The buoyancy of water and the force of its movement make it difficult

form and behavior may change with the seasons

and aquatic plants may accommodate to different water levels

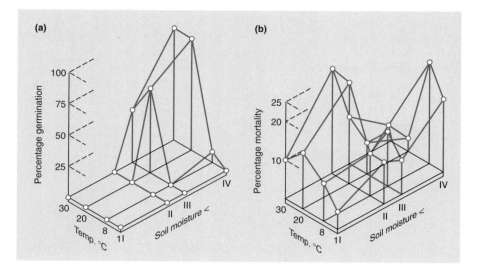

Figure 3.6

(a) In a simple experiment, corn was sown in pots of dry soil that remained unwatered or received three levels of watering—the wettest had been waterlogged and then allowed to drain. The pots were then held at temperatures of 1, 8, 20, or 30°C for 15 days. After that time many of the seeds had germinated and the seedlings had emerged. But none germinated in the soils with the two lowest water contents nor in soils held at 1°C. In soil held at 8°C no seeds germinated except in the wettest soil. This showed that the conditions of soil moisture and temperature interacted in their influence on germination. (b) At the end of the experiment, the ungerminated seeds were tested to see how many were still alive. Mortality was very low in warm damp soil, where germination had been widespread, but many seeds were also alive in the very dry cold soil, where, however, they remained dormant. But there was heavy mortality of seeds in the cooler and wetter soils, and soil-borne fungal diseases were responsible for most of this mortality. (Harper, 1955.)

Figure 3.7

The incidence of southern corn leaf blight (*Helminthosporium maydis*) on corn growing in rows at various distances from trees that shaded them. (Redrawn from Lukens & Mullany, 1972.)

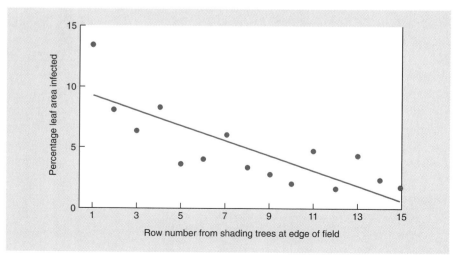

for rigid structures to survive. This is particularly the case in seaweeds and kelps, which tolerate the ferocity of wave action. Fresh waters also provide a range of environmental conditions that are exploited by species with specialized tolerances. They include free-floating species such as the water hyacinth (*Eichhornia crassipes*) and the duckweeds (*Lemna* sp.), which move with the flow of water; species with thin submerged leaves (like *Vallisneria* sp.); and those with floating leaves such as the water lilies (*Nuphar* and *Nymphaea* sp.) or rigid emergent leaves like those of many land plants. Several species even bear remarkably different leaf forms on the same plant when it extends itself across quite different conditions (Figure 3.8).

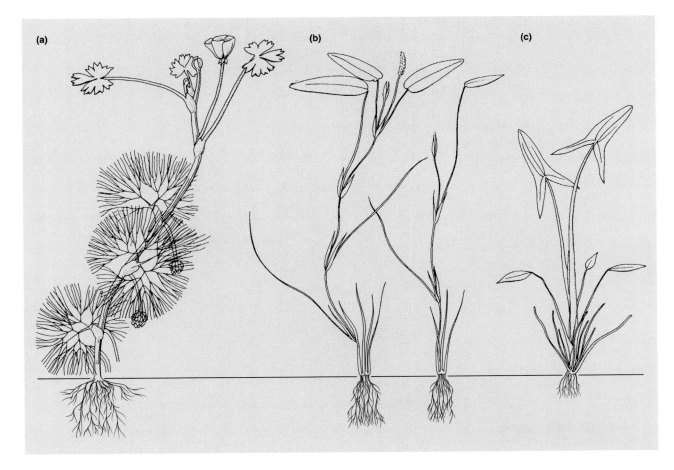

Figure 3.8
Differences in the form of submerged, floating and emergent leaves on some aquatic plants. The submerged leaves are often thin and grass-like or even filamentous. They are flexible and tolerate the force of flowing water. Floating leaves are broader and thicker and the upper surface has a strong water-repellent cuticle. The leaf stalks are long and flexible and move easily in the water. The same plants may produce emergent leaves with stiff leaf blades that have no need to bend and sway with moving water. (a) *Ranunculus aquatilis* with submerged and floating leaves. (b) *Potamogeton natans* with submerged, and floating leaves. (c) *Sagittaria sagittifolia* with submerged, floating and emergent leaves. (Courtesy of S. Ross-Craig and W. Keble-Martin.)

Box 3.1

Recording Seasonal Changes

Recording the changing behavior of organisms through the season (*phenology*) was essential before agricultural activities could be intelligently timed. The earliest phenological records were apparently the Wu Hou observations made in the Chou and Ch'in (1027–206 B.C.) dynasties. The date of first flowering of cherry trees has been recorded at Kyoto, Japan, since A.D. 812.

A particularly long and detailed record was started in 1736 by Robert Marsham at his estate near the city of Norwich, England. He called these records "Indications of the Spring." Recording was continued by his descendants until 1947. Marsham recorded 27 phenological events every year: the first flowering of snowdrop, wood anemone, hawthorn, and turnip; the first leaf emergence of 13 species of tree, and various animal events such as the first appearance of migrants (swallow, cuckoo, nightingale), the first nest building by rooks, croaking of frogs, and toads and the appearance of the brimstone butterfly (*Gonopteryx rhamni*).

Long series of measurements of environmental temperature are not available for comparison with the whole period of Marsham's records, but they are available from 1771 for Greenwich, about 100 miles away. There is surprisingly close agreement between many of the flowering and leaf emergence Marsham events and the mean January–May temperature at Greenwich (Figure 3.9).

Figure 3.9

The relationships between mean January–May temperatures and the annual mean dates of ten flowering and leafing events from the classic Marsham records started in 1736. (From redrawn figures of Margary in Ford, 1982.)

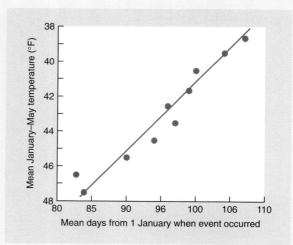

3.2.6 Animal responses to environmental temperature

ectotherms and endotherms

Most species of animals, like plants, are ectotherms: they rely on external sources of heat to determine the pace of their metabolism. This includes the invertebrates and also fish, amphibians, and lizards. Others, mainly birds and mammals, are endotherms: they regulate their body temperature by producing heat within their body.

The distinction between ectotherms and endotherms is not absolute. Some typical ectotherms, some insects, for example, can control body temperature through muscle activities (e.g., shivering flight muscles). Some fish and reptiles can generate heat for limited periods of time, and even some plants can use metabolic activity to

However, not surprisingly, events such as the time of arrival of migrant birds bears little relationship to temperature. Analyzes that made use of a shorter but more reliable series of temperature records and the Marsham data for the emergence of leaves on six species of tree produced the relationship shown in Figure 3.10, which indicates that the mean data of leafing are advanced by 4 days for every 1°F increase in the mean temperature from February to May.

Similarly, for the eastern United States, Hopkins' *bioclimatic law* states that the indicators of spring such as leafing and flowering occur 4 days later for every 1° latitude northward, 5° longitude westward, or 400 feet of altitude.

Collecting phenological records has now been transformed from the pursuit of gifted amateurs to sophisticated programs of data collection and analysis. At least 1500 phenological observation posts are now maintained in Japan alone. The vast accumulations of data have suddenly become exciting and relevant as we try to estimate the changes in floras and faunas that will be caused by global warming.

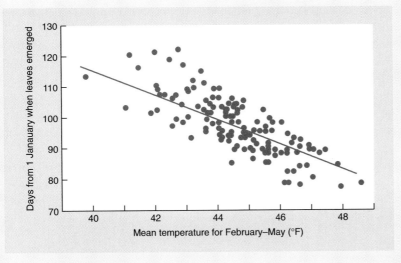

Figure 3.10
The relationship between the mean temperature in the 4-month period February–May and the average date of six leafing events. The correlation coefficient is −0.81. (From redrawn figures of Kington in Ford, 1982.)

raise the temperature of their flowers. Some typical endotherms, on the other hand, such as dormice, hedgehogs, and bats, allow their body temperature to fall and become scarcely different from that of their surroundings when they are hibernating (Figure 3.11).

Despite these overlaps, endothermy is inherently a different strategy from ectothermy. Over a certain narrow temperature range, an endotherm consumes energy at a basal rate. But at environmental temperatures further and further above or below that zone, endotherms expend more and more energy maintaining their constant body temperature. This makes them relatively independent of environmental conditions and allows them to stay longer at or close to peak performance. It makes them

Figure 3.11
Changes in the body temperature of the European
hamster during a 3-day bout of hibernation. (After
Nedergaard & Cannon, 1990.)

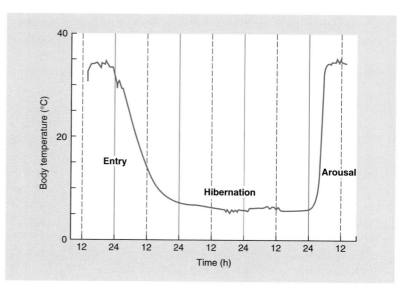

more efficient in both searching for food and escaping from predators. However, there is a cost—a high requirement for food to fuel this strategy.

Endotherms have morphological modifications that reduce their energetic costs. In cold climates most have low surface area to volume ratios (short ears and limbs), and this reduces heat loss through surfaces. Typically, endotherms that live in cold temperate or arctic environments are insulated from the cold with extremely dense fur (polar bears, mink, and foxes) or feathers and extra layers of fat. In contrast, desert endotherms often have thin fur, and long ears and limbs, which help dissipate heat.

temperatures that vary seasonally pose special problems

Variability of conditions can set biological challenges as great as extremes. Seasonal cycles, for example, can expose an animal to summer heat close to its thermal maximum, and winter chill close to its thermal minimum. Some of the responses to these changing conditions include the laying down of different coats

The thick white winter coat and the thinner, browner summer coat of the Arctic fox.

in the fall (thick and underlain by a thick fat layer) and in the spring (a thinner coat and loss of the dense fat layer) (Figure 3.12). Some animals also take advantage of each other's body heat as a means to cope with cold weather (i.e., huddling). Hibernation—relaxing temperature control—allows some vertebrates to stay sheltered and survive periods of winter cold and food shortage (Figure 3.11), avoiding the difficulties of finding sufficient fuel over these periods. Migration is another "avoidance" strategy: the Arctic tern, to take an extreme example, travels from the Arctic to the Antarctic and back each year, avoiding the polar winters.

Despite the quite profound differences in the physiological characteristics of ectotherms and endotherms, they often have very similar ecological features. Large marine endotherms such as whales, porpoises, and dolphins survive broadly the same environmental conditions as large ectotherms such as sharks. Very small desert endotherms such as small seed-eating rodents share very similar conditions with small desert ectotherms such as lizards and snakes. On the other hand, there are now no very large ectotherms on land that match the endotherm elephant, rhinoceros, or hippopotamus. Such ectotherms would have large volumes to heat up—but surface areas over which to absorb heat that were small relative to that volume. (There is still some dispute whether the large dinosaurs were endo- or ectotherms.) Similarly, there are plenty of species of very small fish but no comparably small aquatic endotherms (which would have a relatively large suface area over which to lose heat but only a small body volume from which to generate it).

3.2.7 Microorganisms in extreme environments

Microorganisms survive and grow in all the environments that are lived in or tolerated by animals and plants. They show the same range of strategies, which carry them through extreme conditions—avoid, tolerate, or specialize. Many microorganisms produce resting spores that survive drought, high temperature, or cold. There are

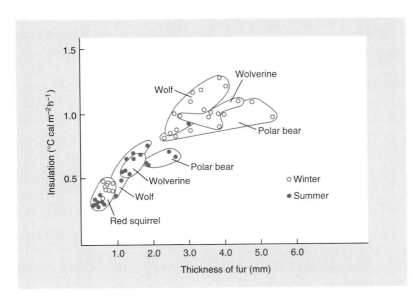

Figure 3.12
Seasonal changes in the thickness of the insulating fur coats of some arctic and northern temperate mammals.

also some that are capable of growth and multiplication in conditions far outside the range of tolerance of higher organisms: they inhabit some of the most extreme environments on Earth. Temperatures maintained higher than 45°C are lethal to almost all plants and animals, but thermophilic, "temperature-loving," microbes grow at much higher temperatures. Although similar in many ways to heat-intolerant microbes, the enzymes of these thermophiles are stabilized by especially strong ionic bonds.

Microbial communities that not only tolerate but grow at low temperatures are also known; these include photosynthetic algae, diatoms, and bacteria that have been identified on Antarctic sea ice. Microbial specialists have also been identified from other rare or peculiar environments. These include *acidophiles*, which thrive in environments that are highly acidic. One of them, *Thiobacillus ferroxidans*, is found in the waste from industrial metal-leaching processes and tolerates pH 1.0; *T. thiooxidans* can grow at a pH of 0! At the other end of the pH spectrum, the cyanobacterium *Plectonema nostocorum* from soda lakes can grow at pH 13. As noted previously, these oddities may be relicts from environments that prevailed much earlier in the Earth's history. Perhaps they are the kind of organism that we might look for on other planets.

3.3 ▷ Plant Resources

Resources may be either biotic or abiotic components of the environment: they are whatever an organism uses or consumes in its growth and maintenance, leaving less available for other organisms. When a photosynthesizing leaf intercepts radiation, it deprives some of the leaves or plants beneath it. When a caterpillar eats a leaf, there is less leaf material available for other caterpillars. By their nature, resources are both critical for survival, growth, and reproduction and also inherently a potential source of conflict and competition between organisms.

resource requirements of nonmotile
organisms

If an organism can move about, it has the potential to search for its food. Organisms that are fixed and "rooted" in position cannot search. They must rely on growing toward their resources (like a shoot or root) or catching resources that move to them. The most obvious examples are green plants, which depend on (1) energy that radiates to them, (2) atmospheric carbon dioxide that diffuses to them, (3) mineral cations that they obtain from soil colloids in exchange for hydrogen ions, and (4) water and dissolved anions that the roots absorb from the soil. In the following sections, we concentrate on green plants. But it is important to remember that many of the nonmobile animals, like corals, sponges, and bivalve molluscs, depend on resources that are suspended in the watery environment and are captured by filtering the water or even just waiting for them open-mouthed.

3.3.1 Solar radiation

sun and shade plants

Solar radiation is a critical resource for green plants. We often refer to it loosely as "light," but green plants actually use only about 44 percent of that narrow part of the spectrum of solar radiation that is visible to us between infrared and ultraviolet. The rate of photosynthesis increases with the intensity of the radiation that a leaf receives, but with diminishing returns. The shape of the curve that relates the rate of photosynthesis by a leaf to the intensity of radiation varies greatly between species (Figure 3.13), especially between those species that usually live in shaded habitats (and reach saturation at low radiation intensities) and those that normally experience full sunlight and can take advantage of it (sun herbs and common cereal grains).

The solar radiation that reaches a plant is forever changing. Its angle and intensity change annually and diurnally in a regular and systematic way. There are also irregular, unsystematic variations due to changes in cloud cover or shadowing by the leaves of neighboring plants. There are gaps in a leafy canopy, and leaves beneath it are not exposed to uniform shading. As light flecks pass over them they receive seconds or minutes of direct bright light and then plunge back into shade. The daily photosynthesis of a leaf integrates these various experiences; the whole plant integrates the diverse exposure of its various leaves.

There is enormous variation in the shapes and sizes of leaves that in a way parallels the diversity in the mouthparts of insects described later. But whereas insects search for widely different resources, all leaves garner the same two simple resources: radiation and carbon dioxide. Most of the heritable variation in the shapes of leaves has probably evolved under selection not primarily for high photosynthesis but rather for optimal efficiency of water use (photosynthesis achieved per unit of water transpired) and minimization of the damage done by foraging herbivores.

Not all the variations in leaf shape are heritable, though: many are programmed responses by the individual to its immediate environment. The genotype does not specify a shape—rather, it specifies a series of alternatives. Many trees, especially, produce different types of leaf in positions exposed to full sunlight ("sun leaves") and in places lower in the canopy where they are shaded ("shade leaves"). Sun leaves are thicker, with more densely packed chloroplasts (which process the incoming radiation) within cells and more cell layers. The more flimsy shade leaves intercept diffused and filtered radiation low in the canopy and may supplement the main photosynthetic activity of the sun leaves high in the canopy.

Among herbaceous plants and shrubs, specialist "sun species" or "shade species" are much more common than species with the ability to produce sun and shade

sun and shade leaves

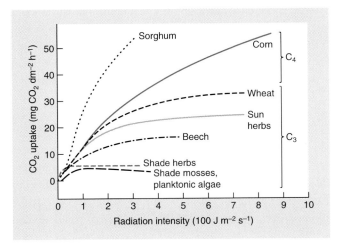

Figure 3.13

The response of photosynthesis by the leaves of various types of green plant (measured as carbon dioxide uptake) to the intensity of solar radiation at optimal temperatures and with natural supply of carbon dioxide. (After Larcher, 1980; and other sources.)

leaves. Leaves of sun plants are commonly exposed at acute angles to the midday sun and are typically superimposed into a multilayered canopy where even the lower leaves may have a positive rate of net photosynthesis. The leaves of shade plants are typically arranged in a single-layered canopy and angled horizontally, maximizing their ability fully to capture the available radiation.

3.3.2 Water

water is lost from plants that photosynthesize

Most plant parts are largely composed of water. In some soft leaves and fruits, as much as 98 percent of the volume may be water. Yet this is a minute fraction of the water that passes from the soil through a plant to the atmosphere during plant growth. Photosynthesis depends on the plant's absorbing carbon dioxide. This can only happen across surfaces that are wet—most notably the walls of the photosynthesizing cells in leaves. If a leaf allows carbon dioxide to enter, it is almost impossible to prevent water vapor from leaving. Likewise, any mechanism or process that slows down the rate of water loss, such as closing the stomata (pores) on the leaf surface, is almost bound to reduce the rate of photosynthesis.

wilting

Green plants serve as wicks that conduct water from the soil and release it to the atmosphere. If the rate of uptake falls below the rate of release, the body of the plant (the wick) starts to dry out. The cells lose their turgidity and the plant wilts. This may just be temporary (though it may happen every day in summer), and they may recover and rehydrate at night. But if the deficit accumulates, the leaf is killed and the whole plant may die.

When a plant has absorbed water from the soil, less remains for others. Water that a plant absorbs from the soil may keep its leaves turgid and in a state in which it can photosynthesize—whereas a neighor that it has deprived may have wilted and been unable to photosynthesize or even died. Water is certainly something for which plants may compete.

plant life in water deficit: avoiders and tolerators

Species of green plant differ in the ways in which they survive in dry environments. One strategy is to avoid the problems. Avoiders such as desert annuals, annual weeds, and most crop plants have a short life span: their photosynthetic activity is concentrated during periods when they can maintain a positive water balance. For the remainder of the year, they remain dormant as seeds, a stage that requires neither photosynthesis nor transpiration. Some perennial plants shed their photosynthetic tissues during periods of drought. Some species then replace them with new leaf forms that are less extravagant of water or spend the driest season with no leaves at all—just green stems.

Other plants, tolerators, have evolved a different compromise, producing long-lived leaves that transpire slowly (for example, by having few and sunken stomata). They tolerate drought, but of course their photosynthesis is slower. These plants have sacrificed their ability to achieve rapid photosynthesis when water is abundant but gained the insurance of being able to photosynthesize at a low level throughout the seasons. This is not only a property of plants from arid areas but also of the pines and spruces that survive where water may be abundant but is usually frozen and therefore inaccessible.

The evaporation of water lowers the temperature of the body with which it is in contact. For this reason, if plants are prevented from transpiring they may overheat. This, rather than water loss itself, may be lethal. The desert honeysweet (*Tidestromia oblongifolia*) grows vigorously in Death Valley, California despite the fact that its leaves are killed if they reach 50°C—a temperature that is commonly reached in the surrounding air. Transpiration cools the surface of the leaf to a tolerable 40–45°C. Most desert plants bear hairs, spines and waxes on the leaf surface. These reflect a high proportion of incident radiation and help to prevent overheating. Other more general modifications in desert plants include the characteristic "chunky" shape of succulents with few branches, giving a low surface area to volume ratio over which radiant heat is absorbed.

Specialized biochemical processes may increase the amount of photosynthesis that can be achieved per unit of water lost. The majority of plants on Earth photosynthesize using what is termed the *C3 pathway*. Although these plants are highly productive photosynthesizers, they are relatively wasteful of water, reach their maximum rates of photosynthesis at relatively low intensities of radiation, and are less successful in arid areas. Alternative pathways of photosynthesis—termed the *C4 pathway* and *CAM*—are more economical in their water use. C4 plants have a particularly high affinity for carbon dioxide—and so absorb more per unit of water lost. CAM plants open their stomata at night and absorb carbon dioxide and fix it as malic acid. They close their stomata during the day and release the CO_2 internally for photosynthesis. C4 and CAM plants are most common in arid and, in particular, hot arid areas. They are restricted in range because the associated costs of their systems apparently make them less competitive under less arid conditions. For example, C4 plants' photosynthesis is inefficient at low radiation intensities (Figure 3.13) and so they are poor shade plants. CAM plants must store their accumulated malic acid every night. Most of them are succulents with extensive swollen water-storage tissues that cope with this problem.

Many of the problems in the life of terrestrial plants arise because the resources of radiation and water behave so differently. Solar radiation may be intercepted by leaves and a small fraction may be fixed in photosynthesis. The products are subsequently respired and the energy is released as heat. The remainder of the incident solar radiation is directly dissipated as heat. *It cannot be stored and can never be used again.* The behavior of water is in total contrast. Although plants may receive water as rain directly onto their leaves, they absorb scarcely any of it. Instead it passes to the soil and (apart from what drains through the soil under the force of gravity) is stored. Plants obtain virtually all their water from this stored reserve. It is held against gravity by capillary forces and as colloids. Sandy soils have wide pores: these do not hold much water but what is there is held with weak forces and plants can withdraw it rather easily. Clay soils are close textured with very fine pores. They retain more water against the force of gravity than sandy soils, but surface tension in the fine pores makes it more difficult for the plants to withdraw it.

The primary water absorbing zone on roots is covered with root hairs that make intimate contact with soil particles (Figure 3.14). As water is withdrawn from the soil, the first to be released is from the wider pores, where capillary forces retain it

water and overheating

increasing the efficiency of water use: C4 and CAM

obtaining water from the soil

Figure 3.14
Highly diagrammatic picture of a root hair withdrawing water from pores in a very wet soil. Even the widest pores shown are full of water. As water is withdrawn, the wider pores become emptied and water flows only along the twisted pathways through narrower pores.

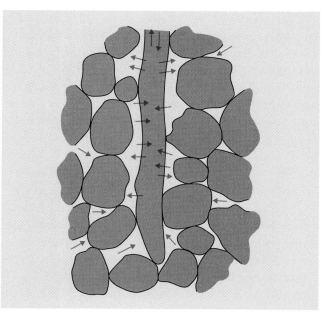

only weakly. Subsequent water is withdrawn from narrower paths in which the water is more tightly held. Consequently, the more the soil around the roots is depleted of water, the more resistance there is to water flow. As a result of this withdrawal, roots create water depletion zones (or, more generally, resource depletion zones [RDZs]) around themselves. The faster the roots draw water from the soil, the more sharply defined the RDZs, and the more slowly water will move into that zone. In a soil that contains abundant water, rapidly transpiring plants may still wilt because water does not flow fast enough to replenish the RDZs around their root systems (or because the roots cannot explore new soil volumes fast enough).

The shapes of root systems are much less tightly programmed than those of shoots. The root architecture that a plant establishes early in its life can determine its responsiveness to later events. Plants that develop under waterlogged conditions usually set down only a superficial root system. If drought develops later in the season these same plants can suffer because their root system did not tap the deeper soil layers. A deep tap root, however, will be of little use to a plant in which most water is received from occasional showers on a dry substrate. Figure 3.15 illustrates some characteristic differences between the root systems of plants from damp temperate and dry desert habitats.

Figure 3.15
Profiles of root systems of plants from contrasting environments. (a–d) Northern temperate species of open ground: (a) *Lolium multiflorum,* an annual grass; (b) *Mercurialis annua,* an annual weed; (c) *Aphanes arvensis* and (d) *Sagina procumbens,* both ephemeral weeds. (From Fitter, 1991.) (e–i) Desert shrub and semishrub species, Mid Hills, eastern Mojave Desert, California. (Redrawn from a variety of sources.)

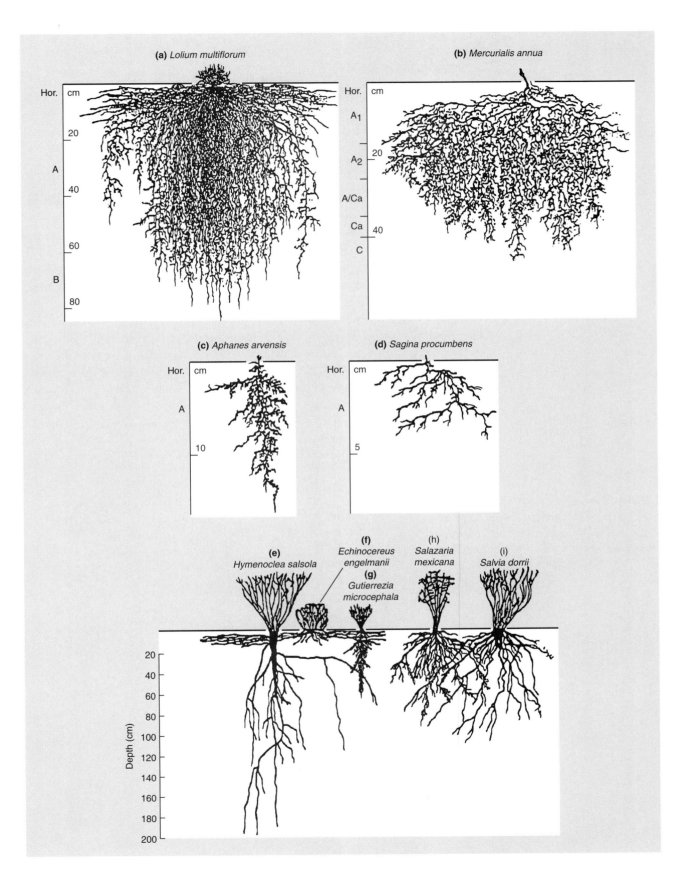

(a) *Lolium multiflorum*

Hor. cm
20
A
40
60
B
80

(b) *Mercurialis annua*

Hor. cm
A₁
A₂ 20
A/Ca
Ca 40
C

(c) *Aphanes arvensis*

Hor. cm
A
10

(d) *Sagina procumbens*

Hor. cm
A
5

(e)
Hymenoclea salsola

(f)
Echinocereus engelmanii

(g)
Gutierrezia microcephala

(h)
Salazaria mexicana

(i)
Salvia dorrii

Depth (cm)
20
40
60
80
100
120
140
160
180
200

3.3.3 Mineral nutrients

Roots extract water from the soil, but they also extract key minerals (Figure 3.16). Plants require mineral resources of nitrogen (N), phosphorus (P), sulfur (S), potassium (K), calcium (Ca), magnesium (Mg), and iron (Fe), together with traces of manganese (Mn), zinc (Zn), copper (Cu), and boron (B). All of these must be obtained from the soil (or directly from the water in the case of free-living aquatic plants).

Root architecture is particularly important in this process because different nutrients are held in the soil by different forces. Soils are patchy and heterogeneous, and as roots grow through them, they may meet regions that vary in nutrient and water content. They tend to branch profusely in the richer patches (Figure 3.17).

the architecture of roots determines their foraging efficiency

The significance of root branching becomes clearer when we consider the different behavior of mineral nutrients in soil. Nitrate ions diffuse rapidly in soil water, and rapidly transpiring plants may bring nitrates to the root surface faster than they are accumulated in the body of the plant. However, other key nutrients such as phosphate are tightly bound in the soil (have low diffusion coefficients). Plants obtain the vast majority of their phosphate from within 0.1 mm of their root surface. The phosphate RDZs of two roots 0.2 mm apart scarcely overlap, and the parts of a finely branched root system scarcely compete with each other. Consequently, if phosphate is in short supply, a highly branched surface root will greatly improve phosphate absorption. A more widely spaced extensive root system, in contrast, will tend to maximize access to nitrate.

Figure 3.16
Seedlings of mustard were grown in tubes of soil that contained radioactively labeled phosphate. The figure is an autoradiograph; the zones from which phosphate has been withdrawn by the seedlings show up white. (After Nye & Tinker, 1977.)

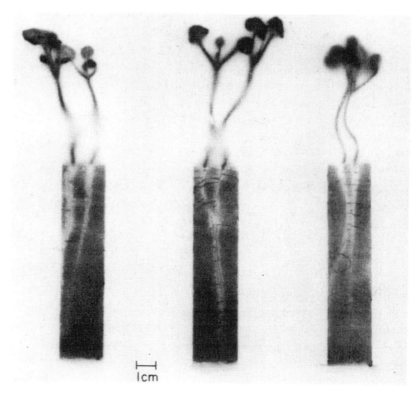

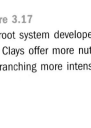

Figure 3.17
The root system developed by a young plant of wheat growing through a sandy soil with a layer of clay. Clays offer more nutrient resources and hold more water than sand and the roots respond by branching more intensively in the clay. (Courtesy of J. V. Lake.)

3.3.4 Carbon dioxide

Plants take in carbon dioxide through the stomatal pores on leaf surfaces and, using the energy of sunlight, capture the carbon atoms and release oxygen. Not only is carbon the base of all plant growth, and consequently the building block of every food chain, but life as we currently know it on Earth was not possible until photosynthetic organisms had substantially altered the atmosphere through the release of oxygen.

Carbon dioxide varies in its concentration at a variety of scales. In 1750, atmospheric carbon dioxide concentrations were approximately 280 µl/l. Currently, the figure is over 350 µl/l and is increasing by 0.4–0.5 percent per year, mainly as a result of the burning of fossil fuels. Plants have responded to even larger fluctuations of carbon dioxide over geological history. During the Triassic, Jurassic, and Cretaceous periods, atmospheric concentrations of carbon dioxide were four to eight times greater than at present.

Not only is carbon dioxide an essential resource for green plants, but increasing carbon dioxide above atmospheric levels in greenhouses can increase crop yields. It might be thought that global increases in atmospheric carbon dioxide will lead to global increases in crop production. The reality seems to be much more complicated, and it is not easy to see a general trend in the behavior of either crop plants or natural vegetation (Bazzaz, 1996).

In a terrestrial community, carbon dioxide increases and moves upward through the canopy at night as plants respire and there is no photosynthesis. The release of carbon dioxide from decomposing organic matter in the soil also contributes to the balance sheet. During the day, carbon dioxide is absorbed by plants within the canopy and the flux is downward as plants actively remove carbon dioxide from the air. It is very doubtful whether plants in the field ever compete with each other for carbon dioxide. Gaseous diffusion and mixing are extremely rapid, and active uptake of carbon dioxide by one plant is very unlikely to create resource depletion zones that overlap with those of its neighbors.

UNANSWERED QUESTION:

what will be the repercussions of global increases in atmospheric carbon dioxide?

Global Warming? Can We Risk it?

Carbon dioxide is one of several global "greenhouse gases" (Section 13.4.3), increases in the concentration of which are believed by many scientists to be leading increasingly to rises in global mean temperatures, to a growth in the number of "extreme" and "record" weather events, and to the threat of the major biomes of the earth substantially changing their distribution (see Box 4.3).

Not all scientists, however, believe these threats to be significant. Politicians, governments, and policy-makers in general are therefore faced with different groups of scientific "experts" offering different projections into the future, and with many interest groups, including a number of industries, resisting attempts to force them to change their behavior in order to reduce emissions of greenhouse gases.

Even though the majority of scientists believe the problem to be a very real one, the truth is that predictions of the future can never be made with absolute certainty.

Put yourself in the position of a policy-maker. Would it be reasonable of you to demand major changes of significant sectors of the national economy, in order to avert a disaster that may never happen in any case? Or, since the consequences of the "worst case" and even some of the "middle-of-the-road" scenarios are so profound, is the only responsible course of action to minimize risk?—to behave as if disaster is certain if we do not change our collective behavior, even though it is not? One alternative might be to wait for better data. But suppose that by the time better data are available it is too late . . .

3.4 ⟩ Animals and Their Resources

autotrophs and heterotrophs

Green plants are *autotrophs*: their resources are quanta of radiation, ions, and simple molecules. Plants assemble them into complex molecules (carbohydrates, fats, and proteins) and then package them into cells, tissues, organs, and whole organisms. It is these packages that form the food resources for virtually all other organisms, the *heterotrophs* (decomposers, predators, grazers, and parasites). These consumers unpack the packages, metabolize and excrete some of the contents, and reassemble the remainder into their own bodies. They in turn may be consumed, unpacked, and reconstituted in a chain of events in which each consumer becomes, in turn, a resource for some other consumer.

Heterotrophs can generally be grouped as follows:

1. *Decomposers,* which feed on already dead plants and animals
2. *Parasites,* which feed on one or a very few host plants or animals while they are alive but do not (usually) kill their hosts, at least not immediately
3. *Predators,* which, during their life, eat many prey organisms, typically (and in many cases always) killing them
4. *Grazers,* which, during their life, consume parts of many prey organisms, but do not (usually) kill their prey, at least not immediately

The usual mental image of a predator-prey relationship is something akin to a lion eating a gazelle, but the relationship encompasses a much wider array of consumer–resource relationships than this picture brings to mind. For example, a squirrel is a predator when it eats an acorn (it kills the acorn embryo); a whale is a predator as it feeds on krill; a fungus can be regarded as a predator when it feeds on and kills a growing seedling; and carnivorous plants are of course predators of their (usually) insect prey. In each case, the predator kills its food resource as it consumes all or part of it. Here, we concentrate on animal consumers (and take the subject further still in Chapter 8).

An important distinction between animal consumers is whether they are special- **monophagy and polyphagy** ized or generalized in their diet. Generalists (*polyphagous* species) take a wide variety of prey species though they very often have clear preferences and a rank order of what they will choose when there are alternatives available. Specialists may specialize on particular parts of their prey but range over a number of species. This is most common among herbivores because, as we shall see, different parts of plants are quite different in their composition. Thus, many birds specialize on eating seeds though they are seldom restricted to a particular species. Finally, a consumer may specialize on a single species or a narrow range of closely related species (when it is said to be *monophagous*). Examples are caterpillars of the cinnabar moth (which eat leaves, flower buds, and very young stems of species of ragwort, *Senecio*) and many species of host-specific parasites.

Many of the resource-use patterns found among animals reflect the different life **the importance of life spans** spans of the consumer and what it consumes. Individuals of long-lived species are likely to be generalists: they cannot depend on one food resource's being available throughout their life. Specialization is increasingly likely if a consumer has a short life span. Evolutionary forces can then shape the timing of the consumer's food demands to match the timetable of its prey. Specialization allows the evolution of very efficiently designed structures that make it possible to deal efficiently with some resources—this is particularly the case with mouthparts. The stylet of the aphid (Figures 3.18a and b) can be interpreted as an exquisite product of the evolutionary process that has given the aphid access to a uniquely valuable food resource—or as an example of the ever-deepening rut of specialization that has constrained what aphids can feed on. The more specialized the food resource required by an organism, the more it is constrained to live in patches of that resource *or* to spend time and energy in searching for it among a mixture of resources. This is one of the costs of specialization.

A number of these points can be illustrated by reference to a population of wild raspberries in a temperate woodland. Flowering brings only a short period of nectar production. Bees feed from this but also make use of nectar from many other species and hence continue to feed throughout the summer. By contrast, the raspberry beetle lays eggs in, and its larvae feed on, the flowers of *only* this species, and the larva completes its life cycle within the developing fruit. It then remains, inactive, as a pupa until the next raspberry flowering season in 10–11 months' time. The larva of the raspberry moth, on the other hand, has a much longer active period because it feeds on a consistently present resource: the pith within the woody stems. Thus, a single resource (the raspberry plant) can be used by a variety of types of consumer, or, to look at it another way, a single species may provide many different resources.

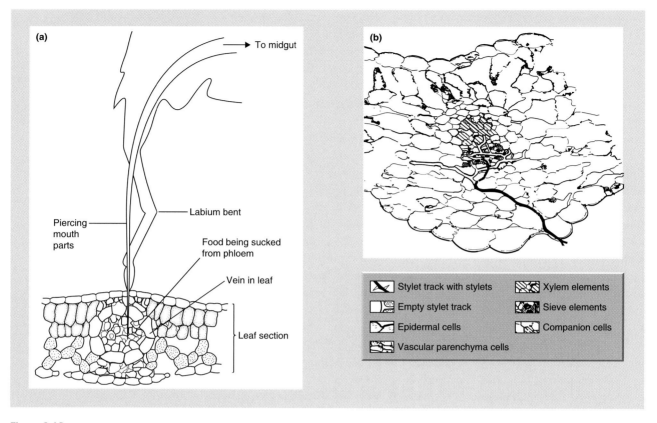

Figure 3.18
The stylet of an aphid penetrating the host tissues and reaching the sugar-rich phloem cells in the leaf veins. (a) Aphid mouthparts and cross section of a leaf. (b) A stylet, showing its circuitous path through a leaf. (After Tjallingii & Ogen Esch, 1993.)

3.4.1 Nutritional needs and provisions

plants as (a variety of) foods

The various parts of a plant have very different composition (Figure 3.19) and so offer quite different resources. Bark, for example, is largely composed of dead cells with corky and lignified walls, packed with defensive phenolics, and is quite useless as a food for most herbivores. Species of bark beetle specialize on the nutritious cambium layer just beneath the bark, rather than on the bark itself. The richest concentrations of plant proteins (and hence of nitrogen) are in the meristems in the buds at shoot apices and in leaf axils. Not surprisingly, these are usually heavily protected with bud scales and defended from herbivores by prickles and spines on the shoots that bear them. Seeds are packaged and usually dried resources that are usually rich in starch or oils as well as specialized storage proteins. The very sugary and fleshy fruits are resources provided by the plant as "payment" to animals that disperse the seeds. Very little of the plants' nitrogen is "spent" on these rewards.

The diversity of different food resources offered by plants is matched by the diversity of specialized mouthparts and digestive tracts that have evolved to consume

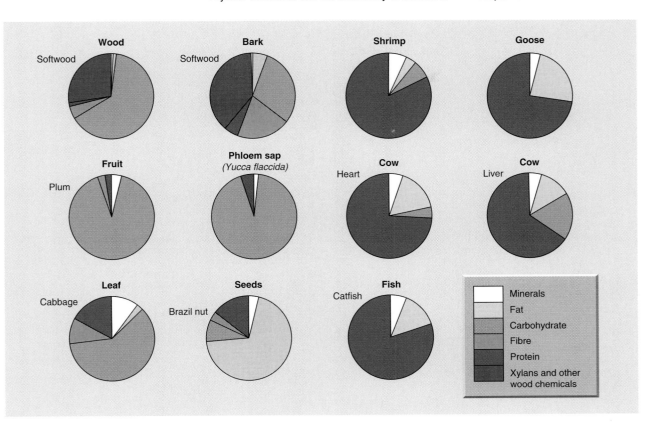

Figure 3.19
The composition of various plants and animals that may serve as food resources for herbivores or carnivores. Note that the different parts of a plant have very different composition whereas different species of animal (and their parts) are remarkably similar.

them. This diversity is especially developed in the beaks of birds and the mouthparts of insects (Figure 3.20).

For a consumer, the body of a plant is a quite different package of resources from the body of an animal. First, plant cells are bounded by walls of cellulose, lignin, and other structural materials that give plants their high fiber content and contribute to the high ratio of carbon to other important elements. The large amounts of fixed carbon in these structural plant materials means that they are potentially rich sources of energy. Yet the overwhelming majority of animal species lack cellulolytic and other enzymes that can digest these compounds: they are quite useless as a direct energy resource for most herbivores. Moreover, the cell wall material of plants hinders the access of digestive enzymes to the plant cell contents. The acts of chewing by the grazing mammal, cooking by humans, and grinding in the gizzard of birds are necessary precursors to digestion of plant food because they allow digestive enzymes access to the cell contents. The carnivore, by contrast, can more safely gulp its food.

Many herbivores have made up for their own lack of cellulolytic enzymes by entering into a *mutualistic* (beneficial to both parties) association with cellulolytic bacteria and protozoa in their alimentary canal. The rumen (sometimes the cecum)

from plants into animals

UNANSWERED QUESTION:

why so few herbivores with their own cellulases?

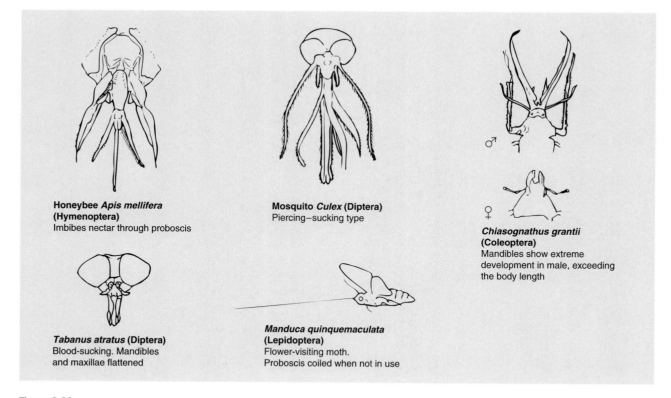

Figure 3.20
Examples of the variety of specialized mouthparts in herbivorous insects.

of many herbivorous mammals is a temperature-regulated culture chamber for bacteria into which already partially fragmented plant tissues and cells flow continually, like the chemostat of an industrial fermentation or brewery. The bacteria receive a home and a supply of food. The "hosts" benefit by absorbing many of the major by-products of this bacterial fermentation, especially fatty acids.

Unlike plants, animal tissues contain no structural carbohydrate or fiber component but are rich in fat and protein. The C:N ratio of plant tissues commonly exceeds 40:1, in contrast to ratios that rarely exceed 10:1 in bacteria, fungi, and animals. Thus herbivores, which undertake the first stage of making animal bodies out of plants, are involved in a massive burning off of carbon as the C:N ratio is lowered. The main waste products of herbivores are therefore carbon-rich compounds (carbon dioxide and fiber). Carnivores get most of their energy from the protein and fats of their prey, and their main excretory products are in consequence nitrogenous.

Even if the cell wall fraction is excluded from consideration, the C:N ratio is high in plants compared with other organisms. Aphids, which gain direct access to cell contents by driving their stylets into the phloem transport system of plants and extracting phloem sap (Figure 3.18), gain a resource that is rich in soluble sugars (Figure 3.19). They use only a fraction of this energy resource and excrete the rest in sugary rich honeydew that may drip as a rain from an aphid-infested tree. For most herbivores and decomposers, the body of a plant is a superabundant source of energy and carbon; it is other components of the diet, especially nitrogen, that are more usually limiting.

The bodies of different species of animal have remarkably similar composition (Figure 3.19). In terms of protein, carbohydrate, fat, water, and minerals per gram, there is very little to choose between a diet of caterpillars or cod, or of earthworms, shrimps, or venison. The packages may be differently packaged (and the taste may be different), but the contents are essentially the same. Moreover, the different parts of an animal have very similar nutritional content. Carnivores are not faced with difficult problems of digestion (and they vary very little in their digestive apparatus) but rather with difficulties in finding, catching, and overcoming the defenses of their prey.

animals as food

3.4.2 Defense

The value of a resource to a consumer is determined not only by what it contains but by how well its contents are defended. Not surprisingly, organisms have evolved physical, chemical, morphological, and behavioral defenses against being attacked. These serve to reduce the chance of an encounter with a consumer and/or increase the chance of survival in such encounters. The spiny leaves of holly are not eaten by larvae of the oak eggar moth, but if the spines are removed the leaves are eaten quite readily. No doubt similar results would be achieved in equivalent experiments with foxes as predators and despined hedgehogs or porcupines as prey. On a smaller scale, many plant surfaces are clothed in epidermal hairs (trichomes) that may keep the smaller predators (such as thrips) away from the leaf surface.

Any feature of an organism's form or life-style that increases the energy that a consumer spends in discovering or handling it is a defense if, as a consequence, the consumer eats less of it. The thick shell of a nut increases the time that an animal spends to extract a unit of effective food, and this may reduce the number of nuts that are eaten. We have already seen that most green plants are relatively overprovided with energy resources in the form of cellulose and lignin. It may therefore be cheap to build husks and shells around seeds (and woody spines on stems) if these defense tissues contain rather little protein, and if what is protected are the real riches.

some resources are protected

Both plants and animals have a battery of chemical defenses. The plant kingdom is very rich in "secondary" chemicals that apparently play no role in the normal of plant biochemical pathways. A defensive function is generally ascribed to these chemicals and a defensive role has been demonstrated unequivocally in some cases. Populations of white clover, for example, commonly contain some individuals that release hydrogen cyanide when their tissues are attacked (*cyanogenic* forms) and others that do not. Those that cannot are eaten by slugs and snails. The cyanogenic forms, however, are nibbled but then rejected (Table 3.1). Noxious plant chemicals have been broadly classified into two types: toxic (or qualitative) chemicals, which are poisonous even in small quantities, and quantitative chemicals, such as tannins, which act by binding proteins and make the tissues that contain them, such as mature oak leaves, relatively indigestible. The growth of caterpillars of the winter moth decreases with an increase in the concentration of tannin included in the diet.

or defended

Table 3.1

Slugs (*Agriolimax reticulatus*) graze on the leaves of clover (*Trifolium repens*). There are forms of clover that release hydrogen cyanide when the cells are damaged. Slugs nibble clover leaves and reject cyanogenic forms but continue to consume the leaves of noncyanogenic forms. Two plants, one of each form, were grown together in plastic containers and slugs were allowed to graze for seven successive nights. The table shows the numbers of leaves in different conditions after slug grazing. $\pm$ indicates deviation from random expectation. The difference from random expectation is significant at $P < 0.001$.

	CONDITIONS OF LEAVES AFTER GRAZING			
	NOT DAMAGED	**NIBBLED**	**UP TO 50% OF LEAF REMOVED**	**MORE THAN 50% OF LEAF REMOVED**
Cyanogenic plants (AcLi)	160 (+)	22 (+)	38 (−)	9 (−)
Acyanogenic plants (acli)	87 (−)	7 (−)	30 (+)	65 (+)
(After Dirzo & Harper, 1982.)				

chemical defense in animals

Animals have more options than plants when it comes to defending themselves, but some still make use of chemicals. For example, defensive secretions of sulfuric acid of pH 1 or 2 occur in some marine gastropod groups, including the cowries. Other animals, which can tolerate the chemical defenses of their plant food, may actually be able to store the plant toxins and use them in their own defense. A classic example is the monarch butterfly, whose caterpillars feed on milkweeds, which contain cardiac glycosides, which are poisonous to mammals and birds. These caterpillars can store the poison, and it is still present in the adult. Thus, a bluejay will vomit violently after eating one, and, once it recovers, will reject all others on sight. In contrast, monarchs reared on cabbage are edible.

Chemical defenses are not equally effective against all consumers. Indeed, what is unacceptable to some animals may be the chosen, even unique diet of others. Many herbivores, particularly insects, specialize on one or a few plant species whose particular defense they have overcome. For example, females of the cabbage root fly, with eggs to lay, home in on a brassica crop from distances as far as 15 m downwind of the plants. It is probably hydrolyzed glucosinolates (toxic to many other species) that provide the attractive odor.

crypsis, aposematism, and mimicry

An animal may be less obvious to a predator if it matches its background, or possesses a pattern that disrupts its outline, or resembles an inedible feature of its environment. A straightforward example of such *crypsis* is the green coloration of many grasshoppers and caterpillars. Cryptic animals may be highly palatable, but their morphological traits and color (and their choice of the appropriate background) reduce the likelihood that they will be used as a resource. In contrast, noxious or dangerous animals often seem to advertize the fact by bright, conspicuous colors and patterns (*aposematism*). The monarch butterfly (see earlier), for example, is aposematically colored. One attempt by a bird to eat an adult monarch is so

Lepidopterous caterpillars illustrate a range of defense strategies. Clockwise from top-left: the irritating hairs of the gypsy moth, aposematism (advertising distastefulness) in the black swallowtail, a cryptic (camouflaged) noctuid (it looks like bark), and another swallowtail rearing and hence possibly startling a potential predator.

memorable that others are subsequently avoided for some time. There is also a strong selection pressure for a predator to remember what its food looks like. The adoption of memorable body patterns by distasteful prey, moreover, immediately opens the door for deceit by other species—there will be a clear advantage to a palatable prey if it *mimics* an unpalatable species. Thus, the palatable viceroy butterfly mimics the distasteful monarch, and a bluejay that has learned to avoid monarchs will also avoid viceroys.

By living in holes, animals (millipedes, moles) may avoid stimulating the sensory receptors of predators, and by "playing dead" (opossum, African ground squirrel) animals may fail to stimulate a killing response. Animals that withdraw to a prepared

behavior

retreat (rabbits and prairie dogs to their burrows, snails to their shells) or roll up and protect their vulnerable parts by a tough exterior (armadillos, hedgehogs) reduce their chance of capture. Other animals seem to try to bluff themselves out of trouble by threat displays. The startle response of moths and butterflies, which suddenly exposes eye spots on their wings, is one example. No doubt the most common behavioral response of an animal in danger of becoming a used resource is to run away.

3.5 ▶ The Effect of Intraspecific Competition for Resources

Resources are consumed. The consequence is that there may not be enough of a resource to satisfy the needs of a whole population of individuals. Individuals may then compete with each other for the limited resource. *Intraspecific competition* is competition between individuals of the same species.

exploitation: competitors depleting each other's resources

In many cases, competing individuals do not interact with one another directly. Rather, they deplete the resources that are available to each other. Grasshoppers may compete for food, but a grasshopper is not directly affected by other grasshoppers, but by the level to which they have reduced the food supply. Two grass plants may compete and each be adversely affected by the presence of close neighbors, but this is most likely to be because their resource depletion zones overlap—each may shade its neighbors from the incoming flow of radiation, and water or nutrients may be less accessible than they would otherwise be around the plants' roots. The data in Figure 3.21, for example, show the dynamics of the interaction between a single-celled aquatic plant, a diatom, and one of the resources it requires, silicate. As diatom density increases over time, silicate concentration decreases: there is then less available for the many than there had previously been for the few. This type of competition—in which competitors interact only indirectly, through their shared resources—is termed *exploitation*.

direct interference

On the other hand, competing individual vultures may fight one another over access to a newly found carcass. Individuals of other species may fight for ownership

Figure 3.21

A population of the freshwater diatom *Asterionella formosa* was grown in flasks of culture medium. The diatom consumes silicate during growth and the population of diatoms stabilizes when the silicate has been reduced to a very low concentration. (After Tilman et al., 1981.)

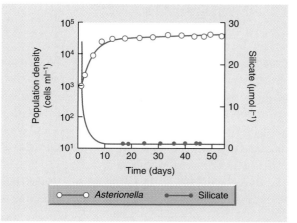

of a "territory" and access to the resources that a territory brings with it. A barnacle that settles on a rock denies the space to another barnacle. This is called *interference* competition.

Whether competition occurs through exploitation, interference or a combination of the two, its ultimate effect is on the *vital rates* of the competitors—their survival, growth, and reproduction—compared with what they would have been if resources had been more abundant. Competition typically leads to decreased rates of resource intake per individual, and thus to decreased rates of individual growth or development, perhaps to decreases in amounts of stored reserves or to increased risks of predation. Figure 3.22a shows how the mortality rate of flour beetles, as they develop from eggs to adults, increases as the number of competing beetles rises; Figure 3.22b shows how the birth rate of the sand dune grass *Vulpia* declines as individuals become increasingly crowded.

In practice, intraspecific competition is often a very one-sided affair: a strong early seedling will shade and suppress a stunted, late one; a large vulture is likely to fight off a smaller one. Some of the competitive strength of individuals is related to timing (the early seedling) or to random events (one seed may germinate in a depression where it obtains more water than its neighbors). Sometimes the winner and loser may be genetically different and then competition will be playing a role in natural selection.

The effects of intraspecific competition on any individual are typically greater the more crowded the individual is by its neighbors—the more the resource depletion zones of other individuals overlap its own. This often translates into saying that the greater the density of a population of competitors the greater is the effect of competition. Hence, the effects of intraspecific competition are often said to be *density-dependent*. But it is doubtful that any organism has a way of detecting the density of its population! Rather it responds to the effects of being crowded.

On the other hand, at low densities in the preceding examples, the *per capita* death rate or mortality rate (Figure 3.22a) and the per capita birth rate or fecundity (Figure 3.22b) are *independent* of density (where *per capita* means literally "per head" or "per individual"). That is, the risk of an individual's dying is effectively

competition and vital rates

density dependence

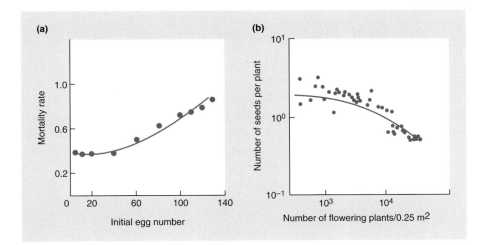

Figure 3.22

(a) The rate of mortality in population of the flour beetle *Tribolium confusum* in cultures established from different numbers of eggs. (After Bellows, 1981.) (b) The average number of seeds produced per plant of the dune grass *Vulpia fasciculata*. (After Watkinson & Harper, 1978.)

the same at a density of 40 beetle eggs per dish as it is at a density of 10 in the first case; and fecundity is effectively the same at a density of 1000 plants/0.25 m² as it is at a density of 500/0.25 m² in the second. Thus, there is no evidence at these densities that individuals are affected by the presence of other individuals and hence no evidence of intraspecific competition. But as density increases further, the per capita death rate progressively increases—the higher the density, the higher the death rate (Figure 3.22a)—or the birth rate progressively decreases (Figure 3.22b). These effects are now density-dependent, and this may be taken as an indication that at these densities, individuals are suffering directly from interference with each other or that there is a shortage of resources as a result of intraspecific competition.

competition and the total number of survivors

The patterns in Figure 3.22 make the point that as crowding (or density) increases, the fecundity per individual is likely to decline and the mortality per individual likely to increase (which would mean that the survival rate per individual would *decrease*). But what can we expect to happen to the *total* number of seeds or eggs produced by populations at different densities—or to the total number of survivors? In some cases, although the rate per individual declines with increasing density, the total fecundity or total number of survivors in the population continues to increase. This can be seen, in Figure 3.23a, to have been the case for the plant populations in Figure 3.22b—at least over the range of densities examined. In other cases, the rate per individual declines so rapidly with increasing density that the total fecundity or total number of survivors in the population actually gets smaller the greater the number of contributing individuals. This can be seen, in Figure 3.23b, to have been the case for the beetle populations in Figure 3.22a.

In yet further cases, the mortality risk per individual declines with increasing density such that the total fecundity or total number of survivors is the same irrespective of the number of contributing individuals. This is referred to as *exactly compensating density dependence*, and the competition leading to it is sometimes referred to as "contest-like." This is a pattern you would expect to see in a competition if there was a fixed number of winners and all the other competitors were doomed to lose. Examples are shown for fecundity in Figure 3.23c and for survivors in Figure 3.23d.

3.6 ▶ Conditions, Resources, and the Ecological Niche

Finally, many of the ideas in this chapter can be brought together in the concept of the ecological niche. The term *niche*, though, is frequently misunderstood and misused. It is often used loosely to describe the sort of place in which an organism lives, as in the sentence "Woodlands are the niche of woodpeckers." However, more strictly, where an organism lives is its *habitat*. A niche is not a place but an idea: a summary of the organism's tolerances and requirements. The habitat of a digestive-tract microorganism would be an animal's gut; the habitat of an aphid might be a garden; and the habitat of a fish could be a whole lake. Each habitat, however, provides many different niches: many other organisms also live in the gut, the garden, or the lake—and with quite different life-styles. The word *niche* began to gain its present scientific meaning when Charles Elton wrote in 1933 that the niche of an

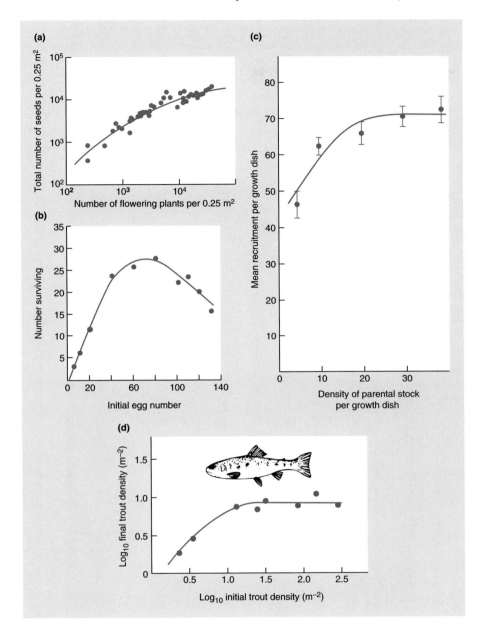

Figure 3.23
Undercompensating, overcompensating, and exactly compensating effects of intraspecific competition.
(a) An undercompensating effect on fecundity: the total number of seeds produced by *Vulpia fasciculata* continues to rise as density increases. (After Watkinson & Harper, 1978.)
(b) An overcompensating effect on mortality: the number of surviving *Tribolium confusum* actually becomes lower as initial numbers increase to high levels. (After Bellows, 1981.)
(c) An exactly compensating effect on fecundity: the numbers produced per dish in the fingernail clam (*Musculium securis*) is independent of the number of parents at higher densities. (After Mackie et al., 1978.) (d) An exactly compensating effect on mortality: the number of surviving trout fry is independent of initial density at higher densities. (After Le Cren, 1973.)

organism is its mode of life "in the sense that we speak of trades or jobs or professions in a human community." The niche of an organism started to be used to describe how, rather than just where, an organism lives.

The modern concept of the niche was proposed by Evelyn Hutchinson in 1957 to address the ways in which tolerances and requirements interact to define the conditions and resources needed by an individual or a species in order to practice its way of life. Temperature, for instance, is a condition that limits the growth and reproduction of all organisms, but different organisms tolerate different ranges of temperature. This range is one *dimension* of an organism's ecological niche: Figure

the niche of an organism is defined by its needs and tolerances

3.24a shows how species of plants vary in the temperature dimension of their niche. But there are many such dimensions for the niche of a species: its tolerance of various other conditions (relative humidity, pH, wind speed, water flow, and so on), and its need for various resources (nutrients, water, food, and so on). Clearly the real niche of a species must be multidimensional.

It is easy to visualize the early stages of building such a multidimensional niche. Figure 3.24b illustrates the way in which two niche dimensions (temperature and salinity) together define a two-dimensional area that is part of the niche of a sand shrimp.

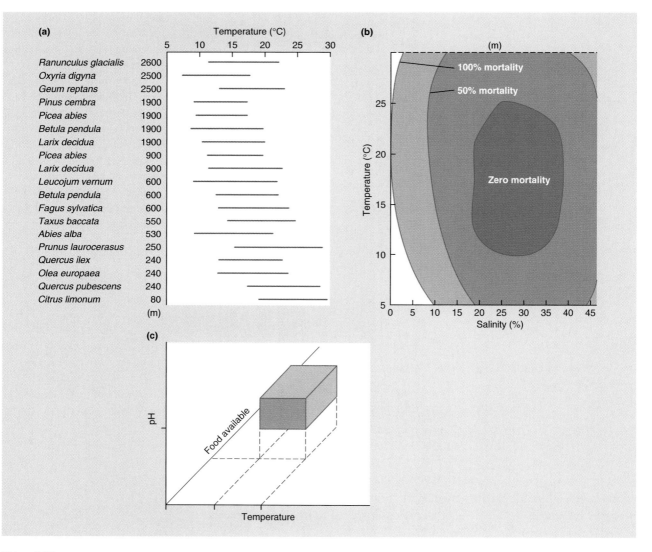

Figure 3.24

(a) A niche in one dimension. The range of temperatures at which a variety of plant species from the European Alps can achieve net photosynthesis of low intensities of radiation (70 W m⁻²). (After Pisek et al., 1973.) (b) A niche in two dimensions for the sand shrimp (*Crangon septemspinosa*) showing the fate of egg-bearing females in aerated water at a range of temperatures and salinities. (After Haefner, 1970.) (c) A diagrammatic niche in three dimensions for an aquatic organism showing a volume defined by the temperature, pH, and availability of food.

Three dimensions, such as temperature, pH, and the availability of a particular food, may define a three-dimensional niche volume (Figure 3.24c). It is hard to imagine and impossible to make a diagram of the more realistic niche that has to have many more than three dimensions (technically an *n-dimensional hypervolume*, where *n* is the number of dimensions that make up the niche). But the simplified three-dimensional version nonetheless captures the idea of the ecological niche of a species—defined by the boundaries that limit where it can live, grow, and reproduce. The niche concept has become a cornerstone of ecological thought, as we shall see in later chapters.

Summary

Conditions and resources

Conditions are physicochemical features of the environment such as its temperature and humidity. They may be altered but are not consumed. Environmental resources are consumed by living organisms in the course of their growth and reproduction.

Environmental conditions

There are three basic types of response curve to conditions. Extreme conditions may be lethal, with, between the two extremes, a continuum of more favorable conditions; or a condition may be lethal only at high intensities; or a condition may be required by organisms at low concentrations but become toxic at high concentrations.

These responses are accounted for, in part, by changes in metabolic effectiveness. But at extremely high temperatures, for example, enzymes and other proteins become unstable and break down, and the organism dies, and at high environmental temperatures, terrestrial organisms may encounter serious, perhaps lethal, problems of dehydration. At temperatures a few degrees above zero, organisms may be forced into extended periods of inactivity, or ice may form between cells and draw water from within them. The timing and duration of temperature extremes, however, may be as important as absolute temperatures.

In practice, the effects of conditions may be determined largely by the responses of other community members, through food consumption, disease, or competition.

Many conditions are important stimuli for growth and development and prepare an organism for conditions that are to come.

Plant resources

Solar radiation, water, minerals, and carbon dioxide are all critical resources for green plants. The shape of the curve that relates the rate of photosynthesis to the intensity of radiation varies greatly among species. The radiation that reaches a plant is forever changing; the plant integrates the diverse exposures of its various leaves.

Most variation in leaf shape has probably evolved under selection to optimize the photosynthesis achieved per unit of water transpired. Any mechanism or process that slows the rate of water loss, such as closing of the stomata, reduces the photosynthetic rate. If the rate of water uptake falls below the rate of release, the body of the plant starts to wilt. If the deficit accumulates, the whole plant may die. Plants may avoid or tolerate water shortage. Specialized biochemical processes may increase the amount of photosynthesis that can be achieved per unit of water lost in C4 and CAM (as opposed to C3) plants.

The primary water absorbing zone on roots is covered with root hairs that make intimate contact with soil particles. Roots create water depletion zones around themselves. Root architectures are much less tightly programmed than those of shoots, and those established early in a plant's life can determine its responsiveness to later events. Roots also extract key minerals from the soil. Root architecture is particularly important here because different nutrients are held in the soil by different forces.

Animals and their resources

Green plants are autotrophs. Decomposers, predators, grazers, and parasites are heterotrophs. The various parts of a plant have very different compositions and so offer quite different resources. This diversity is matched by the diversity of mouthparts and digestive tracts that have evolved to consume them. The body of a plant is a quite different package of resources from the body of an animal. Many herbivores enter into a mutualistic association with cellulolytic bacteria and protozoa in their alimentary canal.

The C:N ratio of plant tissues greatly exceeds those of bacteria, fungi, and animals. Thus, herbivores typically have a superabundant source of energy and carbon, but nitrogen is often limiting; their main waste products are carbon dioxide and fiber. The bodies of different species of animal have remarkably similar compositions. Carnivores are not faced with problems of digestion, but rather with difficulties in finding, catching and overcoming the defenses of their prey. Carnivores' main excretory products are nitrogenous.

The effect of intraspecific competition for resources

Individuals may compete indirectly, via a shared resource, through exploitation, or directly, through interference. The ultimate effect of competition is on survival, growth, and reproduction of individuals. Typically, the greater the density of a population of competitors, the greater is the effect of competition (density dependence). As a result, though, the total number of survivors, or of offspring, may increase, decrease, or stay the same as initial densities increase.

Conditions, resources, and the ecological niche

Where an organism lives is its habitat. A niche is a summary of an organism's tolerances and requirements. The modern concept, proposed by Hutchinson in 1957, is an n-dimensional hypervolume.

Review Questions

⚠ = Challenge Question

1. ⚠ Explain, referring to a variety of specific organisms, how the amount of water in different organisms' habitats may define either the conditions for those organisms, or their resource level, or both.

2. Discuss whether you think the following statement is correct: "A layperson might describe Antarctica as an extreme environment, but an ecologist should never do so."

3. What is an endotherm, and why might endothermy be described as a "high-cost, high-benefit" strategy?

4. ⚠ Drawing examples from a variety of both animals and plants, contrast the responses of tolerators and avoiders to seasonal variations in environmental conditions and resources.

5. Describe how plants' requirements to increase the rate of photosynthesis and to decrease the rate of water loss interact. Describe, too, the strategies used by different types of plants to balance those requirements.

6. ⚠ Describe and account for the differences in both root and shoot architecture exhibited by different plants.

7. Describe and account for the consequences of the contrasting carbon-to-nitrogen ratios in the tissues of plants and animals.

8. Define crypsis, aposematism, and mimicry. Why are mimics themselves unlikely to be more common than the organisms they mimic?

9. Explain, with examples, what exploitation and interference intraspecific competition have in common and how they differ.

10. What is meant when an ecological niche is described as an n-dimensional hypervolume?

The interplay of conditions and resources profoundly influences the composition of the world's communities. At the global scale, patterns of climate circulation are largely responsible for distinctive terrestrial biomes, such as deserts and rain forests, with their characteristic assemblages of plants and animals. Distinct types of marine and freshwater communities can sometimes also be identified at a broad geographical scale. Within each biome or aquatic category, however, there are enormous variations in conditions and resources that are reflected in community patterns viewed at a smaller scale.

4

Conditions and Resources as Forces Shaping the World's Communities

Key Concepts

In this chapter you will

- understand that conditions and resources interact to help determine the composition of whole communities.

- appreciate that climatic conditions over the surface of the Earth cause a mosaic of dry, wet, cool, and warm climates that, in turn, are responsible for the large-scale pattern of distribution of terrestrial biomes (such as tropical rain forest, desert, and tundra).

- recognize that biomes are not homogeneous because local topography, geology, and soil influence the communities of plants and animals that occur.

- understand that in most aquatic environments it is difficult to recognize anything comparable to terrestrial biomes; the communities of streams, rivers, lakes, estuaries, and open oceans reflect local conditions and resources rather than global patterns in climate.

- appreciate that conditions and resources at a location may change over time scales of hours to millennia, and that these changes are paralleled by temporal patterns in the composition of communities.

4.1 ⟩ Introduction

Having examined in Chapter 3 the ways in which individual organisms are affected by conditions and resources, we now turn to the larger question of how the interplay of conditions and resources influences whole communities (the assemblages of species that occur together). The answer to this question depends fundamentally on the scale at which we choose to study communities; this will be a pervasive theme throughout the chapter.

scale and patchiness—central themes of this chapter

Not surprisingly, because of its influence on both conditions and resources, climate plays a major role in determining the large-scale pattern of distribution of different types of community across the face of the Earth. However, local factors, such as soil type in terrestrial environments and water chemical composition in aquatic environments, are responsible for patchiness in community composition on much smaller scales. We discuss some of the causes of spatial patterns in community distribution in Section 4.2 before turning, in Section 4.3, to temporal patterns in conditions and resources whose consequences are changes in community composition over time scales from days to millennia. Section 4.4 describes the characteristics of the Earth's major terrestrial biomes; Section 4.5 deals with the diversity of aquatic community types.

4.2 ⟩ Geographic Patterns at Large and Small Scales

4.2.1 Large-scale climatic patterns

solar radiation

At the largest scale, the geography of life on Earth is mainly a consequence of the planet's movement through space. The tilt of the Earth as it makes its annual orbit around the sun causes solar radiation to strike the Earth's surfaces with different intensities at different latitudes (Figure 4.1). Because the equator is tilted toward the sun, equatorial and tropical areas receive more direct sunlight and are warmer than other latitudes. Warm air holds more moisture than cold air, increasing the water holding capacity of air around the tropics. Solar radiation draws water from the vegetation by evaporation, but because the air is so moist, much of the water condenses and falls back as rain. Thus, the air that cycles to the atmosphere from the tropics is relatively dry, having lost most of its moisture as local rainfall before it ascends to the lower atmosphere.

The rotation of the Earth causes the air masses from the tropics to curve to the north and south. Air that was warmed at the tropics (and which lost moisture as local rain) cools in the atmosphere and descends again at approximately 30° latitude (N and S). The air mass warms as it descends, increasing its capacity to hold water and causing the descending air mass to "soak up" available water from the land. As a result, most of the major deserts, including the Sahara, Kalahari, Mojave, and Sonoran, are found at approximately this latitude. Another smaller evaporation/precipitation system occurs between 30° and 60° latitude as warm, now moist, air rises and is blown farther north or south, respectively. As it cools, the air descends again and rains, producing wetter environments.

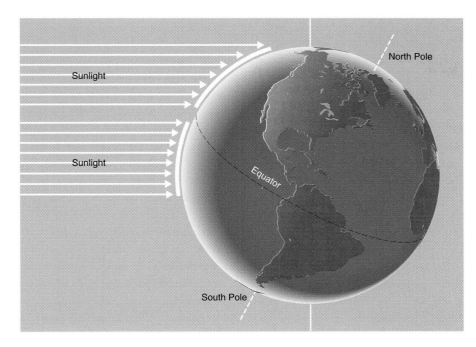

Figure 4.1
The tilt of the Earth on its axis and its rotation around the sun define the amount of radiation striking the atmosphere around the Earth's surface. This, in combination with the daily spin of the Earth on its axis, is responsible for the large-scale patterns of rainfall and solar radiation that define the pattern of global climates. This diagram shows winter in the Northern hemisphere with radiation falling almost vertically south of the equator, but the same amount of radiation is spread over greater areas north of the equator; less is therefore received and there is less heating per unit area. (Adapted from Audesirk & Audesirk, 1996.)

Ocean currents have further powerful effects on climatic patterns. Southern waters circulate counter-clockwise; they carry cold Antarctic waters up along the western coasts of continents and distribute warmer waters from the tropics along their eastern coasts (Figure 4.2). In the northern hemisphere, currents circulate clockwise, carrying cold arctic waters along the eastern coasts of continents and returning warm tropical currents along western coasts. The cool, dry climate of

ocean currents

Figure 4.2
The movements of the major ocean currents. (Adapted from Audesirk & Audesirk, 1996.)

eastern South America is an effect of the Antarctic Humboldt current; the relatively dry climate of California is a result of arctic currents. Conversely, on the eastern side of North America the strong tropical Gulf Stream carries with it warm and moist air far into the Atlantic Ocean, affecting even the climate of western Europe.

and mountain ranges

The topography of the land has consequences at an intermediate scale for the pattern of terrestrial climates. As winds meet mountain ranges they are forced up and become cooler as they rise. The cooler air holds less moisture so that water is released (as rain and snow) on the windward slopes of the mountains (the Rockies and Himalayas provide striking examples of this effect). As the air passes over to the leeward sides of the mountains it descends, becomes warmer, and now absorbs water. This produces a desiccating effect and causes a *rain shadow* along the leeward slopes (Figure 4.3).

produce a mosaic of dry, wet, cool, and warm climates over the face of the Earth

The variety of influences on climatic conditions over the surface of the globe has given rise to a mosaic of dry, wet, cool, and warm climates. In the patches of this mosaic, distinctive terrestrial associations of vegetation and animals have formed. A world traveler sees repeatedly what can be recognized as characteristic types of vegetation, which ecologists call *biomes* (such as coniferous forest, savanna, and rain forest). Figure 4.4 provides an example of a set of biomes that can be recognized and used to construct a global map. Figure 4.5 shows the ranges of rainfall and mean monthly minimum temperature recorded within some major biomes of the world. We describe the characteristics of the communities inhabiting biomes in Section 4.4.

that, in turn, is responsible for the large-scale distribution of terrestrial biomes

4.2.2 Small-scale patterns in conditions and resources

We need to be wary of seduction by cartographers who draw sharp lines on maps to show geographic boundaries. Neat pigeonholes, sharp categories, and tidy boundaries are a convenience for cartographers, not a reality of nature. Moreover, biomes are not homogeneous within their hypothetical boundaries; every biome has gradients of physicochemical conditions related to features of local topography and geology. The communities of plants and animals that occur in these different regions may be quite distinct.

Figure 4.3
The influence of topography on rainfall. Moisture-laden westerlies are forced higher by a mountain range. As they rise they become cooler and release the moisture as rain or snow. This leaves a drier rain shadow on the eastern slopes. (Adapted from Audesirk & Audesirk, 1996.)

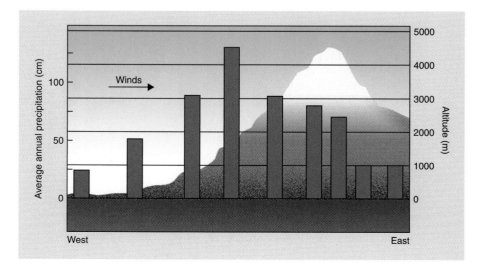

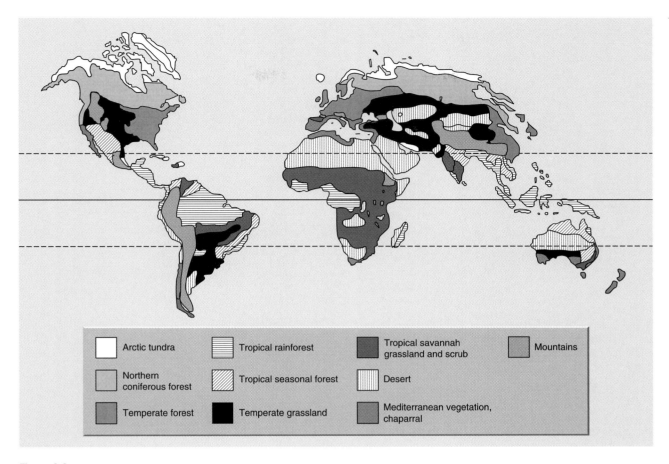

Figure 4.4
The world distribution of the major biomes of vegetation. (Adapted from Audesirk & Audesirk, 1996.)

Legend:
- Arctic tundra
- Northern coniferous forest
- Temperate forest
- Tropical rainforest
- Tropical seasonal forest
- Temperate grassland
- Tropical savannah grassland and scrub
- Desert
- Mediterranean vegetation, chaparral
- Mountains

Local variations in topography can override elements of the broad climatic pattern described in Section 4.2.1. For example, temperature falls with increasing elevation and one effect is that vegetation high on a mountain in the tropics tends to resemble vegetation at low altitudes in northern latitudes. Traveling up a mountain in the tropics involves passing along a very similar ecological gradient as traveling northward (Figure 4.6).

local topography

It is worth remembering that the Earth's surface would consist of a mosaic of different environments even if climate were identical everywhere. Geological history has provided a variety of rocks that differ in their mineral composition. When the surfaces of these rocks are decomposed by heat, frost, and thaw they give rise to a variety of types of soil that reflect their geological origin. Without soil, it is impossible for significant terrestrial vegetation to grow. Soils provide a source of stored water, a reserve of mineral nutrients, a medium in which atmospheric nitrogen can be fixed for plant use, and the support that allows plants to stand up and expose their leaves to the sunlight.

local geology and soil

Limestone rocks and chalk have formed as marine deposits of calcium carbonate, often containing some magnesium and other carbonates. Where these deposits have

acidic and calcareous soils bear very different vegetation

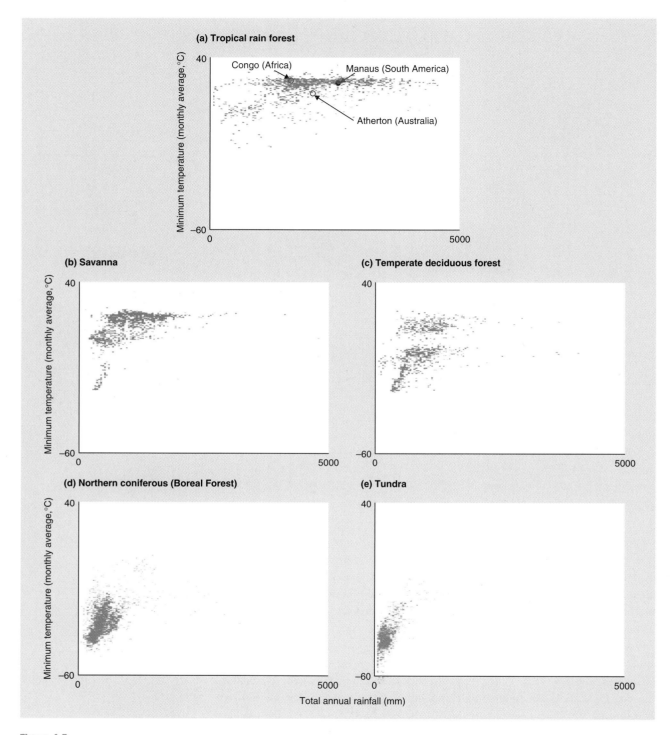

Figure 4.5

The variety of environmental conditions experienced in terrestrial environments can be described in terms of their annual rainfall and mean monthly minimum temperatures. The range of conditions experienced in (a) tropical rain forest, (b) savanna, (c) temperate deciduous forest, (d) northern coniferous forest (taiga), and (e) tundra.

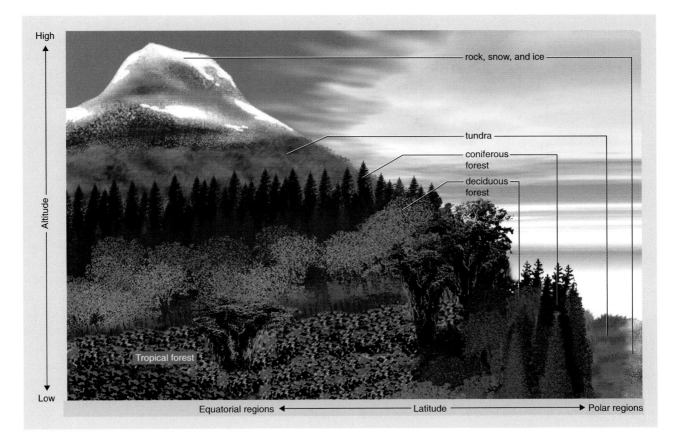

Figure 4.6
The effect of altitude and latitude on the distribution of biomes. (Adapted from Audesirk & Audesirk, 1996.)

been raised and exposed as land surfaces they become the basis for neutral or slightly alkaline *calcareous* soils, which bear a characteristic *calcicole* (calcium-loving) flora. On the other hand, plants normally found on more acid soils (*calcifuges*), such as *Rhodendron, Azalea,* and species of *Kalmia* (sheep laurel and calico bush), commonly suffer nutrient deficiency on calcareous soils. Strict calcicoles, in contrast, suffer on acidic soils, where they are intolerant of aluminum ions released at low pH. In the United States, for example, yellow poplar (*Liriodendron tulipifera*) and northern white cedar (*Thuja occidentalis*) are found on neutral or alkaline soils, and balsam fir (*Abies balsamea*) and eastern hemlock (*Thuja canadensis*) are usually confined to highly acidic soils.

Organic matter accumulates at different rates in different soils; local variations in the balance between mineral and organic material in the soil contribute to the complexity of the environmental mosaic within biomes. In extreme conditions, especially where the rocks are acidic, the temperatures are low, and/or the soil is waterlogged, the decomposition of organic matter is so impeded that bogs of peat form, bearing very specialized plants and animals.

variations in the rate of decomposition of organic matter

When ecologists refer to a *patch* in a community, they generally have in mind an area in which a single variable distinguishes it from its surroundings. Thus, a fallen tree in a forest leaves a gap in the canopy and a patch on the forest floor

patchiness is in the eye of the observer

where sufficient radiation may penetrate to allow seedlings to grow and eventually fill the gap. A tide pool is a patch on a rocky shore, but within that pool snails may graze and clear a patch of algae. It is often useful to think of patches as the scale at which particular organisms experience the environment around them. For an aphid in a forest, an individual leaf of a particular species of tree is a patch—it provides both the conditions and the resources necessary for the insect. For a warbler feeding on caterpillars, the canopies of individual trees are patches that it encounters in its daily life. But owls or hawks hunt over a large part of the forest, and for them a patch can be thought of as the zone that each bird defends.

and all communities are patchy

4.2.3 Patterns in conditions and resources in aquatic environments

In most aquatic environments it is difficult to recognize anything comparable to terrestrial biomes. The exceptions occur at the ocean's edge; mangrove swamps, coral reefs and kelp forests have a flora and fauna that are as distinctive as any of the various terrestrial biomes, but this is largely due to their link with major terrestrial climates. In contrast, the open oceans form a continuum in which there is flow of water and dissolved chemicals across the globe. We have seen how variation in the intensity of solar radiation from place to place and between the seasons has dramatic effects on the temperature and water relations of terrestrial environments. But this is not the case in the oceans. The high thermal capacity of water makes the oceans slow to heat and slow to cool. One effect is that the temperature of the water at one point on the globe is a better reflection of where the water has come from than of the heat exchanged locally.

The world's large lakes can be distinguished and classified according to their physical conditions. For example, large lakes in lowland equatorial regions generally experience permanent stratification (distinct layers of water at particular temperatures), whereas seasonal patterns of stratification and mixing are the rule in temperate regions. Within the polar circles, permanent ice cover with no mixing is characteristic of large lakes. However, local geological conditions and basin size and shape have strong influences on conditions and resources in lakes, particularly relating to water chemical composition. Consequently, a broad geographic classification of lake communities has only limited merit. In the cases of streams, rivers, and estuaries, we will see that local conditions and resources are paramount in determining community patterns (Section 4.5).

4.3 ▷ Temporal Patterns in Conditions and Resources—Succession

The composition of communities can change over time scales of hours to millennia, as conditions and resources themselves change. For example, the microbial community that colonizes and decomposes a dead worm or a fragment of a leaf may change from hour to hour (like a newly inoculated culture of yogurt). At the other extreme, we can trace patterns in community composition related to the ice ages. Thus, changes in climate during the Pleistocene ice ages bear much of the responsibility for present patterns of distribution of plants and animals (Figure 4.7). In the 20,000 years since the peak of the last glaciation, global temperatures have risen by about 8°C. Many species continue, even today, to migrate northward, following the retreat of the glaciers. Climatic changes that occurred 3200 years ago are still evident in persistent patches of hemlocks among the dominant hardwood species in the forests of the Upper Peninsula of Michigan state (Davis et al., 1994).

At intermediate temporal scales, predictable successions of plant species may occur over periods of decades to centuries. For example, the so-called raised peat bogs display a succession of species of moss of the genus *Sphagnum* that grow on the peat. As their dead parts accumulate, they form hummocks, raised above the surface of the peat. Each species of *Sphagnum* occupies a zone at a particular height on the hummocks and they replace each other through the decades as the hummocks grow. Ultimately the hummocks are invaded by heather, *Calluna vulgaris*, and when this dies the whole hummock breaks down and the sequence of *Sphagnum* species starts again (Figure 4.8). The peat bog is a mosaic of various stages in this successional process. To the pilot of an aircraft, the peat bog may look very uniform but observers on their knees see a finely structured and repetitive patchwork.

plant successions—"raised" peat bogs

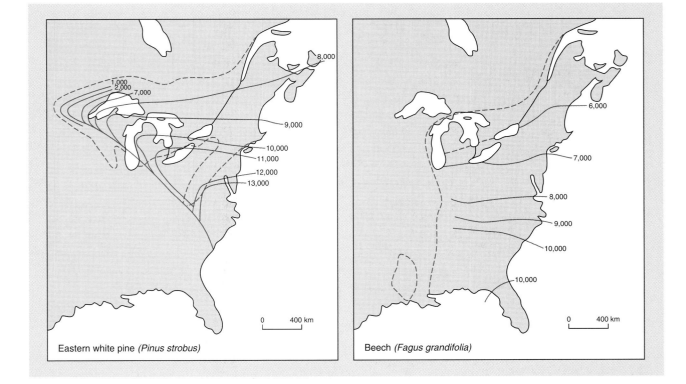

Figure 4.7

A map showing the spread of two species of forest tree in eastern North America after the retreat of the last ice age glaciation. Note that the two species (a) eastern white pine (*Pinus strobus*) and (b) beech (*Fagus grandifolia*) have not followed the same invasion path. The lines on the maps (isochrones) define the time of arrival of each species at 1000-year intervals. The numbers on the maps refer to thousands of years. The dotted lines show current distribution. (From Davis, 1976.)

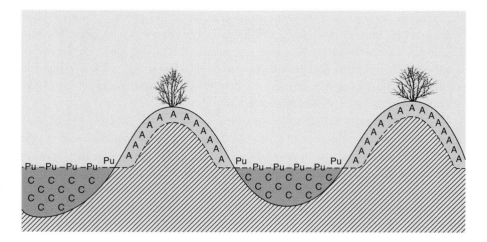

Figure 4.8

The structure of the hummock-hollow cycle on an actively growing raised bog. The moss *Sphagnum cuspidatum* (C) invades small pools of water, and its slowly decaying parts accumulate in them, creating a habitat suitable for *Sphagnum pulchrum* (Pu). This in its turn is replaced by *Sphagnum papillosum* (A) and in its growth this species builds up hummocks of its dead parts. The hummocks rise above the level of standing water and become invaded by small woody shrubs such as *Calluna vulgaris* and then by the lichen *Cladonia arbuscula*. When the *Calluna* dies the hummocks erode and break down and new pools are formed—starting the whole sequence again.

and old field successions

Old field successions, which occur over periods of decades to centuries, are particularly associated with eastern parts of the United States (see Chapter 1, Section 1.3.2). When the western frontier was opened up in the 19th century, farmers moved west and cultivated land in the east was abandoned (see Chapter 1). In the typical succession, annual weeds were first on the scene but were later replaced by herbaceous perennials which, in turn, were succeeded by shrubs, by early successional trees (e.g., eastern red cedar *Juniperus virginiana*) and finally by late successional species (e.g., sugar maple *Acer saccharum*). Animals and fungi are in many cases simply passive followers of successions among plants. Similar successions also occur in established communities, for example, after the death of a single tree. All vegetation is patchy on its own successional time scale. The important topic of community succession is discussed in more detail in Section 9.4.

4.4 ▷ The Terrestrial Biomes

the patterns that we recognize in nature depend on how we focus our attention

Different geographers recognize different numbers of biomes; some make do with just five biomes and others find they need many more. The perspective of the scientist is as important as the system being studied; "splitters" tend to distrust broad generalizations and emphasize the diversity of the natural world, whereas "lumpers" force diversity into a minimum of easily mapped categories. Seven will be adequate for our purposes—tropical rain forest, savanna, temperate grassland, desert, temperate deciduous forest, northern or boreal coniferous forest (taiga), and tundra. This chapter section is largely about global patterns that can be represented, without too many frills, on a map of the world. If we were dealing with California and not the whole world, we would be able to recognize more subtle variations and classify vegetation and map the land on a finer scale: biomes would be far too crude. What we see depends on the way we look; we can recognize patterns in nature whether we look through a telescope or a microscope!

4.4.1 Describing and classifying biomes

We pointed out in Chapter 2 the crucial importance of geographic isolation in allowing populations to diverge under selection. The geographical distributions of species, genera, families, and even higher taxonomic categories of plants and animals often reflect this geographic divergence. All species of lemurs, for example, are found on the island of Madagascar and nowhere else. Similarly, 230 species in the genus *Eucalyptus* (gum tree) are found in Australia (and 2 or 3 in Indonesia and Malaysia). The lemurs and the gum trees occur where they do because they evolved there—not because these are the only places where they could survive and prosper. Indeed, many *Eucalyptus* species grow with great success and spread rapidly when they have been introduced to California or to Kenya. A map of the natural world distribution of lemurs tells us quite a lot about the evolutionary history of this group. But as far as its relationship with a biome is concerned, the most we can say is that lemurs happen to be one of the constituents of the tropical rain forest biome in Madagascar.

Another theme of Chapter 2 concerned the way species with quite different evolutionary origins have been selected to converge in their form and behavior (convergent evolution, as in the penguins, dolphins, and sharks: Figure 2.18). There were also examples of taxonomic groups that have radiated into a range of species with strikingly similar form and behavior (parallel evolution, as in the marsupial and placental mammals: Figure 2.19). Examples like these reveal much about the ways in which the form and behavior of organisms match and restrict them to the conditions and resources in their environments. Thus, particular biomes in Australia include certain marsupial mammals and the same biomes in other parts of the world are home to their placental counterparts.

A map of biomes is not usually a map of the distribution of species: it does not describe where to find forests of *Eucalyptus* or marshes dominated by reed-mace (*Typha*) or papyrus (*Cyperus*). Instead, it shows where we could find areas of land dominated by plants with characteristic shapes, forms, and physiological processes. These are the types of vegetation that can be recognized from an aircraft passing over them or from the windows of a fast car or train. It does not require a botanist to identify them—indeed, by nature and training the botanist may not be able to see the wood for the trees.

describing and classifying vegetation

Moreover, because of parallel and convergent evolution, the list of species present in a particular biome will usually be quite different in different parts of the world. The scrubby maquis or garrigue vegetation of countries around the coast of the Mediterranean Sea provides a striking example. The spectrum of plant forms that gives this vegetation its distinctive nature also occurs in similar environments in California (as chaparral) and in Australia—but the species and genera of plants are

(a)

(b)

(c)

(d)

(e)

(f)

(g)

We illustrate each biome by means of two photographs, one focusing on the detail of the vegetation and the other providing a distant view and emphasizing the great structural variation to be found among the world's terrestrial communities. We will see in our discussion of each of these biomes (Sections 4.4.2–4.4.7) that the animals cannot be ignored. Animals are obvious in the savanna photo, but invertebrate and vertebrate animals are busy behind the scenes in all the biomes. **a.** (top) Carrizo Badlands, Anza-Bonnego Desert State Park, California; (bottom) Red Rock Canyon, Las Vegas, Nevada. **b.** (top) Ozark Forest and Current River, Ozark National Scenic Riverways, Missouri; (bottom) mature eastern deciduous forest. **c.** (top) Fir tree forest, Jasper National Park, Alberta, Canada; (bottom) foggy coniferous forest, Sierras. **d.** (top) Maasai Mara Game Preserve at dawn; (bottom) African savanna with zebra and buffalo. **e.** (top) Rain forest, western slope of Andes, Ecuador; (bottom) lake in mixed dipterocarp forest, Mulu National Park, Sarawak, Borneo. **f.** (top) Lone pronghorn antelope looks tiny in vast mixed-grass prairie, Stanley County, central South Dakota; (bottom) view of prairie in flower with Blazing Star and Black-Eyed Susan. **g.** (top) Green tundra with glacial morraine and Alaska mountain range, Denali National Park, Alaska; (bottom) wet summer tundra.

quite different. We recognize different biomes and different types of aquatic community from the *types* of organisms that live in them—not from lists of the species present. How can we describe their similarities so that we can classify them, compare them and map them?

It was not until 1934 that the Danish biogeographer Raunkiaer developed his idea of life-forms, a deep insight into the ecological significance of plant form (Figure 4.9). He then used the spectrum of life-forms present as a means of describing the ecological character of different types of vegetation. Plants grow by developing new shoots from the buds that lie at the apices (tips) of shoots and in the leaf axils. Within the buds the meristematic cells are the most sensitive part of the whole shoot. Raunkiaer recognized that the buds are the "Achilles heels" of plants. He argued that the ways in which these buds are protected in different plants is a powerful indicator of the hazards in their environments (Figure 4.9).

We recognize a *Mediterranean* type of vegetation when we see it in Chile, Australia, California, or Crete because the life-form spectrums are similar, though the lists of the species present do not tell us that these communities have anything in common. Similarly, a Raunkiaer spectrum of the plant life-forms in Amazon rain forest will tell us how ecologically similar they are to the spectrum of life-forms in rain forests of the Congo or Borneo (Figure 4.9). The detailed taxonomy of their floras and faunas would only emphasize how different they are.

Trees expose their buds high in the air, fully exposed to wind, cold, and drought (Raunkiaer called them *phanerophytes*: Greek *phanero* = "visible"; *phyte* = "plant"). The least protected tree buds are those of tropical rain forest where the dominant species are very tall and evergreen; their buds are freely exposed and produce flushes of new shoots throughout the year. In contrast, the buds of the dominant trees in almost all other forests are dormant during the coldest or driest seasons of the year and are strongly protected by sheaths of bud scales.

Many perennial herbs form cushions or tussocks in which buds are borne above ground but are protected from drought and cold in the dense mass of old leaves and shoots (*chamaephytes*: "on the ground"). Buds are even better protected when they are formed at or in the soil surface (*hemicryptophytes*: "half hidden") or on buried dormant storage organs (bulbs, corms, and rhizomes) that allow the plants to make rapid growth and flower in a short growing season before they die back to a dormant state (*geophytes*: "earth plants"). Such life-forms are characteristic of steppe grasslands and Mediterranean regions but also feature in the ground vegetation of deciduous forests where they use their reserves for growth early in spring, before being shaded by tree foliage.

Another major category consists of annual plants that depend wholly on dormant seed to carry their populations through seasons of drought and cold (*therophytes*: summer plants). These are the plants of deserts (they make up nearly 50 percent of the flora of Death Valley), sand dunes, and repeatedly disturbed habitats. They also include the annual weeds of arable lands, gardens, and urban wastelands—the botanical "camp followers" of humanity.

Of course there is no vegetation that consists entirely of one growth form. All vegetation contains a mixture, a spectrum, of Raunkiaer's life-forms. The spectrum on tropical islands, for example, is predominantly of phanerophytes, but these are

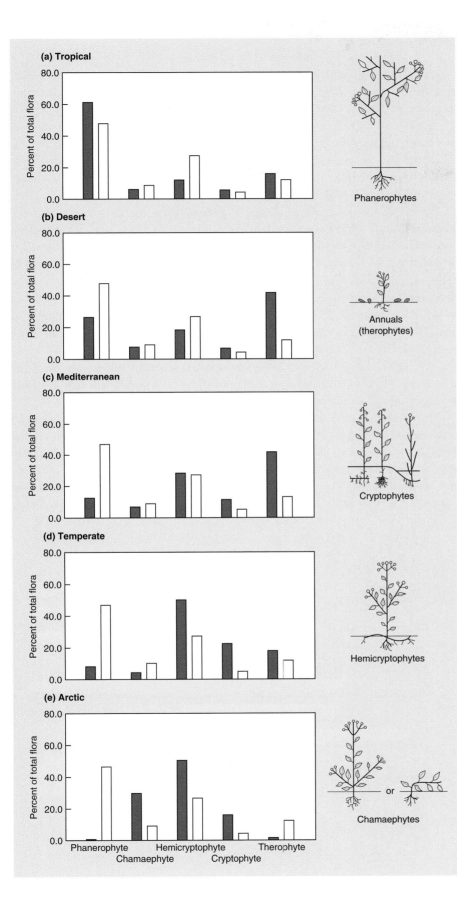

Figure 4.9
On the right drawings depict the variety of plant forms distinguished by Raunkiaer on the basis of where they bear their buds. On the left are life-form spectrums for five different biomes. The black bars show the percentage of the total flora that is composed of species with each of the five different life-forms. The white bars are the proportions of the various life-forms in the world flora for comparison. (From Crawley, 1986.)

completely absent from the spectrum of arctic islands. The composition of the spectrum in any particular habitat is as good a shorthand description of its vegetation as ecologists have yet managed to devise.

describing and classifying faunas

Faunas are bound to be closely tied to floras—if only because most herbivores are choosy about their diet and the diet of a great many, especially among insects, is restricted to one plant species. Terrestrial carnivores range more widely than their herbivore prey, but the distribution of herbivores still gives the carnivores a broad vegetational allegiance (Figure 4.10).

In the development of the science of ecology, plant scientists have been more interested in classifying floras than animal scientists in classifying faunas. Nevertheless, it is interesting to ask whether there is, for the fauna of an area, any one feature (or set) of form and behavior with a role comparable to the protection and position of buds in vegetation. The proportions in the community of herbivores, carnivores, and detritivores might be a candidate. Such a zoological spectrum could allow us to make a quantitative comparison of faunas and test hypotheses (for example, that species of mammalian carnivores are particularly abundant in the fauna of arctic regions or that very small insect herbivores are abundant in the tropics).

An interesting attempt to classify faunas used a particularly animal character, namely, the mode of locomotion, and combined this with dietary habits, to compare the mammals of forests in Malaya, Panama, Australia, and Zaire (Andrews et al., 1979). They classified the mammals into carnivores; herbivores, including fruit eaters; insectivores; and mixed feeders and subdivided each of these into those that were aerial (mainly bats and flying foxes), arboreal (tree dwellers), scansorial (climbers), and small ground mammals (Figure 4.11). The comparison here was only for the mammalian fauna, but it reveals some strong contrasts and similarities. For example, the ecological diversity spectrums for the Australian and Malayan forests

Figure 4.10

The distribution of bird species along a gradient of plant succession in the piedmont region of Georgia, United States. The variation in intensity of shading indicates the relative abundance of the species. As a bare field is taken over by grassland and then by scrub and eventually forest, the community of associated birds also changes. (After Johnston & Odum, 1956.)

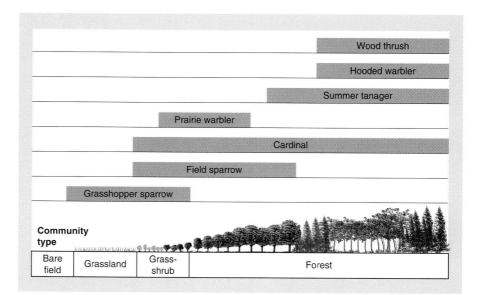

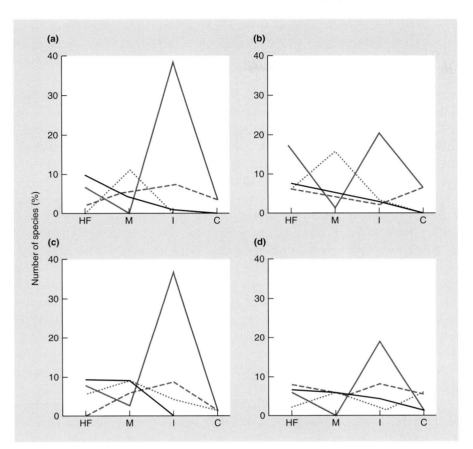

Figure 4.11
A comparison of the forest mammals present in various communities of tropical forest. The mammals were classified according to their diet, and the categories were subdivided according to the zones in which they forage. Dietary categories: carnivores (C), herbivores that are leaf and fruit feeders (HF), insect feeders (I), mixed feeders or omnivores (M). Foraging categories are _____, aerial; _____, arboreal; . . . . . , scansorial; _____ , ground feeders. (a) 161 Species present in forested areas of Malaysia; (b) 70 species in Panama dry forest; (c) 50 species in Cape York forest, Australia; (d) 96 species in Irangi forest, Zaire, Africa. (After Andrews et al., 1979.)

were very similar despite the fact that their faunas are taxonomically very distinct—the Australian mammals are marsupials and the Malaysian mammals are placentals.

There are splendid questions to be asked and answered about comparisons of birds or insects and indeed about whole faunas. Would the categories appropriate for mammals be appropriate for these other groupings? What groupings might be appropriate for comparisons of tropical, Mediterranean, temperate and arctic faunas? Perhaps the presence and character of dormancy (e.g., hibernation) might be as valuable for an ecological classification of faunas as they were for Raunkiaer in plants. Perhaps it will be spectrums of relative abundances of body sizes, endotherms versus ectotherms, or life history traits that will form the basis for some future comparison and classification of whole faunas.

UNANSWERED QUESTION:

how can ecologists most effectively define life-form spectrums of faunas?

4.4.2 Tropical rain forest

We have chosen to discuss tropical rain forest in greater depth than the other biomes because it represents the global peak of evolved biological diversity: all the other biomes suffer from a relative poverty of resources or more strongly constraining conditions.

Tropical rain forest is the most productive of the Earth's biomes with a photosynthetic productivity that can exceed 1000 grams of carbon fixed per square meter per year (see Section 11.2.1). Such exceptional productivity results from the coincidence of high solar radiation received throughout the year and regular and reliable rainfall (illustrated in Figure 4.5). The production is achieved, overwhelmingly, high in the dense forest canopy of evergreen foliage (i.e., Raunkiaer's phanerophytes). It is dark at ground level except where fallen trees create gaps in the canopy so that some radiation reaches the ground. A characteristic of this biome is that often many tree seedlings and saplings remain in a suppressed state from year to year and only leap into action if a gap forms in the canopy above them.

Almost all the action in a rain forest (not just photosynthesis but also flowering, fruiting, predation and herbivory) happens high in the canopy. Apart from the trees, the vegetation is largely composed of plant forms that reach up into the canopy vicariously; they either climb and then scramble in the tree canopy (vines and lianas, including many species of fig) or grow as epiphytes, rooted on the damp upper branches. The epiphytes depend on the sparse resources of mineral nutrients that they extract from crevices and pockets of humus on the tree branches. The rich floras and faunas of the canopy are not easy to study; even to gain access to the flowers in order to identify the species of tree is difficult without the erection of tree walks. It is a measure of the problems of doing research in rain forest that botanists have trained monkeys to collect and throw down flowers and a research team has used hot air balloons to move over the canopy and work in it.

Most species of both animals and plants in tropical rain forest are active throughout the year, though the plants may flower and ripen fruit in sequence. In Trinidad, for example, the forest contains at least 18 trees in the genus *Miconia*, whose combined fruiting seasons extend throughout the year (Figure 4.12).

Dramatically high species richness is the norm for tropical rain forest (Box 4.1, and see Section 10.5.2), and communities rarely if ever become dominated by one or a few species—a very different situation from the biomonotony of northern coniferous forests. This raises some fundamental questions that have proved very difficult to resolve.

UNANSWERED QUESTION:

what permitted high species richness to evolve in tropical rain forests?

First, what is it about the evolutionary history of tropical rain forest that has allowed such diversity to evolve? Part of the answer relates to the comparative stability of patches of rain forest during the ice ages. It is thought that during these periods, drought forced tropical rain forests to contract into "islands" (in a "sea" of savanna), and these expanded and coalesced again as wetter periods returned. This would have promoted genetic isolation of populations, a phenomenon that is so important for speciation to occur (Section 2.4). Moreover, biotic interactions among mutualists (such as ants and their host plants), and between plants and their pathogens, animals and their parasites, and predators and their prey, are particularly obvious in tropical rain forests. These are powerful forces of natural selection that can be expected to have contributed to the diversity of tropical floras and faunas.

UNANSWERED QUESTION:

and what prevents domination by one or a few species?

We may also ask why it is that among the diversity of species in tropical rain forest a few have not dominated and suppressed the rest in a struggle for existence. We will see later (Section 10.5.2) that at least part of the answer is that populations

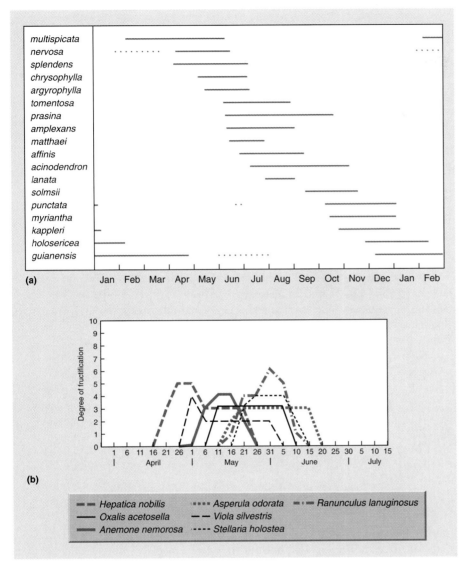

Figure 4.12
The seasonal patterns of fruit or seed production by a rain forest flora and by the flora of a temperate deciduous forest. (a) The fruiting seasons of 18 species of *Miconia* in the rain forest of Arima valley, Trinidad. (From Snow, 1964, in Harper, 1977.) (b) The seasonal production of fruit and seeds by herbs in a deciduous temperate forest (lime [*Tilia*] and hornbeam [*Carpinus*]) at Bialowieza, Poland. (From Falinska in Harper, 1977.)

of specialized pathogens and herbivores develop near mature trees and attack new recruits of the same tree species nearby. Thus, the chance that a new seedling will survive can be expected to increase with its distance from a mature tree of the same species, reducing the likelihood of dominance by one or a few species in the forest.

The diversity of rain forest trees provides for a corresponding diversity of resources for herbivores (Figure 4.13). A variety of fresh young leaves is available throughout the year, and a procession of seed and fruit production through the year provides reliable food for specialists such as fruit-eating bats. Moreover, a diversity of flowers, such as epiphytic orchids with their specialized pollinating mechanisms, require a parallel specialized diversity of pollinating insects. Rain forests are the

tropical rain forest is also associated with high animal diversity

Box 4.1

The Barro Colorado Island Study

Stephen Hubbell and a group of colleagues made a detailed census and inventory of the vegetation on a 0.5-km² plot of tropical rain forest on Barro Colorado Island (Panama) that had been isolated in 1913 when part of the Panama canal was filled in (see figure). Every individual tree or shrub of more than 1-cm diameter at breast height was censused and its position was recorded in 1982, 1985, and 1990. These records were vital because they made it possible at the time of each census to discover which had died and how much the survivors had grown and also to identify new recruits. The whole exercise was enormous and shows just how much effort may be needed to answer quite straightforward ecological questions. On each occasion 240,000 trees and large shrubs were censused and these included 300 species.

This very large data set has been analyzed to detect associations or exclusions of species that might suggest that there were competitive interactions between them. Analyses were also made that might detect effects of the density of a species on its own growth rate and the relationships between its abundance and that of other species. The analyses were made by dividing the sampled area into quadrats (plots) of different sizes so as to

detect whether the behavior of particular species was determined by large-scale effects in the study area or by fine-scale effects such as might arise from the very local crowding of individuals.

The results of statistical analysis of these data are best summarized by the authors. "Intraspecific density-dependent effects in the Barro Colorado Island (Panama) study area are far stronger, and involve far more species, than had previously been suspected. Significant effects on recruitment, many extremely strong, are seen for 67 out of the 84 most common species in the plot, including the 10 most common. Significant effects on the intrinsic rate of increase are seen in 54 of the 84 species. These effects are far more common than interspecific [between different species] effects, and are predominantly of the type that should maintain tree diversity" (Wills et al., 1997).

This is a classic ecological study, and many of the findings are highly relevant to issues discussed elsewhere in the book, especially intraspecific competition (Chapter 3), interspecific competition (Chapter 6), host–parasite interactions (Chapters 7 and 8), and species richness (Chapter 10). (Wills et al., 1997).

center of diversity for ants—43 species have been recorded on a single tree in a Peruvian rain forest. And there is even more diversity among the beetles; Erwin (1982) estimated that there are 18,000 species of beetle in 1 hectare of Panamanian rain forest (compared with only 24,000 in the whole of the United States and Canada!).

and intense soil activity   There is intense biological activity in the soil of tropical rain forests. Leaf litter decomposes faster than in any other biome and as a result the soil surface is often almost bare. The mineral nutrients in fallen leaves are rapidly released, and, as rainfall seeps down the soil profile, nutrients may be carried well below the levels at which roots can recover them. Almost all the mineral nutrients in a rain forest are held in the plants themselves, where they are safe from leaching. When such forests are cleared for agriculture, or the timber is felled or destroyed by fire, the nutrients are released and leached or washed away: on slopes the whole soil may go too. The full regeneration of soil and of a nutrient budget in a new body of forest

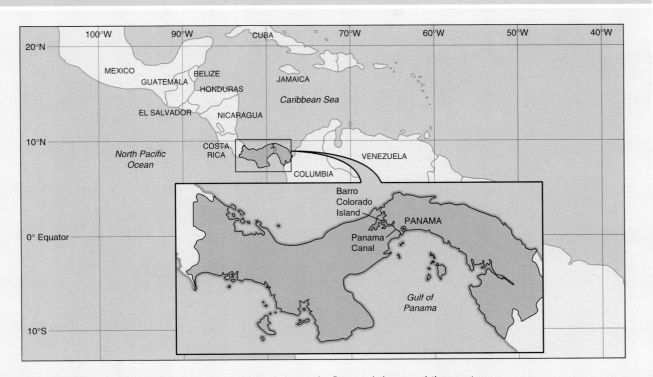

Map showing the position of Barro Colorado Island in relation to the Panama isthmus and the equator.

biomass may take centuries. Evidence of cultivated patches within rain forest can still be seen clearly from the air 40 years or more after they have been deserted.

All the other terrestrial biomes can be seen as the poor relations of tropical rain forest. They are all colder or dryer and all are more seasonal. They have had prehistories that prevented the evolution of a diversity of fauna and flora that approaches the remarkable species richness of tropical rain forest. Moreover, they are generally less suited to the lives of extreme specialists, both plant and animal.

4.4.3 Savanna

The vegetation of savanna characteristically consists of grassland with scattered small trees, but extensive areas have no trees. Herds of grazing herbivores (e.g., zebra, wildebeest in Africa) have a profound influence on the vegetation, favoring grasses (which protect their regenerating meristems in buds at or just below ground level)

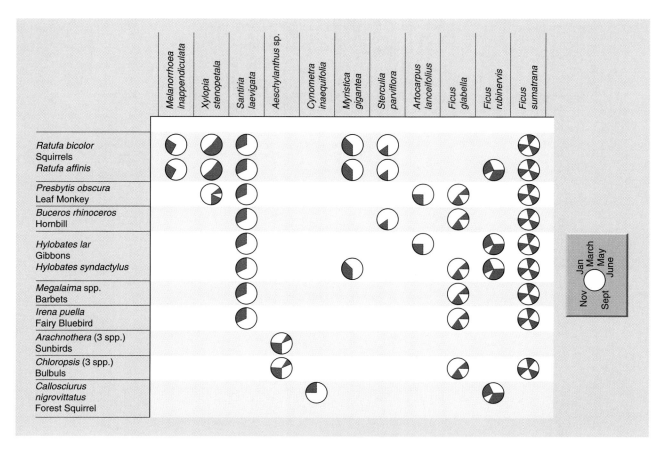

Figure 4.13
Animals that feed on the fruit of trees at various times of the year at Selangor, Malaysia. Each circle is a calendar in which the feeding season is shown in black. (From Harper, 1977; after data of McClure.)

seasonal glut and food shortage are characteristic of savanna

and hindering the regeneration of trees (which expose their meristems to browsing animals and to fire). Fire is a common hazard in the dry season and, like grazing animals, tips the balance in the vegetation against trees and favors grassland.

Seasonal rainfall places the most severe restrictions on the diversity of plants and animals in savanna. Plant growth is limited for part of the year by drought, and there is a seasonal glut of food, alternating with shortage; as a consequence, the larger grazing animals suffer extreme famine (and mortality) in drier years. The strong seasonality of savanna ecology is well illustrated by its bird populations. An abundance of seeds and insects supports large populations of migrating birds, but only a few species can find sufficiently reliable resources to be resident year round.

4.4.4 Temperate grasslands

Temperate grassland is the natural vegetation over large areas in every continent—the steppes of Asia, the prairies of North America, the pampas of South America, and the veldt of South Africa. Typically, these grasslands experience seasonal drought, but the role of climate in determining vegetation is usually overridden by the effects of grazing animals. Populations of invertebrates, such as grasshoppers, are often very large and their biomass may exceed that of grazing vertebrates.

Many of these natural grasslands have been cultivated and replaced by arable annual "grasslands" of wheat, oats, barley, rye, and corn. Such annual grasses of temperate regions, together with rice in the tropics, provide the staple food of human populations worldwide. In fact, the vast increase in the size of the human population in historical times (see Section 12.2) has depended on the domestication of grasses for human food or feed for domestic animals. At the drier margins of the biome, where cultivation is not economical, many of the grasslands are "managed" for meat or milk production, sometimes requiring a nomadic human life-style. The natural populations of grazing animals, especially bison and pronghorn antelope in North America and ungulates in Africa, have been driven back in favor of cattle, sheep, and goats. Of all the biomes, this is the one most coveted, used, and transformed by humans.

of all the biomes, temperate grassland has been most transformed by humans

4.4.5 Desert

In their most extreme form, the hot deserts are too arid to bear any vegetation; they are as bare as the cold deserts of Antarctica. Where there is sufficient rainfall to allow plants to grow in arid deserts, its timing is always unpredictable.

Desert vegetation falls into two sharply contrasted patterns of behavior. Many species have an opportunistic life-style, stimulated into germination by the unpredictable rains (physiological clocks are useless in this environment). They grow fast and complete their life history by starting to set new seed after a few weeks. These are the species that can occasionally make a desert bloom; the ecophysiologist Fritz Went called them "belly plants" because only someone lying on the ground can appreciate their individual charm.

contrasting patterns of behavior of desert plants

A different pattern of behavior of arid desert plants is to be long-lived with sluggish physiological processes. Cacti and other succulents, and small shrubby species with small, thick, and often hairy leaves, can close their stomata (pores through which gas exchange takes place) and tolerate long periods of physiological inactivity. In arid deserts, freezing temperatures are common at night and tolerance of frost is almost as important as tolerance of drought.

The relative poverty of animal life in arid deserts reflects the low productivity of the vegetation and the indigestibility of much of it. Desert perennials including species of wormwood (*Artemisia*) and creosote plant (*Larrea mexicana*) in the southwestern United States, and mallee species of *Eucalyptus* in Australia carry high concentrations of chemicals that are repellent to herbivores. Ants and small rodents rely on seeds as a relatively reliable perennial resource, whereas bird species are largely nomadic, driven by the need to find water. Only desert carnivores can survive on the water they obtain from their food. In the deserts of Asia and Africa, camels, donkeys, and sheep are managed for transport and food by migrant groups of humans.

animal diversity is low in deserts

4.4.6 Temperate forest

Like all biomes, temperate forest includes under one name a variety of types of vegetation. At its southern limits in Florida and New Zealand, winters are mild, frosts and droughts are rare, and the vegetation largely consists of broad-leaved evergreen trees. At its northern limits in the forests of Maine and the Upper Midwest of the United States, the seasons are strongly marked, winter days are short, and there may be six months of freezing temperatures. Deciduous trees, which dominate in most temperate forests, lose their leaves in the fall and become dormant after transferring much of their mineral content to the woody body of the tree. On the forest floor, diverse floras of perennial herbs often occur, particularly those that grow quickly in the spring (Raunkiaer's geophytes), before the new tree foliage has developed.

All forests are patchy because old trees die, providing open environments for new colonists. This patchiness is on an especially large scale after hurricanes fell the older and taller trees or after fire kills the more sensitive species. In temperate forests the canopies are often composed of a mixture of long-lived species, such as red oaks in the midwest of the United States, and colonizers of gaps, such as sugar maple.

Temperate forests provide a seasonally limited sequence of food resources for animals (compare Figure 4.12b with 4.12a) and only species with short life cycles, such as leaf-eating insects, can be dietary specialists. Many of the birds of temperate forests are migrants that return in spring but spend the remainder of the year in warmer biomes.

temperate forest soils are rich in organic matter

Soils are usually rich in organic matter that is continually added to, decomposed and churned by earthworms and a rich community of other detritivores (organisms that feed on dead organic matter). Only waterlogging and low pH inhibit the decomposition of organic matter and force it to accumulate as peat or "raw humus."

Large swathes of deciduous forest in Europe and the United States have been cut down to provide for agriculture, but these have sometimes been allowed to regenerate as farmers abandoned the land (a conspicuous feature in New England).

4.4.7 Northern coniferous forest (taiga) grading into tundra

Northern coniferous forest (also known as *taiga*) and tundra occur in regions where the short growing season and the cold of winter limit the vegetation and its associated fauna.

Coniferous forest consists of a very limited tree flora. In areas with less severe winters, the forests may be dominated by pines (*Pinus* species, which are all evergreens), and deciduous trees such as larch (*Larix*), birch (*Betula*), or aspens (*Populus*), often as mixtures of species. Farther north, these species give way to monotonous single-species forests of spruce (*Picea*) over immense areas of North America, Europe and Asia. This is *biomonotony*, an extreme contrast to the biodiversity of tropical rain forests!

The areas of vegetation now occupied by tundra and northern coniferous forests (and much of northern deciduous forest) were occupied by the ice sheet during the last ice age, which only started to withdraw 20,000 years ago. Temperatures are now as high as they have ever been since that time, but the vegetation has not yet caught up with the changing climate and the forests are still spreading north. The very low diversity of northern floras and faunas is in part a reflection of a slow recovery from the catastrophes of the ice ages.

Biomonotonous communities provide ideal conditions for the development of disease and epidemics of pests. For example, the spruce budworm (*Choristoneura fumiferana*) lives at low densities in immature northern forests of spruce. As the forests mature, the budworm populations explode in devastating epidemics. These wreck the old forest, which then regenerates with young trees. This cycle takes about 40 years to run its course.

the low diversity of northern coniferous forest provides ideal conditions for pest outbreaks

The overriding environmental constraint in northern spruce forests is the presence of permafrost: the water in the soil remains frozen throughout the year, creating permanent drought except when the sun warms the very surface. The root system of spruce can develop in the superficial soil layer, from which the trees derive all their water during the short growing season.

To the north of the spruce forest, the vegetation changes to tundra, and the two often form a mosaic in the Low Arctic. In the colder areas, grasses and sedges disappear, leaving nothing rooted in the permafrost. High winds exaggerate the aridity of the environment, and ultimately vegetation that consists only of lichens and mosses gives way, in its turn, to the polar desert. The number of species of higher plants (i.e., excluding mosses and lichens) that form the vegetation of the arctic decreases from the Low Arctic (600 species in North America) to 100 species in the High Arctic (north of 83°) of Greenland and Ellesmere Island. In contrast, the flora of Antarctica contains only two native species of vascular plant and some lichens and mosses that support a few small invertebrates. The biological productivity and diversity of Antarctica are concentrated at the coast and depend almost entirely on resources harvested from the sea.

The faunas of northern coniferous forests and tundra have intrigued ecologists because populations of lemmings, mice, voles, and hares (herbivores), and the fur-bearing carnivores (e.g., lynx and ermine) that feed on them, pass through remarkable cycles of expansion and collapse (Section 8.5.2). These were first recorded in the statistics of the Hudson's Bay fur trading company. Lemmings (*Lemmus*) are famous for their population cycles and the role they play in the tundra. When the snow melts during a period when the lemming cycle is at a high point, the animals are exposed and they support large migratory populations of predatory birds (owls, skuas, and gulls) and mammals, such as weasels. Reindeer and caribou (they are the same species, *Rangifer tarandus*) occur in migrant herds capable of foraging on lichens of the tundra, which they can reach through snow cover.

dramatic animal population cycles are characteristic of northern biomes

The leaves of conifers are hard and resistant to decomposition: low temperatures further slow their rate of decomposition. Fallen leaves accumulate in layers on the forest floor as an acid humus, and there is little animal activity in the soil to churn and mix the organic matter into the mineral soil. In the tundra, the decomposition of organic matter is severely limited by the low temperature of the soil and the absence of any forest protection allows ferocious winds to develop so that dead organic matter is blown from place to place.

4.5 ▷ Aquatic Environments

The dominating characteristics of aquatic environments result from the physical properties of water. A water molecule is composed of an oxygen atom, which is slightly negatively charged, bonded with two hydrogen atoms, which are slightly positively charged. This *dipolar structure* enables water molecules to attract and dissolve more substances than any other liquid on Earth. Consequently, water can hold mineral ions in solution, providing the nutrient resources required for the growth of algae and higher plants.

the special properties of water as a medium in which to live

On the other hand, the solubility of oxygen, an essential resource for both plants and animals, decreases rapidly with increasing temperature, and it diffuses only slowly in water. This problem can place major limits on life in water. Many aquatic animals maintain access to oxygen by forcing a continual flow of water over their respiratory surfaces (e.g., the gills of fish) or have very large surface areas relative to their body volume. Oxygen is rapidly used up when dead organic matter decomposes. In places where tree leaves accumulate or untreated sewage is discharged into a river or lake, decomposition can create anaerobic conditions, which are lethal for fish and other animals that have a high biological oxygen demand.

Water is viscous, and moving water transports whole living organisms, such as small plants and animals. It offers resistance to the movement of motile animals such as fish, otters, and aquatic birds; not surprisingly, many motile aquatic animals are streamlined. Many plants that live in moving water depend on rooting in the substratum to hold them against water currents, and many smaller animals are attached to the plants or hide in crevices or under rocks where they are protected from the drag of moving water.

Water is unusual in remaining liquid over a wide range of temperatures. It requires a lot of energy to heat it (i.e., it has *high thermal capacity*), but retains heat efficiently. One consequence is that the temperature of large bodies of water (oceans and large lakes) varies little over the seasons. A further peculiar physical property of water is that it is less dense when frozen than when liquid. Like most liquids, water becomes denser and sinks as it cools. However, at temperatures below 4°C, water becomes less dense and when ice forms (at 0°C), it floats. Ice on a water surface insulates the water beneath; lakes and streams can remain liquid, free flowing, and inhabitable under a layer of ice.

4.5.1 Stream ecology

Streams and rivers contain a minute portion of the world's water (0.006 percent), but an enormous proportion of the fresh water that can be used by people. Consequently, they have been tapped, dammed, straightened, rerouted, dredged, and polluted since the beginning of civilization. Understanding the impacts and sustainability of some of these practices begins with understanding the basics of stream ecology.

Streams and rivers are characterized by their linear form, unidirectional flow, fluctuating discharge, and unstable beds. The narrow nature of river channels means that they are very intimately connected to the surrounding terrestrial environment. Thus, a proper understanding of river ecology requires us to consider the river and its drainage basin as a unit.

Oxygen concentration is often high in turbulent, upstream locations and low farther downstream, where higher temperatures cause reduced solubility. This is reflected in river fish communities, with active upstream species such as brown trout (*Salmo trutta*) having a high oxygen demand, whereas more sluggish species such as pike (*Esox lucius*) can tolerate the lower concentrations in their habitats downstream.

the importance of oxygen concentration

A variety of other chemical and physical conditions vary from stream to stream, or down the length of a given river. Figure 4.14 illustrates how the species composition

pH and temperature

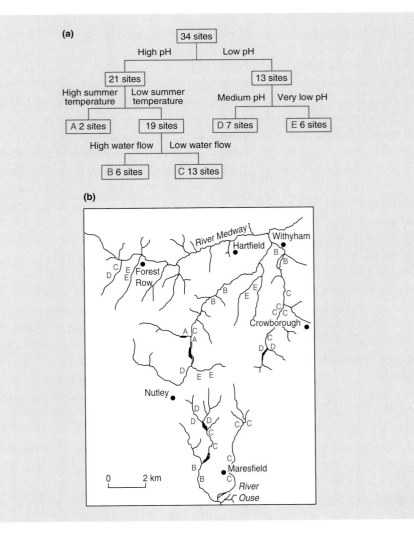

Figure 4.14
The species composition of stream invertebrate communities varies with conditions such as pH, summer temperature, and water flow.
(a) Classification of 34 stream communities. (b) Distribution of community classes A to E, derived from community classification in (a). (After Townsend et al., 1983.)

of stream invertebrate communities varies with conditions. There were 30–40 species at each site, with much overlap in the list of species present. The data were subjected to an analysis called *community classification*, which is conceptually similar to taxonomic classification. In taxonomy, similar individuals are grouped together in species, similar species in genera, and so on. In community classification, communities with similar species compositions are grouped together in sets. These sets are then, in turn, grouped into more inclusive sets, and so on. In this case, the conditions that were most influential in determining the pattern of grouping—and thus were most influential in determining community composition—were pH, stream temperature, and the volume of water flowing per unit time (discharge).

and disturbance of the streambed

Because stream discharge responds to events such as thunderstorms and snow-melt, streams are highly disturbed systems. Stream ecologists have recently been focusing attention at this particular temporal scale, looking at ways in which different regimes of disturbance of the streambed are reflected in the composition of the community. For example, the disturbance regimes of 54 stream sites in New Zealand were assessed by painting particles (pebbles, cobbles, boulders) representative of their bed and determining the percentage that moved during several periods; this varied from 10 to 85 percent. The insect inhabitants of the streams were categorized according to properties that might help them deal with highly disturbed conditions, including small size (small species generally have short life cycles and their populations can rapidly rebuild), a streamlined or flattened body (less prone to being dislodged), and good powers of flight of the adult insects that emerge from the stream to mate (more likely to recolonize after a disturbance). The representation of these traits was higher in the more disturbed streams, testifying to the ecological importance of disturbance regime (Figure 4.15). This relationship between animal traits and a predominant condition in their stream environment parallels the relationships between plant form and conditions in their terrestrial environments identified by Raunkiaer (Figure 4.9).

interactions between the stream and surrounding land

The terrestrial vegetation surrounding a stream (the *riparian* vegetation) has two influences on the resources available to its inhabitants. First, by shading the streambed it may reduce primary production of attached algae and other plants. Second, by shedding leaves it can contribute directly to the food supply of animals and microorganisms. Rivers that begin their course in forested regions are often dominated by the external supply of organic matter, and many of the invertebrates have mouthparts that can handle large particles (*shredders*) (Vannote et al., 1980). Farther downstream, where the stream is wider and where shading is less intense, invertebrates that graze or scrape algae from stones (grazer–scrapers) may be more abundant. As a result of the shredding of large particles into small organic particles (and also physical processes that break up leaves), food for *collector-gatherers* and *collector-filterers* may also increase downstream (Figure 4.16).

may be disrupted by human activities

When riparian vegetation is changed, for example when forest is converted to agriculture, there can be far-reaching effects. Less particulate organic matter enters the stream, but there is less shading and more nutrient runoff from farmland. Results are an increase in productivity of stream plants and a corresponding change in the

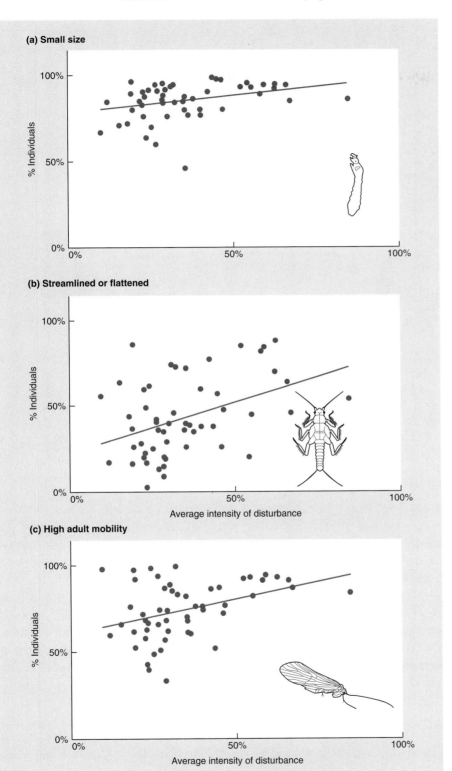

Figure 4.15
Disturbances are important determinants of stream ecology and, in particular, of communities of insects. Disturbed streams contained proportionately more larval insects that (a) were small, (b) had streamlined bodies, and (c) became adults that were strong fliers: characteristics that would enable these organisms to withstand a disturbance and recolonize afterward. $P < 0.01$ in each case. (After Townsend et al., 1997.)

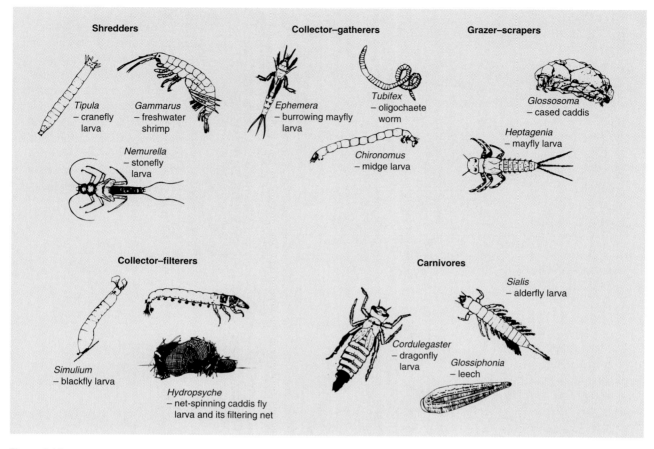

Figure 4.16

Examples of the various categories of invertebrate consumers in stream environments.

stream food web. There may also be effects on discharge (increased when trees are removed), water temperature (higher if shading removed), and streambed characteristics (increased input of fine mineral particles).

The intimate relationship between land and water is also obvious on the floodplains of rivers such as the Amazon, where seasonal floods inundate huge areas of surrounding forest and provide massive inputs of nutrients and organic matter to the river. Many of the world's floodplains have been deliberately drained or cut off from their associated river channels. The effects of this dislocation of spatial continuity are only beginning to be understood (Townsend, 1996).

4.5.2 Lake ecology

Just as river ecology is defined by the unidirectional flow of water, lake ecology is defined by the relatively stationary nature of water within its basin. A critical component of lake ecology is the way in which water can stratify vertically in response to temperature (as mentioned in Section 4.2.3). As water sits in a lake basin, the upper layer is exposed to the sun and heats up. Because warm water is less dense than cold, and because there is resistance of water to heating, the top layer stratifies. This layer, the *epilimnion*, is warm and well illuminated and has a high oxygen content because surface waters exchange oxygen with the atmosphere. It is usually extremely productive, with high densities of plant and animal life.

The importance of vertical spatial scales becomes apparent in deeper lakes, where two further layers may form. Below the epilimnion is a transitional layer, the *thermocline*, in which temperature, oxygen concentration, and light all decrease. The deepest layer, the *hypolimnion*, is cold and often poor in oxygen. It is here that sunken dead organic matter is decomposed and its mineral nutrients released. In temperate regions of the Earth, stratification of lake water breaks down in the fall when the upper layer cools. Currents then mix the water layers and the minerals released in the hypolimnion become available at the lake surface.

Lake ecologists are increasingly turning their attention to the larger spatial scale of whole lake districts. Lakes high in a landscape (such as those in northern Wisconsin) receive a greater proportion of their water from direct precipitation, whereas lakes at lower altitude receive more water as an input from groundwater (Figure 4.17). This is reflected in the higher concentrations of important ions in lakes low in the landscape. The contrasting ion concentrations can be expected to affect the ecology and distribution of freshwater sponges, whose skeletons require silica, and crayfish and snails, which have a particular need for calcium.

Nutrient-rich lakes may support a rich flora of microscopic floating phytoplankton (microscopic plants), together with a diversity of invertebrates and fish species, but a rooted flora of flowering plants is confined to shallow waters near the shore, the *littoral* zone. This zone is usually rich in oxygen, light, food resources, and hiding places. However, some fish and invertebrates specialize in the deeper colder waters of lakes. Lake trout and walleye are two popular sport fish whose habitat is restricted to the colder regions of lakes.

Many lakes in arid regions, lacking a stream outflow, lose water only by evaporation and become rich in sodium and other ions. These saline lakes should not be considered as oddities; globally, they are just as abundant as freshwater lakes. They are usually very fertile and have dense populations of blue–green algae, and some, such as Lake Nkuru in Kenya, support huge aggregations of plankton-filtering flamingoes.

lakes may become thermally stratified, with major consequences for their ecology

saline lakes are common in some parts of the world

4.5.3 The oceans

The oceans cover the major part of the Earth's surface and receive most of the Earth's income of solar radiation. However, much of this radiation is reflected at the water surface or absorbed by water itself and by particles in suspension. Even in clear water the intensity of radiation falls off exponentially with depth and photosynthesis mainly restricted to the upper 100 m—the *euphotic zone*. In most waters the

Figure 4.17
Lakes at different positions in the landscape differ in the source of their water and the concentrations of chemicals important to their inhabitants. (a) Map of Wisconsin Lake District: study lakes are shaded black and contours are shown (meters above sea level). (b) Relationship between landscape position and concentrations of (Ca + Mg) and silica in the five lakes. (Based on Kratz et al., 1997.)

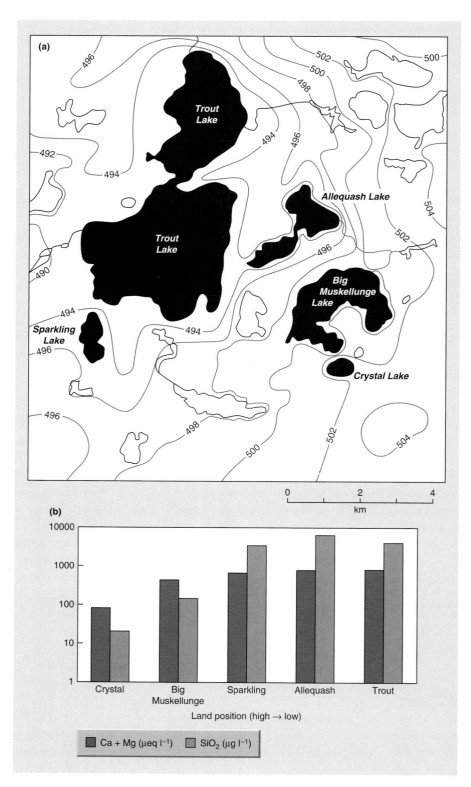

euphotic zone is much shallower, especially where water is more turbid close to coasts and estuaries.

The green plants that photosynthesize in the open oceans are planktonic, mainly single-celled algae that we know from experimental cultures are capable of using solar radiation very efficiently. But, in the real world, many areas of ocean that receive the greatest intensity of solar radiation have the lowest biological activity— because they are limited by shortage of mineral nutrients. The great tropical parts of the Atlantic and Pacific oceans have a biological productivity of less than 35 gC m^{-2} year^{-1}. This compares with more than 800 gC m^{-2} year^{-1} in terrestrial communities at the same latitudes.

The areas of greatest marine productivity (exceeding 90 gC m^{-2} year^{-1}) occur where there is a reliable supply of minerals (especially nitrogen and phosphorus and perhaps iron). This occurs via leaching from the land through rivers and estuaries or where deep currents in the oceans well up to the surface and bring dissolved nutrients into the euphotic zone (see Section 11.2.2). In areas where upwellings occur, the desert of the ocean becomes transformed to a productive environment, as, for example, off the coast of Peru. Dense populations of planktonic algae support small crustacea, which in turn are eaten by schools of anchovies (*Engraulis ringens*). The fish support bonitos, sea lions, and flocks of cormorants, pelicans, and gannets.

We saw earlier in this chapter that the distribution of terrestrial communities depends largely on the intensity of solar radiation and its effects on temperature and water availability. In complete contrast, variations between oceanic communities are ruled mainly by the availability of mineral nutrients.

Below the euphotic zone is increasing darkness and the ocean floor is in total darkness, intensely cold, and under great pressure. This abyssal environment supports the very slow biological activity of a community of extraordinary biological diversity (including worms, crustaceans, mollusks, and fish found nowhere else), which depends on the rain of dying and dead organisms falling from the euphotic zone above. Many of the invertebrate animals are tiny, have very low metabolic rates, and possess a life span that may last for decades. Yet further diversity has been discovered recently in hydrothermal vents that occur at a number of isolated places 2000–4000 m deep. In these remarkable environments, there are high sulfide concentrations and very high temperatures, up to 350°C, where superheated fluid emerges from "chimneys," and there is a sharp gradient down to 2°C, the temperature normally encountered in abyssal depths close by. The vent areas are inhabited by productive thermophilic (heat-loving) bacteria and a unique fauna of polychaete worms, crabs, and very large mollusks. Unfortunately, from the point of view of their inhabitants, these vents are also sites rich in valuable minerals (Box 4.2).

unique communities occur in the abyssal depths of the oceans

4.5.4 Coasts

Marine environments change dramatically near land. Not only are they enriched by nutrients from the land, they are also affected by waves and tides that bring new physical forces to bear. In particular, there are now surfaces to which organisms can attach; indeed if they do not do so they are liable to be washed out to sea or stranded

Box 4.2   Topical ECOncerns

Deep-Sea Vent Communities at Risk

One of the latest controversies to pit environmentalists against industrialists concerns deep sea vents, unique in the communities they harbor (see figure), but now known to be sites rich in minerals. This newspaper article by William J. Broad appeared in the San Jose Mercury News, *January 20, 1998.*

With miners staking claim to valuable metals lying in undersea lodes in the South Pacific, questions surface about how to prevent disasters in these fragile, little understood ecosystems.

The volcanic hot springs of the deep sea are dark oases that teem with blind shrimp, giant tube worms, and

A deep-sea vent community.

on the shore. At a broad scale, coastal communities are strongly influenced by waves and tides and the topography of the coast. Within a single stretch of coast, we can recognize a zonation in the flora and fauna marked by high and low tide marks and differing between areas with heavy or light wave action (Figure 4.18).

waves and tides are key influences in coastal ecology

The extent of the intertidal zone depends on the height of tides and the slope of the shore. Away from the shore, the tidal rise and fall are rarely greater than 1 m, but, closer to shore, the shape of the land-mass can funnel the ebb and flow of the water to produce extraordinary spring tidal ranges of, for example, nearly 20 m in the Bay of Fundy (between Nova Scotia and New Brunswick, Canada). In contrast, the shores of the Mediterranean Sea experience scarcely any tidal range.

On steep shores and rocky cliffs the intertidal zone is very short and zonation is compressed. Both plants and animals are profoundly affected by the physical force of wave action. Anemones, barnacles, and mussels attach themselves securely and other bizarre creatures, sometimes in profusions great enough to rival the chaos of rain forests. And they are old.

Scientists who study them say these odd environments, first discovered two decades ago, may have been the birthplace of all life on Earth, making them central to a new wave of research on evolution.

Now, in a moment that diverse ranks of experts have feared and desired for years, miners are invading the hot springs, possibly setting the stage for the last great battle between industrial development and environmental preservation.

The undersea vents are rich not just in life but in valuable minerals such as copper, silver, and gold. Indeed, their smoky chimneys and rocky foundations are virtual foundries for precious metals. . . . The fields of undersea gold have long fired the imaginations of many scientists and economists, but no mining took place, in part because the rocky deposits were hard to lift from depths of a mile or more.

Now, however, miners have staked the first claim to such metal deposits after finding the richest ores ever. The estimated value of copper, silver, and gold at a South Pacific site is up to billions of dollars. Environmentalists, though, want to protect the exotic ecosystem by banning or severely limiting mining.

Consider the following options and debate their relative merits:

1. *Allow the mining industry free access to all deep sea vents, since the wealth created will benefit many people.*

2. *Ban mining and other disruption of all deep sea vent communities, recognizing their unique biological and evolutionary characteristics.*

3. *Carry out biodiversity assessments of known vent communities and prioritize according to their conservation importance, permitting mining in cases that will minimize overall destruction of this category of community.*

permanently to the substrate and filter planktonic plants and animals from the water when the tides cover them. Other animals, such as limpets, move to graze, and crabs move with the tides and use rock crevices as refuges. The flora in a rocky subtidal zone is usually dominated by the large brown seaweeds (kelps), which fix themselves to the rock with specialized "holdfasts."

Environments are quite different on shallow sloping shores on which the tides deposit and stir up sand and mud. Here the dominant animals are mollusks and polychaete worms, living buried in the substrate and feeding by filtering the water when they are covered by the tides. This environment is completely free of large seaweeds, whose holdfasts can find no anchorage. Flowering plants are almost, but not completely, absent from intertidal environments. The exceptions occur where it is possible for them to be anchored by their roots and this requirement limits them to the more stable and muddy areas colonized by "sea

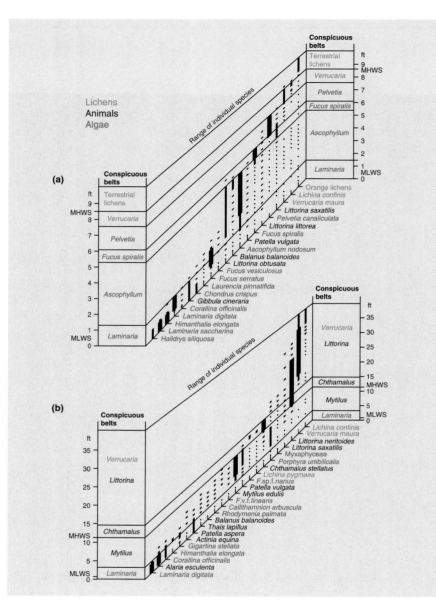

grasses" such as *Zostera* and *Posidonia* or tussocks of *Spartina*. In the tropics, mangroves occupy this kind of habitat, adding a shrubby, woody dimension to the marine littoral zone.

4.5.5 Estuaries

Estuaries occur at the confluence of a river (fresh water) and a tidal bay (salt water). They provide an intriguing mix of the conditions normally experienced in rivers, shallow lakes, and tidal communities. Salt water, more dense than fresh water, tends to enter along the bottom of an estuary as a salt wedge. As it mixes with the outflowing fresh water, a brackish middle layer is created, then returns downstream. The shape of the wedge of salt water is determined in large part by the size of the discharge of the river flowing into the estuary; high discharge tends to create a smaller wedge of salt water and less mixing. The strong gradients in salinity, in both space and time, are reflected in a specialized estuarine fauna; some animals cope through particular physiological mechanisms; many avoid the variable salt concentrations by burrowing, closing protective shells, or moving away.

4.6 ▷ Final Words about Terrestrial Biomes and Aquatic Communities

Ecologists accept that broad categories of terrestrial community exist, and that these biomes are primarily related to global patterns of climate. The broad-scale boundaries of biomes are not immutable. In particular, predicted changes in global climate over the next few decades can be expected to result in dramatic changes to the distribution of biomes over the face of the Earth (Box 4.3).

It is important to remember that the world's vegetation does not, in reality, have the neat boundaries implied by global maps—biomes grade into each other and the boundaries that we draw on maps ought all to be fuzzy. No mapmakers have yet found ways to express this truth adequately. The outlines of biomes on vegetation maps are rough generalizations. The vegetation of the real world is a patchy and mottled carpet, and the fauna largely follows suit. Southern slopes on hillsides within the tundra biome, for example, often carry tussocks of leafy flowering plants and the occasional stunted tree. Similarly, north facing hillsides within a northern conifer forest may, in fact, consist of pure tundra, bare of all but moss and lichens.

The topographical and geological features of a landscape superimpose their own rules on the broader scale of pattern that geographers can display on their maps. Nowhere is this more dramatic than on major mountain ranges in the tropics. The journey up a mountain in Kenya or in Mexico can pass from rain forest or savanna through grasslands, deciduous forest, coniferous forest, and tundra to snow-capped peaks. A journey in time would also see the vegetation of an area changing because continents have moved, ice sheets have advanced and retreated, and climates have changed. In the past 50 million years, for example, central Australia has changed through four biomes, from tropical rain forest, through savanna, to grassland, to the present desert.

Aquatic ecologists observe patterns in communities as they descend from the surface to the depths of lakes and oceans, and these are just as profound as those seen as a mountain is scaled. Changes of a somewhat smaller magnitude are evident as we move from the headwaters to the mouth of a large river. For most aquatic communities, there is little evidence of broad geographic patterns in community types, such as those observed in terrestrial biomes. This is because, for the most part, it is local conditions and resources that govern the nature of aquatic communities, rather than global patterns of temperature and rainfall.

Box 4.3   *Topical ECOncerns*

Predicted Changes in the Distribution of Biomes as a Result of Global Climate Change

As a result of human activities, the atmosphere contains increasing concentrations of gases, particularly carbon dioxide, but also nitrous oxide, methane, ozone and chlorofluorocarbons (CFCs). These changes are predicted to lead to increased temperatures and altered patterns of climate over the face of the Earth (see Section 13.4.3). Given the controlling influence of climate on the distribution of biomes, ecologists expect the biome map of the world to change significantly as carbon dioxide concentrations double over the next 60–70 years.

It is no easy matter to predict the precise details of future climate or its consequences for biome distribution.

Scientists have come up with a number of feasible scenarios, which differ according to the basic assumptions included in their models. The details of these need not concern us here; it is enough to note that the simulations shown in Figures 4.19 and 4.20 are based on a climate change model that assumes an effective doubling of carbon dioxide concentrations and takes into account the coupling of atmosphere and ocean in determining changes in patterns of temperature and rainfall. This is translated into patterns in the distribution of biomes (using the biogeographic model MAPSS) by simulating the potential mature vegetation that could live under the "average"

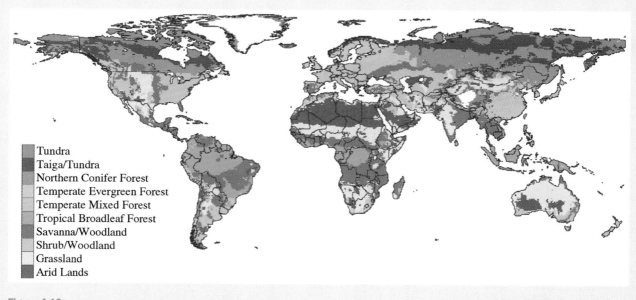

Tundra
Taiga/Tundra
Northern Conifer Forest
Temperate Evergreen Forest
Temperate Mixed Forest
Tropical Broadleaf Forest
Savanna/Woodland
Shrub/Woodland
Grassland
Arid Lands

Figure 4.19
The distribution of major biome types under current climate by the MAPPS biogeography model. (After Neilson et al., 1998.)

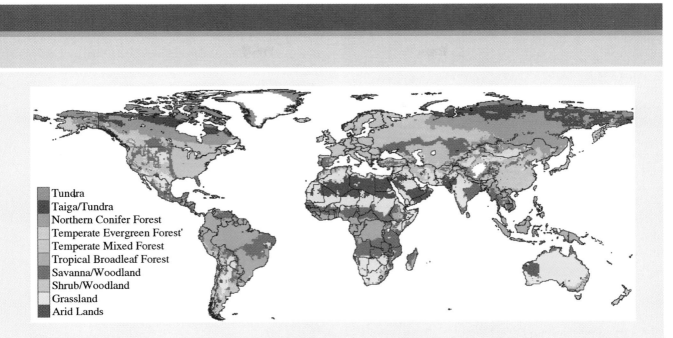

Tundra
Taiga/Tundra
Northern Conifer Forest
Temperate Evergreen Forest'
Temperate Mixed Forest
Tropical Broadleaf Forest
Savanna/Woodland
Shrub/Woodland
Grassland
Arid Lands

Figure 4.20

The potential distribution of major biomes resulting from climate changes associated with an effective doubling of CO_2 concentration, as simulated by MAPPS. (After Neilson et al., 1998.)

seasonal climate prevailing (see Neilson et al., 1998, for further details).

The distribution of biomes shown in Figure 4.19 is as simulated by the model for current climate (Neilson et al., 1998). In other words, it is the model's picture of the way biomes are distributed now (and reflects reality well; note that the biome categories are not exactly the same as those we discuss elsewhere in the chapter). The map in Figure 4.20, by contrast, is the predicted distribution of biomes in 60–70 years' time (Neilson et al., 1998).

This model predicts a reduction in area of the northern biomes of tundra and taiga/tundra (the open woodland that occurs between the treeless taiga and the dense

northern coniferous forest). It also predicts a decrease in arid lands and an increase in temperate forest. These conclusions are in broad agreement with a variety of models that incorporate different starting assumptions.

Consider each of the terrestrial biomes in turn and try to assess what you think might be the major problems associated with an expansion in the area they cover, a contraction, or simply a movement from one global zone to another. Will the problems arise in responding to changes? Or resisting them?

(Thanks to Dr. Ron Neilson for help in compiling this box.)

Summary

Geographic patterns at large and small scales

The variety of influences on climatic conditions over the surface of the globe causes a mosaic of dry, wet, cool, and warm climates. This, in turn, is responsible for the large-scale pattern of distribution of terrestrial biomes. However, biomes are not homogeneous within their hypothetical boundaries; every biome has gradients of physicochemical conditions related to local topographical, geological, and soil features. The communities of plants and animals that occur in these different locations may be quite distinct.

In most aquatic environments it is difficult to recognize anything comparable to terrestrial biomes; the communities of streams, rivers, lakes, estuaries, and open oceans reflect local conditions and resources rather than global patterns in climate.

The composition of local communities can change over time scales of hours (during the decomposition of a leaf fragment), through decades (as plant species replace each other after a gap opens in a forest) to millennia (as tree species move over the face of the Earth in response to climate change associated with the ice ages).

The terrestrial biomes

A map of biomes is not usually a map of the distribution of species. Instead, it shows where we could find areas of land dominated by plants with characteristic shapes, forms, and physiological characteristics. The objective description of communities according to these gross features poses a difficult challenge for ecologists.

Tropical rain forest represents the global peak of evolved biological diversity: all the other biomes suffer from a relative poverty of resources or more strongly constraining conditions. Its exceptional productivity results from the coincidence of high solar radiation received throughout the year and regular and reliable rainfall.

Savanna consists of grassland with scattered small trees. Seasonal rainfall places the most severe restrictions on the diversity of plants and animals in savanna; grazing herbivores and fire also influence the vegetation, favoring grasses and hindering the regeneration of trees.

Temperate grassland occurs in the steppes, the prairies, the pampas, and the veldt. Typically, this biome experiences seasonal drought, but the role of climate in determining vegetation is usually overridden by the effects of grazing animals. Humans have transformed temperate grassland the most of all the biomes.

Desert vegetation shows two kinds of behavior. Many species have an opportunistic life-style, stimulated into germination by the unpredictable rains: others, such as cacti, are long-lived and have sluggish physiological processes. Animal diversity is low in deserts, reflecting the low productivity of the vegetation and the indigestibility of much of it.

Temperate forest, at its southern limits, experiences mild winters, and the vegetation consists of broad-leaved evergreen trees. At its northern limits the seasons are strongly marked and the vegetation is dominated by deciduous trees. Soils are usually rich in organic matter, which is continually added to, decomposed, and churned by earthworms and other detritivores.

Northern coniferous forest consists of a very limited tree flora—a *biomonotony* that contrasts strongly with the biodiversity of tropical rain forests. The very low diversity of northern floras and faunas is in part a reflection of a slow recovery from the catastrophes of the ice ages, and the overriding local constraint is frozen soil. To the north of the spruce forest, the vegetation changes to tundra, and the two often form a mosaic in the Low Arctic. The faunas of the northern biomes have intrigued ecologists because their mammal populations often pass through remarkable cycles of expansion and collapse.

Aquatic environments

Streams and rivers are characterized by their linear form, unidirectional flow, fluctuating discharge, and unstable beds. The terrestrial vegetation surrounding a stream has strong influences on the resources available to its inhabitants; the conversion of forest to agriculture can have far-reaching effects.

Lake ecology is defined by the relatively stationary nature of water within its basin. Some lakes stratify vertically in response to temperature, with consequences for the availability of oxygen and plant nutrients. Lakes high in a landscape may receive a greater proportion of their water from rainfall, whereas lakes at lower altitude receive more water as an input from groundwater. These differences are reflected in community composition. Saline lakes in arid regions lack a stream outflow, lose water only by evaporation, and become rich in sodium and other ions.

The oceans cover the major part of the Earth's surface and receive most of the Earth's income of solar radiation. However, many areas of ocean have very low biological activity—because they are limited by shortage of mineral nutrients. Below the surface zone is increasing darkness, and the ocean floor is in total darkness, intensely cold, and under great pressure, an abyssal environment that supports the very slow biological activity of a diverse community.

Coastal communities are enriched by nutrients from the

land, but they are also affected by waves and tides that bring new physical forces to bear. Within a single stretch of coast, there is a zonation in the flora and fauna marked by high and low tide marks and differing between areas with heavy or light wave action.

Estuaries occur at the confluence of a river (fresh water) and a tidal bay (salt water). Strong gradients in salinity, in both space and time, are reflected in a specialized estuarine fauna.

Review Questions

⏶ = Challenge Question

1. How does solar radiation drive global climate patterns on Earth? Why are deserts more likely to be found at around 30° latitude than at other latitudes?

2. How would you expect the climate to change as you crossed from west to east along the Rocky Mountains?

3. ⏶ Biomes are differentiated by gross differences in the nature of their communities, not by the species that happen to be present. Discuss possible schemes for the objective classification of gross differences among floras and faunas that can be used to distinguish among biomes.

4. ⏶ Which of the Earth's biomes do you think have been most strongly influenced by people? How and why have some biomes been more strongly affected by human activity than others?

5. Discuss how the physical properties of water constrain the life-forms of organisms in aquatic ecosystems.

6. Describe how the logging of a forest may influence the community of organisms inhabiting a stream running through the affected area.

7. Urban runoff from coastal communities has increased the flow of fresh water into many estuaries. What effects might this change have on local aquatic communities?

8. Why is much of the open ocean, in effect, a "marine desert"?

9. Discuss some reasons why community composition changes as one moves (a) up a mountain (b) down the continental shelf into the abyssal depths of the ocean.

10. ⏶ Why are broad geographic classifications of aquatic communities less feasible than broad geographic classifications of terrestrial communities? What characteristics of aquatic ecosystems buffer the effects of climate?

Web Research Questions

1. Global warming is expected to cause dramatic changes to patterns of rainfall and temperature. How could these changes impact on the boundaries of the world's biomes? Do different models of global warming give different results?

2. Hydrothermal vents in the depths of the oceans are home to bizarre and diverse arrays of creatures ("from inner space"). How do these ecosystems differ from those in terrestrial and shallow water locations in terms of their energy base and trophic structure? Evaluate the idea that such vents may be where life originated on earth.

Individuals, Populations, Communities, and Ecosystems

All questions in ecology—however scientifically fundamental, however crucial to immediate human needs and aspirations—can be reduced to attempts to understand the distributions and abundances of organisms, and the processes—birth, death, and movement—that determine distribution and abundance. In this chapter, these processes, methods of monitoring them, and their consequences are introduced.

5

Birth, Death, and Movement

Key Concepts

In this chapter you will

- appreciate the difficulties of counting individuals, but the necessity of doing so for understanding the distribution and abundance of organisms and populations.

- appreciate the range of life cycles and patterns of birth and death exhibited by different organisms.

- understand the nature and the importance of life tables and fecundity schedules.

- understand the role and the importance of dispersal and migration in the dynamics of populations.

- understand the impact of intraspecific competition on birth, death, and movement and hence on populations.

- appreciate that life history patterns can be constructed that link types of organism to types of habitat, but also recognize the limitations of those patterns.

what is a population?

5.1 ▷ Introduction

Ecologists try to describe and understand the distribution and abundance of organisms. They may do so because they wish to control a pest or conserve an endangered species or simply because they are fascinated by the world around them and the forces that govern it. A major part of their task, therefore, involves studying changes in the size of populations. It is usual to use the term *population* to describe a group of individuals of one species under investigation. What actually constitutes a population, though, varies from species to species and from study to study. In some cases, the boundaries of a population are readily apparent: the sticklebacks occupying a small, homogeneous, isolated lake are *the stickleback population of the lake*. In other cases, boundaries are determined by an investigator's purpose or convenience. It is possible to study the population of lime aphids inhabiting one leaf, one tree, one stand of trees or a whole woodland. What is common to all uses of *population* is that it is defined by the number of individuals that compose it and hence grows or declines by changes in those numbers.

birth, death, and movement change
the size of populations

The processes that change the size of populations are birth, death, and movement into and out of that population. Trying to understand the causes of changes in populations is a major preoccupation of ecologists because the science of ecology is not just about understanding nature but often also about how to predict or control it. We might, for example, wish to reduce the size of a population of rabbits, which can do serious harm to crops. We might do this by increasing the death rate, by introducing the myxomatosis virus to the population, or by decreasing the birth rate by offering them food that contains a contraceptive. We might encourage their emigration by bringing in dogs or prevent their immigration by fencing.

A nature conservationist may wish to increase the population of a rare endangered species. In the 1970s, the numbers of bald eagles, ospreys and other birds of prey in the United States began a rapid decline. This might have been because their birth rate had fallen, or their death rate had risen, or because the populations were normally maintained by immigration and this had fallen, or because individuals had emigrated and settled elsewhere. Eventually the decline was traced to reduced birth rates. The insecticide dichlorodiphenyltrichloroethane (DDT) was widely used at the time (it is now banned in the United States) and had been absorbed by fish and other prey of the birds. It accumulated in the bodies of the birds and affected their physiology so that the shells of their eggs became thin and the chicks often died in the egg before they were able to hatch successfully. Conservationists charged with restoring the bald eagle population had to find a way to increase the birds' birth rate. The banning of DDT achieved this end.

5.1.1 What is an individual?

unitary and modular organisms

A population is a number of individuals, but for some kinds of organism it is not always clear what we mean by an individual. For many organisms there is no problem. An individual rabbit has two ears, two eyes, and four legs. An individual spider has eight legs. A spider that lived a long life would not develop more legs and a rabbit would not grow a third eye. The whole form of such organisms, and their program of development from the moment when a sperm fuses with an egg, is predictable and *determinate*. They are called *unitary* organisms.

Birds, insects, reptiles, and mammals are all unitary organisms. We could (in theory, anyway) determine the number of sheep in a flock by counting all their legs and dividing by 4. But there is no way we could do anything similar with a population of modular organisms such as trees, shrubs and herbs, corals, sponges, and very many other marine invertebrates. These grow by the repeated production of modules (leaves, polyps, etc.) and almost always form a branching structure. Such organisms have an architecture: most are rooted or fixed, not motile (Figure 5.1). Both their structure and their precise program of development are not predictable but *indeterminate*. We could count all the leaves in a forest, but this would not allow us to determine the number of trees. We might count all the polyps in a piece of coral reef, but this could not tell us the number of fertilized eggs (zygotes) from which that piece of reef was derived.

In modular organisms, then, we need to distinguish between the genet, the genetic individual, and the module. The *genet* is the individual that starts life as a single-celled zygote and is considered dead only when all its component modules have died. A *module* starts life as a multicellular outgrowth from another module and proceeds through a life cycle to maturity and death even though the form and development of the whole genet are indeterminate. We usually think of unitary organisms when we write or talk about populations, perhaps because we ourselves are unitary, and there are certainly many more species of unitary than of modular organisms. But modular organisms are not rare exceptions and oddities. Most of the living matter (biomass) on Earth and a large part of that in the sea is of modular organisms: the forests, grasslands, coral reefs, and peat-forming mosses. In all of these, and in many other cases, distinguishing individuals can be fraught with difficulty.

modular organisms are themselves populations of modules

5.1.2 Counting individuals, births, and deaths

Even with unitary organisms, ecologists face enormous technical problems when they try to count what is happening to populations in nature. A great many ecological questions remain unanswered because of these problems. Resources, for example, can only be focused on controlling a pest effectively if it is known when its birth rate is highest. But this can only be known by monitoring accurately either births themselves or rising total numbers—neither of which is ever easy.

If we want to know how many fish there are in a pond we might obtain an accurate count by putting in poison and counting the dead bodies. But apart from the questionable morality of doing so, ecologists usually want to continue studying a population after they have counted it. Occasionally it may be possible to trap alive all the individuals in a population, count them, and then release them. With birds, for example, it may be possible to mark nestlings with leg rings and ultimately recognize every individual (except immigrants) in the population of a small woodland. It is not too difficult to count the numbers of large mammals such as deer on an isolated island. But it is very much more difficult to count the numbers of lemmings in a patch of tundra because they spend a large part of the year (and reproduce)

the difficulties of counting

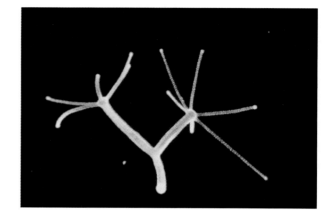

(a)

(b)

(c)

(d)

(e)

Figure 5.1
A range of modular organisms: plants to the left and animals to the right. (a) Modular organisms that fall to pieces as they grow: duckweed (*Lemna* sp.) and *Hydra* sp. (b) Freely branching organisms in which the modules are displayed as individuals on "stalks." A vegetative shoot of a higher plant (*Lonicera*) with leaves (feeding modules) and a flowering shoot (*Tellima*) with flowers (reproductive modules) and a hydroid colony (*Obelia*) bearing both feeding and reproductive modules. (c) Stoloniferous organisms in which colonies spread laterally and remain joined by "stolons" or rhizomes: a single plant of strawberry (*Fragaria*) spreading by means of stolons and a colony of the hydroid *Tubularia crocea*. (d) Tightly packed colonies of modules; tussock of the spotted saxifrage (*Saxifraga bronchialis*) and a segment of the hard coral *Turbinaria reniformis*. (e) Modules accumulated on a long persistent, largely dead support; an oak tree (*Quercus robur*) in which the support is mainly the dead woody tissues derived from previous modules and a gorgonian coral in which the support is mainly heavily calcified tissues from earlier modules.

under thick snow cover! Even a census of the number of humans in a town is difficult when some individuals hide to escape tax collectors or immigration officials!

Ecologists, therefore, are almost always forced to use indirect measures of the number of individuals in a population: they estimate rather than count. They may estimate the numbers of aphids on a crop, for example, by counting the number on a representative sample of leaves, then estimating the number of leaves per square meter of ground, and from this estimating the number of aphids per square meter. Sometimes more complex methods are used (Box 5.1).

estimates from representative samples

Box 5.1

Mark–Recapture Methods for Estimating Population Size

An estimate of the size of a population can sometimes be made by capturing a sample of individuals, marking them in some way (paint spots, leg rings), and then releasing them. Later, another sample is captured, and the proportion that is marked gives some estimate of the size of the whole population (see figure). For example, we might capture and mark 100 individuals from a population of sparrows and release them back into the population. If we later sample a further 100 individuals from the population and find half are marked, we might

conclude that the whole population is composed of about 200 individuals. But this technique of mark and recapture is far less straightforward than it appears at first sight. There are many pitfalls in the sampling process and in interpreting the data. Suppose, for example, that many of the individuals we marked died between our first and second visits. Modifications of the method would be needed to take account of this. For many motile organisms, however, it is the only technique that we have to estimate the size of a population.

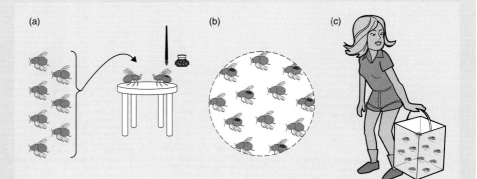

The mark and recapture technique for estimating the size of a population of mobile organisms (in simplified form). (a) On a first visit to a population of unknown total size N, a representative sample is caught (r individuals) and given a harmless mark. (b) These are released back into the population, where they remix with the unknown number of unmarked individuals. (c) On a second visit, a further representative sample is caught. Because it is representative, the proportion of marks in the sample (m out of a total sample of n) should, on average, be the same as that in the whole population (r out of a total of N). Hence N can be estimated.

UNANSWERED QUESTION:

modular organisms: what should we count?

It is *relatively* easy to count the number of plants in a pasture or barnacles on a rock face, because they remain in position, almost waiting to be counted. But what do we count in a forest? When is a tree a tree? Do we include saplings and seedlings? Do we include live seeds in the soil? Do we include individual trees of which all the parts are dead except for one or two shoots that still bear a live leaf or two? A commercial forester may make an arbitrary decision to count as trees only those above a certain size: for example, those more than 20 cm in diameter measured at breast height. To an entomologist interested in a pest that feeds on the trees, however, it may be the number of leaves rather than the number of trees that he needs to know.

And how are births to be counted? Students of genetics usually regard the formation of the zygote as the starting point in the life of an individual. This is when the genetic blueprint for development is first assembled. It is also the stage at which organisms start to die, but it is a stage that is often hidden and extremely hard to study. We simply do not know, for most animals and plants, how many embryos die before "birth," though in the rabbit at least 50 percent of embryos are thought to die in the womb, and in many higher plants it seems that about 50 percent of embryos abort before the seed is fully grown and mature.

the special difficulties of counting births

For ecologists, it is almost always impossible in practice to treat the formation of a zygote as the time of birth. Some other, less logical starting point is usually forced upon us. A student of birds may count the moment that an egg hatches as the time of birth. A student of mammals may take the anthropocentric view of birth as when an individual ceases to be supported within the mother on her placenta and starts to be supported outside her as a suckling. The botanist may regard the germination of a seed as the birth of a seedling, although it is really only the moment at which a developed embryo restarts into growth after a period of dormancy. We need to remember that half or more of a population will probably have died before they can be recorded as born!

Counting deaths in a population poses as many problems as counting births. Dead bodies do not linger long in nature. Only the skeletons of large animals persist long after death. Seedlings may be counted and mapped one day and gone without trace the next. Mice, voles, and soft bodied animals such as caterpillars and worms are digested by predators or rapidly consumed by decomposing organisms (see Chapter 10). They leave no carcasses to be counted and no forensic evidence of the cause of death.

and deaths

Only a tiny fraction of the population of any species completes the full cycle from birth to reproduction to senescence and death. Those individuals that do so are anomalies—highly unusual exceptions. There are many more caterpillars than butterflies, many more tadpoles than frogs.

5.2 ▷ Life Cycles

5.2.1 Life cycles and reproduction

If we wish to understand the forces determining the abundance of a population of organisms, we need to know the important phases of those organisms' lives: that is, the phases when these forces act most significantly. For this, we need to understand the sequences of events that occur in those organisms' life cycles.

There is a point in the life of an individual when, if it survives that long, it will start to reproduce and leave progeny. A highly simplified, generalized life history (Figure 5.2) comprises birth, followed by a prereproductive period, a period of

reproduction, a postreproductive period, and then death as a result of senescence
(though of course other forms of mortality may intervene at any time). The life
histories of all unitary organisms can be seen as variations around this simple pattern,
though a postreproductive period (as seen in humans) is probably rather unusual.

Some organisms fit several or many generations within a single year, some have
just one generation each year (annuals), and others (perennials) have a life cycle
extended over several or many years, as individuals repeatedly experience the cycle
of the seasons. For all organisms, though, a period of growth occurs before there
is any reproduction, and growth usually slows down (and in some cases stops
altogether) when reproduction starts. Growth and reproduction both require resources and there is clearly some conflict between them. Thus, as the annual plant
groundsel *Senecio vulgaris* enters its reproductive stage, flowers and seeds can be
seen to have been produced at the expense of roots and leaves (Figure 5.3). There
are also many biennial plants (e.g., foxgloves) which spend their first year in vegetative
growth, and then flower and die in the second or a later year. If the flowers of these
species are removed before their seeds begin to set, however, the plants usually
survive to the following year, when they flower and set seed even more vigorously.
It seems to be the cost of provisioning the offspring rather than flowering itself that
is lethal. Similarly, pregnant humans are advised to increase their caloric intake by
as much as half their normal consumption: when nutrition is inadequate, pregnancy
can harm the health of the mother.

Among both annuals and perennials, there are some—*iteroparous species*—that
breed repeatedly, devoting some of their resources during a breeding episode not
to breeding itself, but toward survival to further breeding episodes (if they manage
to live that long). We ourselves are examples. There are others, *semelparous species*,
like the biennial plants described above, in which there is a single reproductive
episode, with no resources set aside for future survival, so that reproduction is
inevitably followed quickly by death.

the conflict between growth and reproduction

iteroparous and semelparous species

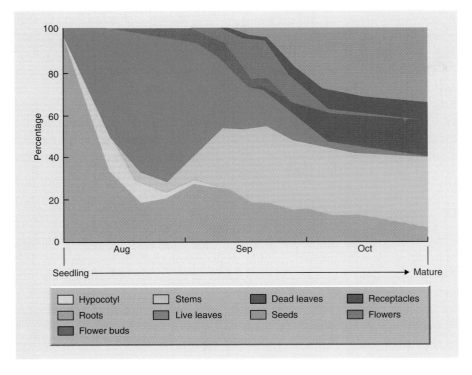

Figure 5.3
Percentage allocation of energy to different structures throughout the life cycle of the annual weed *Senecio vulgaris*. Note the development of reproductive parts at the expense of roots and leaves toward the end of the growing season. (After Harper & Ogden, 1970.)

5.2.2 Annual life cycles

In strongly seasonal, temperate latitudes, most annuals germinate or hatch as temperatures start to rise in spring, grow rapidly, reproduce, and then die before the end of summer. The European common field grasshopper *Chorthippus brunneus* is an example of an iteroparous annual. It emerges from its egg in late spring and passes through four juvenile stages of nymph before becoming adult in midsummer and dying by mid-November. During their adult life, the iteroparous females reproduce repeatedly, each time laying egg pods containing about 11 eggs, and recovering and actively maintaining their bodies between each burst of reproduction.

Many annual plants, by contrast, are semelparous: they have a sudden burst of flowering and seed set, and then die. This is commonly the case among the weeds of arable crops. Others, such as groundsel, are iteroparous: they continue to grow and produce new flowers and seeds through the season until they are killed by the first lethal frost of winter. They die with their buds on.

Most annuals spend part of the year dormant as seeds, spores, cysts, or eggs. In many cases these dormant stages may remain viable for many years; there are reliable records of seeds of the annual weeds *Chenopodium album* and *Spergula arvensis* remaining viable in soil for 1600 years. Similarly, the dried eggs of brine shrimps remain viable for many years in storage. This means that if we measure the length of life from the time of formation of the zygote, many so-called annual animals and plants live very much longer than a single year! Large populations of dormant seeds form a *seed bank* buried in the soil. As many as 86,000 viable seeds per square

seed banks

meter have been found in cultivated soils. Species of annuals that had been thought to have become locally extinct may suddenly reappear after the soil is disturbed and these seeds germinate.

the ephemeral "annuals" of deserts

Dormant seeds, spores or cysts are also necessary to the many ephemeral plants and animals of sand dunes and deserts that complete most of their life cycle in less than 8 weeks. They then depend on the dormant stage to persist through the remainder of the year and survive the hazard of low temperatures in winter and the droughts of summer. In desert environments, in fact, the rare rains are not necessarily seasonal and it is only in occasional years that sufficient rain falls and stimulates the germination of characteristic and colorful floras of very small ephemeral plants.

5.2.3 Longer life cycles

repeated, seasonal breeders

There is a marked seasonal rhythm in the lives of many long lived plants and animals, especially in their reproductive activity: a period of reproduction once per year (Figure 5.4a). Mating (or the flowering of plants) is commonly triggered by the *photoperiod*—the daily light–dark cycle, which varies continuously through the year— and usually makes sure that young are born, eggs hatch, or seeds are ripened when seasonal resources are likely to be abundant.

In populations of perennial species, the generations overlap and individuals of a range of ages breed side by side. The population is maintained in part by surviving adults and in part by new births. A study of great tits *Parus major*, for example, showed that of 50 eggs that were laid by a breeding population of 10 birds in one season, only 30 hatchlings survived to become fully fledged, and only 3 of these survived to adulthood the following year. These 3 1-year-old birds were joined in that second year, though, by a further 5 birds aged between 2 and 5 years—the survivors of the previous year's 10 (Figure 5.5).

continuous breeders

In wet equatorial regions, on the other hand, where there is very little seasonal variation in temperature and rainfall and scarcely any variation in photoperiod, we

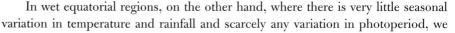

Desert in bloom. In desert areas where rainfall is rare and seasonally unpredictable a dense and spectacular flora of very short-lived annuals commonly develops after rain storms. They often complete their life cycle from germination to seed set in little more than a month.

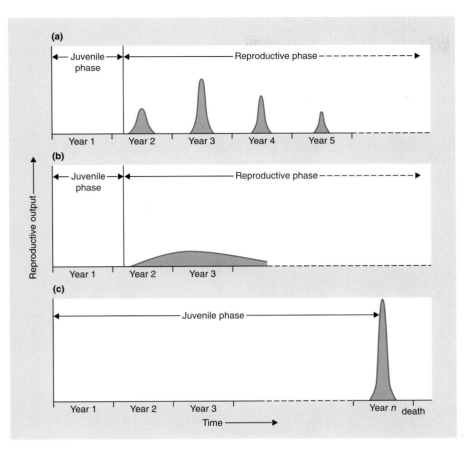

Figure 5.4
Simplified life histories for organisms living more than one year. (a) An iteroparous species breeding seasonally once per year. Death tends not to occur predictably after any given time, though a decline toward senescence is often observed. (b) An iteroparous species breeding continuously throughout the year. The pattern of death and decline is similar to that in (a). (c) A semelparous species passing several or many years in a prereproductive juvenile phase, followed by a burst of reproduction, followed in turn by inevitable death.

find species of plant that are in flower and fruit throughout the year—and continuously breeding species of animal that subsist on this resource (Figure 5.4b). There are several species of fig (*Ficus*), for instance, that bear fruit continuously and form a reliable year-round food supply for birds and primates. In more seasonal climates, humans are unusual in also breeding continuously throughout the year, though numbers of other species, cockroaches, for example, do so in the stable environments that humans have created.

In a contrasting program of lifetime fecundity (Figure 5.4c) the plant or animal spends almost all its life in a long nonreproductive (juvenile) phase and then has one lethal burst of reproductive activity. We saw such semelparity earlier in biennial plants, but it is also characteristic of some species that live much longer than 2 years. The Pacific salmon is a familiar example. Salmon are spawned in rivers. They spend the first phase of their juvenile life in fresh water and then migrate to the sea, often traveling thousands of miles. At maturity they return to the stream in which they were hatched. There they lay their eggs in a terminal bout of reproduction. Some mature and return to reproduce and die after only 2 years at sea; others mature more slowly and return after 3, 4, or 5 years. At the time of reproduction the population of salmon is composed of overlapping generations of semelparous individuals.

semelparous species like salmon and bamboo

Figure 5.5

A diagrammatic life history for a population of great tits near Oxford, England. Population sizes (in rectangles) are per hectare; the proportions surviving from one stage to the next are in triangles; the rate of egg production per female is shown in the diamond. (After Perrins, 1965.)

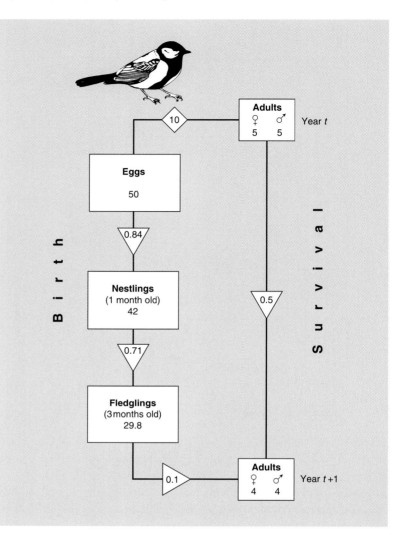

There are even more dramatic examples of species that have a long life but reproduce just once in a lifetime. Many species of bamboo form dense clones of shoots that remain vegetative for many years: in some species, 100 years. The whole population of shoots then flowers simultaneously in a mass suicidal orgy. Even when shoots have become physically separated from each other, the parts still flower synchronously.

size is important   Organisms of long-lived species that are the same age, however, are not necessarily the same size—especially in modular organisms. Some individuals may be very old but have been suppressed in their growth and development by predators or by competition. Age is often a particularly poor predictor of fecundity. Charles Darwin examined a population of 32 dwarfed fir trees in a square yard of heathland and found one that was 26 years old—repeatedly grazed by cattle, still alive, but far too small to reproduce. An analysis that classifies the members of a population according to their size is often more useful (Figure 5.6) than one that classifies them by age.

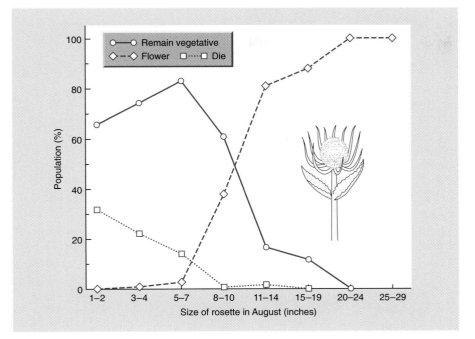

Figure 5.6
The fate of individual teasel plants (*Dipsacus fullonum*) in the following year depends on how large they are in August rather than whether they are 2, 3, or 4 years old. (After Werner, 1975.)

5.3 > Monitoring Birth and Death: Life Tables and Fecundity Schedules

The previous sections have outlined the different patterns of births and deaths in different species. But patterns are just a start. What are the *consequences* of these patterns in particular cases in terms of their effects on how a population might grow to pest proportions, say, or shrink to the brink of extinction? To determine these consequences, we need to monitor the patterns in a quantitative way.

There are different ways of doing so. To monitor and quantify survival, we may follow the fate of individuals from the same *cohort* within a population: that is, all individuals born within a particular period. A *cohort life table* then records the survivorship of the members of the cohort over time (Box 5.2). A different approach is necessary when we cannot follow cohorts but we know the ages of all the individuals in a population. We can then, at one time, describe the numbers of survivors of different ages in what is called a *static life table* (Box 5.2).

The fecundity of individuals also changes with their age, and to understand properly what is going on in a population we need to know how much individuals of different ages contribute to births in the population as a whole: these can be described in *age-specific fecundity schedules.*

5.3.1 Annual life tables

Probably the most straightforward life table to construct is a cohort life table for annuals, because with nonoverlapping generations, it is indeed often possible to follow a single cohort from the first birth to the death of the last survivor. A simplified

a grasshopper life table

Box 5.2

The Basis for Cohort and Static Life Tables

In the figure below, a population is portrayed as a series of diagonal lines, each line representing the life "track" of an individual. As time passes, each individual ages (moves from bottom-left to top-right along its track) and eventually dies (the dot at the end of the track). In some cases, individuals are classified by their age (as here), but in others it is more appropriate to split the life of each individual into the different developmental stages.

In this case, three individuals are born prior to the time period t_0, four during t_0, and three during t_1. To construct a *cohort life table*, we direct our attention to a particular cohort (in this case, those born during t_0) and monitor the subsequent development of the cohort. The life table is constructed by noting the number surviving to the start of each time period. Here, we note that two of the four individuals have survived to the beginning of t_1; only one of these is alive at the beginning of t_2; none survives to the start of t_3. The first data column of the cohort life table thus comprises the series of declining numbers in the cohort: 4, 2, 1, 0.

A different approach is necessary when we cannot follow cohorts but we know the ages of all the individuals in a population (perhaps from some clue such as the condition of the teeth in a species of deer). We can then, as the figure shows, direct our attention to the whole population during a single period (in this case, t_1) and note the numbers of survivors of different ages in the population. These may be thought of as entries in a life table if we assume that rates of birth and death are, and have previously been, constant—a very big assumption. What results is called a *static life table*. Here, of the seven individuals alive during t_1, three were actually born during t_1, two were born in the previous time interval, and two in the interval before that. The first data column of the life table would thus comprise the series 3, 2, 2, 0. This can be thought of as implying, first, that a single life table can typify the survival of individuals during the whole period shown, and, second, that the typical cohort will have started with three and declined over successive time intervals to two, then two again, then zero.

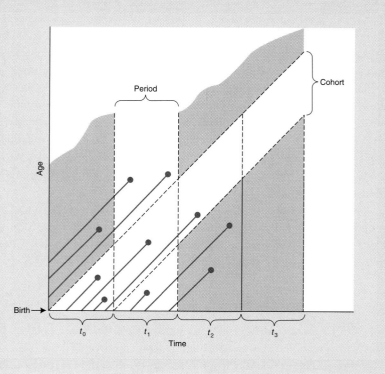

life table is shown in Table 5.1 for the common field grasshopper. The first column is a list of the stages of the grasshopper's life. The second is then the raw data from the study, collected in the field. It reports the number of individuals surviving to the beginning of each stage.

Since ecologists will typically be interested not just in examining populations in isolation but in comparing the dynamics of two or more perhaps rather different populations (in the presence and absence of a pollutant, for instance), it is necessary to standardize the raw data so that comparisons can be made. This is done in the third column of the table, which is said to contain l_x values, where l_x is defined as the proportion of the original cohort surviving to the start of stage x (or in other cases, age class x). The first value in the third column, l_0 (spoken: L-zero), is therefore the proportion surviving to the beginning of this original stage. Obviously, here and in every life table, l_0 is 1.00 (the whole cohort is there at the start). In fact, the grasshopper population had an initial number of 44,000 eggs. The l_x values for subsequent stages are therefore expressed as proportions of this number. Only 3513 individuals survived to reach the first nymphal stage. Thus, the second value in the third column, l_1, is the proportion 3513/44,000 = 0.08 (that is, only 0.08 or 8 percent of the original cohort survived this first step), and so on. In a full life table, subsequent columns would then use these same data to calculate the proportion of the original cohort that died at each stage and also the mortality rate for each stage, but for brevity these columns have been omitted here.

Table 5.1

A simplified cohort life table for the common field grasshopper *Chorthippus brunneus*.
The columns are explained in the text.

STAGE (x)	NUMBER OBSERVED AT START OF EACH STAGE a_x	PROPORTION OF ORIGINAL COHORT SURVIVING TO START OF EACH STAGE l_x	EGGS PRODUCED IN EACH STAGE F_x	EGGS PRODUCED PER SURVIVING INDIVIDUAL IN EACH STAGE m_x	EGGS PRODUCED PER ORIGINAL INDIVIDUAL IN EACH STAGE $l_x m_x$
Eggs (0)	44,000	1.000	—	—	—
Instar I (1)	3513	0.080	—	—	—
Instar II (2)	2529	0.058	—	—	—
Instar III (3)	1922	0.044	—	—	—
Instar IV (4)	1461	0.033	—	—	—
Adults (5)	1300	0.030	22,617	17	0.51

$$R_0 = \Sigma l_x m_x = \frac{\Sigma F_x}{a_0} = 0.51.$$

(After Richards & Waloff, 1954.)

and fecundity schedule

Table 5.1 also includes a fecundity schedule for the grasshopper (columns 4 and 5). Column 4 shows the total number of eggs produced by each stage F_x (though in this case only the adult stage produced eggs). In fact, the 1300 surviving adults produced 22,617 eggs. The fifth column is then said to contain m_x values: the mean number of eggs produced per surviving individual. In the present case there is only one such value, for the adults: $m_5 = 22,617/1300 = 17$.

combined to give the basic reproductive rate

In the final column of a life table, the l_x and m_x columns are brought together to express the overall extent to which a population increases or decreases over time— reflecting the dependence of this on both the survival of individuals (the l_x column) and the reproduction of those survivors (the m_x column). In this case, 0.03 of the original cohort survived to the adult stage, and those survivors produced 17 eggs on average. Hence, the mean number of eggs produced at the end of the cohort per egg present at the beginning was $0.03 \times 17 = 0.51$ eggs. This is therefore a measure of the overall extent by which this population has (in this case) decreased in a generation. We call this *the basic reproductive rate* and denote it by R. In this case, then, $R = 0.51$.

a life table for an annual plant

Data from annual plants can be treated in the same way. Thus an initial cohort of 996 seeds of the annual plant *Phlox drummondii* was followed from seed germination, with the life cycle broken down into successive periods of 30–60 days (Table 5.2). In contrast to the grasshoppers, whose population clearly declined over the annual cycle, this population set approximately two and a half times more seed at the end of the season than was present at the beginning. Notice, though, that these seeds were produced not by just one stage (as among the grasshoppers) but by several. It is apparent therefore that R (= 2.41 in this case) is generally the sum of all the $l_x m_x$ values, denoted by $\Sigma l_x m_x$, where the symbol Σ means "the sum of."

logarithmic survivorship curves

It is also possible to study the detailed pattern of decline in the *Phlox* cohort. Figure 5.7a, for example, shows the numbers surviving relative to the original population—the l_x values—plotted against the age of the cohort. However, this can be misleading. If the original population is 1000 individuals, and it decreases by half to 500 in one time interval, then this decrease looks more dramatic on a graph like Figure 5.7a than a decrease from 50 to 25 individuals later in the season. Yet the risk of death to individuals is the same on both occasions. If, however, l_x values are replaced by $\log(l_x)$ values, that is, the logarithms of the values, as in Figure 5.7b (or, effectively the same thing, if l_x values are plotted on a log scale), then it is a characteristic of logs that the reduction of a population to half its original size will always look the same. *Survivorship curves* are, therefore, conventionally plots of $\log(l_x)$ values against cohort age. Figure 5.7b shows that there was a relatively rapid decline in the size of the cohort over the first 6 months, but that the death rate thereafter remained steady and rather low until the very end of the season, when the survivors all died.

It is possible to see, therefore, even from these two examples, how life tables can be useful in characterizing the "health" of a population—the extent to which it is growing or declining—and identifying where in the life cycle, and whether it is survival or birth, that is apparently most instrumental in determining that rate of increase or decline. Either or both of these may be vital in determining how best to conserve an endangered species or control a pest.

Table 5.2

A simplified cohort life table for the annual plant *Phlox drummondii.* The columns are explained in the text.

AGE INTERVAL (DAYS) $x - x'$	NUMBER SURVIVING TO DAY x a_x	PROPORTION OF ORIGINAL COHORT SURVIVING TO DAY x l_x	F_x	m_x	$l_x m_x$
0–63	996	1.000	—	—	—
63–124	668	0.671	—	—	—
124–184	295	0.296	—	—	—
184–215	190	0.191	—	—	—
215–264	176	0.177	—	—	—
264–278	172	0.173	—	—	—
278–292	167	0.168	—	—	—
292–306	159	0.160	53.0	0.33	0.05
306–320	154	0.155	485.0	3.13	0.49
320–334	147	0.148	802.7	5.42	0.80
334–348	105	0.105	972.7	9.26	0.97
348–362	22	0.022	94.8	4.31	0.10
362–	0	0.000	—	—	—
			2408.2		2.41

$$R_0 = \Sigma l_x m_x = \frac{\Sigma F_x}{a_0} = 2.41.$$

(After Leverich & Levin, 1979.)

5.3.2 Life tables for populations with overlapping generations

Many of the species for which we have important questions, and for which life tables may provide an answer, have repeated breeding seasons (or continuous breeding, as in the case of humans), but constructing life tables here is more complicated, largely because these populations have individuals of many different ages living together. A cohort life table is sometimes possible, but this is relatively uncommon. Apart from the mixing of cohorts in the population, it can be difficult simply because of the longevity of many species.

Another approach is to construct a "population snapshot" in a static life table (Box 5.2). Superficially, the data look like a cohort life table: a series of different numbers of individuals in different age classes. But great care is required: they can

a static life table—useful if used with caution

Figure 5.7
Following the survival of a cohort of
Phlox drummondii (Table 5.2).
(a) When l_x is plotted against cohort
age, it is clear that most individuals are
lost earlier in the life of the cohort, but
there is no clear impression of the risk
of mortality at different ages. (b) By
contrast, a survivorship curve plotting
log (l_x against age shows that an initial
6 months of moderate survivorship was
followed by an extended period of
higher survivorship (less risk of
mortality) and then by very low
survivorship in the final weeks of the
annual cycle.

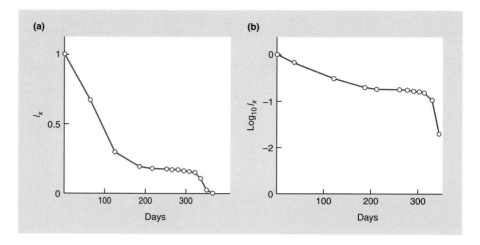

only be treated and interpreted in the same way if patterns of birth and survival in
the population have remained much the same since the birth of the oldest individuals—and this will happen only rarely. Nonetheless, useful insights can sometimes be
gained by combining the data from a static life table (an *age structure*: the numbers
in different age classes) with corresponding background information. This is illustrated by a study of two populations of the long-lived tree *Acacia burkittii* in South
Australia (Figure 5.8). Although differences in age structure between the populations
are obvious, the reasons are not. Fortunately, background information provides
important clues.

5.3.3 A classification of survivorship curves

Life tables provide a great deal of data on specific organisms. But ecologists search
for generalities: patterns of life and death that we can see repeated in the lives of
many species. A useful set of survivorship curves was developed in 1928 by Raymond
Pearl, a human demographer, whose three types generalize what we know about the
way in which the risks of death are distributed through the lives of different organisms
(Figure 5.9).

- Type I describes the situation in which mortality is concentrated toward the end
  of the maximum life span. It is perhaps most typical of humans in developed
  countries and their carefully tended zoo animals and pets.
- Type II is a straight line and describes a constant mortality rate from birth to
  maximum age. It describes, for instance, the survival of buried seeds in a seed
  bank.
- Type III indicates extensive early mortality, but a high rate of subsequent survival.
  This is typical of species that produce many offspring. Few survive initially, but
  once individuals reach a critical size, their risk of death remains low and more or
  less constant. This appears to be the most common survivorship curve among
  animals and plants in nature.

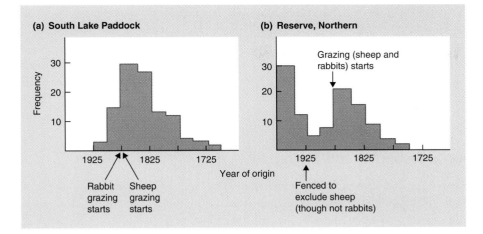

Figure 5.8

Age structures (and hence static life tables) of *Acacia burkittii* populations at two sites in South Australia. South Lake Paddock populations had been grazed by sheep from 1865 to 1970 and by rabbits from 1885 to 1970, whereas the Reserve population had been fenced in 1925 to exclude sheep (but did not exclude rabbits). With this information in hand, the effect of grazing from 1865 onward is evident in the decreased numbers of new recruits to both populations. However, the effects of fencing after 1925 are equally obvious in the Reserve population, where the proportion of new recruits increased dramatically. The effects of rabbit grazing on recruitment after fencing in the Reserve population can, however, still be detected, since, for example, the 1925–1940 age class was much smaller than the (pregrazing) 1845–1860 class, even though the latter had survived an additional 75 years.

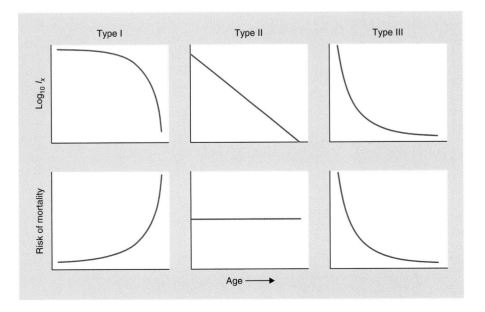

Figure 5.9

A classification of survivorship curves plotting log (l_x) against age, above, with corresponding plots of the changing risk of mortality with age, below. The three types are discussed in the text. (After Pearl, 1928; Deevey, 1947.)

These types of survivorship curve are useful generalizations, but in practice, patterns of survival are usually more complex. Thus, in a population of *Erophila verna*, a very short-lived annual plant inhabiting sand dunes, survival can follow a type I curve when the plants grow at low densities; a type II curve, at least until the end of the life span, at medium densities; and a type III curve in the early stages of life at the highest densities (Figure 5.10).

5.4 ❯ Dispersal and Migration

patterns of distribution

Birth is only the beginning. If we were to stop there in our studies, many crucial ecological questions would remain unanswered. From their place of birth, all organisms move to where we eventually find them. Plants grow where their seeds fall, but seeds may be moved by the wind, water, animals or shifting soil. Animals move

Figure 5.10

Survivorship curves for the sand dune annual plant *Erophila verna* monitored at three densities: high (initially 55 or more seedlings per 0.01-m² plot); medium (15–30 seedlings per plot); and low (1–2 seedlings per plot). The horizontal scale (Plant age) is standardized to take account of the fact that each curve is the average of several cohorts, which lasted different lengths of time (around 70 days on average). (After Symonides, 1983.)

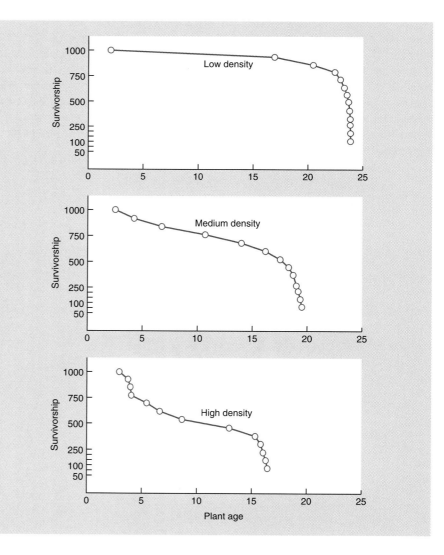

in search of food and safe havens, whether it is only to move 1 cm along a leaf from where their egg was deposited, or to move like some migratory birds halfway around the globe. The effects of those movements are varied. In some cases they aggregate members of a population into clumps; in others they continually redistribute and shuffle them; and in still others they spread the individuals out and "dilute" their density. Three generalized spatial patterns that result from this movement—aggregated (clumped), random, and regular (evenly) spaced—are illustrated in Figure 5.11. Clearly, movement and spatial distribution (the latter sometimes, confusingly, called "dispersion") are intimately related.

Technically, the term *dispersal* describes the way individuals spread away from each other, such as when seeds are carried away from a parent plant or young lions leave the pride in search of their own territory. *Migration* refers to the mass directional movement of large numbers of a species from one location to another. Migration therefore describes the movement of locust swarms but also includes the smaller-scale movements of intertidal organisms, back and forth twice a day as they follow their preferred level of immersion or exposure.

Our view of dispersal and migration, and of the resulting distributions, is determined by the scale on which we are working. For example, consider the distribution of an aphid living on a particular species of tree in a woodland. On a large scale, the aphids appear to be aggregated in the woodlands and nonexistent in the open fields. If the samples we took were smaller, and taken only in woodlands, the aphids would still appear to be aggregated, but now aggregated on their host trees rather than on trees in general. However, if samples were collected at an even smaller scale—the size of a leaf within a canopy—the aphids might appear to be randomly distributed over the tree as a whole. And on the scale experienced by the aphid itself (1 cm²), the distribution might appear regular as individuals on a leaf spread out to avoid one another (Figure 5.12).

This example also illustrates the difference between the "average density" and the crowding experienced by individuals in a population. The *average density* is simply the total number of individuals divided by the total size of the habitat—but it depends very much on how we define the habitat. For the aphids, if it includes everything, woodland and nonwoodland, then average density will be low. It will

the perception of pattern depends on the spatial scale

density and crowding

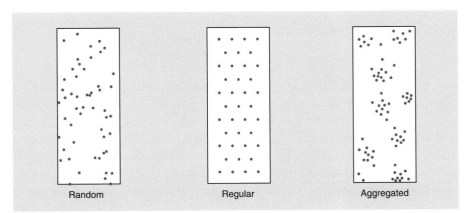

Figure 5.11
Three generalized spatial patterns that may be exhibited by organisms across their habitat.

Random   Regular   Aggregated

Figure 5.12
Are aphids distributed evenly, randomly,
or in an aggregated fashion? It all
depends on the spatial scale at which
they are viewed.

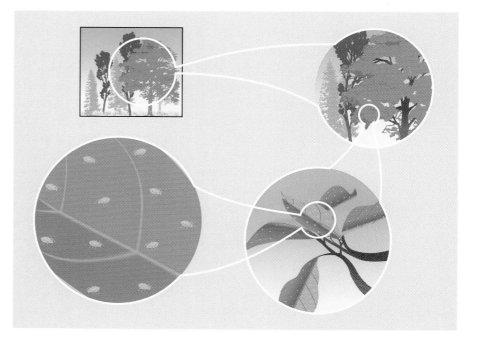

be higher, but still quite low, if we include only woodland but every species of tree. It will be much higher, however, if we include only the aphids' host trees.

The average density of individuals in the United States is about 75 persons/km². Yet there are vast areas of the United States—rural and wilderness areas—within which the density is low, and also crowded cities and towns within which the density is much higher. And because the majority of people live in urban and suburban settings, the density actually experienced by people, on average, has been calculated at 3630 persons/km². There may be little impetus for dispersal, or migration, at the relatively low population pressure of 75 individuals/km². At 3630 people/km², however, individuals are much more likely to find ways to escape from their neighbors. Real measures of crowding as experienced by individuals are likely to be more important forces driving dispersal and migration than some average value of population density.

5.4.1 Dispersal determining abundance

a sand dune plant example

Dispersal can have a profound effect on the dynamics of a population. This was seen in a study of *Cakile edentula*, a summer annual plant growing on the sand dunes of Martinique Bay, Nova Scotia. The population was concentrated in the middle of the dunes and declined toward both the sea and the land. Only in the area toward the sea, however, was seed production high enough and mortality sufficiently low for the population to maintain itself year after year. At the middle and landward sites, mortality exceeded seed production. Hence, one might have expected the population to become extinct (Figure 5.13). But the distribution of *Cakile* did not change over time. Instead, large numbers of seeds from the seaward

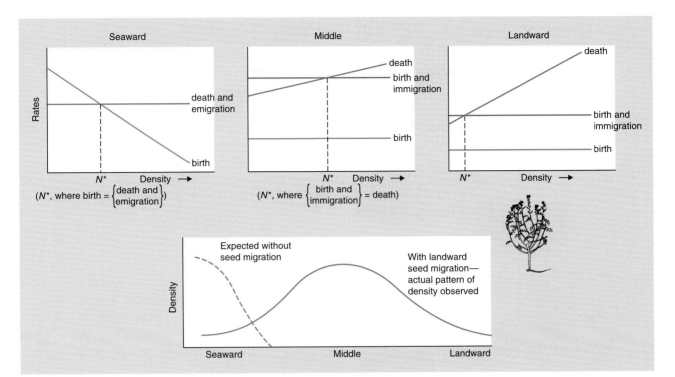

Figure 5.13
Diagrammatic representation of variations in mortality and seed production rate of *Cakile edentula* in three areas along an environmental gradient from open sand beach (seaward) to densely vegetated dunes (landward). In contrast to that in other areas, seed production was prolific at the seaward site. Births, however, declined with plant density, and where births and [deaths + emigration] were equal, an equilibrium population density can be envisaged (N^*) (see also Section 5.5). In the middle and landward sites, deaths always exceeded births resulting from local seeds, but populations persisted there because of the landward drift of the majority of seed produced by plants on the beach (seaward site). Thus, the sum of local births plus immigrating seeds can balance mortality in the middle and landward sites, resulting in equilibria at appropriate densities. (After Keddy, 1982; Watkinson, 1984.)

zone dispersed to the middle and landward zones. Indeed, more seeds were dispersed into and germinated in these two zones than were produced by the residents. The distribution and abundance of *Cakile* were directly due to the dispersal of seeds in the wind and the waves.

The dynamics of few populations are weighted so heavily by immigration as is that of *Cakile*. Yet most populations are more affected by immigration and emigration than is commonly imagined. Within the United States, for example, over 40 percent of U.S. residents, over 100 million people, can trace their roots to the 12 million immigrants who entered the United States through the Ellis Island port from 1870 to 1920.

It is hard to count the births and deaths in a population. It is even harder to determine the number of individuals that immigrate or emigrate. It is possible to determine that of 5000 seeds produced by a ragwort plant, 3000 fell within a square meter of the parent, but it is next to impossible to determine the fate of the other 2000 seeds. Yet the importance of dispersal cannot be overemphasized. In some

UNANSWERED QUESTION:

where do all the dispersers go?

cases, such as the *Cakile* example, it is repeated dispersal that maintains the abundance of a population. In other cases, it may be the extreme distances dispersed by a very few individuals that are responsible for the distribution of a species. It takes only one seed, carried to a new area, to begin a new population. Often ecologists have made the rash assumption that the individuals that leave a population are balanced by those that join it. We now know that this can be grossly misleading. Moreover, immigrants and emigrants not only influence the numbers in a population—they also affect its composition. A higher proportion of individuals usually escape from a crowded than from a sparse population, and those that leave are not a random sample. The dispersers are often the young, and males frequently do more moving about than females. The behavior of humans provides plenty of examples to illustrate these generalizations!

5.4.2 The role of migration

The mass movements of populations that we call migration are almost always from regions where the food resource is declining to regions where it is abundant (or where it will be abundant for the progeny). Planktonic plants live in the upper layers of the water in lakes where the light needed for photosynthesis is brightest. But at night they migrate to lower, nutrient-rich depths. Crabs migrate along the shore with the tides, following the movement of their food supply as it is washed up in the waves. At longer time scales, some shepherds still follow the ages-old practice of transhumance, moving their flocks of sheep and goats up to mountain pastures in summer and down again in the fall to track the seasonal changes in climate and food supply.

The long-distance migrations of terrestrial birds in many cases involve movement between areas that supply abundant food, but only for a limited time. They are areas in which seasons of comparative glut and famine alternate, and which cannot support large all-year-round resident populations. For example, swallows (*Hirundo rustica*) migrate seasonally from northern Europe in the fall, when flying insects start to become rare, to South Africa when they are becoming common. In both areas the food supply that is reliable throughout the year can support only a small population of resident species. The seasonal glut supports the populations of invading migrants, which make a large contribution to the diversity of the local fauna.

5.5 ▷ The Impact of Intraspecific Competition on Populations

Intraspecific competition was introduced in Section 3.5 because its intensity is typically dependent on resource availability. It reemerges here because its effects are expressed through the focal topics of this chapter—rates of birth, death, and movement. Competing individuals that fail to find the resources they need may grow more slowly or even die; survivors may reproduce later and less; or, if they are mobile, they may move farther apart or migrate elsewhere. Examples in which the dynamics of a species can be understood without a firm grasp of the effects of competition on those dynamics are exceedingly rare.

One measure of the pressure that a population may put on limiting resources is its density, though, as we have seen, the straightforward density of a population need not be a good measure of the extent to which its individuals are crowded. Modular sessile organisms are particularly sensitive to competition from their immediate neighbors: they cannot withdraw from each other and space themselves more evenly or escape by dispersal or migration. Thus, when silver birch trees (*Betula pendula*) were grown in small groups, there were more suppressed and dying branches on the sides of individual trees where their branches shaded each other, whereas on the sides away from neighbors there was more vigorous growth (Figure 5.14).

crowding not density—especially in modular organisms

The inexact relationship between density and crowding notwithstanding, we saw in Section 3.5 that, over a sufficiently large density range, competition between individuals generally reduces the per capita birth rate as density increases, and increases the death rate, and that this effect is described as *density-dependent*. Thus, when birth and death rate curves are plotted against density on the same graph,

density-dependent birth and death and the carrying capacity

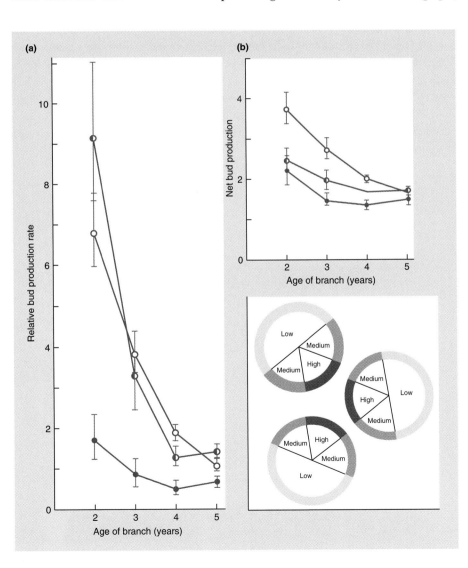

Figure 5.14
Mean relative bud production (new buds per existing bud) for silver birch trees (*Betula pendula*), expressed (a) as gross bud production and (b) as net bud production (birth minus death), in different interference zones. These zones are themselves explained in the inset. ●, high interference; ◐, medium; ○, low. Bars represent standard errors. (After Jones & Harper, 1987.)

and either or both are density-dependent, the curves must cross (Figure 5.15a–c). They do so at the density at which birth and death rates are equal, and because they are equal, there is no overall tendency at this density for the population either to increase or to decrease (ignoring, for convenience, both emigration and immigration). The density at the crossover point is called the *carrying capacity* and is denoted by the symbol K. At densities below K, births exceed deaths and the population increases. At densities above K, deaths exceed births and the population decreases. There is therefore an overall tendency for the density of a population under the influence of intraspecific competition to settle at K.

In fact, because of the natural variability within populations, the birth rate and death rate curves are best represented by broad lines, and K is best thought of not as a single density, but as a range of densities (Figure 5.15d). Thus, intraspecific competition does not hold natural populations to a single, predictable and unchanging level (K), but it may act upon a very wide range of starting densities and bring them to a much narrower range of final densities. It therefore tends to keep density within certain limits, and may thus be said to play a part in *regulating* the size of populations.

Of course, graphs like those in Figure 5.15 are generalizations on a grand scale. Many organisms, for example, have seasonal life cycles. For part of the year births vastly outnumber deaths, but a period of high juvenile mortality typically comes soon after the period of peak births. Most plants, for example, die as seedlings soon after germination. Thus, although births may balance deaths over the year, a population that is "stable" from year to year will often oscillate wildly over the

population regulation by competition—but not to a single carrying capacity

Figure 5.15
Density-dependent birth and mortality rates lead to the regulation of population size. When both are density-dependent (a), or when either of them is (b, c), their two curves cross. The density at which they do so is called the *carrying capacity* (K). However, the real situation is closer to that shown by thick lines in (d), where mortality rate broadly increases, and birth rate broadly decreases, with density. It is possible, therefore, for the two rates to balance not at just one density, but over a broad range of densities, and it is toward this broad range ('K') that other densities tend to move.

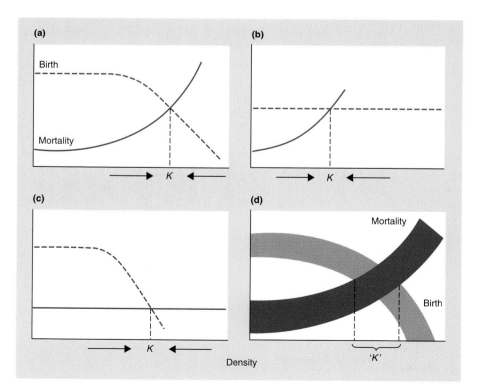

Box 5.3   Topical ECOncerns

Intraspecific Competition in Wild Oats

The wild oat *Avena fatua* is an important weed of cereal crops including winter wheat. As the negative slopes of the three lines in the figure below show, in experimental treatments it is subject to intraspecific competition: at higher initial densities in the soil, its rate of increase over a season decreases. A rate of "increase" of 1 means, in fact, that the population remains at the same size, neither increasing nor decreasing. The density at which this rate of 1 occurs is therefore the density to which the wild oat population is regulated.

The three lines (three different treatments) suggest three different stable, regulated densities. The highest is the *carrying capacity* (wild oats grown alone, subject

only to intraspecific competition). The stable density would be lower in the presence of the crop winter wheat (*interspecific* competition discussed in Chapter 6). It would be much lower still in the presence of winter wheat and treatment from a herbicide selective against wild oat. Of course, the more wild oats there are, the more of an adverse effect they will have on the winter wheat.

Is it sensible to attempt to eliminate wild oats entirely from crop fields? Under what circumstances do these results suggest that herbicide treatment would be sensible? What further information would you require to allow you to give more definite answers to these questions?

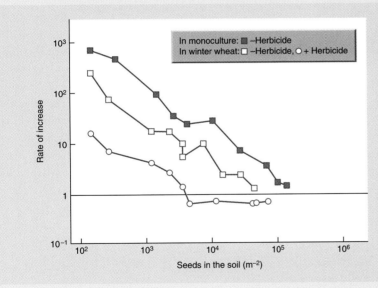

Density-dependent reductions in the rate of increase of wild oat *Avena fatua*, subject to three different experimental treatments. (After Manlove, 1985.)

seasons. Nonetheless, such generalizations can offer insights that are useful both fundamentally and also from a practical point of view (Box 5.3).

5.5.1 Patterns of population growth

When populations are sparse and uncrowded they may grow rapidly—it is only as crowding increases that density-dependent changes in birth and death rates start to take effect. In essence, populations at these low densities grow by simple multiplica-tion over successive intervals of time. This is *exponential* growth (Figure 5.16; see also Box 5.4) and the rate of increase is the population's *intrinsic rate of natural increase* (denoted by r). Of course, any population that behaved in this way would soon run out of resources, but as we have seen, the rate of increase tends to become reduced by competition as the population grows, and it falls to zero when the population reaches its carrying capacity (since birth rate then equals death rate). A steady reduction in the rate of increase as densities move toward the carrying capacity gives rise to population growth that is not exponential but S-shaped (Figure 5.16). The pattern is also often called *logistic* growth after the so-called logistic equation (Box 5.4).

The S-shaped curve can best be seen in action in laboratory studies of microor-ganisms or animals with very short life cycles such as grain beetles (Figure 5.17a). In these kinds of experiment it is easy to have experimental control of environmental conditions and resources. In the real world, outside the laboratory and the mind of the mathematician, the world is less simple. The complex life cycles of organisms, changing conditions and resources through the seasons, and the patchiness of habitats introduce many complications. In nature, populations often follow a very bumpy ride along the path of perfect logistic growth (Figure 5.17b), though not always (Figure 5.17c).

Another way to summarize the ways in which intraspecific competition affects populations is to look at *net recruitment*—the number of births minus the number of deaths in a population over a period of time. When densities are low, net recruitment will be low because there are few individuals available either to give birth or to die. Net recruitment will also be low at much higher densities as the carrying capacity is approached. Net recruitment will be at its peak, then, at some intermediate density. The result is a "humped" curve (Figure 5.18). Again, of course, as with the ideal logistic curve, real data from nature never fall on a single line. But the humped curve reflects the essence of net recruitment patterns when density-dependent birth and death are the result of intraspecific competition.

Figure 5.16

Exponential and S-shaped or *sigmoidal* increase in the size of a population over time. These patterns describe the growth to be expected in general in populations in the absence (exponential) and under the influence (sigmoidal) of intraspecific competition but are also generated, specifically, by the exponential and logistic equations shown (see also Box 5.4).

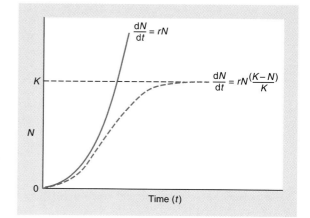

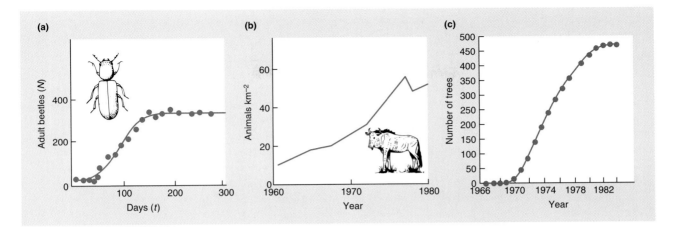

Figure 5.17

Real examples of S-shaped population increase. (a) The beetle *Rhizopertha dominica* in 10 g of wheat grains replenished each week. (After Crombie, 1945.) (b) The population of wildebeest *Connochaetes taurinus*, of the Serengeti region of Tanzania and Kenya seems to be leveling off after rising from a low density caused by the disease rinderpest. (After Sinclair & Norton-Griffiths, 1982; Deshmukh, 1986.) (c) The population of the willow tree (*Salix cinerea*) in an area of land after myxomatosis had effectively prevented rabbit grazing. (After Alliende & Harper, 1989.)

Wildebeest. Also known as gnu. An African antelope of the genus *Connochaetes* that forms large herds that graze on open plains and savannas and are continually on the move as they search for new areas to graze.

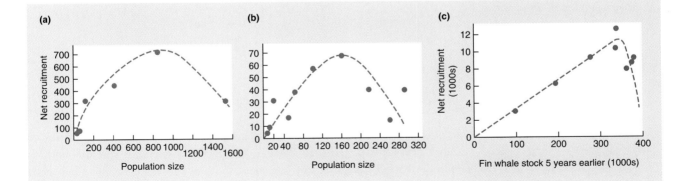

Figure 5.18

Some "humped" net-recruitment curves, drawn by eye through the data points shown. (a) The ring-necked pheasant on Protection Island after its introduction in 1937. (After Einarsen, 1945.) (b) An experimental population of the fruit fly *Drosophila melanogaster*. (After Pearl, 1927.) (c) Estimates for the stock of Antarctic fin whales. (After Allen, 1972.)

Box 5.4

The Exponential and Logistic Equations of Population Growth

In this box, simple mathematical models are derived for populations in the absence of, and under the influence of, intraspecific competition. These and other mathematical models play an important part in ecology (see Chapter 1). They help us to follow through the consequences of assumptions we may wish to make, and to explore the behavior of ecological systems that we may find it hard to observe in nature or construct in the laboratory. These particular models themselves form the basis for more complex models of *interspecific* competition and predation: they are important building blocks. It is essential to appreciate, however, that a pattern generated by such a model—for example, the S-shaped pattern of population growth under the influence of intraspecific competition—is not of interest, or important, because it is generated by the model. There are, no doubt, many other models that could generate very similar (indistinguishable) patterns. Rather, the point about the pattern is that it reflects important, underlying ecological processes—and the model is useful in that it appears to capture the essence of those processes.

We start with a model of a population in which there is no intraspecific competition and then incorporate that competition later. Our models are in the form of *differential equations,* describing the net rate of increase of a population, which will be denoted by dN/dt (referred to as dN by dt). This represents the speed at which a population increases in size, N, as time, t, progresses.

The increase in size of the whole population is the sum of the contributions of the various individuals within

it. Thus, the average rate of increase per individual, or the per capita rate of increase (*per capita* means per head) is given by $dN/dt(1/N)$. In the absence of intraspecific competition (or any other force that increases the death rate or reduces the birth rate) this rate of increase is a constant and as high as it can be for the species concerned. It is called the *intrinsic rate of natural increase* and denoted by r. Thus:

$$\frac{dN}{dt}\left(\frac{1}{N}\right) = r$$

and the net rate of increase for the whole population is therefore given by

$$\frac{dN}{dt} = rN$$

This equation describes a population growing *exponentially* (Figure 5.16). Intraspecific competition can now be added. This is done by deriving the *logistic equation*: a simple mathematical model developed by Pierre-François Verhulst (1804–1849), in a series of papers on "the law that a population follows during its growth." It was almost completely ignored until it was rediscovered in 1920 by Raymond Pearl, who used it to model the growth of the population of the United States.

The equation can be derived by the method set out in the figure on page 195. The net rate of increase per individual is unaffected by competition when N is very close to zero. It is still therefore given by r (point A). When

5.6 ▷ Life History Patterns

One of the ways in which we can try to make sense of the world around us is to search for repeated patterns. In doing so, we are not pretending that the world is simple or that all categories are watertight, but we can hope to move beyond a description that is no more than a series of unique special cases. This final section of the chapter decribes some simple, useful, though by no means watertight patterns linking different types of life history and different types of habitat.

N rises to K (the carrying capacity) the net rate of increase per individual is zero (point B). For simplicity, we assume a straight line between A and B; that is, we assume a linear reduction in the per capita rate of increase, as a result of intensifying intraspecific competition, between $N = 0$ and $N = K$.

Thus, on the basis that the equation for any straight line takes the form $y =$ intercept $+$ slope x, where x and y are the variates on the horizontal and vertical axes, here we have

$$\frac{dN}{dt}\left(\frac{1}{N}\right) = r - \frac{r}{K} \cdot N$$

or, rearranging,

$$\frac{dN}{dt} = rN\left[1 - \left(\frac{N}{K}\right)\right]$$

This is the logistic equation, and a population increasing in size under its influence is shown in Figure 5.16. It describes a sigmoidal growth curve approaching a stable carrying capacity, but it is only one of many reasonable equations that do this. Its major advantage is its simplicity. Nevertheless, it has played a central role in the development of ecology.

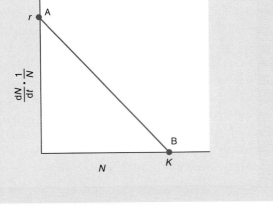

An ideal linear decline in the net rate of increase per individual with increasing population (N) (see text).

Earlier we made the point that in all types of life history there is typically a limited total amount of energy or some other resource available to an organism for growth and reproduction. Some trade-off may therefore be necessary: either grow more and reproduce less or reproduce more and grow less. Specifically, there may be an observable cost of reproduction in that when reproduction starts, or increases, growth may slow or stop completely, as resources are diverted from growth to reproduction. In many forest trees, for example, there are years when very heavy crops of seeds are produced: *mast years*. The trunks of trees are formed by annual

the "cost" of reproduction—a life history trade-off

increments: rings of growth. In mast years these are conspicuously narrower (Figure 5.19). We can, of course, look at this trade-off between growth and reproduction the other way around: an organism that makes vigorous growth, and so thrives in competition with its neighbors, may do so by delaying or reducing reproductive activity. Furthermore, the diversion of resources to present reproduction may also jeopardize subsequent survival (as in the salmon and foxgloves described earlier), or simply reduce the capacity for future reproduction.

But early reproduction can bring some striking rewards, particularly because the progeny themselves start reproduction earlier. The effect is shown by considering the life cycle of fruit flies (*Drosophila*). The number of eggs produced by a female in her lifetime is about 780. Doubling that number would clearly boost the intrinsic rate of increase, but such a massive increase in reproductive output is asking a great deal of an individual. So what other changes in the life history of *Drosophila* would have a similar effect? In fact, the same rise in the rate of increase would be attained by shortening the juvenile period by just over 1.5 days (reproducing sooner, rather than growing longer) in a total period of around 10 or so days. Populations of individuals that reproduce early in their lives can grow extremely fast—even if this means producing many fewer total offspring over their life than they would otherwise. Conversely, the rate of growth of populations can be slowed by delaying the onset of reproduction. One way in which the growth rate of human populations can be slowed down, for example (Chapter 11), is by discouraging early marriage and childbearing.

r and *K* species

The potential of a species to multiply rapidly—producing large numbers of progeny early in the life cycle—is advantageous in environments that are short-lived, allowing the organisms to colonize new habitats quickly and exploit new resources. This rapid multiplication is a characteristic of the life cycles of terrestrial organisms that invade disturbed land (for example, many annual weeds) or colonize newly opened habitats such as forest clearings and of the aquatic inhabitants of temporary puddles and ponds. These are species whose populations are usually found expanding after the last disaster or exploiting the new opportunity. They have the life cycle properties that are favored by natural selection in such conditions. They have been called *r* species, because they spend most of their lives in the near-exponential,

Figure 5.19

The negative correlation between cone crop size and annual growth increment for a population of Douglas fir *Pseudotsuga menziesii*. (After Eis et al., 1965.)

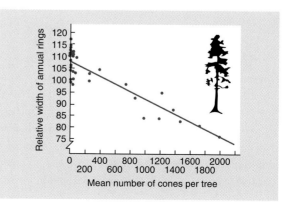

r-dominated phase of population growth (Box 5.4), and the habitats in which they are likely to be favored have been called *r*-selecting.

Organisms with quite different life histories survive in habitats where there is often intense competition for limited resources. The individuals that are successful in leaving descendants are those that have captured, and often held on to, the larger share of resources. Their populations are usually crowded and those that win in a struggle for existence do so because they have grown faster (rather than reproducing) or have spent more of their resources in aggression or some other activity that has favored them in competition with others. They are called *K* species because their populations spend most of their lives in the *K*-dominated phase of population growth (Box 5.4)—"bumping up" against the limits of environmental resources—and the habitats in which they are likely to be favored have been called *K*-selecting.

A further common distinction between *r* and *K* species is whether they produce many small progeny (characteristic of *r* species) or few large progeny (characteristic of *K* species). This is another example of a life history trade-off: an organism has limited resources available for reproduction, and natural selection will influence how these are packaged. In environments where rapid population growth is possible, those individuals that produce large numbers of small progeny will be favored. The size of progeny can be sacrificed because they will usually not be in competition with others. However, in environments in which the individuals are crowded and there is competition for resources, those progeny that are well provided with resources by the parent will be favored. Producing progeny that are well endowed requires the trade-off of fewer progeny (see, for example, Figure 5.20).

The *r/K* concept can certainly be useful in describing some of the general differences among different organisms. For instance, among plants it is possible to describe a number of very broad and general relationships (Figure 5.21). Trees in a forest are splendid examples of *K* species. They compete for light in the canopy, and survivors are those that put their resources into early growth and overtopping

r, *K*, and progeny size and number

evidence for the *r/K* scheme?

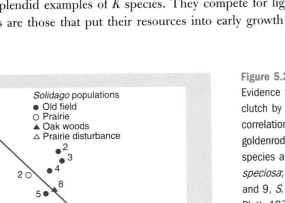

Figure 5.20

Evidence for a trade-off between the number of offspring produced in a clutch by a parent and the individual fitness of those offspring: a negative correlation between the number of propagules per stem produced by goldenrod plants (*Solidago*) and the weight of single propagules. The species are 1, *S. nemoralis*; 2, *S. graminifolia*; 3, *S. canadensis*; 4, *S. speciosa*; 5, *S. missouriensis*; 6, *S. gigantea*; 7, *S. rigida*; 8, *S. caesia*; and 9, *S. rugosa*, taken from a variety of habitats as shown. (After Werner & Platt, 1976.)

Figure 5.21
Figure 5.21
Broadly speaking, plants show some conformity with the *r/K* scheme. For example, trees in relatively *K*-selecting woodland habitats (a) have a relatively high probability of being iteroparous and a relatively small reproductive allocation; (b) have relatively large seeds; and (c) are relatively long lived with relatively delayed reproduction. (After Harper, 1977; from Salisbury, 1942; Ogden, 1968; Harper & White, 1974.)

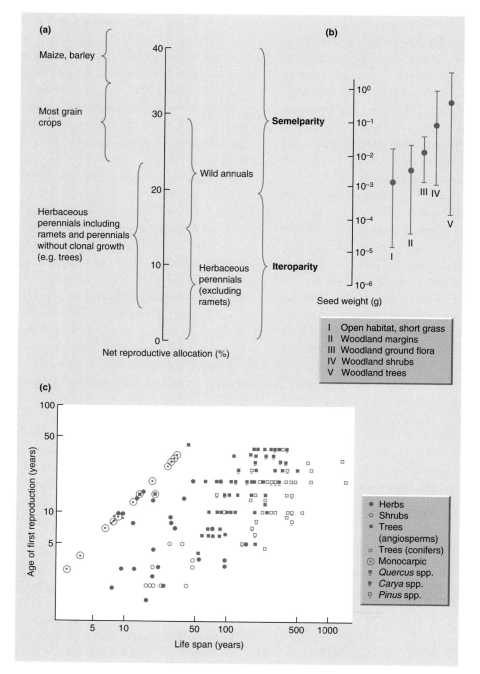

their neighbors. They usually delay reproduction until their branches have an assured place in the canopy of leaves. Once established they hold on to their position and usually have a very long life, with a relatively low allocation to reproduction overall but large individual seeds. By contrast, in more disturbed, open, *r*-selecting habitats, the plants tend to conform to the general syndrome of *r* characteristics: a greater reproductive allocation, but smaller seeds, smaller size, earlier reproduction, and a shorter life (Figure 5.21).

On the other hand, broad reviews of the studies that have been carried out tend to find as many examples that fail to fit the *r/K* scheme as examples that correspond. One might regard this as a damning criticism of the *r/K* concept, since it undoubtedly demonstrates that the explanatory powers of the scheme are limited. But it is equally possible to regard it as very satisfactory that a relatively simple concept can help make sense of a large proportion of the multiplicity of life histories. Nobody, though, can regard the *r/K* scheme as the whole story. Like all attempts to classify species and their characteristics into pigeonholes, the distinction between *r* and *K* species has to be recognized as a convenient (and hopefully useful) human creation rather than a fundamental statement about the living world.

Summary

Counting individuals, births, and deaths

Ecologists try to describe and understand the distribution and abundance of organisms. The processes that change the size of populations are birth, death, and movement. A population is a number of individuals, but for some kinds of organism, especially *modular* organisms, it is not always clear what we mean by an individual. Ecologists face enormous problems when they try to count what is happening to populations in nature. They almost always estimate rather than count. There are particular problems in counting modular organisms and the numbers of births and deaths.

Life cycles and reproduction

The life histories of all unitary organisms can be seen as variations around a simple, sequential pattern. Some organisms fit several or many generations within a single year, some breed predictably just once each year (annuals), and others (perennials) have a life cycle extended over several or many years. Some, *iteroparous* species, breed repeatedly; others, *semelparous* species, have a single reproductive episode followed quickly by death.

Most annuals germinate or hatch in spring, grow rapidly, reproduce, and then die before the end of summer. Most spend part of the year dormant. There is a marked seasonal rhythm in the lives of many long-lived species. Where there is very little seasonal variation, some reproduce throughout the year; others have a long nonreproductive phase and then one lethal burst of reproductive activity.

Monitoring birth and death: life tables and fecundity schedules

Life tables can be useful in identifying what in a life cycle is apparently most instrumental in determining rates of increase or decline. A cohort life table records the survivorship of members of a single cohort. When we cannot follow cohorts, it may be possible to construct a static life table, but great care is required. The fecundity of individuals also changes with age, described in age-specific fecundity schedules. Ecologists search for patterns of life and death that we can see repeated in the lives of many species. A useful set of survivorship curves (types 1–3) has been developed, but in practice patterns of survival are usually more complex.

Dispersal and migration

Dispersal is the way individuals spread away from each other. Migration is the mass directional movement of large numbers of a species from one location to another. Movement and spatial distribution are intimately related. Dispersal and migration can have a profound effect on the dynamics of a population and on its composition.

The impact of intraspecific competition on populations

Over a sufficiently large density range, competition between individuals generally reduces the birth rate as density increases and increases the death rate (i.e., is density-dependent). Intraspecific competition therefore tends to keep density within certain limits and may thus be said to play a part in regulating the size of populations.

When populations are sparse and uncrowded they tend to exhibit *exponential* growth, but the rate of increase tends to become reduced by competition as the population grows, giving rise to population growth that is not exponential but S-shaped or *logistic*. Intraspecific competition also affects *net recruitment*, typically resulting in a humped curve.

Life history patterns

There is typically a limited total amount of energy or some other resource available to an organism for growth and reproduction. There may be an observable cost of reproduction. But populations of individuals that reproduce early in their lives can grow extremely fast.

The potential of a species to multiply rapidly is favored by natural selection in environments that are short-lived, allowing the organisms to colonize new habitats quickly and exploit new resources. Such species have been called *r* species. Where there is often intense competition for limited resources, the individuals that are successful in leaving descendants are those that have captured the larger share of resources, often because they were born larger and/or have grown faster (rather than reproducing): so-called *K* species. The *r/K* concept can be useful in interpreting many of the differences in form and behavior of organisms, but of course it is not the whole story.

Review Questions

▲ = Challenge Question

1. Contrast the meaning of the word *individual* for unitary and modular organisms.

2. In a mark–recapture exercise during which a population of butterflies remained constant in size, an initial sample provided 70 individuals, each of which was marked and then released back into the population. Two days later, a second sample was taken, totaling 123 individuals of which 47 bore a mark from the first sample. Estimate the size of the population. State any assumptions that you have had to make in arriving at your estimate.

3. ▲ Define *annual, perennial, semelparous,* and *iteroparous.* Try to give an example of both an animal and a plant for each of the four possible combinations of these terms. In which cases is it difficult (or impossible) to come up with an example and why?

4. Contrast the derivation of *cohort* and *static life tables* and discuss the problems of constructing and/or interpreting each.

5. The following is an outline life table and fecundity schedule for a cohort of a population of sparrows. Fill in the missing values (wherever there is a question mark).

6. Describe what are meant by *aggregated, random,* and *regular* distributions of organisms in space, and outline, with actual examples where possible, some of the behavioral processes that might lead to each type of distribution.

7. ▲ Why is the average density of people in the United States lower than the density experienced by people, on average, in the United States? Is a similar contrast likely to apply to most species? Why? Under what conditions might it not apply?

8. ▲ Compare unitary and modular organisms in terms of the effects of intraspecific competition both on individuals and on populations.

9. What is meant by the *carrying capacity* of a population? Describe where it occurs in (a) S-shaped population growth, (b) the logistic equation, and (c) humped net-recruitment curves.

10. ▲ What is meant by the cost of reproduction? Contrast the role that it is likely to have played in the evolution of *r*-selected and *K*-selected species.

Stage (x)	Numbers at start of stage (a_x)	Proportion of original cohort alive at start of stage (l_x)	Mean no. of eggs produced per individual in stage (m_x)
Eggs	173	?	0
Nestlings	107	?	0
Fledgelings	64	?	0
1-year-olds	31	?	2.5
2-year-olds	23	?	3.7
3-year-olds	8	?	3.1
4-year-olds	2	?	3.5
$R = ?$			

Interspecific competition is one of the most fundamental phenomena in ecology, affecting not only the current distribution and success of species but also their evolution. Yet the existence and effects of interspecific competition are often remarkably difficult to establish and demand an armory of observational, experimental, and modeling techniques.

6

Interspecific Competition

Key Concepts

In this chapter you will

- appreciate the difficulty of distinguishing between the power and importance of interspecific competition in principle and in practice.

- distinguish between fundamental and realized niches.

- define the Competitive Exclusion Principle and understand its limitations.

- appreciate the potential role of the evolutionary effects of competition in species coexistence and the difficulty of proving that role.

- understand the nature of niche complementarity and the importance of scale in communities ddstructured by interspecific competition.

- appreciate the difficulties of determining the prevalence of current competition in nature, and of distinguishing between the effects of competition and mere chance.

6.1 ▶ Introduction

The discussion of *intraspecific* competition in previous chapters contributes to an understanding of *interspecific* competition. Its essence is that individuals of one species suffer a reduction in fecundity, survivorship, or growth as a result of exploitation of resources or interference by individuals from another species. These competitive effects on individuals are likely to affect the population dynamics of the competing species. These, in turn, can influence the species' distributions, and also their evolution. The distributions and abundances of species, of course, determine the compositions of the biological communities of which they are part. And evolution, in *its* turn, can itself influence the species' distributions and dynamics.

two separate questions—the possible and actual consequences of competition

This chapter, then, is about both the ecological and the evolutionary effects of interspecific competition on individuals, on populations, and on communities. But it also addresses a more general issue in ecology and indeed in science—that there is a difference between what a process can do and what it does do: a difference between what, in this case, interspecific competition is capable of doing and what it actually does in practice. These are two separate questions, and we must be careful to keep them separate.

The way these different questions can be asked and answered will be different, too. To find out what interspecific competition is capable of doing is relatively easy. Species can be *forced* to compete in experiments, or they can be examined in nature in pairs or groups chosen precisely because they seem most likely to compete. On the other hand, to discover how important interspecific competition is, actually, is much more difficult. It will be necessary to ask how realistic our experiments were, how typical they were of the way species interact in nature, and how typical of pairs and groups of species in general those singled out for special attention were. As will become evident, this is just one of those many areas of ecology, and of science in general, where clear-cut answers are as yet unavailable.

We begin, though, with some examples of what interspecific competition *can* do.

6.2 ▶ The Ecological Effects of Interspecific Competition

6.2.1 Competition between diatoms for silicate

Competition was investigated in the laboratory between two species of freshwater diatom (single-celled plants), *Asterionella formosa* and *Synedra ulna*, both of which require silicate in the construction of their cell walls (see Section 3.5). The population densities of the diatoms were monitored, but at the same time their impact on their limiting resource (silicate) was also being recorded. When either species was grown alone in a liquid medium to which resources were continuously being added, it established a steady population density while reducing the silicate to a constant low concentration (Figure 6.1a and b). However, in exploiting this resource, *Synedra*

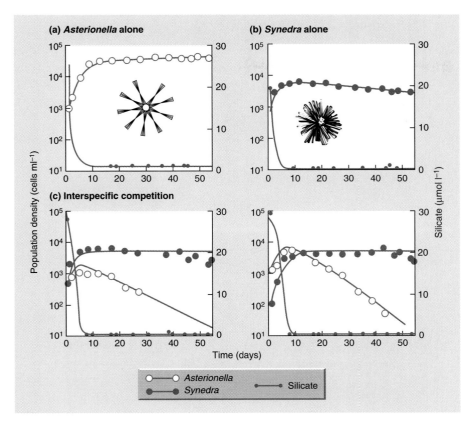

Figure 6.1
Competition between diatoms.
(a) *Asterionella formosa*, when grown alone in a culture flask, establishes a stable population and maintains a resource, silicate, at a constant low level. (b) When *Synedra ulna* is grown alone it does the same, but maintains silicate at an even lower level. (c) When grown together, in two replicates, *Synedra* drives *Asterionella* to extinction. (After Tilman et al., 1981.)

reduced the silicate concentration to a lower level than did *Asterionella*. Hence, when the two species were grown together, *Synedra* maintained the concentration at a level that was too low for the survival and reproduction of *Asterionella* and only *Synedra* survived (Figure 6.1c).

Thus, although both species were capable of living alone in the laboratory habitat, when they competed, *Synedra* excluded *Asterionella*, because it was the more effective exploiter of their shared, limiting resource.

a more effective diatom exploiter excludes a less effective one

6.2.2 Coexistence and exclusion of competing barnacles

Next, a field study in the early 1960s showed that two barnacle species, *Chthamalus stellatus* and *Balanus balanoides*, occurred in distinct zones along the intertidal range on rocky shores in Scotland, though adult *Chthamalus* generally occurred higher up on the shore (Figure 6.2). This raises several questions. Do the distinct distributions of these species reflect their reliance on equally distinct combinations of conditions and resources? Or is each species confined to its own zone by the effects of competition from the other species? Could *Chthalamalus*, for example, survive lower down the shore if *Balanus* were not there? Could *Balanus* survive in the higher zone in the absence of *Chthamalus*, or is it intolerant of the desiccation to

Figure 6.2
The intertidal distribution of adults and newly settled larvae of *Balanus balanoides* and *Chthamalus stellatus*, with a diagrammatic representation of the relative effects of desiccation and competition. (After Connell, 1961.) Photo: rock barnacles below (*Balanus*), encrusting barnacles above (*Chthamalus*).

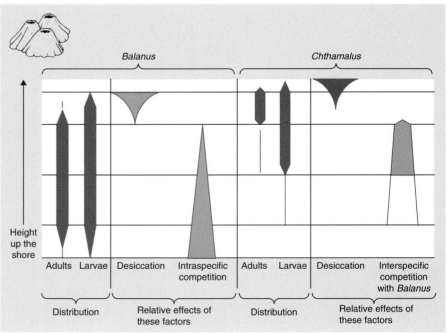

distributions of barnacles—competitive exclusion or intolerance of conditions?

which it would be subjected there? These questions were addressed by both careful observation and experimentation.

Young *Chthamalus* barnacles did settle in the *Balanus* zone, but they rarely survived to adulthood. In order to test whether this was simply an effect of the length of the period for which they were submerged in each tidal cycle, or of competition from *Balanus*, *Balanus* were physically removed from the area around these juvenile *Chthamalus*. Free of the influence of *Balanus*, each individual *Chthamalus* persisted, regardless of the tidal zone. By contrast, *Balanus* appears simply unable to survive the desiccating conditions in the highest tidal zone. Thus, it seems that the distribution of *Chthamalus* is the result of a competitive exclusion from the lower zone by *Balanus*, but the distribution of *Balanus* in the lower zone is a result

not of interspecific competition from *Chthamalus* but simply of its intolerance to
desiccation.

Further observation showed that *Balanus* actually grew over and crushed the
smaller *Chthamalus* individuals when the latter did become established in the *Balanus*
zone. The action of the competition in this case, then, was direct interference:
Balanus physically eliminated *Chthamalus* from the lower zones.

6.2.3 Some general observations

This example, and the previous one, illustrate several points of general importance.

- Competing species often coexist at one spatial scale but are found to have distinct
  distributions at a finer scale of resolution. Here, the barnacle species coexisted
  on the same shore but each was confined to its own zone within the shore.
- Species are often excluded by interspecific competition from locations at which
  they could exist perfectly well in the absence of interspecific competition. Here,
  Chthamalus can live in the *Balanus* zone—but only when there are no *Balanus*
  there. (Similarly *Asterionella* can live in laboratory cultures—but only when there
  are no *Synedra* there.)
- We can describe this by saying that the conditions and resources provided by the   **fundamental and realized niches**
  Balanus zone are part of the *fundamental* ecological niche of *Chthamalus* (see
  Section 3.6 for an explanation of ecological niches) in that the basic requirements
  for the existence of *Chthamalus* are provided there. But the *Balanus* zone does
  not provide a *realized* niche for *Chthamalus* when *Balanus* itself is present. (And
  the laboratory cultures provided the requirements of the fundamental niches of
  both *Synedra* and *Asterionella*, but those of the realized niche only of *Synedra*.)
- Thus, a species' fundamental niche is the combination of conditions and resources
  that allows that species to exist, grow, and reproduce when considered in isolation
  from any other species that might be harmful to its existence, whereas its realized
  niche is the combination of conditions and resources that allows it to exist, grow,
  and reproduce in the presence of specified other species that might be harmful
  to its existence—especially interspecific competitors.
- Competing species can therefore coexist when both are provided with a realized
  niche by their habitat (in the present case, the shore as a whole provided a realized
  niche for both *Balanus* and *Chthamalus*), but even in locations that provide a
  species with the requirements of its fundamental niche, that species may be
  excluded by another, superior competitor that denies it a realized niche there.
- Interspecific competition may, moreover, be very one-sided. In the present case,
  Chthamalus was excluded by *Balanus* from a zone that could provide *Chthamalus*
  with the requirements of its fundamental niche, but there was no competitive
  exclusion of *Balanus* by *Chthamalus*—the *Chthamalus* zone did not provide
  Balanus with the requirements of even its fundamental niche.
- Finally, the barnacle study illustrates the importance of experimental manipulation
  if we wish to discover what is really going on in a natural population—"nature"
  needs to be prodded to reveal its secrets.

6.2.4 Coexistence and exclusion of competing bumblebees

A related study, but one that focuses on species competing not through direct interference but through exploitation of shared resources, concerns two bumblebee species, *Bombus appositus* and *B. flavifrons*, in the Rocky Mountains of Colorado (Inouye, 1978). The bees live in the same area, but *B. appositus* forages primarily from the flowers of larkspur, *Delphinium barbeyi*, and *B. flavifrons* forages primarily from monkshood, *Aconitum columbianum*. However, each species appears to specialize only in the presence of the other species. When one or the other was removed, the remaining species quickly modified its behavior to include much more feeding from the previously less-preferred flower and it stayed longer at each visit (an indication of increased foraging success) (Figure 6.3).

The experiment therefore demonstrated that each bee's fundamental niche includes both species of flower: it is the presence of the competitor that restricts each bee's range to a realized niche of one flower type. Once again, therefore, the species coexist in the habitat as a whole but exclude one another from particular parts of it. This reciprocal competitive effect is almost certainly related to differences in the length of each species' proboscis. Although capable of foraging on either flower, *B. appositus* has a longer proboscis, making it better able to forage from larkspur, whose flower has a longer corolla. *B. appositus* therefore diminishes the nectar in larkspur

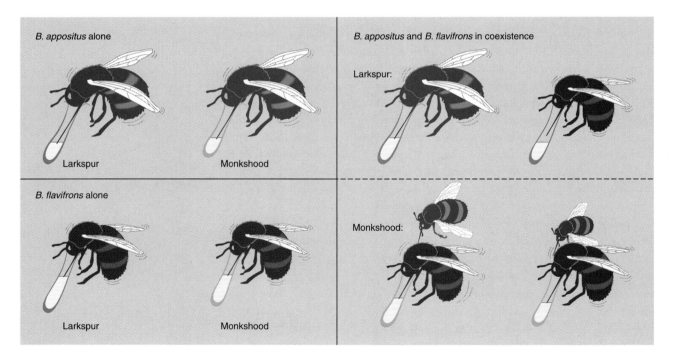

Figure 6.3

Bombus appositus and *B. flavifrons* both forage from both larkspur and monkshood if they are the only bumblebee present, but when they coexist, *B. appositus*, with a longer proboscis, excludes *B. flavifrons* from larkspur, whereas *B. flavifrons* specializes on monkshood, making it less attractive for *B. appositus*.

to a level that makes it unattractive for *B. flavifrons*. *B. flavifrons* consequently specializes on monkshood, making it less attractive for *B. appositus*.

In the case of the diatoms exploiting silicate, described previously, *Synedra*, which held the resource at the lower level when exploiting it alone, outcompeted *Asterionella*. Here with the bumblebees, we see that two species exploiting two resources (nectar in each of two flower species) can compete but still coexist when each species holds one of the resources at a level that is too low for effective exploitation by the other species.

coexisting competitors holding different resources at levels too low for effective exploitation by others

6.2.5 Coexistence of competing diatoms

In fact, a further experimental study of diatoms also shows two competing species coexisting on two shared, limiting resources. The two species were *Asterionella formosa* (again) and *Cyclotella meneghiniana*, and the resources that were both capable of limiting the growth of both diatoms were silicate and phosphate. However, whereas *Cyclotella* was the more effective exploiter of silicate, *Asterionella* was the more effective exploiter of phosphate. Thus, as we might now predict, in cultures with relatively balanced supplies of silicate and phosphate, the two diatoms coexisted, whereas *Cyclotella* excluded *Asterionella* when there were especially low supplies of silicate, and *Asterionella* excluded *Cyclotella* when there were especially low supplies of phosphate (Figure 6.4).

6.2.6 Coexistence of competing rodents and ants

The examples described so far have all involved pairs of closely related species—diatoms, barnacles, or bumblebees. This is potentially misleading in at least two important respects. First, competition may occur among larger groups of species

competition among groups of unrelated species

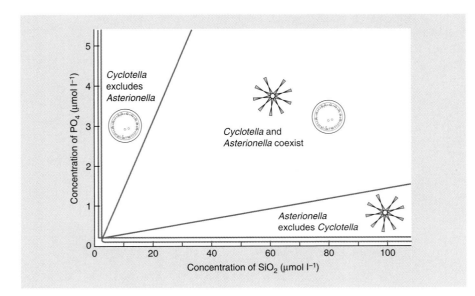

Figure 6.4

Asterionella formosa and *Cyclotella meneghiniana* coexist when there are roughly balanced supplies of silicate and phosphate, but *Asterionella* excludes *Cyclotella* when there are especially low supplies of phosphate, whereas *Cyclotella* excludes *Asterionella* when there are especially low supplies of silicate. (Simplified from Tilman, 1982.)

than just a pair—where it is sometimes, therefore, called "diffuse" competition. And second, competition may occur between completely unrelated species.

Both points are illustrated by a study of interspecific competition involving seed-eating ants and seed-eating rodents in deserts of the southwestern United States. At the study sites, only two *guilds* (groups of species that feed on similar foods in a similar fashion) fed on seeds: the rodents and the ants. In studying the size of the seeds harvested by each guild, it became apparent that the two exhibited significant overlap in the size of the seeds they ate (Figure 6.5). Ants did eat a larger proportion of the smallest seeds, but overall the potential for resource competition between them was very high.

As indicated earlier, however, the only true test for whether competition occurs between them would be to manipulate the abundance of each competitor and observe the response of its counterpart. Consequently, eight plots were established in similar habitats. In two, rodents were trapped and excluded by fencing, to ensure that only ants now had access to the seeds within. In another two, ants were eliminated by repeated applications of pesticides. In two further plots both ants and rodents were excluded, and finally two plots were maintained as unmanipulated controls.

When either rodents or ants were removed, there was a statistically significant increase in the numbers of the other guild: the depressive effect of interspecific competition from each guild on the abundance of the other was apparent. Also, when rodents were removed, the ants ate as many seeds as the rodents and ants had previously eaten together—as did the rodents when the ants were removed; only when both were removed did the amount of resource increase. In other words, under normal circumstances both guilds eat less and achieve lower levels of abundance than they would if the other guild were absent. This clearly indicates that rodents and ants, although they coexist in the same habitat, compete interspecifically with one another.

Figure 6.5

The diets of ants and rodents overlap: sizes of seeds harvested by coexisting ants and rodents near Portal, Arizona. (After Brown & Davidson, 1977.)

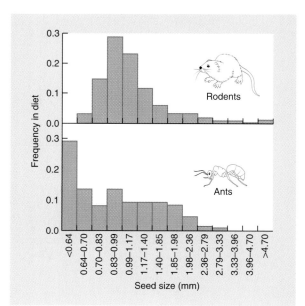

6.2.7 The Competitive Exclusion Principle

The patterns that are apparent in these examples have also been uncovered in many others and have been elevated to the status of a principle: the *Competitive Exclusion Principle* or *Gause's Principle* (named after an eminent Russian ecologist). It can be stated as follows:

- If two competing species coexist in a stable environment, then they do so as a result of niche differentiation, i.e., differentiation of their realized niches.
- If, however, there is no such differentiation, or if it is precluded by the habitat, then one competing species will eliminate or exclude the other.

Although the principle has emerged here from a contemplation of patterns evident in real sets of data, its establishment was—and many modern discussions of interspecific competition still are—bound up with a simple mathematical model of interspecific competition, usually known by the names of its two (independent) originators as the Lotka–Volterra model (Box 6.1).

There is no question that there is some truth in the principle: that competitor species *can* coexist as a result of niche differentiation, and that one competitor species *may* exclude another by denying it a realized niche. But it is crucial also to be aware of what the Competitive Exclusion Principle does *not* say.

The Competitive Exclusion Principle—what it does and does not say

It does *not* say that whenever we see coexisting species with different niches it is reasonable to jump to the conclusion that this is the principle in operation. All species, on close inspection, have their own unique niches. Niche differentiation does not prove that there are coexisting competitors. The species may not be competing at all and may never have done so in their evolutionary history. We require *proof* of interspecific competition. In the preceding examples, this was provided by experimental manipulation—remove one species (or one group of species) and the other species increases its abundance or its survival. But most of even the more plausible cases for competitors coexisting as a result of niche differentiation have not been subjected to experimental proof. So just how important is the Competitive Exclusion Principle in practice? We return to this question in Section 6.5.

Part of the problem is that although species may not be competing now, their ancestors may have competed in the past, so that the mark of interspecific competition is left imprinted on the niches, the behavior, or the morphological characteristics of their present-day descendants. This particular question is taken up in Section 6.3.

Finally, the Competitive Exclusion Principle, as stated earlier, includes the word *stable*. That is, in the habitats envisaged in the principle, conditions and the supply of resources remain more or less constant—if species compete, then that competition runs its course, either until one of the species is eliminated or until the species settle into a pattern of coexistence within their realized niches. Sometimes this is a realistic view of a habitat, especially in laboratory or other controlled environments where the experimenter holds conditions and the supply of resources constant. However, most environments are not stable for long periods. How does the outcome of

(continued on page 214)

Box 6.1

The Lotka–Volterra Model of Interspecific Competition

The Lotka–Volterra model of interspecific competition (Volterra, 1926; Lotka, 1932) is an extension of the logistic equation described in Box 5.4. Like those of the logistic, its advantages are its simplicity and its capacity, in this case, to shed light on the factors that determine the outcome of a competitive interaction.

The logistic equation

$$\frac{dN}{dt} = rN\frac{(K - N)}{K}$$

contains, within the parentheses, a term responsible for the incorporation of intraspecific competition. The basis of the Lotka–Volterra model is the replacement of this term by one that incorporates both intra- and interspecific competition. In it, the population size of one species can be denoted by N_1, and that of a second species by N_2. Their carrying capacities and intrinsic rates of increase are K_1, K_2, r_1, and r_2.

Suppose that 10 individuals of species 2 have, between them, the same competitive, inhibitory effect on species 1 as does a single individual of species 1. The total competitive effect on species 1 (intra- and interspecific) will then be equivalent to the effect of $(N_1 + N_2/10)$ species 1 individuals. The constant (1/10 in the present case) is called a *competition coefficient* and is denoted by α_{12} (alpha one–two). Thus multiplying N_2 by α_{12} converts it to a number of N_1-equivalents. (Note that $\alpha_{12} < 1$ means that individuals of species 2 have less inhibitory effect on individuals of species 1 than individuals of species 1 have on others of their own species, and so on.)

The crucial element in the model is the replacement of N_1 in the parentheses of the logistic equation with a term signifying "N_1 plus N_1-equivalents," i.e.

$$\frac{dN_1}{dt} = r_1N_1\frac{(K_1 - N_1 - \alpha_{12}N_2)}{K_1}$$

and in the case of the second species

$$\frac{dN_2}{dt} = r_2N_2\frac{(K_2 - N_2 - \alpha_{21}N_1)}{K_2}$$

These two equations constitute the Lotka–Volterra model.

To appreciate its properties, we ask the question,

Under what circumstances does each species increase or decrease in abundance? In order to answer, it is necessary to construct diagrams, in Figure 6.6, in which all possible combinations of N_1 and N_2 can be displayed. Certain combinations give rise to increases in species 1 and/or species 2, whereas other combinations give rise to decreases. There must also therefore be a *zero isocline* for each species (a line along which there is neither increase nor decrease) with combinations leading to increase on one side of it and combinations leading to decrease on the other.

In order to draw a zero isocline for species 1, we can use the fact that on the zero isocline $dN_1/dt = 0$ (by definition). Rearranging the equation, this gives us, as the zero isocline for species 1,

$$N_1 = K_1 - \alpha_{12}N_2$$

Below and to the left of this, numbers of both species are relatively low, and species 1, subjected to only weak competition, increases in abundance (arrows in the figure, representing this increase, point from left to right, since N_1 is on the horizontal axis). Above and to the right of the line, numbers are high, competition is strong, and species 1 decreases in abundance (arrows from right to

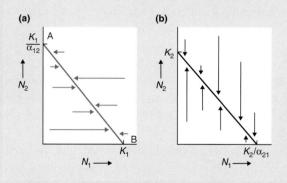

Figure 6.6
The zero isoclines generated by the Lotka-Volterra competition equations. (a) The N_1 zero isocline: species 1 increases below and to the left of it, and decreases above and to the right of it. (b) The equivalent N_2 zero isocline.

left). Based on an equivalent derivation, the figure also shows the species 2 zero isocline, with arrows, like the N_2-axis, running vertically.

In order to determine the outcome of competition in this model, it is necessary to ascertain, at each point on a figure, the behavior of the joint species 1–species 2 population, as indicated by the pair of arrows. There are, in fact, four different ways in which the two zero isoclines can be arranged relative to one another, and the outcome of competition will be different in each case. The different cases can be defined and distinguished by the intercepts of the zero isoclines (Figure 6.7).

In Figure 6.7a, for instance,

$$\frac{K_1}{\alpha_{12}} > K_2 \qquad \text{and} \qquad K_1 > \frac{K_2}{\alpha_{21}}$$

or:

$$K_1 > K_2\alpha_{12} \qquad \text{and} \qquad K_1\alpha_{21} > K_2$$

The first inequality ($K_1 > K_2\alpha_{12}$) indicates that the inhibitory intraspecific effects that species 1 can exert on itself are greater than the interspecific effects that species 2 can exert on species 1. The second inequality, how-

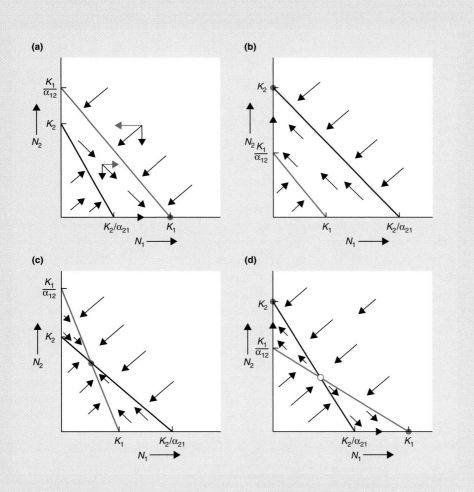

Figure 6.7

The outcomes of competition generated by the Lotka–Volterra competition equations for the four possible arrangements of the N_1 and N_2 zero isoclines. Vectors generally refer to joint populations, and are derived as indicated in (a). The solid circles show stable equilibrium points. The open circle in (d) is an unstable equilibrium point. For further discussion, see text.

(continues)

Box 6. 1

The Lotka–Volterra Model of Interspecific Competition (continued)

ever, indicates that species 1 can exert more of an effect on species 2 than species 2 can on itself. Species 1 is thus a strong interspecific competitor, whereas species 2 is a weak interspecific competitor, and as the arrows in Figure 6.7a show, species 1 drives species 2 to extinction and attains its own carrying capacity. The situation is reversed in Figure 6.7b. Hence Figures 6.7a and b describe cases in which the environment is such that one species invariably outcompetes the other.

In Figure 6.7c,

$$K_1 > K_2\alpha_{12} \qquad \text{and} \qquad K_2 > K_1\alpha_{21}$$

In this case, both species have less competitive effect on the other species than they have on themselves. This would happen, for example, if there were niche differentiation between the species. The outcome, as Figure 6.7c shows, is a stable, equilibrium combination of the two

species, which all joint populations tend to approach: that is, stable coexistence of competitors.

Finally, in Figure 6.7d:

$$K_2\alpha_{12} > K_1 \qquad \text{and} \qquad K_1\alpha_{21} > K_2$$

Thus individuals of both species compete more strongly with individuals of the other species than they do among themselves. This will occur, for instance, when each species is more aggressive toward individuals of the other species than toward individuals of its own species. The consequence is that there are two alternative stable points. At the first, species 1 reaches its carrying capacity with species 2 extinct; at the second, species 2 reaches its carrying capacity with species 1 extinct. Which of these is actually attained is determined by the initial densities: the species that has the initial advantage will drive the other species to extinction.

competition change when environmental heterogeneity in space and time is take into consideration? This is the subject of the next section.

6.2.8 Environmental heterogeneity

competition may only rarely "run its course"

As explained in previous chapters, spatial and temporal variation in environmen is the norm rather than the exception. Environments are usually a patchwork o favorable and unfavorable habitats; patches are often only available temporarily; an patches often appear at unpredictable times and in unpredictable places. Under suc variable conditions, competition may only rarely "run its course," and the outcom cannot be predicted simply by application of the Competitive Exclusion Principl A species that is a "weak" competitor in a constant environment might, for exampl be good at colonizing open gaps created in a habitat by fire, or a storm, or th hoofprint of a cow in the mud—or it may be good at growing rapidly in such gap immediately after they are colonized. It may then coexist with a strong competito as long as new gaps occur frequently enough. Thus, a realistic view of interspecif competition must acknowledge that it often proceeds not in isolation, but under th influence of, and within the constraints of, a patchy, impermanent, or unpredictab world.

The following examples illustrate just two of the many ways in which environmen tal heterogeneity ensures that the Competitive Exclusion Principle is very far from being the whole story when it comes to determining the outcome of an interactio between competing species.

The first concerns the coexistence of the sea palm *Postelsia palmaeformis* (a brown alga) and the mussel *Mytilus californianus* on the coast of Washington in the United States (Paine, 1979; Figure 6.8). *Postelsia* is an annual plant, which must reestablish itself each year in order to persist at a site. It does so by attaching to the bare rock, usually in gaps in the mussel bed created by wave action. However, the mussels themselves slowly encroach on these gaps, gradually filling them and precluding colonization by *Postelsia*. In other words, in a stable environment, the mussels would outcompete and exclude *Postelsia*. But their environment is not stable—gaps are frequently being created. It turns out that these species coexist only at sites in which there is a relatively high average rate of gap formation (at least 7 percent of surface area per year), and in which this rate is approximately the same each year. Where the average rate is lower, or where it varies considerably from year to year, there is (either regularly or occasionally) a lack of bare rock for colonization by *Postelsia*. At the sites of coexistence, on the other hand, although *Postelsia* is eventually excluded from each gap, these are created with sufficient frequency and regularity for there to be coexistence in the site as a whole. In short, there is coexistence of competitors—but not as a result of niche differentiation.

mussels, sea palms, and the frequency of gap formation

Figure 6.8

On shores in which gaps are not created, mussels are able to exclude the brown alga *Postelsia*, but where gaps are created regularly enough, the two species coexist, even though *Postelsia* is eventually excluded by the mussels from each gap. Photo: *Postelsia palmaeformis*.

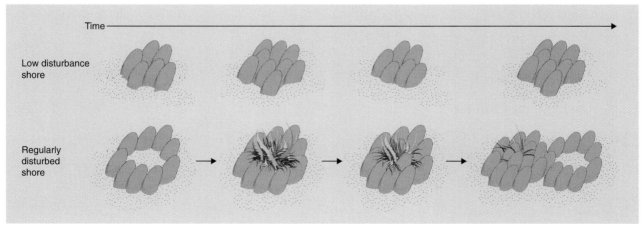

Many environments, by their very nature, are not simply variable but ephemeral (short-lived). Among the more obvious examples are decaying corpses (carrion), dung, rotting fruit and fungi, and temporary ponds. But note too that a leaf or an annual plant can be seen as an ephemeral patch, especially if it is palatable to its consumer for only a limited period. Often these ephemeral patches have an unpredictable life span: a piece of fruit and its attendant insects, for instance, may be eaten at any time by a bird. In these cases, it is easy to imagine the coexistence of two species: a superior competitor and an inferior competitor that reproduces earlier than the former.

coexistence of the good and the fast

One example concerns two species of pulmonate snail, *Physa gyrina* and *Lymnaea elodes,* living in ponds in northeastern Indiana, in the United States (Brown, 1982; Figure 6.9). Artificially altering the density of one or the other species in the field showed that the fecundity of *Physa* was significantly reduced by interspecific competition from *Lymnaea,* but the effect was not reciprocated. *Lymnaea* was clearly the superior competitor when competition continued throughout the summer. Yet *Physa* reproduced earlier and at a smaller size than *Lymnaea,* and in the many ponds that dried up by early July, it was often the only species to have produced drought-resistant eggs by that time. The species therefore coexisted in the area as a whole, in spite of *Physa*'s apparent inferiority, and as a result not of niche differentiation, but of the differing effects on the species of a world composed of ephemeral patches.

These studies, and others like them, go a long way toward explaining the co-occurrence of species that in constant environments would probably exclude one another. The environment is rarely unvarying enough for competitive exclusion to

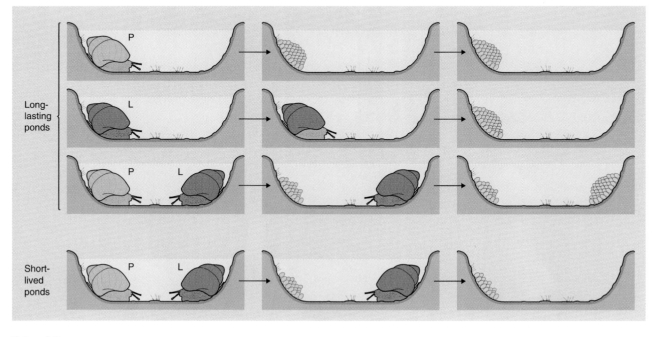

Figure 6.9
Lymnaea is the better competitor, but only *Physa* reproduces at all in ponds that dry up quickly in the summer. Photo: *Lymnaea.*

run its course. Rather, competitive balances are often shifting, with one species and then another being favored by daily, seasonal, and longer-term fluctuations in conditions and resources. Or, when new patches are created or old ones destroyed, competition may simply, for a time, be unimportant.

6.3 ▶ Evolutionary Effects of Interspecific Competition

Putting to one side the fact that environmental heterogeneity ensures that the forces of interspecific competition are often much less profound than they would otherwise be, it is nonetheless the case that the potential of interspecific competition to affect individuals adversely is considerable, and that this potential is often realized. We have seen in Chapter 2 that natural selection in the past will have favored those individuals that, by their behavior, physiological characteristics, or morphological features, have avoided adverse effects acting on other individuals in the same population. The adverse effects of extreme cold, for example, may have favored individuals with an enzyme capable of functioning effectively at low temperatures. Or, in the present context, the adverse effects of interspecific competition may have favored those individuals that, by their behavioral, physiological, or morphological characteristics, avoided those competitive effects.

the evolutionary avoidance of competition

We can, therefore, expect species to have evolved characteristics that ensure that they compete less, or not at all, with members of other species. How will this look to us at the present time? Coexisting species, with an apparent potential to compete, will exhibit differences in behavioral, physiological, or morphological features that ensure that they compete little or not at all. Connell (1980) has called this line of reasoning to account for differences between coexisting species "invoking the ghost of competition past."

invoking the ghost of competition past

Yet the pattern it predicts is precisely the same pattern supposed by the Competitive Exclusion Principle to be a prerequisite for the coexistence of species that *do* still compete. Coexisting present-day competitors, and coexisting species that have evolved an avoidance of competition, can look, at least superficially, the same.

The question of how important either past *or* present competition is, actually, as a force structuring natural communities will be addressed in the last section of this chapter (6.5). For now, we examine some examples of what interspecific competition *can* do as an evolutionary force. Note, however, that in invoking something that cannot be observed directly (evolution), it may be impossible to *prove* an evolutionary effect of interspecific competition, in the strict sense of "proof" that can be applied to mathematical theorems or carefully controlled experiments in the laboratory. Nonetheless, examples will be considered in which an evolutionary (rather than an ecological) effect of interspecific competition is the most reasonable explanation for what is observed.

UNANSWERED QUESTION: the difficulty of distinguishing ecological and evolutionary effects

6.3.1 Israeli rodents

In Israel, two small rodents have colonized the coastal region, but one (*Meriones tristrami*) approached from the north and the other (*Gerbillus allenbyi*) approached from the south. The Mount Carmel ridge divides the coastal region of Israel. Although it has not proved a barrier to *Meriones*, which now occupies the coastal area both north and south of the ridge, it has done so to *Gerbillus*, which only occupies the dunes south of Mount Carmel. Thus, although the dune habitats are nearly identical north and south of the divide, there is now a natural "experiment" in which in one series of study sites (to the north) *Meriones* occurs without its potential competitor, whereas in another (to the south) the two rodents occur together.

North of Mount Carmel, *Meriones* is found on a wide variety of sand and soil types, including the coastal sand dunes. South of Mount Carmel, however, although it occurs on several soil types, it does not occupy the coastal dunes. Only *Gerbillus* occurs there. Is *Meriones* excluded by its presence? Apparently not. When *Gerbillus* was removed from experimental plots, the abundance of *Meriones* remained unchanged. It seems that south of Mount Carmel, *Meriones* has evolved to select those habitats in which it avoids competition with *Gerbillus*, and that even in the absence of *Gerbillus* it retains this preference. There is a differentiation of the realized niches of the two species to the south of Mount Carmel, but this is also a differentiation of their fundamental niches—an evolutionary rather than an ecological effect of interspecific competition (Abramsky & Sellah, 1982; Figure 6.10).

6.3.2 Canadian sticklebacks

The three-spined stickleback (*Gasterosteus aculeatus*) is a marine species, but populations were "left behind" in several freshwater lakes in British Columbia either when land in the area uplifted after deglaciation (about 12,500 years ago) or after the subsequent rise and fall of sea levels (about 11,000 years ago). Following these two opportunities for invasion, some lakes now support two "species" (though they have not been given separate names) while others support only one. Wherever there are two species, one is always associated with the open water (it is limnetic), the other with the lake bottom (benthic). The first concentrates its feeding on plankton and has correspondingly long (and closely spaced) gill rakers that sieve the plankton from the flow of ingested water. The second, with much shorter gill rakers, concentrates on larger prey, which it consumes largely from vegetation or sediments. Wherever there is only one species in a lake, however, this exploits both kinds of food resource and is morphologically intermediate (Figure 6.11).

It is most unlikely to be by chance that whenever there have been two successful invasions of a lake the species just happen to be so different, whereas whenever there has been one invasion, the species is morphologically and behaviorally intermediate. There are, however, two plausible explanations for these patterns—though both are "historical" and both involve interspecific competition. First, it may be that a second successful invasion was only possible when the lake resident from the first invasion was either a limnetic or a benthic specialist, whereas if the resident was a generalist there was insufficient empty niche space for a second species to invade and avoid being competitively excluded.

On the other hand, why should the first invaders have evolved into generalists in some lakes and specialists in other, similar lakes? A more likely alternative, therefore, is that following a second invasion, small initial differences between species became exaggerated over evolutionary time in response to interspecific competition,

character displacement in sticklebacks—and Darwin's finches

Observation:

Experiment: **NORTH**   **SOUTH**

Figure 6.10
The distribution across habitats of *Meriones* is different south of Mount Carmel in the presence of *Gerbillus* from that in the north, in its absence, and this does not change if *Gerbillus* is excluded artificially to the south. Photo: *Meriones*.

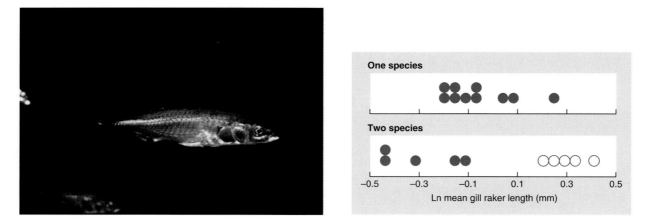

Figure 6.11

Character displacement in three-spined sticklebacks (*Gasterosteus aculeatus*). In small lakes in coastal British Columbia supporting two stickleback species, the gill rakers of the benthic species (filled circle, below) are significantly shorter than those of the limnetic species (open circle, below), whereas those of species of sticklebacks that occupy comparable lakes alone (above) are intermediate in length. Lengths of gill rakers have been adjusted to take account of species differences in size overall. (After Schluter & McPhail, 1993.) Photo: 3-spined stickleback (*Gasterosteus*).

until now there is clear separation both behaviorally and morphologically (called *character displacement*). Moreover, because the differentiation here is associated with character displacement (morphological change), it is particularly suggestive of an *evolutionary* rather than an ecological response to competition. (Another strong candidate for the evolutionary effects of interspecific competition, especially because of its association with character displacement, is provided by "Darwin's" finches of the genus *Geospiza* living on the Galapagos Islands, discussed in Chapter 2.)

6.4 > Interspecific Competition and Community Structure

Interspecific competition, then, has the potential either to keep apart (Section 6.2) or to drive apart (Section 6.3) the niches of coexisting competitors. How can these forces express themselves when it comes to the role of interspecific competition in molding the shape of whole ecological communities—who lives where and with whom?

6.4.1 Granivorous ants

We begin by returning to the study of seed-eating rodents and ants in the deserts of the southwestern United States, but this time concentrating on competition among some of the ant species. In an experiment, eight of the species were presented with artificially produced seeds and seed fragments of various sizes, and the sizes of worker ants and the sizes of the seeds they were carrying were noted. It was apparent that species tend to specialize (though not exclusively) on different sizes of seeds, depending on the size of their own bodies. Thus, the largest ants tend to specialize on the largest seeds, and so on (Figure 6.12).

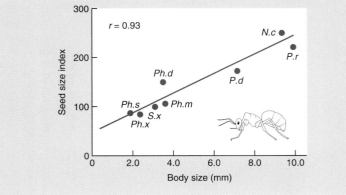

Figure 6.12
The relationship between worker body length and seed size index for experiments with eight species of seed-eating ants near Rodeo, New Mexico. Species are as follows: *N.c., Novomessor cockerelli; Ph.d., Pheidole desertorum; Ph.m., Pheidole militicida; Ph.s., Pheidole sitarches; Ph.x., Pheidole xerophila; P.d., Pogonomyrmex desertorum; P.r., Pogonomyrmex rugosus; S.x, Solenopsis xyloni.* (After Davidson, 1977.)

In addition, the ants follow one or the other of two different foraging strategies: group and individual foraging. Group foragers move together in well-defined columns, searching exhaustively but only in a small portion of the area surrounding the nest. Individual foragers search for their food independently of one another, and, as a result, a larger area surrounding the colony is searched, but less exhaustively. It was possible to demonstrate that group foraging is a more efficient strategy when seeds are clumped and densities are high. But when seeds are at low density and are more evenly distributed, individual foragers are more efficient. In other words, different types of foraging strategy are most effective in different types of habitat.

The occurrence of ant species at five different field sites can be examined (Figure 6.13), bearing these differences in size and foraging strategy in mind. Almost without exception, when species of similar size coexist at a site, they differ in foraging strategy, and when species of similar foraging strategy coexist at a site, they differ in size. The only apparent exception is the coexistence of *Pogonomyrmex desertorum* and *P. maricopa* at site A, and of these the latter occurs only rarely. This, though, does not detract from the general conclusion that when several species of granivorous ant coexist at a site, each specializes in a different way in its utilization of the food resource, and that the guild as a whole demonstrates *niche complementarity:* that is, niche differentiation involves several niche dimensions, and coexisting species that occupy a similar position along one dimension tend to differ along another dimension. Thus, it appears that although in the guild of granivorous ants there is overlap in resource utilization and interspecific competition for food, there is coexistence only between species that differ in size or foraging strategy or both. Species that do not differ in at least one of these respects are apparently unable to coexist.

"niche complementarity," demonstrated by seed-eating ants

6.4.2 Bumblebees in Colorado

Another study of niche differentiation and coexistence examined a number of species of bumblebee in Colorado (two of which were discussed in Section 6.2.4) in which differences in resource utilization appeared to be actively maintained by current competition. Visits were made every 8 days during the summer to 17 sites along an altitudinal gradient (2860 to 3697 m), and on each visit the numbers of each of seven common bumblebee species visiting the flowers of various plant species were

Figure 6.13
Mean worker body lengths of seed-eating ants at five sites in New Mexico and Arizona. Species designations as for Figure 6.12, plus *P.b.,* *Pogonomyrmex barbatus;* *P.c.,* *Pogonomyrmex californicus;* *P.m.,* *Pogonomyrmex maricopa;* *P.p.,* *Pogonomyrmex pima;* *V.p.,* *Veromessor pergandei.* G, group forager; I, individual forager. Species designated by open circles occur only rarely. (After Davidson, 1977.)

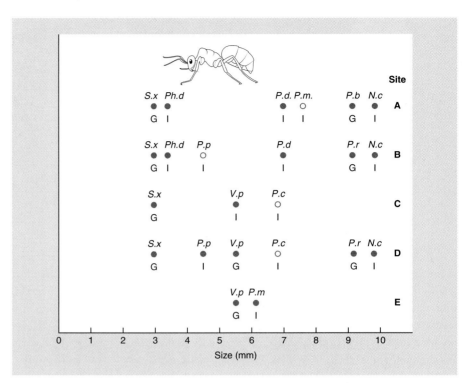

recorded. The bumblebee species fell into four groups in terms of both their proboscis length and the corolla length of the plants they visited preferentially (Figure 6.14). The long-proboscis bumblebees (*Bombus appositus* and *B. kirbyellus*) clearly favored plants with long corollas, particularly *Delphinium barbeyi*. The short-proboscis species (*B. sylvicolla,* *B. bifarius,* and *B. frigidus*) fed most frequently on various flower species of the family Compositae and on *Epilobium angustifolium*, all of which possess quite short corollas. The medium-proboscis species, *B. flavifrons*, fed over the entire range of corolla lengths. Finally, another short-proboscis species, *B. occidentalis*, fed as expected on plants with short corollas but was also able to obtain nectar from long corollas by using its large, powerful mandibles to bite through the bases of the corollas and "rob" them of their nectar. For this reason it was placed in a group of its own.

niche complementarity in nectar-feeding bumblebees

There was a clear tendency, in any single locality on the altitudinal gradient, for the bumblebee community to be dominated by one long-proboscis species (*B. appositus* at low altitude, *B. kirbyellus* at high), one medium-proboscis species (*B. flavifrons*), and one short-proboscis species (*B. bifarius* at low, *B. frigidus* at intermediate, and *B. sylvicola* at high altitude). In addition, the nectar-robbing *B. occidentalis* was present at sites where its preferred and exclusive nectar source was available (the plant *Ipomopsis aggregata*). Thus, the bumblebees also demonstrate niche complementarity. *B. occidentalis* differs from the other six species in terms of food-plant species and feeding method (nectar robbing). Among the others, species differ from each other either in the corolla length (and thus proboscis length), or in the altitude (i.e., habitat) favored, or in both.

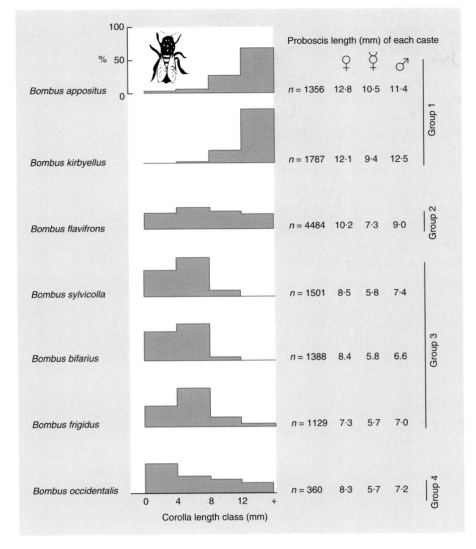

Figure 6.14
Percentage of all observations of each bumblebee species on plants in four classes of flower size. The total number of observations (*n*) is indicated in each case. Also shown for each species are the mean proboscis lengths of each caste (♀, queen; ☿, sterile female worker; ♂, male). The seven species can be divided into three groups according to proboscis length. *Bombus occidentalis* is placed into a fourth group because of its unique mandible structure. (After Pyke, 1982.)

These studies, of ants and bumblebees, illustrate an important point: insofar as interspecific competition plays a role in structuring communities, it tends to do so not by affecting some random sample of the members of that community, nor by affecting every member, but by acting within guilds—groups of species exploiting the same class of resource in a similar fashion.

6.4.3 Canadian grasses and legumes

Our perception of the role of interspecific competition in molding community structure depends, as already pointed out for barnacles, on the scale at which it is examined. On a relatively small scale, we tend to see the co-occurrence of only a few species with complementary ecological niches. But if the scale is broadened, the community appears to contain more species, but these occur within a patchwork, with each patch supporting only a few.

coexistence or exclusion? questions of scale

Patterns of co-occurrence and exclusion have been studied on a very local scale among plants in pastures in Ontario, Canada. The frequencies of physical contact between various species pairs were measured and compared with expected frequencies derived simply from the abundances of the different species. Particular attention was paid to six species—three grasses and three legumes—and the results for these six are presented in the first data column of Table 6.1. A plus sign ($+$) indicates that two species occurred in close contact more frequently than expected by chance; a minus ($-$) indicates that contacts were less frequent than expected.

A negative association could mean either that two species exclude one another through competition or that they have quite different habitat requirements. To

Table 6.1

Coexisting plant species that live in similar types of soil in a Canadian pasture tend not to occur together (possible competitive exclusion), but species living in different types of soil do occur together (coexistence with possible niche differentiation)

SPECIES PAIR	DISTRIBUTIONS MORE ($+$) OR LESS ($-$) CONGRUENT THAN EXPECTED BY CHANCE	NUMBER OF SIGNIFICANT CORRELATIONS BETWEEN THE SOIL TYPES OF THE TWO SPECIES EXAMINED FOR FIVE CHARACTERISTICS		
		$+$	$-$	BALANCE OF CORRELATIONS
TR/TP	$-$	3	0	$+$
TR/ML	$-$	2	1	$+$
TR/BI	$+$	1	1	0
TR/DG	$-$	0	0	0
TR/PP	$+$	0	1	$-$
TP/ML	$-$	2	0	$+$
TP/BI	$-$	0	0	0
TP/DG	$+$	1	0	$+$
TP/PP	$+$	1	0	$+$
ML/BI	$+$	1	1	0
ML/DG	$+$	0	0	0
ML/PP	$-$	1	0	$+$
BI/DG	$-$	2	0	$+$
BI/PP	$-$	2	1	$+$
DG/PP	$-$	2	0	$+$

BI, *Bromus inermis;* DG, *Dactylis glomerata;* ML, *Medicago lupulina;* PP, *Phleum pratense;* TP, *Trifolium pratense;* TR, *T. repens.* Note that the soil characteristics reflect the consequences of a species growing in the soil and extracting nutrients differentially.
(After Turkington et al., 1977.)

distinguish between these alternatives, the local habitats of each of the species were examined by analyzing the soil in the immediate vicinity of the roots. Soil samples were tested for their phosphorus, potassium, magnesium, and calcium content and for their pH. These analyses indicate which resources are left behind by the growing plant, and so establish, for each factor separately, whether particular species pairs tend to deplete the soil in the same or in different ways. The fourth and final data column in the table summarizes the overall similarity (+) or dissimilarity (−) of soil types left behind by the various species pairs. Thus, a minus sign in the first data column (do not co-occur) combined with a plus in the fourth (similar niches) suggests competitive exclusion on a small spatial scale despite coexistence within the pasture community as a whole.

Overall, there was indeed a marked tendency for species that were negatively associated in their distribution to deplete soil nutrients in the same way (−, +). In particular, all three grass–grass pairs and all three legume–legume pairs were in this category. This certainly suggests that coexistence within the pasture community as a whole is accompanied by competitive exclusion in more localized community patches. On the other hand, two pairs (*Trifolium pratense* with *Dactylis glomerata* and with *Phleum pratense*) ran counter to the expected pattern: they were positively associated but depleted the soil in similar ways. This, perhaps, reminds us that the soils were analyzed in only a small number of ways; other factors (other niche dimensions) might have suggested relationships different from those observed.

6.4.4 Niche differentiation among animals and among plants

In spite of all the difficulties of making a direct connection between interspecific competition and niche differentiation, there is no doubt that niche differentiation is often the basis for the coexistence of species within natural communities.

There are a number of ways in which niches can be differentiated. One is resource partitioning or differential resource utilization. This can be observed when species living in precisely the same habitat nevertheless utilize different resources. Since the majority of resources for animals are individuals of other species (of which there are millions of types) or parts of individuals, there is no difficulty, in principle, in imagining how competing animals might partition resources among themselves.

less scope for resource partitioning by plants than by animals

Plants, on the other hand, all have very similar requirements for the same potentially limited resources (Chapter 3), and there is much less apparent scope for resource partitioning. Moreover, although mobile animals may compete for food but still be free from competition for space, water, and other needs, for plants and other sessile organisms, competition for one resource often affects the ability of an organism to exploit another resource. If one plant species invades the canopy of another and deprives it of light, for example, the suppressed species will suffer directly from the reduction in light energy that it obtains, but this will also reduce its rate of root growth, and it will therefore be less able to exploit the supply of water and nutrients in the soil. This in turn will reduce its rate of shoot and leaf growth. Thus, when plant species compete, repercussions flow backward and forward between roots and shoots.

roots and shoots in plant competition

A number of workers have attempted to separate the effects of canopy and root competition by an experimental design in which two species are grown (1) alone, (2) together, (3) in the same soil but with their canopies separated, and (4) in separate soil with their canopies intermingling. One example is a study of subterranean clover and skeleton weed (Figure 6.15). The clover was not significantly affected under any circumstances (another example of one-sided competition), but the skeleton weed was affected when the roots intermingled (reduced to 65 percent of the control value of dry weight) and when the canopies intermingled (47 percent of the control), and when both intermingled the effect was multiplicative, dry weight being reduced to 31 percent of the control compared to the 30.6 percent (65 percent × 47 percent) that might have been expected. Clearly, both were important.

niche differentiation through spatial or temporal patterns in resources and/or conditions

In many cases, the resources used by ecologically similar species are separated spatially. Differential resource utilization will then express itself as either a microhabitat differentiation between the species (different species of fish, say, feeding at different depths) or even a difference in geographical distribution. Alternatively, the availability of the different resources may be separated in time; that is, different resources may become available at different times of the day or in different seasons. Differential resource utilization may then express itself as a temporal separation between the species.

The other major way in which niches can be differentiated is on the basis of conditions. Two species may use precisely the same resources, but if their ability to do so is influenced by environmental conditions (as it is bound to be), and if they respond differently to those conditions, then each may be competitively superior in different environments. This too can express itself as either a microhabitat differentiation, or a difference in geographical distribution, or a temporal separation, depending on whether the appropriate conditions vary on a small spatial scale, on a large spatial scale, or over time. Of course, it is not always easy to distinguish between conditions and resources, especially with plants (Chapter 3). Niches may then be differentiated on the basis of a factor (such as water) that is both a resource and a condition.

Figure 6.15
Root and shoot competition between subterranean clover (*Trifolium subterraneum*) and skeleton weed (*Chondrilla juncea*). Above are the experimental designs used; below are the dry weights of skeleton weed produced as a percentage of the production when grown alone. (After Groves & Williams, 1975.)

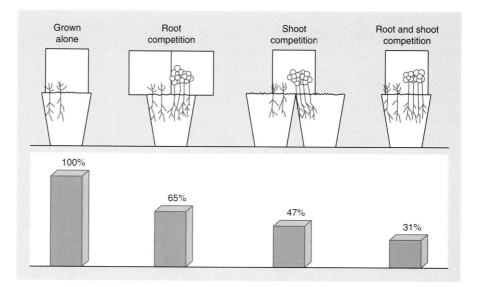

6.5 > How Significant Is Interspecific Competition in Practice?

Competitors may exclude one another, or they may coexist if there is ecologically significant differentiation of their realized niches (Section 6.2). On the other hand, interspecific competition may exert neither of these effects if environmental heterogeneity prevents the process from running its course (Section 6.2.8). Evolution may drive the niches of competitors apart until they coexist but no longer compete (Section 6.3). All these forces may express themselves at the level of the ecological community (Section 6.4). Interspecific competition sometimes makes a high-profile appearance by having a direct impact on human activity (Box 6.2). In this sense, competition can certainly be of practical significance.

In a broader sense, however, the significance of interspecific competition rests not on a limited number of high-profile effects, but on an answer to the question, How widespread are the ecological and evolutionary consequences of interspecific

Box 6.2   *Topical ECOncerns*

Competition in Action

When exotic plant species are introduced to a new environment, by accident or on purpose, they sometimes prove to be exceedingly good competitors and many native species suffer harmful consequences as a result. This is one example from Philadelphia. How severe should be the punishment of those found to have neglectfully introduced weeds? How effective should such punishment be? When is a weed a real problem—when it has an adverse aesthetic effect? Or only when it causes economic damage?

PHILADELPHIA INQUIRER—Sunday 12 April 1998—by Stephanie A. Stanley

WEEDS WREAKING HAVOC ACROSS THE STATE

An alien invader is spreading fast across the hillsides, climbing quickly up tree trunks and creating shaded groves of leafy green towers at the Riverbend Environmental Education Center.

"This is *Akebia*" center director Kevin Peter said as he strayed off Valley View Trail and gestured dejectedly at a sea of fresh, one-inch leaves bristling in the wind. "They do a great job as a ground cover, but they escape and are hard to hunt down."

Weeds. The bane of any caretaker's existence. Peter said at least 85 percent of Riverbend's 30 acres were infiltrated by *Akebia*—which can grow 20 feet in one season—and other exotic vines that go by the names

of mile-a-minute weed, Oriental bittersweet and multiflora rose.

For years, the center has fought the never-ending battle against the growing pests with minor success. But on Saturday, the center will mount its largest offensive when 200 Boy and Girl Scouts strap on their work boots and pick up the clippers for a day of chopping and pulling weeds. The crowd will slay a wide swath of the vines, but, as Peter pointed out, they will not win the war. Indeed, *Akebia* still is sold at gardening centers.

At Riverbend, an indigenous grove of locust trees is blanketed under akebia, choking out light and threatening the naturally brittle branches, Peter said. In addition, the center cannot proceed with plans to plant more indigenous species until the weeds are under control.

In the world of weeds, the government keeps a list of the most egregious. The Pennsylvania Noxious Weed Control List consists of 11 plants, five of which, including mile-a-minute, are growing at Riverbend. Under the state noxious weed control law, it is illegal to "propagate, sell or transport these weeds in the commonwealth." *Akebia* is not on the list.

Box 6.3

Lack's Tits—Alternative Explanations for "Niche Differentiation"

In the 1960s, in an English broad-leaved woodland called Marley Wood, David Lack carried out a landmark study of five bird species: the blue tit (*Parus caeruleus*), the great tit (*P. major*), the marsh tit (*P. palustris*), the willow tit (*P. montanus*), and the coal tit (*P. ater*). Four of these species weigh between 9.3 g and 11.4 g on average (the great tit weighs 20.0 g); all have short beaks and hunt for food chiefly on leaves and twigs, but at times on the ground; all eat insects throughout the year, and also seeds in winter;

and all nest in holes, normally in trees. Nevertheless, despite their similarities (the overlaps between their niches), all five species bred where Lack studied them, and the blue, great, and marsh tits were common. In short, they coexisted but seemed very similar in where they lived, where they bred, and what they ate.

The closer Lack examined these species, however, the stronger became his conclusion that they were separated, at most times of the year, by the precise location

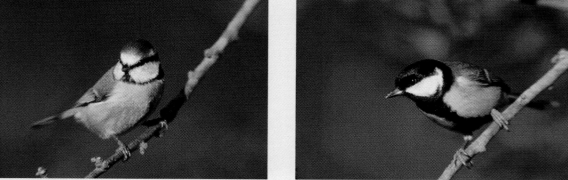

The blue tit (left) and the great tit (right).

competition in practice? The difficulty of answering this question can be illustrated by a study of the coexistence of five closely related woodland bird species: Lack's tits (Box 6.3).

The alternative explanations of patterns revealed by studies of niche differentiation, such as that in Box 6.3, raise a number of important questions, but we limit ourselves here to just two.

- The first, dealt with in Section 6.5.1, is, How prevalent is current competition in natural communities? To demonstrate current competition requires experimental field manipulations, in which one species is removed from or added to the community and the responses of the other species are monitored. It is important to answer this question, because where current competition *is* demonstrable, neither the ghost of competition past nor spatial or temporal variation is likely to have a crucial role. And if current competition *is* prevalent, then interspecific competition is likely to be an important structuring force in nature. However, even if current competition is not prevalent, past competition, and therefore competition generally, may still have played a significant role in structuring communities.

within the trees where they fed, by the size of their insect prey, and by the hardness of the seeds they took. He found, too, that this separation was associated with (often minor) differences in overall size, and in the size and the shape of the birds' beaks. Despite their similarities, these coexisting species exploited slightly different resources in slightly different ways.

Did this, though, have anything to do with competition? Lack thought so. He believed that the species coexisted as a result of evolutionary responses to interspecific competition (Connell's ghost of competition past). But there is a major problem—the absence of any direct proof. We cannot go back in time to check whether the tit species ever competed more than they do now, and it was not part of Lack's study to determine the extent of present-day competition. In fact, with such observational studies of niche differentiation, there are at least five plausible alternative interpretations of the data.

1. The species are competitors now and coexist as a result of the observed niche differentiation.

2. The species competed in the past but do not do so now—but their niche differentiation, which allows their coexistence, evolved in response to that past competition.

3. Competition in the past eliminated a number of other species, leaving behind only those that *were* different in their utilization of the habitat. Thus, we can still see the hand of the ghost of competition past—but acting as an ecological force (eliminating species) rather than an evolutionary one (changing them).

4. The species have, in the course of their evolution, responded to natural selection in different but entirely independent ways. They are distinct species, and they have distinctive niches. But they do not compete now, nor have they ever competed; they simply happen to be different. The coexistence of the tits, and the structure of the community of which they are part, have nothing to do with competition.

5. Finally, although the niches of the species are distinct, they are not different enough to allow all five to coexist in an unvarying environment where competition runs its course. But the environment does vary and competition does not run its course, and the species therefore owe their coexistence to their responses, overall, to a patchy and ever-changing world.

▪ The second problem, dealt with in Section 6.5.2, is to distinguish between interspecific competition (past *or* present) and either "mere chance" (the fourth alternative, Box 6.3) or spatial and temporal variation (the fifth). The many studies in which experimental field manipulations have *not* been possible can be examined to determine whether observed patterns provide strong evidence for a role for competition or are open to alternative interpretations, as in Lack's study (Box 6.3).

6.5.1 The prevalence of current competition

Two surveys of field experiments on interspecific competition were published in 1983. Schoener examined the results of all the experiments he could find—164 studies in all. He found that approximately equal numbers of studies had dealt with terrestrial plants, terrestrial animals, and marine organisms, but that studies of freshwater organisms amounted to only about half the number in the other groups. Among the terrestrial studies, however, he found that most were concerned with temperate regions and mainland populations, and that there were relatively few

surveys of published studies of competition indicate that current competition is widespread . . .

dealing with phytophagous (plant-eating) insects. Any conclusions were therefore bound to be subject to the limitations imposed by what ecologists had chosen to look at. Nevertheless, Schoener found that approximately 90 percent of the studies had demonstrated the existence of interspecific competition, and that the figures were 89, 91, and 94 percent for terrestrial, freshwater, and marine organisms, respectively. Moreover, if he looked at single species or small groups of species (of which there were 390), rather than at whole studies, which may have dealt with several groups of species, he found that 76 percent showed effects of competition at least sometimes, and 57 percent showed effects in all the conditions under which they were examined. Once again, terrestrial, freshwater, and marine organisms gave very similar figures.

Connell's review was less extensive than Schoener's: 72 studies, dealing with a total of 215 species and 527 different experiments. Interspecific competition was demonstrated in most of the studies, more than half of the species, and approximately 40 percent of the experiments. In contrast to Schoener, Connell found that interspecific competition was more prevalent in marine than in terrestrial organisms, and also that it was more prevalent in large than in small organisms.

Taken together, Schoener's and Connell's reviews certainly seem to indicate that active, current interspecific competition is widespread. Its percentage occurrence among species is admittedly lower than its percentage occurrence among whole studies, but this is to be expected, since, for example, if four species were arranged along a single niche dimension and all adjacent species competed with each other, this would still be only three out of six (or 50 percent) of all possible pairwise interactions.

UNANSWERED QUESTION:

. . . but these surveys exaggerate to an unknown extent the true frequency of competition

Connell also found, however, that in studies of just one pair of species, interspecific competition was almost always apparent, whereas with more species the prevalence dropped markedly (from more than 90 percent to less than 50 percent). This can be explained to some extent by the argument outlined previously, but it may also indicate biases in the particular pairs of species studied, and in the studies that are actually reported (or accepted by journal editors). It is highly likely that many pairs of species are chosen for study because they are "interesting" (because competition between them is suspected) and if none is found, this is simply not reported. Judging the prevalence of competition from such studies is rather like judging the prevalence of debauched clergymen from the "gutter press." This is a real problem, only partially alleviated in studies on larger groups of species when a number of "negatives" can be conscientiously reported alongside one or a few "positives." Thus the results of surveys, such as those by Schoener and Connell, exaggerate, *to an unknown extent,* the frequency of competition.

As previously noted, phytophagous insects were poorly represented in Schoener's data, but reviews of this group alone have tended to suggest either that it is relatively rare in this group overall (Strong et al., 1984) or rare in at least certain types of phytophagous insects, for example "leaf-biters" (Denno et al., 1995). On a more general level, it has been suggested that herbivores *as a whole* are seldom food-limited and are therefore not likely to compete for common resources (Hairston et al., 1960; Slobodkin et al., 1967). The bases for this suggestion are the observations that green plants are normally abundant and largely intact, they are rarely devastated, and most herbivores are scarce most of the time. Schoener found the proportion of herbivores exhibiting interspecific competition to be significantly lower than the proportions of plants, carnivores, or detritivores.

Taken overall, therefore, current interspecific competition has been reported in studies on a wide range of organisms, and in some groups its incidence may be particularly obvious, for example, among sessile organisms in crowded situations. However, in other groups of organisms, interspecific competition may have little or no influence. It appears to be relatively rare among herbivores generally, and particularly rare among some types of phytophagous insect.

6.5.2 Competition or mere chance?

When Lack's study of woodland birds is considered with the benefit of hindsight, it is easy to point to a dangerous tendency to interpret "mere differences" between the niches of coexisting species as confirming the importance of interspecific competition, when there may be other, equally plausible explanations. On the other hand, the theory of interspecific competition does more than predict "differences." It predicts, for example, that the niches of competing species should be arranged regularly rather than randomly in niche space; that is, it predicts not simply that their niches differ, but that they differ *more* than would be expected from chance alone.

A more rigorous investigation of the role of interspecific competition, therefore, should address itself to the question, Does the observed pattern, even if it appears to implicate competition, differ significantly from the sort of pattern that could arise

neutral models

A desert lizard of the southwestern United States.

Box 6.4

Neutral Models of Lizard Communities

Lawlor (1980) investigated differential resource utilization in 10 North American lizard communities, consisting of four to nine species, for which there were estimates of the amounts of each of 20 food categories consumed by each species in each community. This pattern of resource use allowed the calculation, for each pair of species in a community, of an index of resource use overlap, which varied between 0 (no overlap) and 1 (complete overlap). Each community was then characterized by a single value: the mean resource overlap for all pairs of species present.

A number of neutral models of these communities were then created. They were of four types. The first type, for example, retained the minimum amount of original community structure. Only the original number of species and the original number of resource categories were retained. Beyond that, species were allocated food preferences completely at random, such that there were far fewer species completely ignoring food in particular categories than in the real community. The niche breadth of

each species was therefore increased. The fourth type, on the other hand, retained most of the original community structure: if a species ignored food in a particular category, then that was left unaffected, but among those categories where food was eaten, preferences were reassigned at random. These neutral models were then compared with their real counterparts in terms of their patterns of overlap in resource use. If competition is or has been a significant force in determining community structure, the niches should be spaced out, and overlap in resource use in the real communities should be less—and statistically significantly less—than predicted by the neutral models.

The results (see figure) were that in all communities, and for all four neutral models, the model mean overlap was higher than that observed for the real community, and that in almost all cases this was statistically significant. For these lizard communities, therefore, the observed low overlaps in resource use suggest that niches are segregated, and that interspecific competition plays an important role in community structure.

in the community even in the absence of any interactions between species? This question was the driving force behind analyses that sought to compare real communities with so-called neutral models. These are hypothetical models of actual communities that retain certain of the characteristics of their real counterparts but reassemble or reconstruct some of the community components in a way that specifically excludes the consequences of interspecific competition. In fact, the neutral model analyses are attempts to follow a much more general approach to scientific investigation, namely, the construction and testing of *null hypotheses*. The idea is that the data are rearranged into a form (the neutral model or null hypothesis) representing what the data *would* look like in the *absence* of interspecific competition. Then, if the actual data show a significant statistical difference from the null hypothesis, the null hypothesis is rejected and the action of interspecific competition is strongly inferred.

The application of null hypotheses to community structure—that is, the reconstruction of natural communities with interspecific competition removed—has not been achieved to the satisfaction of all ecologists. But a brief examination of a study of lizard communities will at least make it possible to understand the potential and rationale of the neutral model approach (Box 6.4).

By contrast to the results for the lizard communities in Box 6.4, where a role for competition appeared to be confirmed, in a similar analysis of five grasshopper

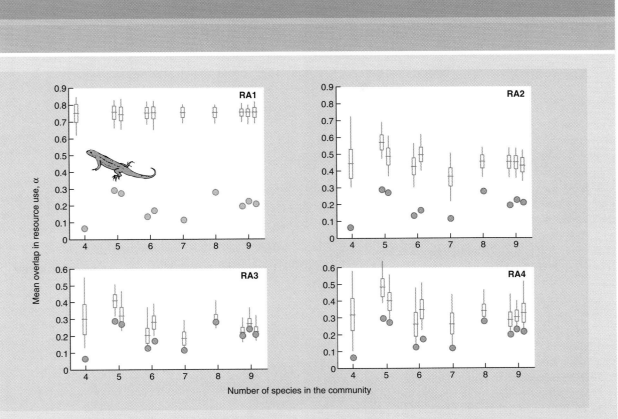

The mean indices of resource-use overlap for each of 10 North American lizard communities are shown as solid circles. These can be compared, in each case, with the mean (horizontal line), standard deviation (vertical rectangle), and range (vertical line) of mean overlap values for the corresponding set of 100 randomly constructed communities. The analysis used four different types of reorganization algorithms (RAs), as described in the text. (After Lawlor, 1980.)

communities, only complete randomization of resource use resulted in mean overlap values larger than those observed in real communities (Joern & Lawlor, 1980). Thus, the evidence for competition structuring grasshopper communities is not convincing, suggesting that this may be a further example of a community of phytophagous insects in which competition plays no significant role.

This pattern—sometimes a role for competition is confirmed, sometimes not—has been the general conclusion from the neutral model approach. What then should be our verdict on it? Perhaps most fundamentally, its aim is undoubtedly worthy. It concentrates the minds of investigators, stopping them from jumping to conclusions too readily: it is important to guard against the temptation to see competition in a community simply because we are looking for it. On the other hand, the approach can never take the place of a detailed understanding of the field ecology of the species in question, or of manipulative experiments designed to reveal competition by increasing or reducing species abundances. It, like so many other approaches, can only be part of the community ecologist's armory.

Summary

The ecological effects of interspecific competition

The essence of interspecific competition is that individuals of one species suffer a reduction in fecundity, survivorship, or growth as a result of exploitation of resources or interference by individuals of another species.

Species are often excluded by interspecific competition from locations at which they could exist perfectly well in the absence of interspecific competition.

With exploitation competition, the more successful competitor is the one that more effectively exploits shared resources. Two species exploiting two resources can compete but still coexist when each species holds one of the resources at a level that is too low for effective exploitation by the other species.

A fundamental niche is the combination of conditions and resources that allows a species to exist when considered in isolation from any other species, whereas its realized niche is the combination of conditions and resources that allows it to exist in the presence of other species that might be harmful to its existence—especially interspecific competitors.

The Competitive Exclusion Principle states that if two competing species coexist in a stable environment, then they do so as a result of differentiation of their realized niches. If, however, there is no such differentiation, or if it is precluded by the habitat, then one competing species will eliminate or exclude the other. However, whenever we see coexisting species with different niches, it is *not* reasonable to jump to the conclusion that this is the principle in operation.

The only true test for whether competition occurs between species is to manipulate the abundance of each competitor and observe the response of its counterparts.

Environments are usually a patchwork of favorable and unfavorable habitats; patches are often only available temporarily; and patches often appear at unpredictable times and in unpredictable places. Under such variable conditions, competition may only rarely run its course.

The evolutionary effects of interspecific competition

Although species may not be competing now, their ancestors may have competed in the past. We can expect species to have evolved characteristics that ensure that they compete less, or not at all, with members of other species. Coexisting present-day competitors and coexisting species that have evolved an avoidance of competition can look, at least superficially, the same.

By invoking something that cannot be observed directly—"the ghost of competition past"—it is impossible to *prove* an evolutionary effect of interspecific competition. However, careful observational studies have sometimes revealed patterns that are difficult to explain in any other way.

Interspecific competition and community structure

Our perception of the role of interspecific competition in molding community structure depends on the scale at which it is examined. On a relatively small scale, we tend to see the co-occurrence of only a few species with complementary ecological niches. But if the scale is broadened, the community appears to contain more species, but these occur within a patchwork, with each patch supporting only a few.

Interspecific competition tends to structure communities by acting within guilds—groups of species exploiting the same class of resource in a similar fashion.

Niche complementarity can be discerned in some communities, where coexisting species that occupy a similar position along one niche dimension tend to differ along another dimension.

How important is interspecific competition in practice?

Surveys of published studies of competition indicate that current competition is widespread, but these exaggerate to an unknown extent the true frequency of competition.

The theory of interspecific competition predicts that the niches of competing species should be arranged regularly rather than randomly in niche space. Neutral models have been developed to determine what the community pattern *would* look like in the *absence* of interspecific competition. Real communities are sometimes structured in a way that makes an influence of competition difficult to deny.

Review Questions

⚠ = Challenge Question

1. Some experiments concerning interspecific competition have monitored both the population densities of the species involved *and* their impact on resources. Why is it helpful to do both?

2. Interspecific competition may be a result of exploitation of resources *or* of direct interference. Give an example of each and compare their consequences for the species involved.

3. Define fundamental niche and realized niche. How do these concepts help us to understand the effects of competitors?

4. With the help of one plant and one animal example, explain how two species may coexist by holding different resources at levels that are too low for effective exploitation by the other species.

5. ⚠ Define the Competitive Exclusion Principle. When we see coexisting species with different niches is it reasonable to conclude that this is the principle in action?

6. Explain how environmental heterogeneity may permit an apparently "weak" competitor to coexist with a species that might be expected to exclude it.

7. ⚠ What is the "ghost of competition past"? Why is it impossible to prove an evolutionary effect of interspecific competition?

8. Provide one example each of niche differentiation involving physiological, morphological, or behavioral properties of coexisting species. How may these differences have arisen?

9. Define niche complementarity and, with the help of an example, explain how it may help to account for the coexistence of many species in a community.

10. ⚠ Discuss the pros and cons of the neutral model approach to evaluating the effects of competition on community composition.

It is impossible to understand the ecology of any species without understanding its relationship with the habitat in which it lives, but for many, indeed most species, this habitat is itself another organism: the host. This chapter examines the intimate interactions between hosts and the organisms that live on and within them.

7

Organisms as Habitats

Chapter Contents

Key Concepts

In this chapter you will

- appreciate that the habitats of most organisms are other organisms and understand the differences between commensalism, symbiosis, parasitism, and mutualism.

- perceive that parasites are very diverse and often highly specialized, whereas mutualists are not diverse or specialized but nevertheless dominate the productivity of many habitats.

- understand that parasites, mutualists, and commensals may inhabit the surface and the body cavity or live within the tissues or cells of their hosts.

- recognize that parasites and mutualists may "manipulate" the growth, form, or behavior of their hosts.

- understand that hosts may make defensive responses to the presence of parasites and mutualists.

- appreciate that the life cycles of parasites are often complex and dominated by sex and dispersal, whereas the lives of mutualists are usually much simpler.

- understand that parasites play a crucial and often subtle role in the evolution of their hosts, and vice versa.

7.1 ▷ Introduction

symbiosis, parasitism, and mutualism

More than half the world's species live in or on the bodies of other organisms, where they find conditions and sometimes also resources that support their growth. Almost every organism is the habitat of others—even bacteria are the habitat of their own viruses, bacteriophages. An intimate association between individuals of different species, in which one lives on or in the other, is known as a *symbiosis*.

More than 60 years ago, Charles Elton (a founding father of animal ecology) and his colleagues surveyed the fauna living on and in wood mice, *Apodemus sylvaticus*, in a small woodland near Oxford, England. They found 47 different species, mainly parasites (including a species of spirochete, though the survey did not include other kinds of bacteria) (Table 7.1). A *parasite* is an organism that lives

Table 7.1

Species found on or in the bodies of wood mice (*Apodemus sylvaticus*) in a woodland near Oxford, England

On skin: 1 tick larva
On fur: At least 12 species of mite
1 tick (adult)
1 beetle
11 fleas
1 louse
On anus and genital organs: 1 mite
Under skin of limbs: 1 mite
In liver: 1 tapeworm larva
In stomach: 1 roundworm
In small intestine: 3 roundworms
2 tape worms
3 flat worms
2 flagellates
1 ciliate
1 amoeba
2 coccidians
In caecum: 1 ciliate
In kidney: 1 spirochaete (no attempt was made to identify other bacteria or viruses)
In blood: 1 trypanosome
(From Elton, C. [1940]. "The Ecology of Animals" 2nd ed. Methuen, London.)

its life intimately associated with one or a very few individuals of another species, its host, deriving resources from its host and doing its host harm, but not predictably killing its host in the short term. We have now come to expect species-rich communities of parasites, not only on mammals like the wood mice, but also on fish, amphibians, reptiles, and birds, and if we survey the types of organism that live on plants in nature, or in cultivation, we find that almost every species of tree, shrub, and herb carries a community of small parasitic animals (mites, aphids, insect larvae) as well as parasitic fungi and bacteria (Figure 7.1). Parasites contribute overwhelmingly to the diversity of plant and animal communities in nature and to the diversity of life on Earth. A useful distinction between microparasites and macroparasites is drawn in Box 7.1.

parasites are the major contributors to the biodiversity of nature

At least as striking as the diversity of parasites they support are the populations of "mutualistic" fungi that live tightly integrated into the tissues of the great majority of plants. A *mutualism* is an association between species (usually a pair of species) that benefits both. A *mycorrhiza*, for example, is a mutualistic association between a fungus and the root of a higher plant, and mycorrhizae are particularly strongly developed in trees and grasses. In addition, in a number of plants there are mutualistic bacteria and actinomycetes living in specialized nodules on the roots, capable of fixing atmospheric nitrogen and supplying it to the plant host. Other familiar mutualisms include those between flowering plants and insect pollinators. There are

but mutualists dominate productivity

Box 7.1

Microparasites and Macroparasites

Microparasites are small, often extremely numerous, and they multiply directly within their host. Many are intracellular. It is always difficult, and usually impossible, to count the number of microparasites in a host, and the number of infected hosts is usually studied, rather than the number of parasites. For example, a study of epidemics of measles will involve counting the number of cases of the disease, rather than the number of particles of the measles virus. Probably the most obvious microparasites are the bacteria and viruses that infect animals and plants (e.g., the yellow net viruses of beet and tomato and the bacterial crown gall disease). The other major group of microparasites affecting animals are the protozoa (e.g., the *Plasmodium* species that cause malaria). In plant hosts some of the simpler fungi behave as microparasites.

Macroparasites grow within their host but do not multiply there. They produce infective stages that are released to infect new hosts. The macroparasites of animals mostly live on the body or in the body cavities (e.g., the gut), rather than within the host cells. In plants,

they are generally intercellular. It is usually possible to count or at least estimate the numbers of macroparasites in or on a host (e.g., worms in an intestine or lesions on a leaf), and the numbers of parasites as well as the numbers of infected hosts can both be studied by the epidemiologist. Parasitic worms are major macroparasites of animals (e.g., tapeworms, flukes, and nematodes). In addition, there are lice, fleas, ticks, and mites and some fungi that attack animals. Plant macroparasites include the higher fungi that give rise to powdery mildews, downy mildews, rusts, and smuts, as well as the gall-forming and mining insects, and such flowering plants as the dodders and broomrapes that are themselves parasitic on other higher plants.

A less useful and certainly less clear distinction is sometimes drawn between parasites and *pathogens*. The term pathogen is more often applied to microparasites than to macroparasites and is usually reserved for parasites that give rise in their host to a "disease" with describable and presumably harmful "symptoms."

Figure 7.1

Higher plants and animals, such as the human, serve as specialized niches occupied by fungi, insects, worms, and protozoa. The two diagrams illustrate a little of this diversity but exclude the nematode worms that parasitize plants. Bacteria and viruses are omitted but comparable diagrams could be drawn for them—especially the diverse species that are parasitic on humans. Several of the worm parasites on humans migrate to different parts of the body during their life cycles. For example, *Schistosoma* occurs in mesenteric veins, but eggs accumulate in and damage the liver and even the lungs. Fungi are shown in brown, worms in blue, insects in green, and protozoa in yellow.

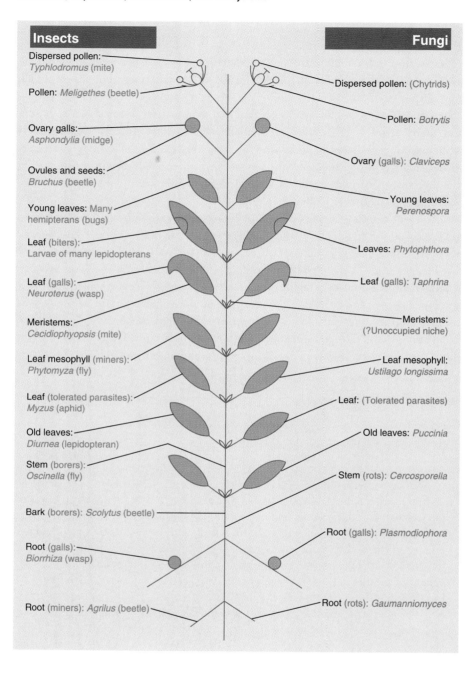

far fewer species of mutualists than of parasites, but mutualistic associations like mycorrhizae dominate the production of biomass in many habitats.

We distinguish parasites and mutualists according to whether they benefit or harm the host. But words like *benefit* and *harm* must be used to mean something more than "what I think I would like or not like if I were that organism." Here, we use *harm* to mean increasing the death rate, reducing the birth rate, decreasing the growth rate, or reducing the carrying capacity, and *benefit* to mean the converse of

a parasite is not always a parasite

Figure 7.1   *(continued)*

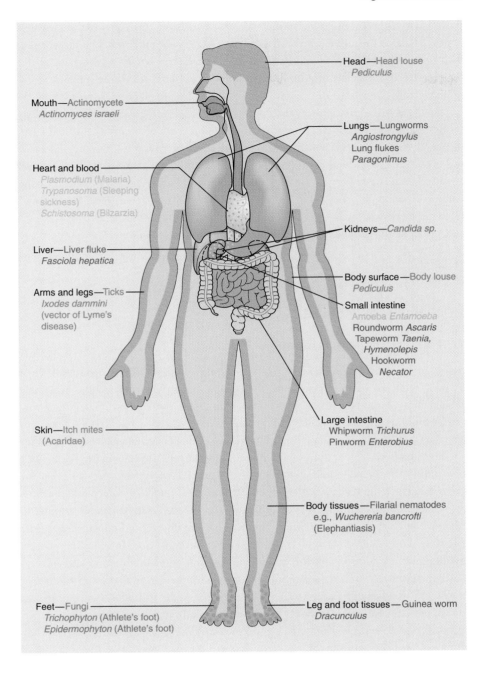

Head—Head louse *Pediculus*

Mouth—Actinomycete *Actinomyces israeli*

Lungs—Lungworms *Angiostrongylus* Lung flukes *Paragonimus*

Heart and blood *Plasmodium* (Malaria) *Trypanosoma* (Sleeping sickness) *Schistosoma* (Bilzarzia)

Kidneys—*Candida sp.*

Liver—Liver fluke *Fasciola hepatica*

Body surface—Body louse *Pediculus*

Small intestine Amoeba *Entamoeba* Roundworm *Ascaris* Tapeworm *Taenia, Hymenolepis* Hookworm *Necator*

Arms and legs—Ticks *Ixodes dammini* (vector of Lyme's disease)

Large intestine Whipworm *Trichurus* Pinworm *Enterobius*

Skin—Itch mites (Acaridae)

Body tissues—Filarial nematodes e.g., *Wuchereria bancrofti* (Elephantiasis)

Feet—Fungi *Trichophyton* (Athlete's foot) *Epidermophyton* (Athlete's foot)

Leg and foot tissues—Guinea worm *Dracunculus*

these. In practice, though, an organism that sometimes harms its host need not always do so. An activity that is harmful in one environment may be tolerated in another. Ticks draw blood from their hosts and might be thought to be obviously parasites. But individuals of the Australian sleepy lizard (*Tiliqua rugosa*) carry populations of ticks that seem not to harm them. An analysis of the relationship between the load of ticks on a lizard and its longevity either showed that there was no correlation or that lizards with more ticks seemed to have lived longer. Perhaps these lizard ticks should be classified as *commensals* or "guests" (neither harmful nor beneficial to the host), bearing in mind that, like guests under other circumstances, they may be unwelcome when the hosts are ill or distressed! It is important not to forget these commensals in a preoccupation with the more dramatic interactions of mutualists or parasites.

In the remainder of this chapter we examine hosts as mosaics of microhabitats for other species. We consider ways in which the diversity of life cycles, growth form, dispersal, and specificity provides the match between colonists and the organism that they colonize. We also show how, in comparison with other kinds of habitats and environments, a living organism is a unique type of habitat. It can respond to being colonized, can repair some of the damage that may be caused, may have a "memory" of past experience, and can evolve. In Chapter 8 we consider the effects of parasites and other natural enemies on the dynamics of host populations.

a host is a habitat that can respond, repair, remember, and evolve

7.2 ▷ Hosts as Habitats

7.2.1 Life on the surface of another organism

organisms may provide physical support for other species

Solid surfaces provide potential support for living organisms and may allow them to gain access to conditions and resources that are otherwise unavailable. Trees and shrubs in particular provide habitats for the many species of birds, bats, and climbing and scrambling animals that are absent from treeless environments. Trees are particularly good examples of what have been called "ecological engineers." By their very presence they create, modify, or maintain habitats for others (Figure 7.2). They provide sites for nesting and roosting and a degree of safety from predators on the ground. Lichens and mosses develop on tree trunks, and climbing plants such as ivy, vines, and figs, though they root in the ground, use tree trunks as support to extend their foliage up into a forest canopy. In warm and humid climates, *epiphytic* plants (plants living on plants), including ferns, orchids, and bromeliads, root in the branches and crevices of the upper canopy. They draw their water and mineral nutrient resources from the excretions of animals in the canopy and rain water that leaches down through the branches. Water that collects in hollows on the trunks and in the cups formed by the leaf bases of bromeliads supports further microcosms of midge larvae and even the larvae of tree frogs. These are all commensals. They draw no resources from the host tree and do it no direct harm. But the community of organisms that make their habitat on trees and shrubs also includes hemiparasites, such as the mistletoes, which penetrate the tissues of the host tree and depend on it for water and mineral nutrients, and true parasitic flowering plants, such as dodders, which depend on the host for photosynthetic products as well (see Section 7.2.3).

In aquatic communities, the solid surfaces of larger organisms are even more important contributors to biodiversity. Seaweeds and kelps normally grow only where they can be anchored on rocks, but their fronds are colonized in turn by filamentous algae and especially by tube-forming worms (*Spirorbis*) and modular animals such

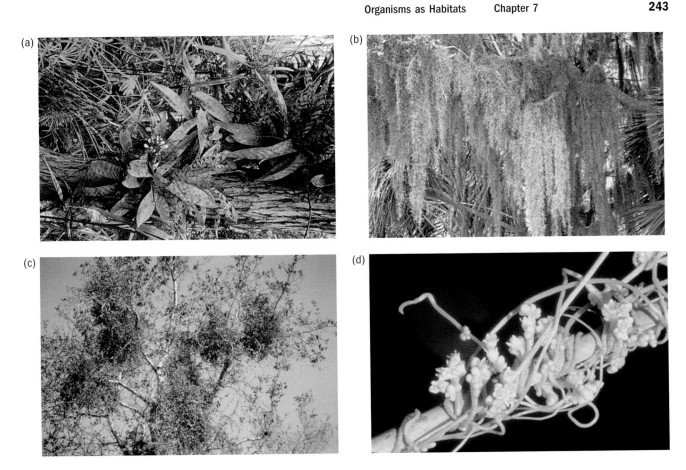

Figure 7.2
A variety of flowering plants use the bodies of other flowering plants as their habitat. (a) Bromeliads are epiphytes, living often high in the canopy on the trunks of trees in tropical forests. They derive only physical support from the host tree. (b) Spanish moss (*Tillandsia usnoides*) is also an epiphytic flowering plant draped over trees in humid environments and with a characteristic lichen-like color. (c) Mistletoe (*Phoraradendron* sp.) on sycamore. This is a hemiparasite, penetrating the host tree and extracting mineral nutrients and water from within the host tree, but capable of independent photosynthesis. (d) Dodder (*Cuscuta* sp.). This is a true parasite that repeatedly penetrates the host plant on which it climbs. It lacks chlorophyll and depends on its host for photosynthesis as well as for supplies of mineral nutrients and water.

as hydroids and bryozoans (Figure 7.3). These depend on the seaweeds and kelps for anchorage and access to resources in the moving waters of the sea. In fresh waters the stems and leaves of aquatic flowering plants are exploited as habitats in the same way.

The shells of mollusks provide an anchorage habitat for a diverse fauna. Most of these surface dwellers again seem to be commensals, but interactions can be complex. For example, the cockle *Austrovenus stutchburyi* is abundant in sheltered shores in New Zealand, and limpets and anemones compete directly for colonization space on their shells. A trematode worm, *Curcuteria australis*, is a common internal parasite of the cockles, which, when infected, are unable to bury themselves properly. They then tend to lie on the surface of the mud, where they are more heavily

(a)

(b)

Figure 7.3

(a) A frond of the brown seaweed *Fucus serratus* bearing the polychaete worm *Spirorbis* (the spiral shells), the zigzag colonies of the hydroid *Obelia geniculata,* and (the brown area in the middle) the bryozoan *Bowerbankia*. (b). A frond of the brown seaweed *Ascophyllum nodosum* bearing tussocks of the red seaweed *Polysiphonia* (which is found almost uniquely living on *Ascophyllum*) and colonies of the hydroid *Obelia dichotoma*. (c) A frond of the brown seaweed *Ascophyllum nodosum* bearing colonies of *Clava multicornis*. (Photos kindly provided by R. N. Hughes.) (d) The water louse *Asellus aquaticus,* like most aquatic organisms, provides the habitat for a variety of other species that use their host as a vantage point from which to filter food from the water that flows past them. The figure shows *Asellus* from beneath (left) and in dorsal view (right) with 12 of the species that live on it shown in their characteristic (often very specialized) habitats. Apart from the rotifers, the species are all protozoa. (From Cook et al., 1998.)

(c)

colonized by limpets than uninfected cockles. The cockle is behaving as an ecological engineer in the community, and the trematode parasite changes its engineering role. Commensals have received far less study than parasites and mutualists, though many of them have ways of life that are quite as specialized and fascinating and their contribution to the diversity of communities may be quite as great.

life on roots and leaves: the rhizosphere and phyllosphere

Plant roots draw water and mineral nutrients from the soil and so modify the local environments on or near their surface. They also shed root cap cells and release organic materials to the soil as the root hairs and outer cortical cells die and decompose. The volume of soil adjacent to and influenced by plant roots is called the *rhizosphere* and it carries an abundant microbial flora, different from that found in the main body of the soil. The rhizosphere flora differs from one host species to

(d)

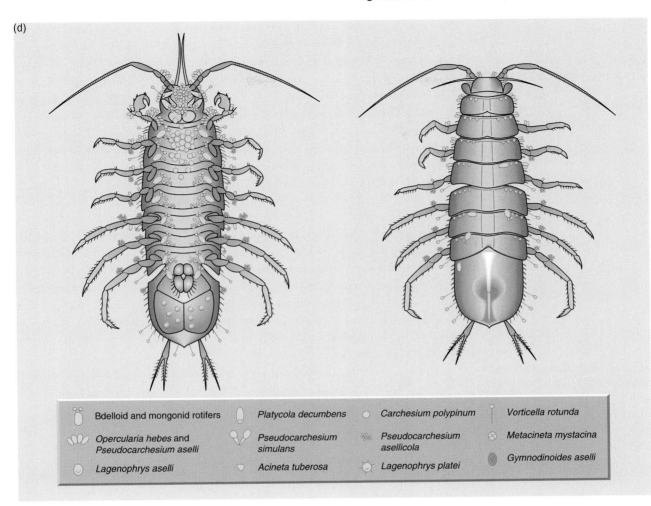

	Bdelloid and mongonid rotifers		*Platycola decumbens*		*Carchesium polypinum*		*Vorticella rotunda*
	Opercularia hebes and *Pseudocarchesium aselli*		*Pseudocarchesium simulans*		*Pseudocarchesium asellicola*		*Metacineta mystacina*
	Lagenophrys aselli		*Acineta tuberosa*		*Lagenophrys platei*		*Gymnodinoides aselli*

another (Westover et al., 1997), and there is evidence that it may sometimes protect the host plant against soil-borne parasites. There is also a distinct *phyllosphere* of species that live on leaf surfaces. Aphids, for example, are *ectoparasites* of plants—they live outside the main body of the plant but tap its phloem with specialized mouthparts and subsist entirely on the extracted sap. It is easy to forget that such behavior is truly parasitism. Very similarly, fungal parasites on leaf surfaces penetrate the tissues of the host with their hyphae. Examples are species of powdery mildews (e.g., *Erysiphe* sp., which cause devastating diseases of grapes) and downy mildews (e.g., *Phytophthora infestans*, which caused the Irish potato famine of the 1840s). The main body of these fungi is carried on the surface of host leaves, where it is easy to disperse their spores. Many of the organisms found on leaf and root surfaces are true parasites and are normally specialized to live on just one species of host and a particular microhabitat on that host (e.g., the root, leaf, or flower; see Figure 7.1). Such habitat specificity is greater among parasites than anywhere else in ecology.

mildews, fleas, and lice: the ectoparasites

Fleas and lice are ectoparasites on the bodies of the larger animals. Most of these are (like the parasitic fungi on leaf surfaces) very tightly specialized to a very few hosts. The more intimate their association with a particular host individual, the more they tend to be restricted to a particular species of host. Thus most species of bird lice, which spend their entire lives on their host, exploit only one species of host, whereas louse flies, which move actively from one host individual to another, can use several species of host (Table 7.2).

The bird lice have chewing mouthparts and feed on the feathers and scales of the skin. Their crawling and gnawing irritate the birds, which may become restless and not feed or digest their food properly. The lice are therefore an economic problem in commercial poultry production. Lice that feed on the feathers of rock doves (*Columba livia*) reduce the protective insulation given by the feathers; infected birds spend more time in preening and less in courtship display and obtain fewer mates than less heavily parasitized males. Indeed there is now much evidence that the parasitized males of many species (especially birds) fail to attract mates—an important way in which parasites can harm their hosts because it translates so directly into effects on male fitness.

the extinction of a host species forces extinction of its specialist parasites

Because parasites are highly specialized to their hosts, the extinction of a host often causes the extinction of its parasites as well. The passenger pigeon became

Table 7.2

Specialization in ectoparasites that feed on birds and mammals

		NUMBER OF SPECIES	PERCENTAGE OF SPECIES RESTRICTED TO		
			1 HOST	2 OR 3 HOSTS	MORE THAN 3 HOSTS
Philopteridae	bird lice (spend whole life on one host)	122	87	11	2
Streblidae	blood sucking flies (parasitize bats)	135	56	35	9
Oestridae	botflies (females fly between hosts)	53	49	26	25
Hystrichopsyllidae	fleas (jump between hosts)	172	37	29	34
Hippoboscidae	louse flies (are highly mobile)	46	17	24	59

(Adapted from Price, 1980.)

extinct in 1914. It was the host for two species of chewing lice, *Columbicola extinctus* and *Campanulotes defectus.* They became extinct at the same time. The hosts of parasites are their habitats: the extinction of a host species is the loss of a unique habitat.

Organisms that specialize in life on the surface of their hosts are usually well placed for dispersal, via spores, eggs, and so on, and the colonization ("infection") of other hosts. However, there are costs to a life spent on host surfaces, such as greater exposure to drought, heat, and cold and usually less intimate contact with the host as a resource. Surface dwellers are dangerously exposed to predators and parasites; for example, aphids on a leaf surface are searched for and attacked by parasitoid wasps and eaten by birds. Organisms that can invade the body cavities of an organism or even its tissues or cells are better placed to extract resources, are better insulated from the outside environment, and on warm-blooded animal hosts are maintained in a constant thermal environment and are better hidden from all but the most specialized enemies.

surface dwelling organisms may disperse easily—but there are costs

7.2.2 Inhabitants of the body cavity

Many species of parasites, commensals, and mutualists exploit the insulation and protection offered by cavities within the body of a host species. The alimentary canal from mouth to anus is a resource-rich habitat. It is largely protected from enemies and drought, is more or less regulated in temperature, and contains a continuous flow of food that has been garnered, and partly processed, by the host.

Many parasites live in the alimentary canal of animals. Gut inhabitants like tapeworms and roundworms, however, do not attack or consume any part of their host. Rather, they live in and on the food that the host has swallowed and is digesting and can be thought of as competing with the host for these digestive products as they travel from mouth to anus. They can certainly cause disease in their host. Some, indeed, cause further damage by releasing toxins: classic examples are *Vibrio cholerae,* the bacterium that causes cholera, and *Amoeba dysenterica,* the protist that causes amoebic dysentery. Note, too, that the diarrhea and dysentery that these parasites induce in their host favors the dispersal and epidemic spread of the *Vibrio* and *Amoeba* species.

life in the gut

The digestive tract is of course not just one environment but a continuum of changing conditions of pH, enzymic character, and aeration along its length (Figure 7.4), and the inhabitants are specialized to, and even migrate to find, local conditions within it. When nematode worms, *Nippostrongylus brasiliensis,* which infect rats, were experimentally transplanted into the anterior and posterior small intestine from their normal position between them in the jejenum, they migrated back to their original habitat.

Organisms that are specialized to use the guts of others as their habitat gain protection and insulation, but dispersal from one host to another is much more difficult than for those that live on the surface. Most of the species of parasite that inhabit the host gut solve the problem of dispersal with complex life histories that carry infection from the feces of the host, disperse it widely, and spread it among others in the host population. The life cycle of the human tapeworm, *Taenia saginata,*

Figure 7.4
The digestive systems of herbivores are commonly modified to provide fermentation chambers inhabited by a rich fauna and flora of microbes. The figure shows the digestive tracts of four different herbivorous mammals with the fermentation chambers highlighted. (a) The rabbit has an expanded caecum. (b) The zebra has fermentation chambers in both the caecum and colon. (c) The sheep has fermentation in an enlarged and four-chambered stomach. (d) The kangaroo has a long fermentation chamber in the first part of the stomach. (Modified from Stevens, 1988.)

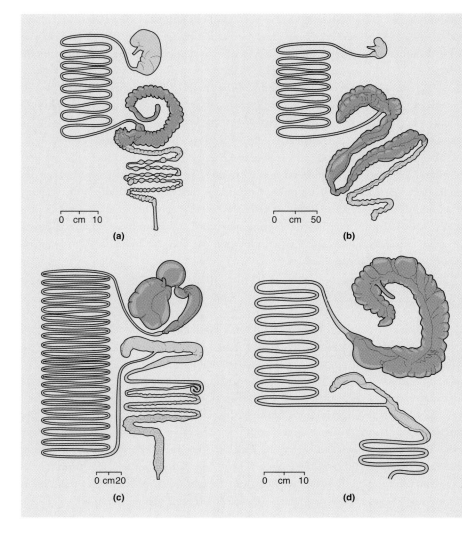

the floras and faunas of the digestive tract are crucial for the digestion of cellulose

involves the passage of parasite eggs in human feces to livestock, for example, to cattle in contaminated food. The parasite develops in the cattle and is then eaten by humans who eat meat from the cattle without cooking it adequately. An even more complex life cycle is illustrated in Figure 7.5.

Immensely diverse communities of mutualistic microorganisms inhabit the alimentary canal of herbivorous animals, playing a crucial role in the digestion of a diet of plants. Ruminants (deer, cattle, and antelopes) have four-chambered "stomachs," and chewed food passes from the first chamber to the second and is then regurgitated and chewed again (rumination) before it passes through to the final chambers and to the small intestine (Figure 7.4). The chambers of the ruminant's stomach provide a fermentation chamber in which huge populations of bacteria and protozoa are stirred in a nutritious anaerobic broth with regulated pH and temperature. In a sheep the volume of the rumen is 4.7 liters, and 20 percent is the bodies of microorganisms. The inhabitants of the rumen are mainly anaerobic; many of them are killed by exposure to oxygen and are unique to this peculiar habitat. The

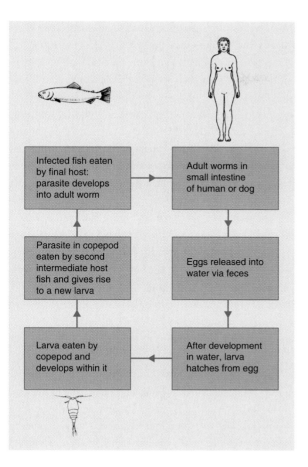

Figure 7.5

The parasitic cestode worm *Diphyllobothrium latum* infects humans who eat infected fish and have not cooked it sufficiently to destroy the parasite. Eggs of the parasite pass in human feces into water, where they hatch and start to develop before they are eaten by copepods (small crustaceans) within which they continue their development. The copepods are eaten in turn by fish, within which they develop to an adult worm and may be eaten by humans (or dogs) in which they produce eggs and start the life cycle again. Notice that the release of eggs into the feces is the only point in the life cycle when the parasite multiplies. The parasite consumes most of the vitamin B_{12} in the host's diet causing severe anemia.

community of species in a rumen is comparable to that of a tropical forest in its diversity and complexity. The protozoa, for example, include species that can decompose cellulose and others that are carnivorous.

The primary importance of this microbial community to the host is that it digests cellulose. Cellulose is the most abundant and bulky of all plant products but is virtually indigestible by animals. Apart from a few insects and protozoa, animals cannot produce cellulases of their own. The anaerobic microbial community ferments cellulose to organic acids that are then absorbed and metabolized by the host. The stomach of ruminants is particularly well studied, but virtually all herbivorous mammals and some lizards and birds rely on a microbial flora to digest cellulose, and specialized fermentation chambers seem to have appeared repeatedly in the evolution of vertebrates (Figure 7.4). Some cellulose fermentation occurs before food reaches the stomach in kangaroos, camels and ruminants, but in horses, rabbits, rodents, elephants and some birds, it is concentrated in the colon and/or the cecum. Moreover, many species, including elephants and rabbits, practice *refection*—they consume their own feces and hence pass material twice through the alimentary canal. This doubles the period for which food is exposed to microbial digestion and, perhaps more important, doubles the period in which the symbiotic mutualists synthesize vitamins that are required by the host.

Many invertebrates also carry a symbiotic microflora that digests cellulose. The stomach of termites, for example, is the habitat of a particularly diverse flora and fauna (Figure 7.6). Some species in this microcosm can even fix atmospheric nitrogen—a valuable property in an associate of a host that spends its life eating something as nonnutritious as wood!

ants inhabit the hollow thorns of acacias

Plants offer a peculiar range of body cavities that become inhabited by microorganisms. For example, flower nectaries are colonized by yeasts, carried from nectary to nectary on the mouthparts of visiting insects. Among the most remarkable inhabited body cavities in plants are the hollow thorns developed by species of *Acacia* in tropical and subtropical America and used as nesting sites by ants. Among the most studied of these are the hollow thorns of the Bull's Horn acacia *Acacia cornigera*, inhabited by the specialized ant *Pseudomyrmex ferruginea* (Figure 7.7). The *Acacia* species produce attractive nectar and also protein-rich *beltian bodies* on their leaves that the ants eat. The ants, for their part, prune off invading shoots from competing neighboring plants and also protect their host from herbivores. *Pseudomyrmex* is just one genus in a subfamily of ants that contains about 230 species. At least 37 of these are specialized inhabitants of plant cavities (domatia) and plant species in 20 genera and 14 families provide these peculiar habitats.

7.2.3 Inhabitants of host tissues and cells

Some species of both parasites and mutualists make their habitat in the tissues of plants. Many of these associations occur within soil, where dispersal from one host to another poses a problem. Organisms that live on roots may form long persistent spores that can await the arrival of a new host, or they persist and grow on a variety of different host species or even on dead organic matter.

the mutualism of fungi and plant roots: mycorrhiza

Most forest trees carry a dense growth of fungi on their root surface. The fungus forms a sheath around the root and encloses its tip. When this happens, the root ceases to elongate and commonly starts to branch. The combined structure of root

Figure 7.6
Gut flora of the termite.

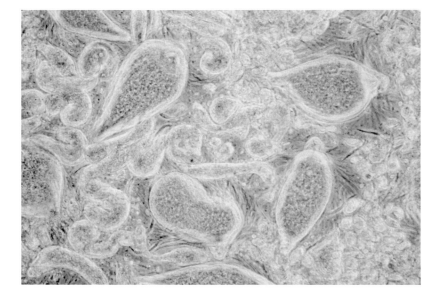

(a)

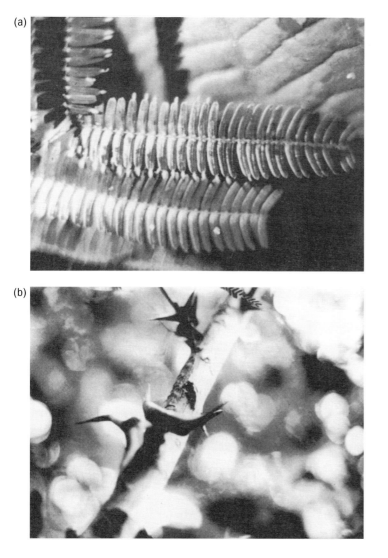

(b)

Figure 7.7
Structures of the Bull's Horn acacia (*Acacia cornigera*) that attract its ant mutualist. (a) Protein-rich Beltian bodies at the tips of the leaflets. (b) Hollow thorns used by the ants as nesting sites. (Courtesy of L. E. Gilbert.)

and fungus has its own morphology and is called a *mycorrhiza*. Filaments of the fungus spread out into the surrounding soil and also penetrate between, though not usually into, the cells of the host root. The fungus obtains energy-rich carbon resources from the host, and phosphate and probably other nutrients pass from the fungal filaments in the soil to the tree via the fungal sheath.

A different type of mycorrhiza is found in most herbaceous plants. In these the fungus spreads from cell to cell of the host root and also out into the surrounding soil. The fungus is tolerated by the host cells and forms highly branched intrusions into them, making intimate intracellular contact between fungus and host. Almost all trees, grasses and perennial herbs are mycorrhizal. Together they account for by far the greatest part of the biomass of terrestrial environments, although, compared to parasites, relatively few species of fungus are involved.

mycorrhizal plants form most of the biomass of terrestrial environments

We have already met organisms that penetrate host cells though the greater part of their body is on the host surface. For many parasites and mutualists, however, active life is spent entirely within host cells, and there is a wide variety of growth

sit-and-wait parasites of higher plants

forms and life cycles all of which solve, in one way or another, the problem of how to disperse from one host to another. A common life cycle among species of parasite that live within the cells of plants involves dormant, often very long-lived spores that germinate in the presence of a host. They release motile single-celled *zoospores,* which penetrate and infect the cells of the host. The parasite multiplies within the host cells, kills them, and releases a new generation of dormant spores. This very simple life cycle is especially common among parasites that persist in the soil and infect plant roots or tubers (club root disease of crucifers [*Plasmodiophora brassicae*] and wart disease [*Synchitrium endobioticum*] of potatoes are examples). Long persistent spores from an infected plant remain dormant, "sitting and waiting" for a new host to disperse to them rather than dispersing in search of a new host. Agriculture and horticulture introduce a quite new element into the dispersal and spread of these parasites, when infected host plants and tubers are carried by people from place to place (often with spores in the attached soil).

flowering plants parasitic on others

There is a specialized flora of flowering plants that find their habitat on other plants, form intimate cellular contact with them, and behave as parasites. Each species has a very limited number of acceptable hosts. The dodders, species of *Cuscuta,* are *holoparasites*: they cannot photosynthesize and are entirely dependent on a host for energy, minerals, and water. When the seed of dodder germinates, the seedling "searches" for a suitable host to colonize. It does this by rotating its shoot tip in a circle and then growing horizontally for a while before becoming erect and rotating again. This continues until either the seedling runs out of its stored resources and dies or it contacts one of a very restricted range of acceptable hosts. It penetrates the host tissues with specialized pegs that make intimate contact within the host tissues. As it branches and spreads over the body of the host plant, the parasite penetrates it repeatedly and soon loses all contact with the ground.

Hemiparasites contain chlorophyll and photosynthesize, but they use other flowering plants as their habitat and obtain water and minerals through intimate connections with the host roots or shoots. Mistletoes are hemiparasites specialized for growth on the branches of trees: their seeds are usually dispersed there by birds. A great variety of mistletoes is found in tropical forests: at least 11 genera and 27 species are found on 126 host species in Sri Lanka alone. Some species of mistletoe (e.g., *Dendrophthoe falcata*) can develop as epidemics in forests and seriously damage forest productivity. *Dendrophthoe*, though, has its own "hyperparasitic" mistletoe (*Viscum capitellatum*), which apparently helps to keep epidemics under control.

the malarial parasite, *Plasmodium*

Many of the parasites of animals that have their habitat within the cells of the host have no dormant persistent stage but depend absolutely on phases of their life cycle that must alternate between two different species of host. The malarial parasite, *Plasmodium,* alternates between humans and mosquitoes and multiplies in both (Figure 7.8).

nitrogen-fixing microbes within the cells of host plants

Many important species of symbiotic mutualists also inhabit and multiply in host cells. Of particular significance are the nitrogen-fixing bacteria of the genus *Rhizobium,* which infect the roots of legumes such as clover, beans, and peas. They stimulate the formation of nodules, which form a specialized habitat in which the bacteria receive resources from the host plant and fix atmospheric nitrogen. They contribute this fixed nitrogen to the host plant, which then becomes independent

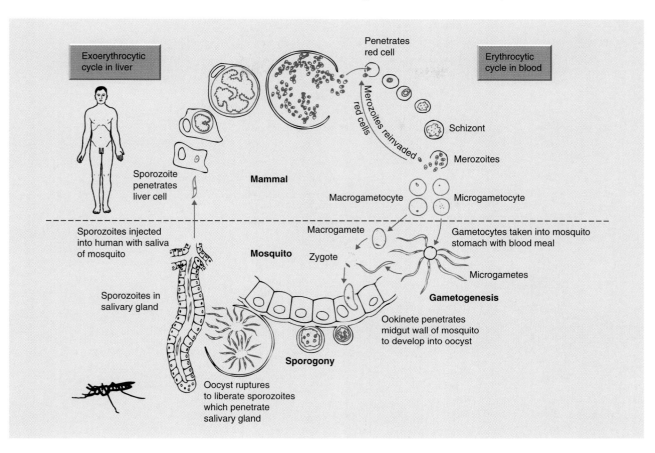

Figure 7.8

The parasite that causes malaria (species of protozoa of the genus *Plasmodium*) lives within the cells of primates and of mosquitoes. It has no life outside its two hosts and must alternate from one to the other. There is no dormant stage in the life cycle. In humans the parasite lives and multiplies in red blood cells, and, as it multiplies, it continually invades new blood cells, causing recurrent bouts of fever in the host. Mosquitoes of the genus *Anopheles* draw blood from the human and *Plasmodium* begins a new intracellular life in the cells of the insect gut and later in the cells of its salivary glands. When the mosquito next draws blood from a human host it infects it with the parasite, which is carried in the insect saliva. The parasite now enters the liver cells of the human and multiplies again. It then escapes from the liver cells and invades the red blood cells. The parasite forms its gametes in the human, but sexual fusion occurs in the mosquito.

of supplies of nitrate or ammonium from the soil. The life cycle is illustrated in Figure 7.9. In some ways the legume behaves as if it is capturing the bacteria and harnessing them to provide its own nitrogen nutrition. The legume even appears to stimulate *Rhizobium* in the rhizosphere by releasing special root exudates. The bacteria multiply to form a colony on a root hair which becomes curled, and the colony stimulates active cell division within the tissues of the root that leads to the formation of a nodule. The cell division starts even before the tissues have become infected. There are mysteries about how exactly the bacteria enter but once inside a root hair they form a tube or "infection thread" which branches and carries the infection to a number of cells, which continue to divide and form much of the body of the nodule. In the infected cells the bacteria multiply further; when the cells are

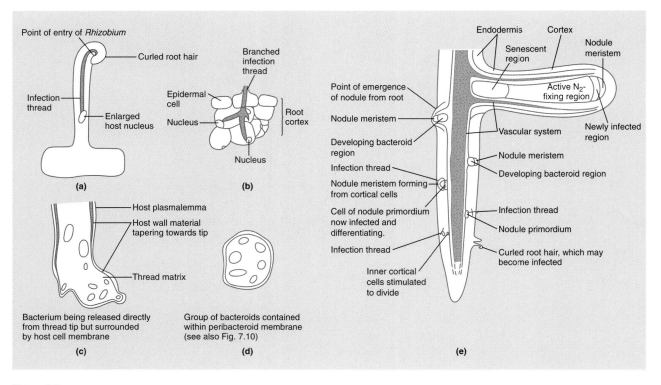

Figure 7.9

A typical life cycle of *Rhizobium*, an intracellular nitrogen-fixing mutualist of legumes such as clover, beans, and peas. (a), (b) and (c) Infection of a host root hair by *Rhizobium*. The hair becomes curled before entry. Subsequently an infection thread carries infection from cell to cell. (d) A mature infected host cell contains tightly packed bacteria ("bacteroids") which have ceased to multiply. (e) The sequence of development of nodules on a host root from infection behind the root tip to the emergence of mature nodules on older parts of the root. (Redrawn from Sprent & Sprent, 1990.)

tightly packed, the bacteria change their shape and become branched *bacteroids* (Figure 7.10), and multiplication stops.

Rhizobial bacteria are widely distributed in soil and grow and multiply in the absence of host plants. But in this free-living condition they do not fix atmospheric nitrogen because this is an anaerobic process. The host provides a paradoxical environment in its nodules: it has rapid diffusive flow of oxygen (allowing the rhizobia to respire aerobically) but extremely low oxygen concentrations (allowing them to fix nitrogen with an effectively anaerobic nitrogenase enzyme).

UNANSWERED QUESTION:

why no nitrogen fixing plants—and so few with nitrogen-fixing microbes?

Enormous research effort has been put into disentangling the details of the evolution and the morphology, anatomy, physiology, and genetics of the interaction between nitrogen-fixing organisms and their hosts, but unanswered questions remain at all these levels. There are many more species of nitrogen-fixing symbionts than previously recognized, and different host species provide quite different environments in their nodules. One important ecological and evolutionary question is why higher plants have never evolved the ability to fix nitrogen for themselves and how it has happened that only a few species have found a solution by providing habitats and resources for nitrogen-fixing microbes. The vast majority of higher plants have to depend on competing among themselves for nitrates or ammonium salts in the soil.

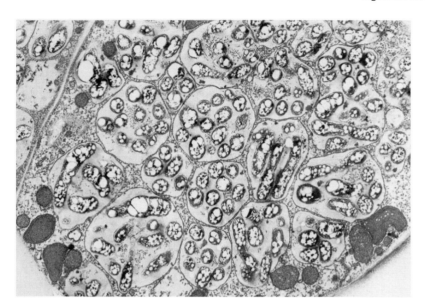

Figure 7.10
Transmission electron micrograph of nitrogen-fixing bacteria (*Rhizobium*) within a cell of a root nodule of soybean (*Glycine max*). The bacteria are grouped in clusters, each cluster enclosed in a peribacterial membrane. Dark bodies at the edge of the cell are host cell inclusions within the cell cytoplasm. (Photograph by courtesy of Janet Sprent.)

Most unicellular green algae are free-living organisms in aquatic or very humid environments. But almost all corals, jellyfish, and sea anemones and many mollusks in shallow tropical waters contain intracellular algae as symbionts—usually flagellates of the genus *Symbiodinium*. In fresh water, too, the sponges, hydras, and some mollusks are hosts for intracellular algae, though here they are usually species of *Chlorella*. As much as 90 percent of the carbon fixed in photosynthesis by symbiotic marine intracellular algae is released to the hosts that provide them with their habitat. Bivalve mollusks that carry algal symbionts capture them directly from free living populations, and flatworms seem to acquire their algae in the same way. However, many corals, anemones, and jellyfish that grow by budding transmit their symbionts from parent to offspring as well as capturing some from free-living forms.

> unicellular algae inhabit specialized cells in many aquatic invertebrates

Intracellular symbionts are usually confined to special cells in the host. An extreme case of this occurs in the many insects that provide a habitat for bacterial symbionts that appear to synthesize B vitamins that the hosts can neither make for themselves nor find in their diet. Aphids carry bacteria of this type (belonging to the genus *Buchnera*) in special cells called *mycetocytes*. The bacteria are transmitted directly from parent to offspring and colonize the mycetocytes in the aphid embryo. If the aphids are freed from their *Buchnera* infection by antibiotic treatment, they cease to reproduce. In effect, *Buchnera* bacteria have become incorporated into the normal life and body of the aphids—neither can live without the other.

> intracellular bacteria inhabit specialized cells in some insects

7.3 ▷ **Hosts as Responsive Habitats**

7.3.1 Changes in the growth and form of hosts caused by their inhabitants

All organisms change the environment in which they live. Often they make it less habitable for themselves (they pollute it! see Chapter 13), though they may transform it into a condition more suitable for others. This generalization is as true for organisms that inhabit living habitats as it is for those that have independent lives.

"parasites" that kill

The most dramatic change that an organism can cause to its living habitat is to kill it. For specialized parasites a dead host is no longer habitable. However, some "parasites" kill their host and then continue to grow and reproduce on the dead body. These are *facultative parasites* (that is, organisms that are not obliged always to be parasitic), but by starting to decompose their food resource before their host is dead they have an advantage over all those that have to wait for death before they can begin to colonize the body. The blowfly, *Lucilia*, is a facultative parasite that commonly kills its host habitat. It may start its life as an egg laid on the body of a mammal such as a sheep. The larvae feed on dung close to the skin or on a wound, and then spread into living tissue, often killing the sheep or making it more likely to die of other causes. They then continue to feed, mature, pupate, and hatch to give further generations, which continue to consume the dead body.

Gray mold (*Botrytis cinerea*) and damping-off (species of *Pythium*) are unspecialized fungal parasites of higher plants. Like blowflies on a sheep they attack and kill their host and then grow and reproduce on the dead tissues. Facultative parasites may use a dead host as a food base from which to invade and kill others.

for obligate parasites a dead host is uninhabitable

Most parasites, however, are obligatorily parasitic. If their host dies, they cease to grow on it. If they persist on the dead host, it is only as an infective dormant residue (e.g., the long-lived spores of *Anthrax* bacilli). Commonly, therefore, though not always, the success of a parasitic way of life depends on keeping the host alive but twisting its growth, form, and behavior in ways that enhance the fitness of the parasite. In response, the host usually reacts to limit the activity of organisms that parasitize it.

parasites and mutualists may induce—or even control—the formation of specialized host tissues

A common host plant reaction to the presence of a parasite is to isolate it with a scab. Potato tubers react to invasion by the parasite *Actinomyces scabies* by forming a layer of cork, which limits it to the tuber surface. Many plants react to leaf parasites by shedding the infected leaves. Some species of parasites and mutualists, however, take control of the local growth of the host and force it to create habitable microenvironments in which the growth of the parasites is supported but confined. The parasitic bacterium *Agrobacterium tumefaciens* induces the formation of galls on plant tissues, especially of woody plants (crown gall disease), and the bacteria then multiply within the enlarged tissues. As in the initiation of nodules on leguminous plants by *Rhizobium* (Section 7.2.3), the host cells divide in anticipation of the entry of the parasite, and gall tissue can be recovered that lacks the parasite but has now been set in its new pattern of morphogenesis: it continues to produce gall tissue—even in tissue culture. The parasite has induced a genetic transformation of the host cells.

Insect galls, caused by agromyzid flies and cecidomyid and cynipid wasps, are some of the most remarkable examples of this takeover of the host's development to provide a habitat for a parasite (Figure 7.11). In the formation of a typical oak gall, the insect lays eggs in the plant tissues which then stimulate it to grow in a quite alien fashion. Just the presence for a time of the parasite's egg appears to be sufficient to start the host into developing a specialized gall. A great many parasites induce galls on species of oak (*Quercus* sp.) and each induces its own specific form of gall. Many of the gall-forming insects have a sexual generation in the spring and

Figure 7.11

Above (a)-(s). Galls formed by wasps of the genus *Andricus* on oaks (*Quercus petraea, Q. robur, Q. pubescens*, or *Q. cerris*). Each figure shows a section through a gall induced by a different species of *Andricus*. The dark shaded areas are the gall tissue and the central unshaded areas are the cavities containing the insect larva. (From Stone & Cook, 1998.) Below. The morphological changes induced in oak trees by gall wasps to create the habitats for their larvae are very precisely controlled by specific interactions of the genes from both the oak and the wasp. (left) Brightly colored "oak apples" are produced by a cynipid wasp on the branches of the oak *Quercus engelmanii*. (right) Flat mushroom-like outgrowths are developed from the underside of oak leaves by the gall wasp *Cecidomyia paculum*.

an asexual generation in the summer and fall, and even these induce quite different shaped galls on quite different parts of the plant. The galls provide a supply of food for the parasitic insect larvae and give them some protection from parasites of their own.

7.3.2 Changes in host behavior caused by their inhabitants

Organisms that use a host as a habitat may manipulate its behavior to their advantage. The human threadworm *Enterobius vermicularis* lives mainly in the cecum, but at nights the adult females migrate to the anus and lay eggs on the skin that surrounds it. This causes the host to itch and scratch: as a result the eggs contaminate the hands and transmit easily to reinfect through the mouth; infection spreads rapidly in families. Plant parasites such as late blight of potatoes cause the leaves to curl in ways that expose fungal spores to dispersal. *Anopheles* mosquitoes infected with the malarial parasite *Plasmodium* take more blood and take it from more individuals than uninfected mosquitoes—thus increasing the chances of parasite transmission. In a field sample of infected mosquitoes, 22 percent contained blood from at least two people compared with only 10 percent of uninfected mosquitoes (Figure 7.12). There can be few more effective ways in which *Plasmodium* could change its host's behavior to its own reproductive ends!

The gordian worm (*Gordius dimorphus*) infects the aquatic larvae of insects such as mayflies, caddis flies, stoneflies, and alderflies. The parasite remains in the aquatic host until it metamorphoses to an adult and leaves the water. The adult insect may then be eaten by large carnivorous insects such as the praying mantis or the New Zealand weta and the parasite develops further at the expense of their internal organs. The parasite grows until it occupies the entire abdomen, coiled up tightly like a watch spring. The infected host now develops a desperate craving for water and jumps in when it finds it. The parasite emerges immediately as a long

Figure 7.12
The effect of infection by the malaria parasite (*Plasmodium falciparum*) on the probability that the mosquito (*Anopheles gambiae*) will have more than one meal. The bars show the percentage of multiple meals in each size group. The light bars represent the uninfected and the dark bars the infected mosquitoes. (From Koella et al., 1998.)

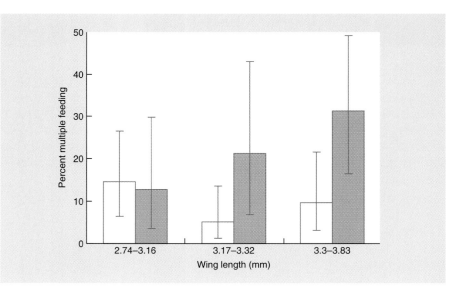

black worm, and the life cycle starts again with infection of aquatic insect larvae. The craving for water that the parasite induces in its terrestrial host is so extreme that mantises rescued from the water will quickly make their way back to it and throw themselves in again.

Many of the organisms that use hosts as habitats use the host's sexual activity to hitch a ride from host to host—most obviously, perhaps, the bacteria that cause venereal diseases of humans (gonorrhea, syphilis). Similarly, though, many viruses of trees are transmitted "sexually" from tree to tree by pollen, and in several species of plant (e.g., the loose smuts of cereals, *Ustilago tritici*, and anther smut of clover, *Botrytis anthophila*) the fungal parasite replaces the pollen or seeds of the host with its own spores.

microbial inhabitants may twist host behavior to their own advantage

7.3.3 Immune and other defensive responses of the host

When an organism reacts to the presence of another in its body it is recognizing and distinguishing between what is self and not-self. In invertebrates, populations of phagocytic cells are responsible for most of a host's response to invaders, even to inanimate particles. Phagocytes may engulf and digest small alien bodies and encapsulate and isolate larger ones. In vertebrates there is also a phagocytic response to material that is not-self, but their armory of defense also includes a much more elaborate process: the immune response (Figure 7.13).

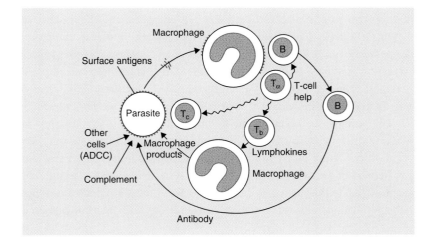

Figure 7.13

A vertebrate host is a reactive habitat in which the immune response of the host converts it from being habitable into being uninhabitable. This sort of reaction distinguishes living organisms as habitats from the sorts of environment more normally considered by ecologists. The response begins when the immune system is stimulated by an antigen that is taken up and processed by a macrophage. The antigen is a part of the parasite, such as a surface molecule. The processed antigen is presented to T and B lymphocytes. T lymphocytes respond by stimulating various clones of cells, some of which are cytotoxic, as others secrete lymphokines, and yet others stimulate B lymphocytes to produce antibodies. The parasite that bears the antigen can now be attacked in a variety of ways. (After Cox, 1982.)

the immune responses of vertebrates allow hosts to eliminate populations of bacterial and viral parasites

The ability of vertebrates to recognize and reject tissue grafts from unrelated members of their own species is just one example of the immune response in action (see Figure 7.13). For the ecology of parasites, an immune response has two vital features. First, it may enable a host to eliminate the population of a parasite and recover from disease. Second, it can give a once-infected host (= habitat) a "memory" that changes its behavior if the parasite (= colonist) strikes again; the host has become immune to reinfection. In mammals, the transmission of immunoglobulins to the offspring in the mother's milk can sometimes even extend protection to the early life of the next generation. The response of vertebrates to infection therefore protects the individual host and increases its potential for survival to reproduction. The response of invertebrates gives poorer protection to the individuals and, when populations recover after disease, this depends more on the high reproductive potential of the survivors than on the recovery of those that had been infected.

transient and persistent infections

For many viral and bacterial infections of higher animals, the colonization of the host is a brief and transient episode. The parasites multiply within the host and elicit a strong immunological response. Assuming the host survives, the parasite population is eliminated and the host acquires lasting, often lifelong immunity to reinfection. By contrast, the immune responses elicited by many of the worms and protozoan parasites, and indeed for example, by herpesviruses, tend to be relatively ineffective. The infections themselves, therefore, tend to be persistent, and hosts are subject to continual reinfection. For the dog or cat lover this means that treatment for fleas and intestinal worms must continually be repeated through the life of the pet, unlike immunization to some viral infections, which may last for life.

plants have no immune system that distinguishes self from not-self

The modular structure of plants, the presence of cell walls, and the absence of a true circulating system (such as blood or lymph) make any form of immunological response an inefficient protection. There is no migratory population of phagocytes in plants that can be mobilized to deal with invaders. Many plants, however, respond even to the earliest stages of infection by "hypersensitive" reactions. The infected cells die, and they and the immediately surrounding cells produce *phytoalexins*—antimicrobial compounds that accumulate to inhibitory levels in the locality of infection. There is also increasing evidence of signaling mechanisms from one part of a plant to another. These may allow one infected part to induce the formation of defenses in the rest of the plant body, and there is some evidence that an infected plant (for example, by aphids) may use gaseous signals to induce defenses in other members of the population. This is an exciting area of biology with many unanswered questions.

7.4 ▷ The Distribution and Regulation of Parasites and Mutualists within Hosts and Their Populations

parasites are usually aggregated

The distribution of parasites within populations of hosts is rarely random. It is usually the case that many hosts harbor few or no parasites, and a few hosts harbor many—the distributions are usually aggregated or clumped (Figure 7.14). Some of the aggregation results when microparasites (Box 7.1) multiply within randomly

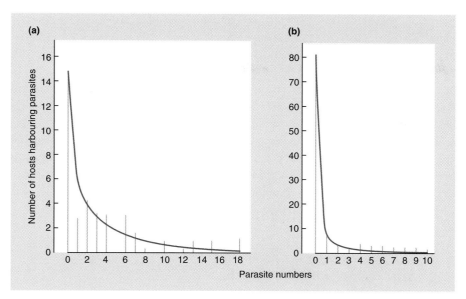

Figure 7.14
Examples of distributions of parasite numbers per host. The vertical bars represent the observed frequencies of different numbers of parasite per host and the solid lines the ideal distribution obtained if every parasite has an equal chance of infecting or failing to infect a host. (a) The distribution of *Toxicara canis,* a gut nematode parasite of foxes and (b) lice (*Pediculus humanis capitis*) on a population of humans. (After Watkins & Harvey, 1942; Williams, 1944.)

infected hosts and quickly generate gross differences between the zero populations of parasites in uninfected hosts and dense populations in others. However, when dispersal is weak or slow, very dense, very local aggregations of infected hosts can occur simply because most transmission occurs between immediate neighbors rather than randomly within a population. Aggregations also arise if individual hosts vary in their susceptibility to infection (whether as a result of genetic, behavioral, or environmental factors).

the prevalence, intensity, and mean intensity of infection

A widely used statistic to describe the distribution of parasites on a population of hosts is the proportion of the host population that is infected. This is termed the *prevalence* of the parasite. It is a statistic with obvious ecological meaning—it is the proportion of habitable patches in an environment that are actually inhabited. The number of parasites in or on a particular host is referred to as the *intensity* of infection. This is comparable to the local density of birds on an island or insects in a patch of vegetation. The *mean intensity* of infection is then the mean number of parasites per host in a population, including those hosts that are not infected, just as the mean density of fish in a series of ponds includes those ponds that contain no fish. In locally aggregated populations, however, the mean density of parasites (mean intensity of infection) may have rather little useful meaning. In a population of humans in which only one person is infected with anthrax, the mean density of *Bacillus anthracis* is a particularly useless piece of information.

the populations of organisms that inhabit a host are regulated

Almost all parasites and symbiotic mutualists are smaller than their hosts and can grow much faster. Yet their growth is constrained within the hosts, and this implies that there are density-dependent processes (crowding effects; see Chapters 3 and 5) at work. It is technically extremely difficult to study the population dynamics of microparasites, which multiply within the host cells (imagine how one might try to follow the growth of a population of measles virus particles in the cells of a child). However, it is quite practicable to study the effects of population size on the behavior of some macroparasites (Box 7.1) and even to control levels of infection experimentally. The results of the experiments shown in Figure 7.15, for example, are remarkably similar to the density-dependent relationships found to result from intraspecific competition and discussed in Chapter 5.

The number of algal cells per host cell in corals and lichens, and of rhizobial bacteria in the cells of legume nodules, must also all be regulated. The algal symbionts of the coral *Stylophora pistillata* can increase 100 times faster than their hosts, but in practice they do not outgrow them. The ways in which such symbiont populations are regulated are obscure, but it is probable that the growth rate of the algae is limited by mineral nutrients and/or carbon—just as in more traditional habitats, density-dependent regulation often involves the effects of resource shortage on birth and death rates.

UNANSWERED QUESTION:

how do symbiotic mutualists regulate one another's populations?

7.5 Life Cycles and Dispersal

The most complicated of all life cycles are found among the parasites; some of these have already been illustrated. Most parasites are extremely specialized and are able to infect, grow, and reproduce on one or a very few, usually closely related species of host. Almost all organisms are totally resistant to almost all parasites! It is therefore all the more remarkable that the life cycles of a number of important parasites require them to spend alternate periods of their lives on two different and usually quite unrelated hosts. Where there is such an alternation, the parasite usually has its sexual phase on one host and most of its multiplication on the other. This means that most genetic recombination and heritable variation are generated on one host and most of the natural selection of its results occurs on the other.

Complex life cycles such as that of the tapeworm in Figure 7.5 imply a quite extraordinary degree of habitat specialization of the parasite. These specialists, in a

Figure 7.15
Density-dependent behavior of macroparasites within their hosts. (a) The relationship between the density of worms and the number of eggs produced by a population of the roundworm *Ascaris* in the human digestive tract and counted in the feces. (b) A similar relationship for the tapeworm *Ancylostoma* in humans.

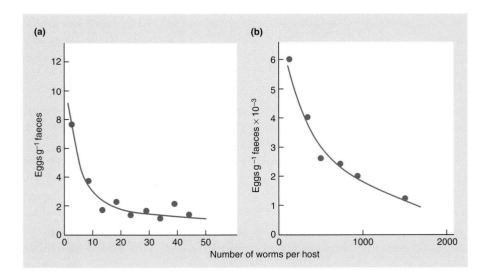

Number of worms per host

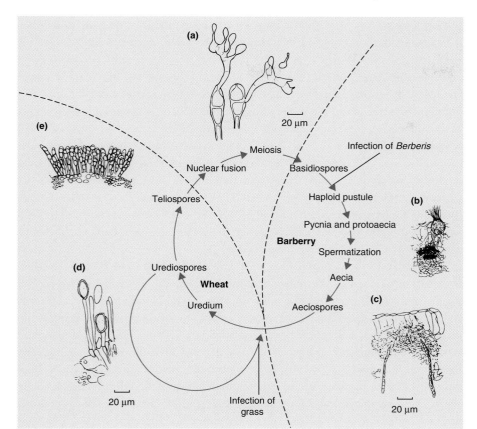

Figure 7.16
The life cycle of black stem rust of wheat (*Puccinia graminis tritici*). The parasite infects woody shrubs of the genus *Berberis* or of the closely related *Mahonia*. The fungus spreads through the cells and tissues of the host, and, after a sexual process has occurred, the fungus forms pustules, which break out from the lower surface of the leaves and release spores (aeciospores). These are quite unable to infect (colonize) *Berberis* or *Mahonia* and can infect only wheat (*Puccinia graminis*). On the wheat host the fungus produces a new type of spore called a *uredospore* which can only infect further wheat plants. Cycles of infection and reinfection of wheat plants can repeat indefinitely. Epidemics develop in crops of wheat if climatic conditions are suitable. At the end of the host's growing season a different type of spore is produced (a teliospore), which cannot infect anything but is thick-walled and can remain alive but dormant. Eventually the teliospores germinate, and, at last, sexual fusion occurs, followed almost immediately by meiosis and the formation of yet a further type of spore (the *basidiospore*) which can infect only *Berberis* or *Mahonia*.

lifetime that extends from parent zygote to offspring zygote, must pass in a defined progression from one defined host to another. Similarly, many of the rust fungi (Uredinales) have life cycles that involve spending phases on two quite unrelated species of host plant (e.g., black stem rust; Figure 7.16). The closest parallels in traditional habitats are the life cycles of amphibians, which must pass from an aquatic habitat for the larvae to a terrestrial habitat for the adults, or of butterflies and moths, which must pass from the larval habitat of a leafy host to the flowers frequented by the adults.

Two relationships are often confused when considering complex life cycles. We refer to *alternate hosts* when there is an obligatory alternation of the parasite between one host and another, as in the tapeworm and black stem rust examples. In contrast, we use the phrase *alternative hosts* to indicate that a particular phase of a parasite's life cycle may occur on different host species. *Berberis* and *Mahonia* are examples of alternative hosts in the life of black stem rust, but *Berberis/Mahonia* and wheat are alternate hosts (Figure 7.16). Similarly, at least 150 species of mammal, including humans, can serve as alternative hosts for the trypanosome *Trypanosoma cruzi*, which causes Chagas' disease in the human host, although the parasite itself multiplies in the gut of 50–60 alternative species of hemipteran insects—but the mammal and the bug are alternate hosts. The parasite must pass from mammal to bug to mammal

alternate and alternative hosts

alternate hosts and vectors

to bug and so on. Note that alternative hosts are usually closely related (often members of the same genus), whereas alternate hosts are never phylogenetically close.

A second easy confusion is between alternate hosts and vectors. Many parasites are spread from one host to another by a vector organism, but the parasite does not draw resources from the vector. A vector only disperses a parasite, often simply as a contaminant, for example, on its mouthparts. This is the most common route of passage of plant viruses from their habitat in one plant to another. Mosaic virus of corn is transmitted by a vector, the plant hopper *Peregrinus maidis*, and the vector of maize bushy stunt virus is a leaf hopper (*Dalbulus* sp.). Aphids are particularly effective vectors of plant viruses. They insert their feeding stylet through tissues of the host plant into the living cells of the plant's phloem transport system, and this places any contaminant virus or bacterium in a position optimal for its spread.

Among animals, probably the most noteworthy groups of vectors are flies, especially mosquitoes. These transmit a number of important human diseases, including malaria and yellow fever. The adult fly takes several blood meals in its short life and transmits infection when consecutive meals are taken from an infectious and then an uninfected, susceptible host (see also Figure 7.12). Ticks are also important vectors. They take only one blood meal in each of their life stages and therefore harbor infections for long periods, often more than a year, while they metamorphose from one stage to the next. Many species of tick are capable of feeding from several alternative host species, and they have recently gained increasing notoriety as the bearers of infections, such as Lyme disease, that normally circulate in wildlife species, because ticks can transmit the disease to humans.

7.6 ▷ Coevolution

Parasites and mutualists act as selective forces on the evolution of the plants and animals that they use as hosts. Between hosts and mutualists there is no conflict—hosts gain in fitness if they allow or encourage invasion by mutualists, which also then benefit—but this is clearly not the case for parasites. Hosts benefit if they evolve resistance or tolerance to parasites and pathogens, but parasites gain fitness by penetrating the defenses of their hosts. One result is to lead parasites deeper and deeper into the rut of extreme specialization: as we have already noted, all species of plant and animal are totally resistant to all but a very few species of parasite. The relationship between hosts and the parasites that use them as habitats involves as precise a match between organism and environment as occurs anywhere in nature.

parasite and host variation: influenza and TB

Indeed, the specialization may go further than that between species. Within species of obligate parasites it is common to find a high degree of genetic variation in the virulence of parasites and/or in the resistance or immunity of hosts. Every few years, a new strain of the influenza virus evolves of sufficient virulence and novelty to generate a widespread epidemic and mortality in human populations that had been relatively resistant to previously circulating strains. No strain has been more devastating than the worldwide epidemic (*pandemic*) of Spanish flu that followed World War I in 1918–1919 and killed 20 million people—many more than died in the war itself. Human diseases can also provide examples of variation in host resistance. When the American Indians of the Canadian Plains were forcibly settled onto reservations in the 1880s, their death rate due to tuberculosis (TB) initially exploded but then gradually declined (Figure 7.17). Environmental factors (inadequate diet, overcrowding, spiritual demoralization) undoubtedly played some part in this, but variation in resistance is also likely to have been significant. The mortality rate among the Indians was often 20 times that of the surrounding white population living in similar conditions but which had previously been exposed to TB. Some Indian families suffered a particularly low mortality in the 1880s epidemic, and many of the survivors in the 1930s were descendants of those families (Ferguson, 1933; Dobson & Carper, 1996).

In some cases, especially in plant diseases—flax and flax rust, powdery mildew of barley, stem rust of wheat—for every gene controlling resistance in the host, there is a corresponding gene in the parasite controlling pathogenicity (a precise gene-for-gene relationship). This may be an oversimplification, but it provides a useful framework within which to think about the perpetual struggle of plants in nature (and in commercial plant breeding) to keep host plants one step ahead of their pathogens.

gene-for-gene relationships and plant breeding

A most dramatic example of the ecological and evolutionary interaction of a parasite and its host involves the rabbit and the myxoma virus, which causes the disease myxomatosis. The disease originated in the South American jungle rabbit *Sylvilagus brasiliensis*, in which it is a mild disease that only rarely kills the host. It is, however, usually fatal when it infects the European rabbit *Oryctolagus cuniculus*. In one of the greatest examples of the biological control of a pest, the myxoma virus

myxomatosis: coevolution in action

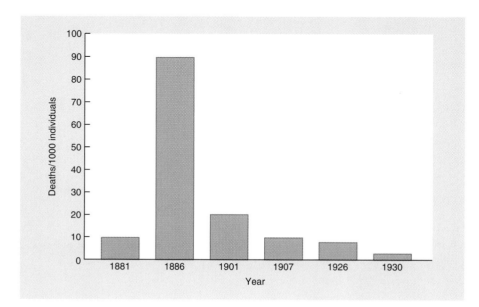

Figure 7.17
The mortality rate due to tuberculosis in three generations of Canadian Plains American Indians after their forced settlement onto reservations. (After Ferguson, 1933; Dobson & Carper, 1996.)

was introduced into Australia in the 1950s to control the European rabbit, which had become a pest of grazing lands. The disease spread rapidly in 1950–1951, and rabbit populations were greatly reduced—by more than 90 percent in some places. A little later, the virus was introduced to England and France, where it also resulted in huge reductions in the rabbit populations. The evolutionary changes that then occurred in Australia were followed in detail by Fenner and his associates, who had the brilliant research foresight to establish baseline genetic strains of both rabbits and virus. They used these to measure subsequent changes in the virulence of the virus and the resistance of the host as they evolved in the field.

When the disease was first introduced to Australia it killed more than 99 percent of infected rabbits. This *case mortality* fell to 90 percent within a year and then declined further (Fenner & Ratcliffe, 1965). The virulence of virus was graded according to the survival time and case mortality of control rabbits. The original, highly virulent virus (1950–1951) was grade I, which killed >99 percent of infected laboratory rabbits. Already by 1952 most of the virus isolates from the field were grade III and IV (Figure 7.18). At the same time the rabbit population in the field was increasing in resistance. When injected with a standard grade III strain of the virus, field samples of rabbits in 1950–1951 had a case mortality of nearly 90 percent, which had declined to less than 30 percent only 8 years later (Marshall & Douglas, 1961).

parasites do not evolve to preserve their hosts

This evolution of resistance in the European rabbit is easy to understand. The case of the virus, however, is more subtle. The contrast between the virulence of the myxoma virus in the European rabbit and its lack of virulence in the American

Figure 7.18

(a) The percentages in which various virulence grades (I–V) of myxoma virus have been found in wild populations of rabbits in Australia at different times from 1951 to 1981. Grade I is the most and grade V the least virulent. (b) Similar data for wild populations of rabbits in Britain from 1953 to 1980. (After May & Anderson, 1983; from data collected by Fenner, 1983.)

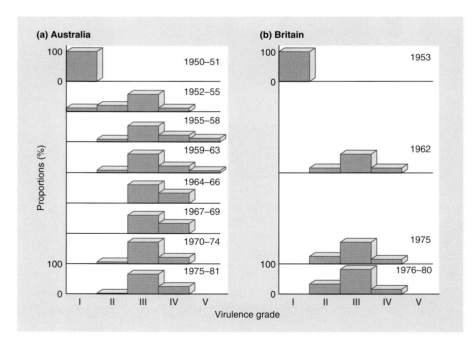

host with which it had coevolved, combined with the attenuation of its virulence in Australia and Europe after its introduction, fit a commonly held view that parasites evolve toward becoming benign to their hosts in order to prevent the parasite eliminating its host and thus eliminating its habitat. This view, however, is quite wrong. The parasites favored by natural selection are those with the greatest fitness (broadly, the greatest reproductive rate). Sometimes this is achieved through a decline in virulence, but sometimes it is not. In the myxoma virus, an initial decline in virulence was indeed favored—but further declines were not.

The myxoma virus is blood-borne and is transmitted from host to host by blood-feeding insect vectors. In Australia in the first 20 years after its introduction, the main vectors were mosquitoes (especially *Anopheles annulipes*), which feed only on live hosts. The problem with grade I and II viruses is that they kill the host so quickly that there is only a very short time in which the mosquito can transmit them. Effective transmission may be possible at very high host densities, but as soon as densities decline, it is not. Hence, there was selection against grades I and II and in favor of less virulent grades, giving rise to longer periods of host infectiousness. At the other end of the virulence scale, however, the mosquitoes are unlikely to transmit grade V of the virus because it produces very few infective particles in the host skin that could contaminate the vectors' mouthparts. The situation was complicated in the late 1960s when an alternative vector of the disease, the rabbit flea *Spilopsyllus cuniculi* (the main vector in England), was introduced to Australia. There is some evidence that more virulent strains of the virus may be favored when the flea is the main vector.

Overall, however, there has been selection in the rabbit-myxomatosis system not for decreased virulence but for *increased transmissibility* (and hence increased fitness)—which happens in this system to be maximized at intermediate grades of virulence. Many parasites of insects rely on killing their host for effective transmission. In these very high virulence is favored. In yet other cases natural selection acting on parasites has clearly favored very low virulence: for example, the human herpes simplex virus may do very little tangible harm to its host but gives it effectively lifelong infectiousness. These differences reflect differences in the underlying host–parasite ecologies, but what the examples have in common is that there has been evolution toward increased parasite fitness.

In this chapter we have focused on organisms as the habitats of both parasites and mutualists. Some of the contrasts between them are summarized in Box 7.2. A susceptible host is a habitable patch (or an inhabitable "island") in the environment. The dynamics of populations of parasites and pathogens depend on the colonization of these patches and on the rate at which they become locally extinct. These in turn depend on the distance apart and longevity of the patches and the dispersibility of the parasite. We discuss the broader ecological issues of patch and island population behavior and metapopulation dynamics in Chapters 9 and 10. But we need to remember that parasites represent a special case: motile animals are motile habitats, moving patches, migrating islands. They pair and mate, they may have social behavior, and they can run away. This adds a delightful subtlety to thinking about hosts as habitats.

hosts as patches—but special patches

Box 7.2 📖

Some Contrasts between Parasites and Mutualists

In comparison with species of parasites, there are remarkably few species of symbiotic mutualists, because they are much less specialized to particular hosts. The greatest diversity is found in the lichens (each lichen is a symbiotic association between a fungus and an alga), but although there are at least 13,500 species of lichen, their algal symbionts are represented by only 30–40 species of green algae and 12 or so species of blue–green algae. More than half the known species of lichen, in fact, carry just one genus of algal symbiont, *Trebouxia*. Similarly, algal symbionts of marine invertebrates are drawn almost entirely from just one genus of dinoflagellate algae (*Symbiodinium*) and one genus of cyanobacteria (*Prochloron*) that inhabits sea squirts. Just three genera of rhizobial bacteria account for all the nitrogen-fixing symbionts of legumes.

Symbiotic mutualists have rather simple life cycles and, unlike parasites, none of them requires an alternation between hosts: there are no alternate hosts but lots of alternative hosts.

The lives of parasites are dominated by dispersal. Their continued existence depends typically on transmission and infection from one unique type of host to another of the same sort or one that is equally unique. But symbi-

otic mutualists lack any structures or behavior specialized for dispersal from host to host and vectors are not involved. Legumes, for example, acquire their rhizobia from free-living forms in the soil. Mollusks and some other marine invertebrates acquire their algal symbionts from free-living cells in the sea. When animal host species are modular and grow by budding or branching (corals, anemones, and so on) algal symbionts pass directly to the tissues of the buds and branches. The intracellular symbiotic bacteria of aphids, *Buchnera*, are vertically transmitted through the mother to her eggs. They have become so completely assimilated into the life of their hosts that they appear to have lost any vestige of a free-living or dispersive stage or even any independent life of their own.

A remarkable feature of the lives of symbiotic mutualists is the almost complete suppression of sex. This is a great contrast with the lives of parasites, and, since sexual reproduction generates so many new variants, the lifeblood of evolution, this presumably reflects the continual evolutionary battle between parasites and their hosts. By definition there is no battle between mutualists and their hosts.

❯ Summary

Symbiosis, parasitism, and mutualism

An intimate association between individuals of different species, in which one lives on or in the other, is known as a *symbiosis*. A *parasite* is an organism that lives its life intimately associated with one or a very few individuals of another species, its host, deriving resources from its host and doing its host harm, but not predictably killing its host in the short term. A *mutualism* is an association between species (usually a pair of species) that benefits both. *Commensals* are neither harmful nor beneficial to their host. Parasites contribute overwhelmingly to the diversity of life on Earth, but mutualistic associations often dominate the production of biomass.

Life on the surface of another organism

The bodies of many organisms provide anchorage for others and give them access to conditions and resources that are otherwise unavailable. These ecological engineers create, modify, maintain, or create habitats for others.

Plant roots shed root cap cells and release organic materials to the adjacent soil. This *rhizosphere* carries its own abundant microbial flora. There is also often a distinct *phyllosphere* of species on leaf surfaces.

Fleas and lice are examples of *ectoparasites* on the bodies of larger animals. Organisms that live on the surface of their hosts are usually well placed to colonize other hosts, but there are costs: ectoparasites are more exposed to the ambient hazards of drought, heat, cold, and predators and make less intimate contact with the host.

Inhabitants of the body cavity

The alimentary canal is a resource-rich, protected, and regulated habitat. Many parasites live there, competing with the host for its digestive products. Most have complex life histories that carry infection from the feces of the host and disperse it widely among others in the host population. Immensely diverse communities of mutualistic microorganisms inhabit the alimentary canal of herbivorous animals, playing a most crucial role in the digestion of cellulose.

Inhabitants of host tissues and cells

The combined structure of root and fungus is called a *mycorrhiza*. The fungus obtains energy-rich carbon resources from the host, and phosphate and other nutrients pass from the fungus to the plant.

Many parasites that live within the cells of plants have dormant, long-lived spores that germinate in the presence of a host. Many of the parasites of animals that have their habitat within the cells of the host have no dormant persistent stage but depend absolutely on alternating phases of their life cycle between two different species of host.

Nitrogen-fixing mutualistic bacteria of the genus *Rhizobium* infect the roots of leguminous plants (though there are other nitrogen-fixing mutualisms). They contribute fixed nitrogen to the host plant, which is then independent of supplies of nitrogen from the soil. Much of the carbon fixed in photosynthesis by marine intracellular algal symbionts is released to hosts that provide their habitat.

Hosts respond to infection

Some "parasites" kill their host and then continue to grow and reproduce on the dead body, preempting those that have to wait for an organism to die. Most parasites, however, are obligatorily parasitic. If their host dies, they cease to grow on it. Other parasites (and mutualists) take control of the local growth of their hosts and force them to create habitable micro-environments. Still others manipulate host behavior.

In invertebrates, phagocytic cells are responsible for most of a host's response to invaders. In vertebrates there is also a much more elaborate process: the immune response. This may enable a host to eliminate a parasite and recover from disease but can also make a once-infected host immune to reinfection. There are no migratory phagocytes in plants, but many respond to infection by producing phytoalexins that are strictly localized and nonspecific in their action.

Parasites are usually aggregated within populations of their hosts. The *prevalence* of a parasite is the proportion of the host population that is infected. The number of parasites in or on a particular host is the *intensity* of infection. The *mean intensity* of infection is the mean number of parasites per host in a population. Almost all populations of parasites and symbiotic mutualists are constrained within their hosts by density-dependent processes (crowding effects).

The life cycles of many parasites require them to spend alternate periods of their lives on two different and usually quite unrelated, "alternate" hosts. Many parasites are spread from one host to another by a vector organism, but the parasite does not draw resources from the vector.

All species of plant and animal are totally resistant to all but a very few species of parasite. In some cases, especially for plant diseases, a *gene-for-gene* relationship has been suggested: for every gene controlling host resistance, there is a corresponding gene in the parasite controlling pathogenicity.

Myxomatosis illustrates coevolution of parasite and host but also indicates that parasites do not evolve to preserve their hosts. Rather, the parasites favored by natural selection are those with the greatest fitness (broadly, the greatest reproductive rate). Sometimes this is achieved through a decline in virulence, but sometimes it is not.

Review Questions

▲ = Challenge Question

1. Define and give examples of symbiosis, parasitism, mutualism and commensalism.

2. Contrast the communities of parasites and mutualists inhabiting the vertebrate alimentary canal.

3. What are mycorrhizae and what is their significance?

4. ▲ Discuss the following propositions: "Most herbivores are not really herbivores but consumers of the by-products of the mutualists living in their gut," and "Most gut parasites are not really parasites but competitors with their hosts for food that the host has captured."

5. ▲ Provide a careful interpretation of Figure 7.19.

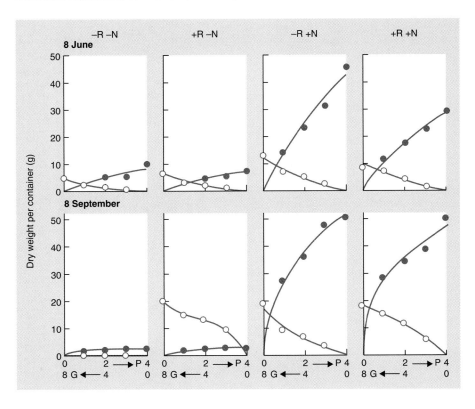

Figure 7.19
Soybeans (*Glycine max*, G—a legume that can form a mutualistic associa- tion with *Rhizobium*) and a grass (*Paspalum*, P) are grown alone and in mixtures with and without nitrogen fertilizer and with and without inoculation with nitrogen-fixing *Rhizobium*. The plants were grown in pots containing zero to four plants of the grass and zero to eight plants of *Glycine*. The hori- zontal scale on each figure shows the proportions of plants of the two spe- cies in each container. −R −N, no *Rhizobium*, no fertilizer; +R −N, inoculated with *Rhizobium* but no fertilizer; −R +N, no *Rhizobium* but nitrate fertilizer was applied; +R +N, inoculated with *Rhizobium* and nitrate fertilizer supplied. (After de Wit et al., 1966.)

6. Account for the fact that the blowfly, *Lucilia*, often kills its host, whereas most parasites do not.

7. ⏶ Give examples of parasites that change (a) the morpho- logical characteristics and (b) the behavior of their host. Can these changes be interpreted as defenses by the host or opportunities gained by the parasite?

8. ⏶ What do we really mean when we say that a parasite harms its host?

9. Compare and contrast the defensive responses to parasites of higher vertebrates, invertebrates, and plants.

10. Account for the decline in virulence of the myxomatosis virus in European rabbits after its initial introductions in Australia and Europe.

Every living organism either is a consumer of other living organisms, or is consumed by other living organisms, or—in the case of most animals—is both. We cannot hope to understand the structure and dynamics of ecological populations and communities until we understand the links between consumers and their prey.

8

Predation, Grazing, and Disease

Key Concepts

In this chapter you will

- distinguish the similarities and differences among "true predators," grazers, and parasites.
- understand the subtleties of predation, including the capacity of prey to compensate.
- appreciate the value of the optimal foraging approach for analyzing predator choices.
- recognize the underlying tendency of populations of predators and prey to cycle and the "damping" effect of crowding and patchy distributions.
- understand the consequences of predation for community composition.

predator: a term extending beyond the
obvious examples

"true" predators, grazers, and
parasites

8.1 ▷ Introduction

Ask most people to name a predator and they are almost certain to say something like a lion, tiger, or grizzly bear—something big, potentially ferocious, instantly lethal. However, from an ecological point of view, predator is a term that can be used to encompass a whole variety of organisms that obtain their resources by consuming other living organisms. They are not all large, aggressive, or instantly lethal—and they need not even be animals. By considering these consumers together, we can more readily understand the part each plays in determining the structure and dynamics of ecological systems.

In its broadest sense, then, a *predator* may be defined as any organism that consumes all or part of another living organism (its prey or host), thereby benefiting itself but, under at least some circumstances, reducing the growth, fecundity, or survival of the prey. Notice that this definition extends beyond the likes of lions and tigers by including those that consume all *or part* of their prey, and those that merely *reduce* their prey's growth, fecundity, or survival. Within this broad definition, however, three main types of predator can be distinguished.

"True" predators
- invariably kill their prey and do so more or less immediately after attacking them
- and consume several or many prey items in the course of their life

True predators therefore include lions, tigers, and grizzly bears, but also spiders, baleen whales that filter plankton from the sea, the zooplanktonic animals within that community that consume the phytoplankton; birds, rodents, and ants that eat seeds (each one an individual organism); cutworm caterpillars that eat through the roots of seedlings and young plants and invariably kill them; carnivorous plants; and so on.

Grazers
- also attack several or many prey items in the course of their life
- but consume only part of each prey item
- and do not usually kill their prey, especially in the short term

Grazers therefore include cattle, sheep, and locusts; mobile, nibbling caterpillars that move from plant to plant; but also, for example, blood-sucking leeches that take a small, relatively insignificant blood meal from several vertebrate prey over the course of their life.

Parasites
- also consume only part of each prey item (usually called their *host*)
- and also do not usually kill their prey, especially in the short term
- but attack one or very few prey items in the course of their life, with which they therefore often form a relatively intimate association

Parasites therefore include some obvious examples: animal parasites and pathogens such as tapeworms and the tuberculosis bacterium, plant pathogens like tobacco mosaic virus, parasitic plants like mistletoes, and the tiny wasps that form galls on oak leaves. But aphids that extract sap from one or a very few plants with which they enter into an intimate association, and even caterpillars that spend their whole life on one host plant, are also, in effect, parasites. A useful distinction, between microparasites and macroparasites, was drawn in Box 7.1.

On the other hand, these distinctions between true predators, grazers, and parasites, as with most categorizations of the living world, have been drawn in large part for convenience—certainly not because every organism fits neatly into one and only one category. We could, for example, have included a fourth class, the parasitoids: little known to nonbiologists, but extensively studied by ecologists (and immensely important in the biological control of insect pests—see Chapter 12). Parasitoids are flies and wasps in which the larvae consume their insect-larva host from within, having been laid there as an egg by their mother. Parasitoids therefore straddle the "parasite" and "true predator" categories (only one host individual, which it always kills), fitting neatly into neither and confirming the impossibility of constructing clear boundaries.

There is, moreover, no truly satisfactory term to describe all the animal consumers of living organisms to be discussed in this chapter. Detritivores and plants are also "consumers" (of dead organisms, or of water, radiation, and so on), whereas the term "predator" inevitably tends to suggest a true predator even after we have defined it to encompass grazers and parasites too. But neither is it very satisfactory to be

parasitoids—and the artificiality of boundaries

A parasitoid wasp, which uses its long ovipositor to insert its eggs into the larvae of other insects, where they develop by consuming their host.

continually using the qualifier "true" when discussing conventional predators like big cats or ladybugs. Thus, throughout this chapter, *predator* will often be used as a shorthand term to encompass true predators, grazers, and parasites, when general points are being made, but it will also be used to refer to predators in the more conventional sense, when it is obvious that this is what is being done.

8.2 ▷ Prey Fitness and Abundance

predators reduce the fecundity and/or survival of individual prey

The fundamental similarity among predators, grazers, and parasites is that each, in obtaining the resources it needs from its prey, reduces either the fecundity or the chance of survival of individual prey, and therefore may reduce prey abundance. The effect of true predators on the survival of individual prey hardly needs illustrating—the prey dies. That the effects of grazers and parasites can be equally profound, if rather more subtle, is illustrated by the following two examples.

When the sand-dune willow, *Salix cordata*, was grazed by a flea beetle in two different years—1990 and 1991—the reduction in the growth rate of the willow was marked in both years, but the consequences were rather different (Figure 8.1). Only in 1991 were the plants also subject to a severe shortage of water; thus it was only in 1991 that the reduced growth rate was translated into plant mortality: 80 percent of the plants died in the high grazing treatment, 40 percent died in the low-grazing treatment, but none of the control, ungrazed plants died.

The pied flycatcher is a bird that migrates early each summer from tropical West Africa to Finland (and elsewhere in northern Europe) to breed. Males that

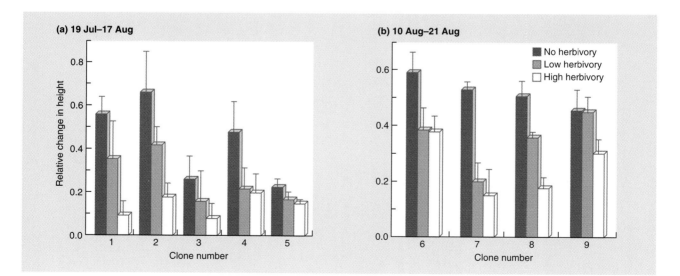

Figure 8.1

Relative growth rates (changes in height, with standard errors) of a number of different clones of the sand-dune willow, *Salix cordata*, (a) in 1990 and (b) in 1991, subjected to either no herbivory, low herbivory (four flea beetles per plant); or high herbivory (eight beetles per plant). (After Bach, 1994.)

arrive relatively early are particularly successful at finding mates. Late arrival therefore has a serious detrimental effect on the expected fecundity of a male: the number of offspring that it can expect to father. Significantly, the later arrivals are disproportionately infected with the blood parasite *Trypanosoma* (Figure 8.2). Infection with the parasite therefore has a profound effect on the reproductive output of individual birds.

It is not so straightforward, though, to demonstrate that reductions in the survival or fecundity of individual prey translate into reductions in prey abundance; we need to be able to compare prey populations in the presence and the absence of predators. As so often in ecology, we cannot rely simply on observation: we need experiments—either those we set up ourselves, or natural experiments set up for us by nature.

For example, Figure 8.3a contrasts the dynamics of laboratory populations of an important pest, the Indian meal moth, with and without a parasitoid wasp, *Venturia canescens*. Ignoring the rather obvious regular fluctuations (cycles) in both moth and wasp, it is apparent that the wasp reduced moth abundance to less than one tenth of what it would otherwise be (notice the logarithmic scale in the figure). A very similar result is shown in Figure 8.3b, though the reduction is not so marked, when another important pest, the flour beetle *Tribolium castaneum*, is infected with the protozoan parasite *Adelina triboli*.

The visual evidence in Figure 8.3c, for a grazer this time, is very different from the previous two examples but nonetheless compelling. The figure contrasts two photographs of Lake Moon Darra, North Queensland, Australia. In 1978 (the first picture) the lake carried an infestation of 50,000 metric tons fresh weight of the aquatic fern *Salvinia molesta*. *Salvinia*, which originated in Brazil, has appeared since 1930 in various tropical and subtropical regions. It was first recorded in Australia in 1952 and spread very rapidly. Under optimal conditions it has a doubling time of only 2.5 days. Significant predators of the plant appear to have been absent. Amongst possible control agents collected from *Salvinia*'s native range in Brazil, the black long-snouted weevil (*Cyrtobagous* sp.) was known to graze only on *Salvinia*.

predators can reduce prey abundance but don't necessarily do so

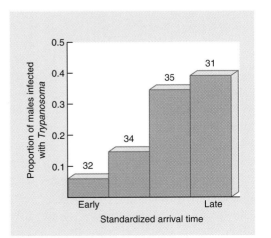

Figure 8.2

The proportion of male pied flycatchers (*Ficedula hypoleuca*) infected with *Trypanosoma* amongst groups of migrants arriving in Finland at different times. (After Rätti et al., 1993.)

Figure 8.3

(a) Long-term population dynamics in laboratory population cages of a host (*Plodia interpunctella*), with and without its parasitoid (*Venturia canescens*). Above: host and parasitoid; below: the host alone. (After Begon et al., 1995a.) (b) The depression of the population size of the flour beetle *Tribolium castaneum*, infected with the protozoan parasite *Adelina triboli*: ———, uninfected; ----, infected. (After Park, 1948.) (c) Lake Moon Darra (North Queensland, Australia). Above, covered by dense populations of the water fern (*Salvinia molesta*); below, after introduction of weevils (*Cyrtobagous* sp.). (Courtesy of P.M. Room.)

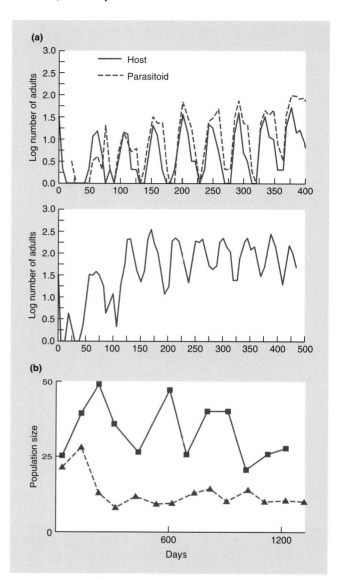

On June 3, 1980, 1500 adults were released at an inlet to the lake and a further release was made on January 20, 1981. By April 18, 1981, *Salvinia* throughout the lake had become dark brown. Samples from the dying weed suggested a total population of 1 billion beetles on the lake. By August 1981, there was estimated to be less than 1 metric ton of *Salvinia* left on the lake (second picture). This has been the most rapid success of any attempted biological control of one organism by the introduction of another. It was a "controlled" experiment in that other lakes continued to bear large populations of *Salvinia*.

All sorts of predators can cause reductions in the abundance of their prey. We shall see as this chapter develops, however, that they do not *necessarily* do so.

(c)

8.3 ▷ The Subtleties of Predation

There is much to be gained by stressing the similarities between different types of
predators. On the other hand, it would be wrong to make this an excuse for
oversimplification (there *are* important differences among true predators, grazers,

and parasites) or to give the impression that all acts of predation can be reduced to the formula (1) prey dies; (2) predator takes one step closer to the production of its next offspring.

8.3.1 Interactions with other factors

Grazers and parasites, in particular, often exert their harm not by killing their prey immediately like true predators, but by making the prey more vulnerable to some other form of mortality. One especially important reason that grazing herbivores have a more drastic effect than is initially apparent is the interaction between grazing and plant competition. This can be seen in the interaction between a chrysomelid beetle, *Gastrophysa viridula*, and two species of dock plant, *Rumex obtusifolius* and *R. crispus*. *R. crispus* is only moderately affected by competition from *R. obtusifolius* and is little affected by light beetle grazing. Indeed, the beetle feeds preferentially on *R. obtusifolius*. Light grazing and competition together, however, have a considerable impact on *R. crispus* (Figure 8.4).

Parasites, too, may disturb competitive relationships. One example, from a natural population, comes from a study of two *Anolis* lizards that live on the Caribbean island of St. Maarten. *A. gingivinus* is the stronger competitor and appears to exclude *A. wattsi* from most of the island. However, the malarial parasite *Plasmodium azurophilum* very commonly affects *A. gingivinus* but rarely affects *A. wattsi*. Wherever the parasite infects *A. gingivinus*, *A. wattsi* is present—but wherever the parasite is absent, even in very local communities, only *A. gingivinus* occurs (Schall, 1992).

Infection may also make hosts more susceptible to predation. For example, cormorants, which feed on fish (including roach), capture a disproportionately large number of roach infected with the tapeworm *Ligula intestinalis*, compared with the prevalence of infected fish in the population as a whole (van Dobben, 1952); postmortem examination of red grouse, a "game" bird, showed that birds killed by natural predators in the spring and summer carried significantly greater burdens of

Figure 8.4

The effects of grazing by the beetle *Gastrophysa viridula* on the dock plant *Rumex crispus* are increased significantly when combined with competition from another dock, *R. obtusifolius*. Plant performance is measured as leaf area, relative to control plants (no grazing, no interspecific competition). (After Bentley & Whittaker, 1979.)

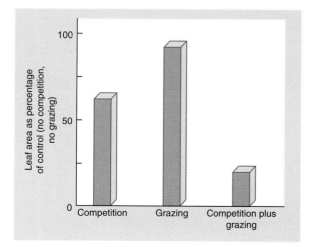

the gut nematode parasite *Trichostrongylus tenuis* than the birds that remained for the autumn shooting season (Hudson et al., 1992a).

8.3.2 Compensation and defense by individual prey

The effects of parasites and grazers, however, are not always more profound than they first seem. They are often *less* profound because, for example, individual plants can compensate for the effects of herbivory. The removal of leaves from a plant may decrease the shading of other leaves and thereby increase their rate of photosynthesis. Or, in the immediate aftermath of an attack from a herbivore, many plants compensate by utilizing reserves stored in a variety of tissues and organs. For example, when two varieties of Italian rye grass (*Lolium multiflorum*) were completely defoliated, the variety that had higher levels of stored carbohydrates in its roots and stubble exhibited higher initial rates of leaf regrowth (Figure 8.5).

Herbivory also frequently alters the distribution of newly photosynthesised material within the plant, the general rule being apparently that a balanced root/shoot ratio is maintained. When shoots are defoliated, an increased fraction of net production is channeled to the shoots themselves, and when roots are destroyed, the switch is toward the roots. The defoliation of grasses often stops root growth altogether.

Often, too, there is compensatory regrowth of defoliated plants when buds that would otherwise remain dormant are stimulated to develop. There is also, commonly, a reduced subsequent death rate of surviving plant parts. For example, the wild

compensatory plant responses

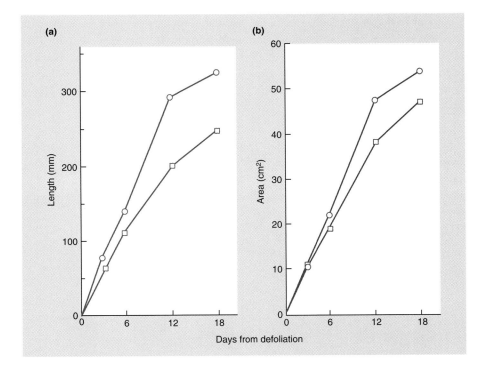

Figure 8.5
After defoliation, the regrowth of the variety "Liscate" of *Lolium multiflorum* (○) is better than that of the variety "S.22" (□) in terms of (a) leaf elongation and (b) leaf area expansion. The fraction of water-soluble carbohydrates in the roots and stubble at the time of defoliation was 27 percent higher in Liscate than in S.22. (After Kigel, 1980.)

parsnip (*Pastinaca sativa*) produces a moderate number of seeds on its primary umbels (flower clusters), a larger number on its secondary umbels, but only a small number on its tertiary umbels because of flower abortion prior to seed production (Figure 8.6). However, when it is attacked by larvae of the parsnip webworm (*Depressaria pastinacella*, a moth), although most of the flowers and fruits on the primary umbels are destroyed, there is little effect on the secondary umbels, and there is a greatly reduced rate of abortion on the tertiary umbels. Overall, therefore, the quantity of seed produced is largely unaltered.

defensive plant responses   Aside from compensating for the attacks of grazers, plants may respond by initiating or increasing their production of defensive structures or chemicals. For example, artificially wounded potato and tomato plants produce increased levels of chemicals that inhibit the protease enzymes of herbivores (Green & Ryan, 1972), and the prickles of dewberries on cattle-grazed plants are longer and sharper than those on ungrazed plants nearby (Abrahamson, 1975).

In at least some cases—though it is not always easy to prove—the plants' responses do seem to be genuinely harmful to the herbivores. When larch trees were defoliated by the larch budmoth, *Zeiraphera diniana*, for example, the survival rate and adult fecundity of the moths were reduced throughout the succeeding 4–5 years as a combined result of delayed leaf production, tougher leaves, higher fiber and resin concentration and lower nitrogen levels (Baltensweiler et al., 1977). Similarly, snowshoe hares (*Lepus americanus*) showed all the adverse signs normally associated with crowding when, under experimental conditions, they were fed birch leaves that had regenerated after severe defoliation (Bryant & Kuropat, 1980).

The defensive responses of hosts to attack by parasites were discussed in Chapter 7. The effectiveness of, for example, the immune response of vertebrates is perhaps most clearly demonstrated by those immunocompromised individuals in which the response is lacking or severely weakened. Many of the symptoms of people who have acquired immunodeficiency syndrome (AIDS) caused by infection with human immunodeficiency virus (HIV) actually result from infections with other pathogens

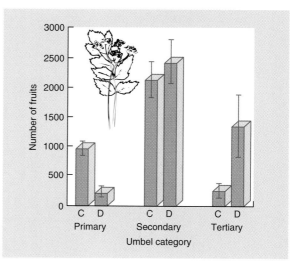

Figure 8.6

Compensation via reduced death rate of flowers. Although most of the flowers and fruits of primary umbels of *Pastinaca sativa* are destroyed by parsnip webworm, damaged plants (D) produce similar numbers of fruits from their secondary umbels and many more fruits from their tertiary umbels than do control plants (C) (means plus the standard error). (After Hendrix, 1979; from Crawley, 1983.)

that are otherwise extremely rare; this provides persuasive testimony to the threat that parasites pose to an undefended host.

8.3.3 From individual prey to prey populations

In spite of these various qualifications, however, the general rule is undoubtedly that predators are harmful to individual prey. But the effects of predation on a population of prey are not always so predictable. In the first place, there may be compensatory changes in the growth, survival, or reproduction of the surviving prey: they may experience reduced competition for a limiting resource or produce more offspring, or other predators may take fewer of the prey. In other words, although predation is bad for the prey that get eaten, it may be good for those that do not.

The impact of predation is most commonly limited by compensatory reactions among the survivors as a result of reduced intraspecific competition. For example, in an experiment in which large numbers of wood pigeons (*Columba palumbus*) were shot, the *overall* level of winter mortality was not increased, and stopping the shooting led to no increase in pigeon abundance (Murton et al., 1974). This was because the number of surviving pigeons was determined ultimately not by shooting but by food availability, and so when shooting reduced density, there were compensatory reductions in intraspecific competition and in natural mortality, as well as density-dependent immigration of birds moving in to take advantage of unexploited food.

compensatory reactions among surviving prey

Predation may also have a negligible impact on prey abundance if an increased loss of prey to predators at one stage of the prey's life simply leads to a decreased loss to predators at some other stage. If, for example, recruitment to a population of adult plants is not limited by the number of seeds produced, then insects that reduce seed production are unlikely to have an important effect on plant population dynamics. The point is illustrated by a study of the shrub *Haplopappus venetus*, which decreased steadily in abundance along a 60-km gradient inland from the coast in California (Louda, 1982, 1983). The level of insect damage to developing flowers and seeds was high. Experimental exclusion of flower and seed predators, therefore, caused an 11 percent decrease in the abortion of flower heads, a 19 percent increase in pollination success and a 104 percent increase in the number of developing seeds escaping damage. This led to an increase in the number of seedlings established. But this was followed by a much greater loss of seedlings, probably to vertebrate herbivores, especially inland. As a consequence, the original abundance gradient was reestablished in spite of the short-term importance of the seed predators.

Compensation, however, is by no means always perfect. Figure 8.7, for example, shows the results of an experiment in which Douglas fir seeds were sown both in open plots and in plots screened from vertebrate seed eaters. The immediate effect of this was an enormous reduction in the loss of seeds to these rodents and birds (though the screens were not totally effective). There were, however, compensatory increases in the losses to fungal infection of seeds prior to germination, of seeds during the germination period, and of seedlings and young plants in the year following germination. Nonetheless, in spite of this compensation, the overall effect of screen-

but compensation is often imperfect

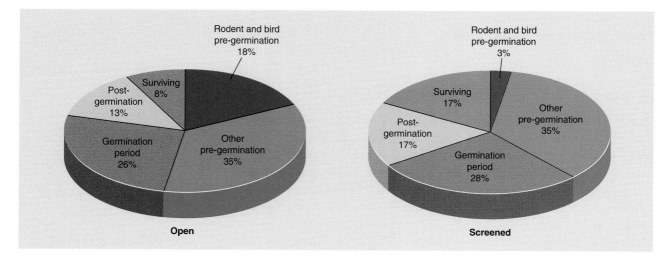

Figure 8.7
When Douglas fir seeds are protected from vertebrate predation by screens, the lowered mortality rate is compensated for (but not *fully* compensated for) by increased mortality rate from other sources. (After Lawrence & Rediske, 1962.)

predators often attack the weakest and most vulnerable

ing was to more than double the number of seedlings still surviving 1 year after germination.

Predators may also have little impact on prey populations as a whole because of the particular individuals they attack. Many large carnivores, for example, concentrate their attacks on the old (and infirm), on the young (and naive), or on the sick. Thus, a study in the Serengeti found that cheetahs and wild dogs killed a disproportionate number from the younger age classes of Thomson's gazelles (Figure

Thomson's gazelle.

8.8a) because: (1) these young animals were easier to catch (Figure 8.8b); (2) they had lower stamina and running speeds; (3) they were less good at outmaneuvering the predators (Figure 8.8c); and (4) they may even have failed to recognize the predators. The effects of predation on the prey population will therefore have been less than would otherwise have been the case, because these young gazelles will have been making no present reproductive contribution to the population, and many would have died anyway, from other causes, before they were able to do so. Similarly, the weevil *Phyllobius argentatus* has been observed to feed mainly on the lower, shaded leaves toward the center of a beech tree, and it therefore has little impact on a tree's overall productivity (Nielsen & Ejlerson, 1977).

It is apparent, then, that the effects of a predator on an individual prey are crucially dependent on the response of the prey, and the effects on prey populations are equally dependent on which prey are attacked and on the responses of other prey individuals and other natural enemies of the prey. The effect of a predator may be more drastic than it appears, or less drastic. It is only rarely what it seems.

8.4 ⟩ Predator Behavior—Foraging and Transmission

So far, we have been looking, in effect, at what happens *after* a predator finds its prey. Now, we take a step back and examine how contact is established in the first place. This is crucially important, because this pattern of contact is critical in determining the predator's "consumption rate," which goes a long way to determining its own level of benefit and the harm it does to the prey, which determines, in turn, the impact on the dynamics of predator and prey populations, and so on.

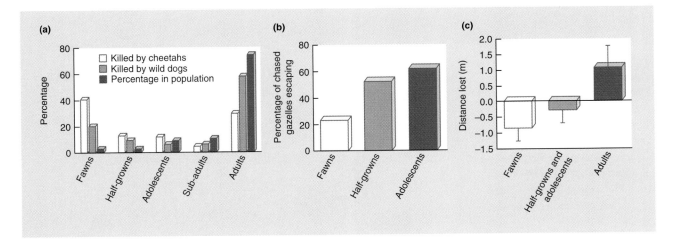

Figure 8.8

(a) The proportions of different age classes (determined by tooth wear) of Thomson's gazelles in cheetah and wild dog kills is quite different from their proportions in the population as a whole. (b) Age influences the probability for Thomson's gazelles of escaping when chased by cheetahs. (c) When prey (Thomson's gazelles) "zigzag" to escape chasing cheetahs, prey age influences the mean distance lost by the cheetahs. (After FitzGibbon & Fanshawe, 1989; FitzGibbon, 1990.)

sit-and-wait predators

True predators and grazers typically "forage." Many move around within their habitat in search of their prey, and their pattern of contact is therefore itself determined by the predators' behavior—and sometimes by the evasive behavior of the prey (Figure 8.9a). This foraging behavior is discussed later. Other predators, web-spinning spiders for instance, sit and wait for their prey, though almost always in a location they (or sometimes their parent) has selected (Figure 8.9b). The carnivorous, net-spinning aquatic larvae of the caddisfly *Plectrocnemia conspersa*, for example, when fed on fly larvae in a laboratory stream, simply abandon unprofitable areas more rapidly than they abandon profitable ones (Figure 8.10a). Furthermore, the likelihood that a larva spins a net in the first place depends on whether it happens to encounter a food item (which it can consume even without a net): larvae that have fed begin net building immediately, whereas unfed larvae continue wandering and are more likely to leave a patch. Overall, therefore, in the natural stream environment, there are most predators sitting and waiting (with their nets) where there are most prey (Figure 8.10b).

parasite transmission

With parasites and pathogens, on the other hand, we usually talk about transmission rather than foraging. This may be direct transmission between infectious and uninfected hosts when they come into contact with one another (Figure 8.9c), or free-living stages of the parasite may be released from infected hosts, so that it is the pattern of contact between these and uninfected hosts that is important (Figure 8.9d). The simplest assumption we can make for directly transmitted parasites—and one that often is made when attempting to understand their dynamics (discussed in Section 8.5)—is that transmission depends on infectious and uninfected hosts' "bumping into one another." In other words, the overall rate of parasite transmission depends both on the density of uninfected, susceptible hosts (since these represent the size of the target) and on the density of infectious hosts (since this represents the risk of the target's being "hit") (Figure 8.9c).

Figure 8.9

The different types of foraging and transmission. (a) Active predators seeking (possibly active) prey. (b) Sit-and-wait predators waiting for active prey to come to them. (c) Direct parasite transmission—infectious and uninfected hosts "bumping into each other." (d) Transmission between free-living stages of a parasite shed by a host and new, uninfected hosts.

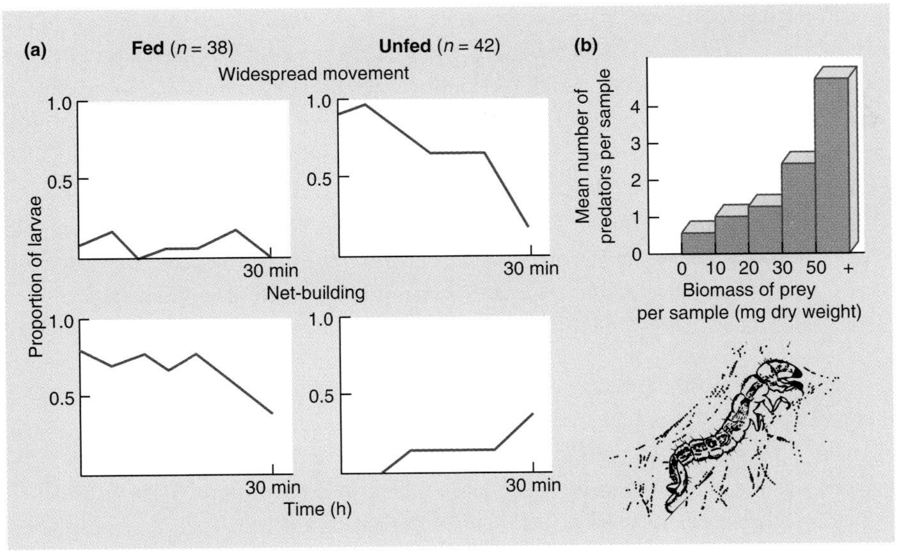

Figure 8.10

(a) On arrival in a patch, fifth-instar *Plectrocnemia conspersa* larvae that encounter and eat a chironomid prey item at the beginning of the experiment ("fed") quickly cease wandering and commence net building. Predators that fail to encounter a prey item ("unfed") exhibit much more widespread movement during the first 30 minutes of the experiment and are significantly more likely to move out of the patch. (b) Directly density-dependent aggregative response of fifth-instar larvae in a natural environment expressed as mean number of predators against combined biomass of chironomid and stonefly prey per 0.0625-m² sample of streambed ($n = 40$). (After Hildrew & Townsend, 1980; Townsend & Hildrew, 1980.)

8.4.1 Foraging behavior

There are many questions we might ask about the behavior of a foraging predator. Where, within the habitat available to it, does it concentrate its foraging? How long does it tend to remain in one location before moving on to another? And so on. Ecologists address all such questions from two points of view. The first is from the viewpoint of the *consequences* of the behavior for the dynamics of predator and prey populations. We turn to this in Section 8.5.

The second is the viewpoint of *behavioral ecology* or *optimal foraging*. The aim is to seek to understand why particular patterns of foraging behavior have been favored by natural selection. Most readers will be familiar with the general approach as applied, for example, to the anatomy of the bird's wing—we may seek to understand why a particular surface area, a particular combination of bone strength and weight, a particular arrangement of feathers, has been favored by natural selection for the effectiveness it imparts to the bird's powers of flight. Quite rightly, in so doing, it would not occur to anybody to suggest that this implies even a basic understanding of aerodynamics theory on the bird's part—only that those birds with the most effective wings have been favored in the past by natural selection and have passed on their effectiveness to their offspring. Likewise, in applying this approach to foraging behavior, there is no question of suggesting conscious decision-making by the predator.

the evolutionary, optimal foraging approach

What, though, is the appropriate measure of effectiveness in foraging behavior, the equivalent of flying ability as a criterion for a successful bird's wing? Usually, the *net* rate of energy intake has been used: that is, the amount of energy obtained per unit time, *after* account has been taken of the energy expended by the predator in carrying out its foraging. The term *optimal foraging* is used, then, because in seeking to understand why particular patterns of foraging behavior have been favored by natural selection, it is assumed that they will have been so favored because they give rise to the highest net rate of energy intake. For many consumers, however, the efficient gathering of energy may be less critical than some other dietary constituent (e.g., nitrogen), or it may be of prime importance for the forager to consume a mixed and balanced diet. The predictions of optimal foraging theory do not apply to all the foraging decisions of every predator.

applying the optimal foraging approach to a range of foraging behavior

A range of the aspects of foraging behavior to which the optimal foraging approach has been applied is illustrated in Figure 8.11. These are elaborated on briefly here, before the whole approach is demonstrated by examining just one of them in detail.

- Where, within the habitat available to it, does a predator concentrate its foraging (Figure 8.11a)? Does it concentrate where the long-term expectation of net energy intake is highest *or* where the risk of extended periods of low intake is lowest?
- Does the location chosen by a predator reflect just the expected energy intake? Or does there appear to be some balancing of this against the risk of being preyed upon by its own predators (Figure 8.11b)?

Figure 8.11
The types of foraging "decisions" considered by optimal foraging theory. (a) Choosing between habitats. (b) The conflict between increasing input and avoiding predation. (c) Patch stay time decisions. (d) The "ideal free" decision—the conflict between patch quality and competitor density. (e) Optimal diets—to include or not to include an item in the diet (when something better might be around the corner).

- How long does a predator tend to remain in one location—one patch, say, of a patchy environment—before moving on to another (Figure 8.11c)? Does it remain for extended periods and hence avoid unproductive trips from one patch to another? Or does it leave patches early, before the resources there are depleted?
- What are the effects of other, competing predators foraging in the same habitat (Figure 8.11d)? The expected net energy intake from a location is now presumably a reflection of both its intrinsic productivity and the number of competing foragers. What is the expected distribution of the predators as a whole over the various habitat patches?

The remaining question, in Figure 8.11e, and the one to which we now turn in Box 8.1 for a fuller illustration of the optimal foraging approach, is that of *diet width*. No predator can possibly be capable of consuming all types of prey. Simple design constraints prevent shrews from eating owls (even though shrews are carnivores) and prevent hummingbirds from eating seeds. Even within their constraints, however, most animals consume a narrower range of food types than they are morphologically capable of consuming.

In summary, Box 8.1 suggests that a predator should continue to add increasingly less profitable items to its diet as long as doing so increases its overall rate of energy intake. This will serve to maximize its overall rate of energy intake.

This optimal diet model, then, leads to a number of predictions.   **predictions of the optimal diet model**

1. Predators with handling times that are typically short compared to their search times should be generalists, because in the short time it takes them to handle a prey item that has already been found, they can barely begin to search for another prey item. In terms of Box 8.1: E_n/h_n is large (h_n small) for a wide range of prey types, whereas $\bar{E}/(\bar{s} + \bar{h})$ is small ($\bar{s}$ large) even for broad diets. This prediction seems to be supported by the broad diets of many insectivorous birds feeding in trees and shrubs. Searching is always moderately time-consuming, but handling the minute, stationary insects takes negligible time and is almost always successful. A bird, therefore, has something to gain and virtually nothing to lose by consuming an item once found, and its overall profitability is maximized by a broad diet.

2. By contrast, predators with handling times that are long relative to their search times should be specialists: maximizing the rate of energy intake is achieved by including only the most profitable items in the diet. For instance, lions live more or less constantly in sight of their prey so that search time is negligible; handling time, on the other hand, and particularly pursuit time can be long (and very energy consuming). Lions consequently specialize on those prey that can be pursued most profitably: the immature, the lame, and the old.

3. Other factors being equal, a predator should have a broader diet in an unproductive environment (where prey items are relatively rare and search times generally are relatively large) than in a productive environment (where search times are generally smaller). This prediction is supported by the two examples shown in 8.12: In experimental arenas, both bluegill sunfish (*Lepomis macrochirus*) and great tits (*Parus major*) had more specialized diets when prey density was higher.

4. The equation in Box 8.1 does not depend on the search time for the next most profitable item in the diet and hence does not depend on their abundance. In other

Box 8.1

Box 8.1

Optimal Diet Width

To obtain food, any predator must expend time and energy, first in searching for its prey and then in handling it (i.e., pursuing, subduing, and consuming it). While searching, a predator is likely to encounter a wide variety of food items. Therefore, diet width—the range of types of food consumed by a predator—may be seen as depending on the responses of predators once they have encountered prey. Generalists, those with a broad diet, pursue (and may then subdue and consume) a large proportion of the prey they encounter. Specialists, those with a narrow diet, continue searching except when they encounter prey of their specifically preferred type.

Generalists clearly have the advantage of spending relatively little time searching—most of the items they find they pursue and, if successful, consume. But they suffer the disadvantage of including relative low-profitability items in their diet. Specialists, on the other hand, have the advantage of only including high profitability items in their diet. But *they* suffer the disadvantage of spending a relatively large amount of their time searching for them. Determining what the optimal foraging stategy for a particular predator should be amounts to determining how these pros and cons should be balanced (MacArthur & Pianka, 1966; Charnov, 1976).

We can start by taking it for granted that any predator will include the single most profitable type of prey in its diet: that is, the one for which the net rate of energy intake is highest. But should it include the next most profitable type of item too? Or, when it comes across such an item, should it ignore it and carry on searching

for the *most* profitable type? And if it does include the second most profitable type, what about the third? and the fourth? and so on.

Consider that "second most profitable food type." It will pay a predator to include it in its diet (pay in energetic terms) if, when it has found it, its expected rate of energy intake over the time spent handling it (pursuing, subduing, and consuming it) exceeds its expected rate of intake if, instead, it searches for *and* handles an item of the *most* profitable type. (The *expected* times are simply the average times for items of a particular type.) Thus, if we call the expected searching and handling times for the most profitable type s_1 and h_1, and its energy content E_1, and the expected handling time for the second most profitable type h_2, and its energy content E_2, then it pays the predator to increase the width of its diet if E_2/h_2 exceeds $E_1/(s_1 + h_1)$ (since the rate of intake is the amount of energy obtained per unit time).

Suppose now that it did pay the predator to expand its diet. What about the third most profitable type? It will pay a predator to include this in its diet if, when it has found it, its expected rate of intake over the time spent handling it, h_3, exceeds the expected rate if it searches for and handles *either* of the two most profitable types, both already included in its diet. Thus, if we call $\bar{s}$, $\bar{h}$, and $\bar{E}$ the searching and handling times and energy content for items already in the diet, it will pay the predator to expand its diet if E_3/h_3 exceeds $\bar{E}/(\bar{s} + \bar{h})$, or, more generally, if E_n/h_n exceeds $\bar{E}/(\bar{s} + \bar{h})$, where n refers generally to the "next" most profitable prey type (not already in the diet).

words, predators should ignore insufficiently profitable food types irrespective of their abundance. Reexamining the examples in Figure 8.12, we can see that these both refer to cases in which the optimal diet model does indeed predict that the least profitable items should be ignored completely. The foraging behavior was very similar to this prediction, but in both cases the animals consistently took slightly more than expected of the less profitable food types. In fact, this sort of discrepancy has been uncovered repeatedly, and there are a number of reasons why it may occur, which can be summarized crudely by noting that the animals are not all-knowing. The optimal diet model, however, like all optimal foraging

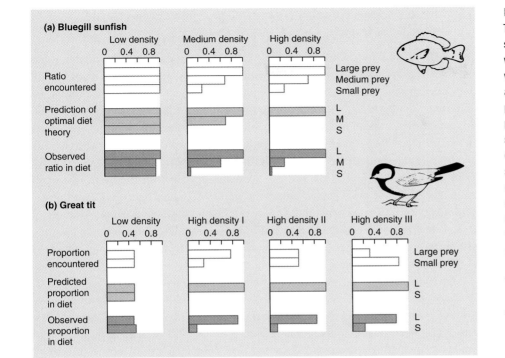

Figure 8.12
Two studies of optimal diet choice that show a clear but limited correspondence with the predictions of the optimal diet width model. Diets are more specialized at high prey densities, but more low-profitability items are included than predicted by the theory. (a) Bluegill sunfish preying on different size classes of *Daphnia*: the histograms show ratios of encounter rates with each size class at three different densities, together with the predicted and observed ratios in the diet. (After Werner & Hall, 1974.) (b) Great tits preying on large and small pieces of mealworm. (After Krebs et al., 1977.) The histograms in this case refer to the proportions of the two types of item taken. (After Krebs, 1978.)

and indeed all evolutionary models, does not predict a perfect correspondence between observation and expectation. It predicts the sort of strategy that will be favored by natural selection and says that the animals that come *closest* to this strategy will be *most* favored. From this point of view, the correspondence between data and theory in Figure 8.12 seems much more satisfactory.

Overall, then, we can see how an evolutionary, optimal foraging approach can help us make sense of predators' foraging behavior—how it makes predictions of what that behavior might be expected to be, and that those predictions may be supported by real examples.

8.5 The Population Dynamics of Predation

What roles do predators play in driving the dynamics of their prey, or prey play in driving the dynamics of their predators? Are there common patterns of dynamics that emerge? The preceding sections should have made it plain that there are no simple answers to these questions. It depends on the detail of the behavior of individual predators and prey, on possible compensatory responses at individual and population levels, and so on. Rather than despair at the complexity of it all, however, we can build an understanding of these dynamics by starting simply and then adding features one by one to construct a more realistic picture.

UNANSWERED QUESTION:

why do different predator–prey systems display different dynamics?

8.5.1 The underlying dynamics of predator–prey interactions: a tendency to cycle

We begin by consciously oversimplifying: ignoring everything but the predator and the prey and asking what underlying tendency there might be in the dynamics of their interaction. It turns out (in this section and Sections 8.5.2 and 8.5.3) that the underlying tendency is to exhibit coupled oscillations—cycles—in abundance. With this established, we can turn to the many other important factors—further predators, shortage of food for prey, environmental patchiness, and so on—which might modify or override this underlying tendency. Rather than explore each and every one of them, however, Sections 8.5.4 and 8.5.5 examine just two of the more important ones: the effects of crowding in either the predator or the prey, and some of the consequences of spatial patchiness. Between them, these two factors cannot, of course, tell the whole story, but they illustrate how the differences in predator–prey dynamics, from example to example, might be explained by the varying influences of the different factors with a potential impact on those dynamics.

Starting simply then, suppose there is a large population of prey (Figure 8.13). Predators presented with this population should do well: they should consume many prey and hence increase in abundance themselves. The large population of prey thus gives rise to a large population of predators. But this increasing population of predators increasingly takes its toll on the prey. The large population of predators therefore gives rise to a small population of prey. Now the predators are in trouble: large numbers of them and very little food. Their abundance declines. But this takes the pressure off the prey: the small population of predators gives rise to a large population of prey, and the populations are back to where they started. There is, in short, an underlying tendency for predators and their prey to undergo coupled oscillations in abundance—population cycles (Figure 8.13)—essentially because of the time delays in the response of predator abundance to that of the prey, and vice versa. (A *time delay* in response means, for example, that a high predator abundance reflects a high prey abundance *in the past*, but it *coincides* with declining prey abundance, and so on.) A simple mathematical model—the Lotka–Volterra model—conveying essentially the same message is described in Box 8.2.

8.5.2 Predator–prey cycles in practice

This underlying tendency for predator–prey interactions to generate coupled oscillations in abundance could produce an expectation of such cycles in real populations. Remember, however, that there were many aspects of predator and prey ecology that had to be ignored in order to demonstrate this underlying tendency, and these can greatly modify expectations. It should come as no surprise, then, that there are rather few good examples of clear predator–prey cycles—albeit examples that have received a great deal of attention from ecologists. And even if a herbivore population, for example, does fluctuate in abundance, that may reflect its interaction with its food *or* with its predators—or it may simply be *tracking* fluctuations in the abundance of its food, which is exhibiting cycles for some quite different reason. Predator–prey interactions *can* generate regular cycles in the abundance of both interacting populations, and they can reinforce such cycles if they exist for some other reason, but

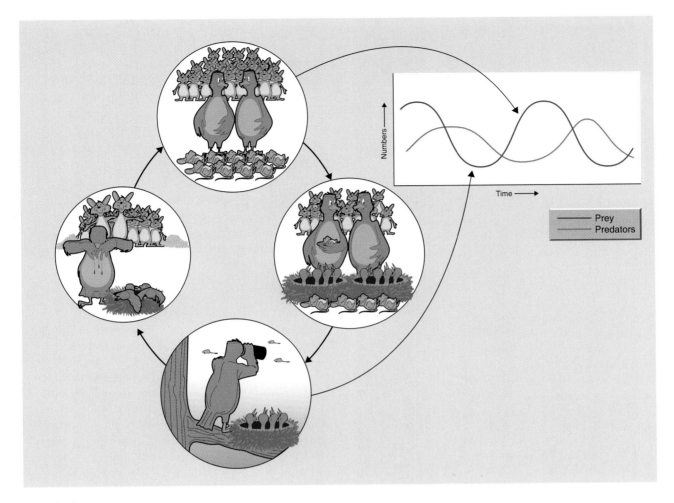

Figure 8.13
The underlying tendency for predators and prey to display coupled oscillations in abundance as a result of the time delays in their responses to each other's abundance.

attributing a cause to regular cycles in nature is generally a difficult task. Nonetheless, in trying to make sense of predator–prey population dynamics, cycles—the underlying tendency—are a good place to start.

Cycles do occur sometimes. It has been possible in several cases, for example, to generate coupled predator–prey oscillations, several generations in length, in the laboratory (Figure 8.17). Amongst field populations, there are a number of examples in which regular cycles of prey and predator abundance can be discerned. Cycles in hare populations, in particular, have been discussed by ecologists since the 1920s and were recognized by fur trappers more than 100 years earlier (Keith, 1983; Krebs et al., 1992). Most famous of all is the snowshoe hare, *Lepus americanus*, which in the boreal forests of North America follows a "10-year cycle" (which in reality varies in length between 8 and 11 years; see Figure 8.18). The snowshoe hare is the

plants, hares, and lynx in North America

(continued on page 297)

Box 8.2

The Lotka–Volterra Predator–Prey Model

Here, as in Boxes 5.2 and 6.1, one of the foundation-stone mathematical models of ecology is described and explained. The model is known (like the model of interspecific competition in Box 6.1) by the name of its originators: Lotka and Volterra (Volterra, 1926; Lotka, 1932). It has two components: P, the numbers present in a predator (or consumer) population, and N, the numbers or biomass present in a prey or plant population.

It is assumed that in the absence of consumers the prey population increases exponentially (Box 5.4):

$$\frac{dN}{dt} = rN$$

But prey individuals are removed by predators at a rate that depends on the frequency of predator–prey encounters. Encounters will increase with increasing numbers of predators (P) and prey (N). However, the exact number encountered and successfully consumed will depend on the searching and attacking efficiencies of the predator, denoted by a'. The consumption rate of prey will thus be $a'PN$, and overall

$$\frac{dN}{dt} = rN - a'PN \qquad (1)$$

In the absence of food, predator numbers are assumed to decline exponentially through starvation

$$\frac{dP}{dt} = -qP$$

where q is their mortality rate. This is counteracted by predator birth, the rate of which is assumed to depend on (1) the rate at which food is consumed, $a'PN$; and (2) the predator's efficiency, f, at turning this food into predator offspring. Overall:

$$\frac{dP}{dt} = fa'PN - qP \qquad (2)$$

Equations 1 and 2 constitute the Lotka–Volterra model.

The properties of this model can be investigated by finding zero isoclines (Box 6.1). There are separate predator and prey zero isoclines, both of which are drawn on a graph of prey density (x-axis) against predator density (y-axis) (Figure 8.14). Each is a line joining those combinations of predator and prey density that lead either to an unchanging prey population, $dN/dt = 0$ in equation (1) (prey zero isocline), or an unchanging predator population, $dP/dt = 0$ in equation (2) (predator zero isocline).

Figure 8.14

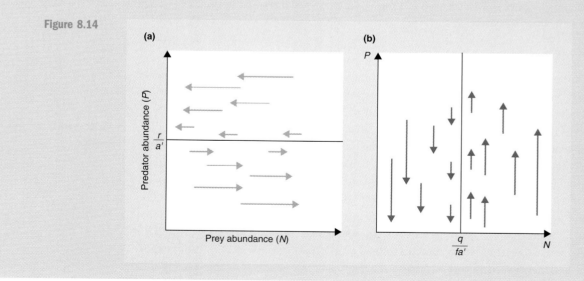

In the case of the prey this is

$$P = r/a'$$

Thus, since r and a' are constants, the prey zero isocline is a line for which P itself is a constant (Figure 8.14a): prey increase when P is less than r/a' but decrease when it is greater.

For the predators the line is

$$N = q/fa'$$

The predator zero isocline is therefore a line along which N is constant (Figure 8.14b): predators decrease when N is less than q/fa' but increase when it is greater.

Putting the two isoclines (and two sets of arrows) together in Figure 8.15 shows the behavior of joint populations: they undergo coupled oscillations in abundance, for the reasons explained in the main text.

The detailed behavior of the model, however, should not be taken too seriously. It exhibits *neutral stability*, which means that the populations follow precisely the same cycles indefinitely unless some external influence shifts them to new values, after which they follow new cycles indefinitely (Figure 8.16). In practice, of course, environments are continually changing, and populations would continually be "shifted to new values." A population following the Lotka–Volterra model would, therefore, not exhibit regular cycles but, because of repeated disturbance, fluctuate erratically. No sooner would it start one cycle than it would be diverted to a new one.

Figure 8.15

Figure 8.16

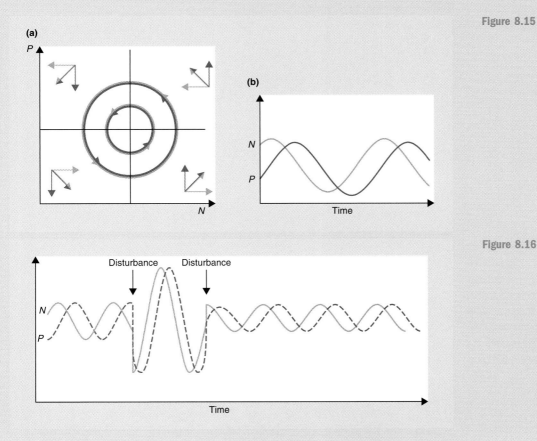

(a)

(b)

The Canada lynx and the snowshoe hare—a predator and prey that may show coupled oscillations in abundance.

Figure 8.17

Coupled predator-prey oscillations in the laboratory: the azuki bean weevil (*Callosobruchus chinensis*) and its braconid parasitoid (*Heterospilus prosopidis*). (After Utida, 1957.)

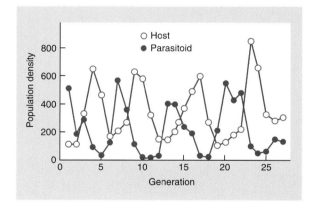

Figure 8.18

Oscillations in abundance of the snowshoe hare (*Lepus americanus*) and the Canada lynx (*Lynx canadensis*) as determined by the number of pelts lodged with the Hudson Bay Company. (After MacLulick, 1937.)

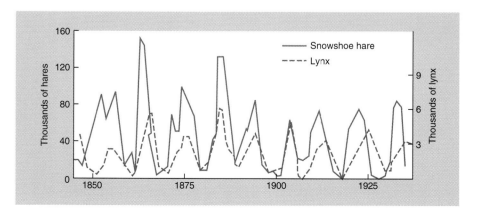

dominant herbivore of the region, feeding on the terminal twigs of numerous shrubs and small trees. A number of predators, including the Canada lynx (*Lynx canadensis*), have associated cycles of similar length. The hare cycles often involve 10- to 30-fold changes in abundance, and 100-fold changes can occur in some habitats. They are made all the more spectacular by being virtually synchronous over a vast area from Alaska to Newfoundland.

The declines in hare abundance are accompanied by low birth rates, low juvenile survivorship and low growth rates or even weight loss. All of these features can be induced experimentally by food shortages. Direct measurements, too, have suggested that there are often shortages of accessible, high-quality food during periods of peak hare abundance (Keith et al., 1984; Smith et al., 1988). Hence, the hares are forced to eat high-fiber (low-quality) food. Indeed, the quantity of relatively palatable, mature shoots does not then recover until after a time lag of another 2–3 years. This is likely to have some direct effect on their body condition, perhaps making them more susceptible to predation. But the hares also spend more time searching for food, exposing them more to predation. As a consequence, the predators concentrate, behaviorally, on hares, and also increase numerically, and so drive hare numbers down again.

Is this, then, a hare–plant cycle or predator–hare cycle? Experiments, as ever, have been instructive in suggesting an answer. Normally, with both plant and predator effects present, there are cycles. But if food is added and predators are excluded from experimental areas (neither effect is active), then hare numbers increase 10-fold and stay there—the cycles are lost. However, if either predators are excluded but no food is added (food effect alone) or food is added in the presence of predators (predator effect alone), numbers of hares double but then drop again—the "cycle" is retained. Both the hare–plant and the predator–hare interactions have some propensity to cycle on their own—but in practice the cycle seems normally to be generated by the interaction between the two.

Interestingly, moreover, a rather sophisticated statistical analysis of the time series in Figure 8.18 suggests that whereas the cycles in lynx abundance reflect their interaction with the hares, the cycles in hare abundance reflect their interaction with both their predators (essentially the lynx) *and* their food (Stenseth et al., 1997). This warns us that even when we have a predator–prey pair both exhibiting cycles (the hare and lynx) we may still not be observing simple predator–prey oscillations. Cycles of predators may also be of more than fundamental scientific interest (Box 8.3).

but how are the cycles generated?

8.5.3 Disease dynamics and cycles

Cycles are also apparent in the dynamics of many parasites, especially microparasites (bacteria, viruses, etc.). To understand the dynamics of any parasite, the best starting point is its basic reproductive rate, R_0. For microparasites, R_0 is the average number of new infected hosts that would arise from a single infectious host in a population of susceptible hosts. An infection will eventually die out for $R_0 < 1$ (each present infection leads to less than one infection in the future), but an infection will spread for $R_0 > 1$. There is therefore a *transmission threshold* when $R_0 = 1$, which must

the basic reproductive rate and the transmission threshold

Box 8.3   Topical ECOncerns

Predation in Action—Rats in Vietnam

Pests fluctuate in abundance from year to year. Recently, as the following article indicates, Vietnamese people have noticed a dramatic increase in rats—"predators" themselves but also preyed upon by many other sorts of animals. What evidence would you need to be confident that the problem is a reduction in the rats' predators? How, in practical terms, do you think recent reductions in the number of rat predators might be reversed?

PHILADELPHIA INQUIRER, Wednesday, March 11, 1998

RODENTS DEFY CONTROL IN VIETNAM
   By Paul Alexander, ASSOCIATED PRESS

Imagine killing 55 million rats in a year—and still losing ground.

Vietnam's vermin plague has gotten so bad that the central government has banned exports of traditional rat predators and closed down restaurants that specialize in serving up cats or snakes.

Some local government officials offer bounties for each tail brought in. Television, radio and newspapers encourage farmers and children alike to go out en masse and use smoke, dogs or digging to flush out rats and kill them.

Losses to Vietnam's critical rice crop amounted to $6 million last year, said Dam Quoc Tru, deputy director general of the plant protection department in the Agriculture and Rural Development Ministry.

The problem has grown almost exponentially. Tru said 220,000 acres of rice paddies were infested in 1995. That rose to 640,000 acres in 1996 and 925,000 last year. In the first two months of 1998, more than 320,000 acres of the winter-spring crop were hit. "If we don't

intervene, the damage could be $30 million a year" Tru said.

At least 55 million rats were killed last year. "We cannot eradicate them, but if we can keep their numbers down, we can reduce the level of damage" Tru said. "The Ministry of Health also is very concerned about rats spreading disease."

Rats have become so prevalent that residents are increasingly blasé about their appearances. Rodents scamper brazenly across the concrete-paved floors of Ben Thanh Market in the center of Ho Chi Minh City, scavenging food.

"The weather is hot, so the ants come out and the rats come out" Vu An Thuy said as a foot-long rat did just that across her food stand. "We're used to it, and the customers don't complain, so it's OK, although there do seem to be more of them in the last couple of weeks."

Part of the increase has been attributed to crop diversification, which provides the rats with plenty of food year-round. But much of the blame is put on the shortage of cats, snakes and barn owls.

A stray cat is a rare sight in Hanoi and Ho Chi Minh City. They either have been caught and served in local restaurants or sold by the thousands each day to China, where they are eaten too. Black cats also are used in traditional medicines.

That has caused an imbalance in the ecosystem, Tru said. "For long-term control, we need to encourage farmers to raise cats" Tru said.

be crossed if a disease is to spread. A derivation of R_b for microparasites with direct transmission (Figure 8.9c) is given in Box 8.4.

Box 8.4 provides us with a crucial insight into disease dynamics: for each directly transmitted microparasite there is a critical *threshold population size*, which needs to be exceeded for a parasite population to be able to sustain itself. For example, measles has been calculated to have a threshold population size of around 300,000 individuals and is unlikely to have been of great importance until quite recently in human biology. However, it has generated major epidemics in the growing cities of the industrialized world in the 18th and 19th centuries, and in the growing concentrations of population in the developing world in the 20th century. Current estimates suggest that around 900,000 deaths occur each year from measles infection in the developing world (Walsh, 1983).

Moreover, the immunity induced by many bacterial and viral infections, combined with the death from the infection, reduces the number of susceptibles in a population, reduces R_b, and therefore tends to lead to a decline in the incidence of the disease itself. In due course, though, there will be an influx of new susceptibles into the population (as a result of new births or perhaps immigration), an increase in R_b, an increase in incidence, and so on. There is thus a marked tendency with such diseases to generate a sequence from high incidence, to few susceptibles, to low incidence, to many susceptibles, to high incidence, and so on—just like any other predator–prey cycle. This undoubtedly underlies the observed cyclic incidence of many human diseases (especially prior to modern immunization programs), with the differing lengths of cycle reflecting the differing characteristics of the diseases: measles with peaks every 1 or 2 years, pertussis (whooping cough) every 3–4 years, diphtheria every 4–6 years, and so on (Figure 8.19).

threshold population sizes and microparasite cycles

Box 8.4

Transmission Threshold for Microparasites

Putting it simply, for microparasites with direct transmission, R_b increases (1) with the average period of time over which an infected host remains infectious, L; (2) with the number of susceptible individuals in the host population, S, because more susceptible hosts offer more opportunities for transmission of the parasite; and (3) with the transmission rate of the disease, β, which itself depends first on the infectiousness of the disease—the probability that contact leads to transmission—but also on the pattern of host behavior, insofar as this affects the likelihood of infectious and susceptible hosts coming into contact (Anderson, 1982). Thus, overall

$$R_b = S\,\beta L.$$

The transmission threshold can now be expressed in terms of a critical *threshold population size*, S_T, where, because $R_b = 1$ at that threshold,

$$S_T = 1/\beta L$$

In populations with fewer susceptibles than this, the disease will die out ($R_b < 1$), but with more than this, the disease will spread ($R_b > 1$).

Figure 8.19
(a) Reported cases of measles in
England and Wales from 1948 to
1968, prior to the introduction of mass
vaccination. (b) Reported cases of
pertussis (whooping cough) in England
and Wales from 1948 to 1982. Mass
vaccination was introduced in 1956.
(After Anderson & May, 1991.)

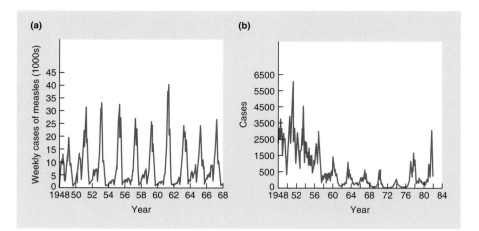

8.5.4 Crowding

**mutual interference among predators
reduces the predation rate**

One fundamental feature that has been ignored so far is the fact that no predator
lives in isolation: all are affected by other predators. The most obvious effects are
competitive; many predators experience exploitation competition for limited amounts
of food when their density is high or the amount of food is small, and this results
in a reduction in the consumption rate per individual as predator density increases
(Chapter 3). However, even when food is not limited, the consumption rate per
individual can be reduced by increases in predator density by a number of processes
known collectively as *mutual interference*. For example, many predators interact
behaviorally with other members of their population, leaving less time for feeding and
therefore depressing the overall feeding rate. Hummingbirds actively and aggressively
defend rich sources of nectar; badgers patrol and visit the "latrines" around the
boundaries between their territories and those of their neighbors; females of *Rhyssa
persuasoria* (a wasp that itself attacks wood wasp larvae) will threaten and, if need
be, fiercely drive away an intruding female from their own area of tree trunk (Spradbery, 1970). Alternatively, an increase in consumer density may lead to an increased
rate of emigration or of consumers' stealing food from one another (as do many
gulls), or the prey themselves may respond to the presence of consumers and become
less available for capture.

In all such cases, the underlying pattern is the same: the consumption rate per
individual predator declines with increasing predator density. This reduction is likely
to have an adverse effect on the fecundity, growth, and mortality rate of individual
predators, which intensifies as predator density increases. The predator population
is thus subject to density-dependent regulation (Chapters 3 and 5).

**a similarly density-dependent effect in
parasites**

With parasites, too, especially because they occupy specific sites within their
host, it is to be expected that individuals will often interfere with each other's
activities, and that there will be intraspecific competition between parasites and
density dependence in their growth and in their birth and/or death rates. In experimental infections of mice with the tapeworm *Hymenolepis microstoma*, for example,
the total weight of worm per mouse rises asymptotically, as if to some ceiling value,
as the number of parasites is increased. The weight of individual parasites decreases

correspondingly. The total number of eggs produced *per host* is much the same at all levels of infection, but the number of eggs produced *per parasite* falls in a strongly density-dependent fashion (Figure 8.20).

Moreover, it is of course not only the predators that may be subject to the effects of crowding. Prey, too, are likely to suffer reductions in growth, birth, and survival rates as their abundance increases and their individual intake of resources declines.

The effect of either predator or prey crowding on their dynamics is, in a general sense, fairly easy to predict. Prey crowding prevents their abundance from reaching as high a level as it would otherwise, and that means in turn that predator abundance is also unlikely to reach the same peaks. Predator crowding, similarly, prevents predator abundance from rising so high, but also tends to prevent them from reducing prey abundance as much as they would otherwise do. Overall, therefore, crowding is likely to have a damping effect on any predator–prey cycles, reducing their amplitude or removing them altogether: not just because crowding chops off the peaks and troughs, but also because each peak in a cycle tends itself to generate the next trough (e.g., high prey abundance → high predator abundance → *low* prey abundance), so that the lowering of peaks in itself tends to raise troughs.

There are certainly examples that appear to confirm the stabilizing effects of crowding in predator–prey interactions. For instance, there are two groups of primarily herbivorous rodents that are widespread in the Arctic: the microtine rodents (lemmings and voles) and the ground squirrels. The microtines are renowned for their dramatic, cyclic fluctuations in abundance, but the ground squirrels have populations that remain remarkably constant from year to year. Significantly, the ground squirrels are strongly self-limited by their aggressive territorial defense of burrows used for breeding and hibernating, and it is to this that their stability has been convincingly attributed (Batzli, 1983).

8.5.5 Predators and prey in patches

The second feature that was ignored initially but will be examined here is the fact that many populations of predators and prey exist not as a single, homogeneous mass, but as a *metapopulation*: an overall population divided, by the patchiness of

crowding tends to dampen or eliminate predator–prey cycles

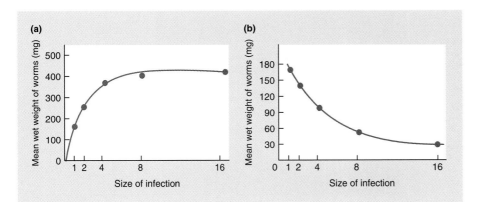

Figure 8.20
Density-dependent responses in populations of parasites. (a) The mean weight of worms per infected mouse after deliberate infection with different levels of infection by the tapeworm *Hymenolepis microstoma*, and (b) the mean weight of individual worms from the same experiment. (After Moss, 1971.)

the environment, into a series of subpopulations, each with its own internal dynamics, but linked to other subpopulations by movement (dispersal) between patches (a topic developed further in Section 9.3).

It is possible to get a good idea of the general effect of this spatial structure on predator–prey dynamics by considering the simplest imaginable metapopulation: one consisting of just two subpopulations. If the patches are displaying the same dynamics, and dispersal is the same in both directions, then the dynamics are unaffected: patchiness and dispersal have no effect in their own right (Figure 8.21a). Differences between the patches, however, either in the dynamics within subpopulations or in the dispersal between them, tend, in themselves, to stabilize the interaction: to dampen any cycles that might exist (Figure 8.21b; Ives, 1992b; Murdoch et al., 1992; Holt & Hassell, 1993). The reason is that any difference leads to asynchrony in the fluctuations in the patches. Inevitably, therefore, a population at the peak of

dispersal and asynchrony dampen cycles

Figure 8.21
Patch separation and dispersal alone, in the absence of asynchrony (a), have no effect on dynamics but with asynchrony (b) tend to dampen cycles. At each dispersal event, half the individuals leave each patch and join the other patch. In (a) there is an equal exchange and hence the overall dynamics are unaltered. In (b), however, the patch near its peak loses more and donates more than the patch near its trough. Hence, peaks tend to be not so high, nor troughs so low—the fluctuations are *dampened*.

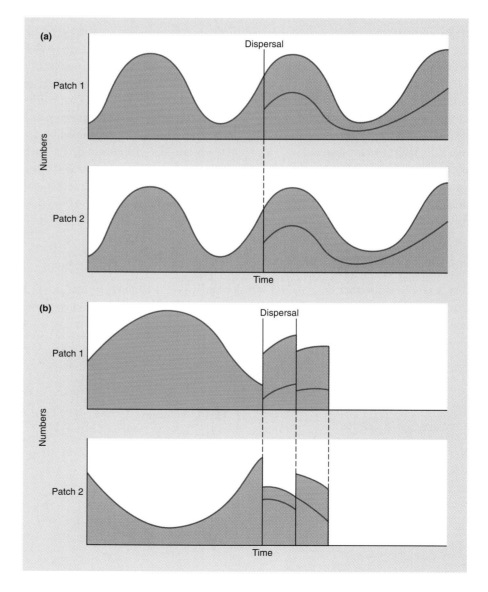

its cycle tends to lose more by dispersal than it gains, a population at a trough tends to gain more than it loses, and so on. Dispersal and asynchrony together, therefore—and some degree of asynchrony is likely to be the general rule—tend to dampen fluctuations in predator–prey dynamics.

Computer simulations allow this theoretical consideration of metapopulation predator–prey dynamics to be taken a stage further (Figure 8.22). In these, the environment consists of a patchwork of squares. In each generation, two processes occur in sequence. First, a fraction of predators and a fraction of prey disperse from each square to the eight neighboring squares. At the same time, predators and prey from the eight neighboring squares disperse into the first square. The second phase then consists of one generation of standard predator–prey dynamics. Simulations are started with random prey and predator populations in a single patch, with all other patches empty.

Within individual squares, if they existed in isolation, the dynamics in these simulations would be unstable, ever-increasing fluctuations. But within the patchwork of squares as a whole, highly persistent patterns can readily be generated (Figure 8.22). The general message is similar to the results that we have already seen: that stability can be generated by dispersal in metapopulations in which different patches are fluctuating asynchronously. The explicitly spatial aspects of this model have, however, quite literally, added another dimension to the results. Depending on the dispersal fractions and the prey reproductive rate, a number of quite different spatial structures can be generated (although they tend to blur into one another) (Figures 8.22a–c). The model, therefore, makes the point very graphically that persistence at the level of a whole population does not necessarily imply either uniformity across the population or stability in individual parts of it.

a computer simulation of a metapopulation leads to a range of persistent dynamics

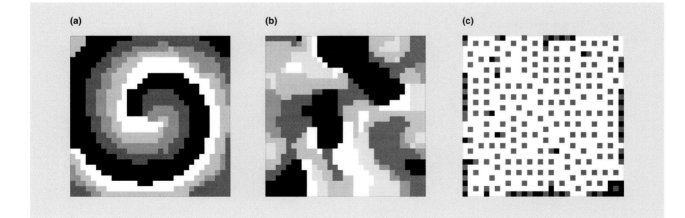

Figure 8.22

Instantaneous maps of population density for simulations of computer model of dispersal between patches in a metapopulation, where the local dynamics are unstable predator–prey cycles. Different levels of shading represent different densities of prey and predators. Black squares represent empty patches; dark shades becoming paler represent patches with increasing prey densities; light shades to white represent patches with prey and increasing predator densities. (a) Spirals, (b) spatial chaos, (c) a crystalline lattice, at different rates of predator and prey dispersal. (After Comins et al., 1992.)

stabilizing metapopulation effects in
Huffaker's mites

Is it possible, though, to see the stabilizing influence of this type of metapopulation structure in practice? One famous example is experimental work on a laboratory system in which a predatory mite *Typhlodromus occidentalis* fed on a herbivorous mite *Eotetranychus sexmaculatus*, which fed on oranges interspersed among rubber balls in a tray. In the absence of its predator, *Eotetranychus* maintained a fluctuating but persistent population (Figure 8.23a), but if *Typhlodromus* was added during the early stages of prey population growth, it rapidly increased its own population size, consumed all of its prey, and then became extinct itself (Figure 8.23b): the underlying predator–prey dynamics were unstable.

The interaction was altered, however, when the habitat was made more patchy: The oranges were spread farther apart and partially isolated from each other by placing a complex arrangement of petroleum jelly barriers in the tray, which the mites could not cross. The dispersal of *Eotetranychus* was facilitated, however, by inserting a number of upright sticks from which they could launch themselves on silken strands carried by air currents. Dispersal between patches was therefore much easier for prey than it was for predators. In a patch occupied by both, the predators consumed all the prey and then either became extinct themselves or dispersed (with a low rate of success) to a new patch. In patches occupied by prey alone, there was rapid, unhampered growth accompanied by successful dispersal to new patches.

Figure 8.23

Predator–prey interactions between the mite *Eotetranychus sexmaculatus* and its predator, the mite *Typhlodromus occidentalis*. (a) Population fluctuations of *Eotetranychus* without its predator. (b) A single oscillation of the predator and prey in a simple system. (c) Sustained oscillations in a more complex system. (After Huffaker, 1958.)

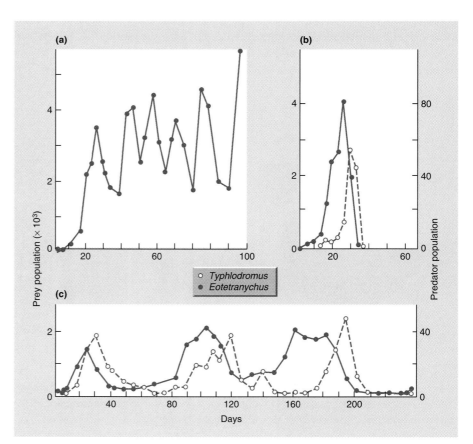

And in a patch occupied by predators alone, there was usually death of the predators before their food arrived. Each patch was therefore ultimately doomed to the extinction of both predators and prey—that is, its dynamics were unstable. But overall, at any one time, there was a mosaic of unoccupied patches, prey–predator patches heading for extinction, and thriving prey patches, and this mosaic was capable of maintaining persistent populations of both predators and prey (Figure 8.23c).

A similar example, from a natural population, is provided by work off the coast of southern California on the predation by starfish of clumps of mussels (Murdoch & Stewart-Oaten, 1975). Clumps that are heavily preyed upon are liable to be dislodged by heavy seas so that the mussels die; the starfish are continually driving patches of their mussel prey to extinction. The mussels, however, have planktonic larvae that are continually colonizing new locations and initiating new clumps, whereas the starfish disperse much less readily. They aggregate at the larger clumps, but there is a time lag before they leave an area when the food is gone. Thus, patches of mussels are continually becoming extinct, but other clumps are growing prior to the arrival of the starfish. As with the mites, the combination of patchiness, the aggregation of predators in particular patches, and a lack of synchrony between the behavior of different patches appears capable of stabilizing the dynamics of a predator–prey interaction.

and in starfish and mussels

In a similar vein, it is significant that populations of the "cyclic" snowshoe hare (described earlier) never show cyclic behavior in habitats that are a mosaic of habitable and uninhabitable areas. In mountainous regions, and in areas fragmented by the incursions of agriculture, the snowshoe hare maintains relatively stable, noncyclic populations (Keith, 1983).

A metapopulation structure, then, like crowding, can have an important influence on predator–prey dynamics. More generally, however, the message of this section is that predator–prey dynamics can take a wide variety of forms, but there are good grounds for believing that we can make sense of this variety through seeing it as a reflection of the way in which the different aspects of predator–prey interactions combine to play out variations on an underlying theme.

an explanation for the variety of predator–prey dynamics begins to emerge

8.6 Predation and Community Structure

What roles can predation play when we broaden our perspective from populations to whole ecological communities? Before this question is addressed, it is worth reflecting that predation, in many of its effects, is just one of the forces acting on communities that can be described as *disturbance*. The result of a predator's opening up a gap in a community for colonization by other organisms, for example, is often essentially indistinguishable from that of battering by waves on a rocky shore or a hurricane in a forest.

It turns out that many of the effects of predation (and other disturbances) on community stucture are the result of its interaction with the processes of competitive exclusion (taking up a theme introduced in Section 6.2.8). In an undisturbed world, the most competitive species might be expected to drive less competitive species to

predation as an interruptor of competitive exclusion: predator-mediated coexistence

extinction. However, this assumes first that the organisms are actually competing, and that in turn implies that resources are limiting. Yet there are many situations in which predation may hold down the densities of populations, so that resources are not limiting and individuals do not compete for them. When predation promotes the coexistence of species among which there would otherwise be competitive exclusion (because the densities of some or all of the species are reduced to levels at which competition is relatively unimportant) this is known as *predator-mediated coexistence*.

the different effects of specialist and generalist predators

The effects of specialist and generalist predators on community structure may be quite different. The effects of predation generally on a group of competing species depend on which species suffer most. If it is subordinate species, then these may be driven to extinction and the total number of species in the community will decline. If it is the competitive dominants that suffer most, however, the results of heavy predation will usually free space and resources for other species, and species numbers may then increase. Generalist predators typically have the effect of increasing the number of species in a community through predator-mediated coexistence, because even if the prey are attacked simply in proportion to their abundance, it will be those species that are competitive dominants that will be most abundant, and will therefore be most severely set back by predation.

unselective rabbits promote the coexistence of plants

Rabbits, for example, although they do not eat every type of plant, are relatively unselective (generalist) grazers. The rabbit is not native to Britain, and its introduction (probably in the 12th century) must have been a major disturbance to the vegetation. Thereafter, though, its presence became part of the normal state of affairs. In 1954, however, the viral disease myxomatosis was introduced to Britain and drastically reduced rabbit populations. The immediate response of the vegetation to this reduction in predation pressure was an increase in the number of flowering perennial plants (e.g., orchids) *observed*. These had previously been present—but not obvious because they had been repeatedly nibbled. Later, however, the total number of species fell as the community became increasingly dominated by a few, taller, more competitive species of grass.

most species occur at intermediate levels of predation?

Grazing by rabbits had apparently kept these aggressive, competitively dominant grasses in check and allowed a greater number of plant species to persist. However, at very high intensities of grazing, species numbers may actually be reduced as the rabbit is forced to turn from heavily grazed, preferred plant species to less preferred species. Plant species may then be driven to extinction (Figure 8.24), so that overall the number of species is greatest at *intermediate* levels of predation.

selective predation on a rocky shore

It can be deduced from these rabbit data that, as a generalization, *selective* predation should favor an increase in species numbers in a community as long as the preferred prey are competitively dominant, although again, species numbers may also be low at very high predation pressures. Along the rocky shores of New England, the most abundant and important herbivore in mid- and low-intertidal zones is the periwinkle snail *Littorina littorea*. The snail will feed on a wide range of algal species but *is* relatively selective: it shows a strong preference for small, tender species and in particular for the green alga *Enteromorpha intestinalis*. The least preferred foods are much tougher (e.g., the perennial red alga *Chondrus crispus* and brown algae).

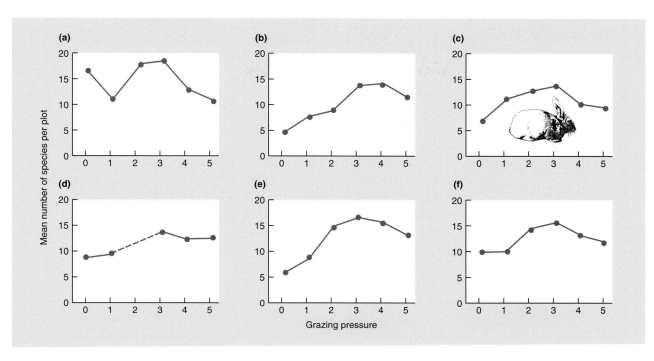

Figure 8.24
Relationship between plant species richness and intensity of rabbit grazing (on a scale of 0–5) in 1-m² plots on five sand dunes (a–e) and all dunes combined (f). (After Zeevalking & Fresco, 1977.)

They are either never eaten by the periwinkles or eaten only if no other food has been available for some time.

Is *Enteromorpha*, the periwinkles' preferred food, a competitive dominant in their absence? In a normal *Chondrus* pool, periwinkles feed on microscopic plants and the young stages of many ephemeral algae that settle on *Chondrus* (including *Enteromorpha*). However, in a *Chondrus* pool from which periwinkles are artificially removed, *Enteromorpha* and several other algae settle or grow from microscopic sporelings or germlings and become abundant. Here, *Enteromorpha* achieves competitive dominance and *Chondrus* individuals become bleached and then disappear. It seems clear that the periwinkles are responsible for the dominance of *Chondrus* in *Chondrus* pools. Addition of periwinkles to *Enteromorpha* pools leads, in a year, to a decline in percentage cover of the green alga from almost 100 percent to less than 5 percent. *Chondrus* colonizes slowly, but eventually comes to dominate pools where periwinkles have eaten out its competitor.

Thus, the algal composition of tide pools in the rocky intertidal region varies from almost pure stands of *Enteromorpha* to almost pure stands of *Chondrus*. Is grazing by the periwinkle responsible for these differences? The preceding argument suggests that when periwinkles are absent or rare, *Enteromorpha* should competitively exclude other species and the number of algal species should be low. A survey of a number of pools with different densities of periwinkles demonstrated just this

(Figure 8.25a). At very high densities of periwinkles, however, all palatable algal species were consumed to extinction and prevented from reappearing, leaving the almost pure stands of the tough *Chondrus*. As with the rabbits, therefore, it was when periwinkles were present in intermediate densities that the abundance of *Enteromorpha* and other ephemeral algal species was reduced, competitive exclusion was prevented, and many species, both ephemeral and perennial, coexisted.

Why then do some pools contain periwinkles while others do not? Predation is again the answer. The periwinkle colonizes pools while it is in an immature, planktonic stage. Although planktonic periwinkles are just as likely to settle in *Enteromorpha* pools as *Chondrus* pools, the crab *Carcinus maenas,* which can shelter in the *Enteromorpha* canopy, feeds on the young periwinkles and prevents them from establishing a new population. The final thread in this tangled web of predator–prey interactions is the effect of gulls that prey on crabs where the dense green algal canopy is absent. Thus there is no bar to continuing periwinkle recruitment in *Chondrus* pools.

the selection of inferiors

The picture is quite different when the preferred prey species is competitively inferior to other prey. Here, increased predation pressure should simply reduce the number of prey species in the community. This can also be illustrated on the rocky shores of New England, where the competitive dominance of the most abundant tide pool plants is actually reversed when the plant species interact on emergent substrata rather than in the tide pools. Here it is perennial brown and red algae that predominate, at least in sites where periwinkles are rare or absent, while a number of ephemeral algal species manage to maintain only a precarious foothold. Any increase in the grazing pressure, however, decreases the algal diversity, as the preferred, ephemeral species are consumed totally and prevented from reestablishing themselves (Figure 8.25b).

Overall, then, predation can have an important role in developing our understanding of the structure of ecological communities, not least in reminding us that the patterns we saw in Chapter 6 when we were focusing on interspecific competition may never get a chance to express themselves because communities in the real world rarely proceed smoothly to an equilibrium state.

Figure 8.25

Effect of *Littorina littorea* density on species richness (a) in tide pools and (b) on emergent substrata. (After Lubchenco, 1978.)

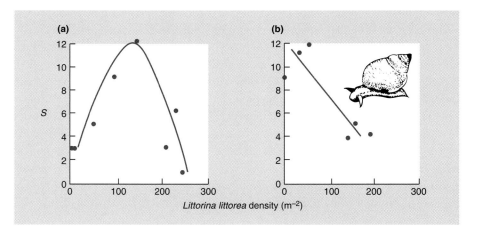

Summary

Predation, true predators, grazers, and parasites

A predator may be defined as any organism that consumes all or part of another living organism (its prey or host) thereby benefiting itself, but, under at least some circumstances, reducing the growth, fecundity, or survival of the prey.

True predators invariably kill their prey and do so more or less immediately after attacking them, and they consume several or many prey items in the course of their life. Grazers also attack several or many prey items in the course of their life but consume only part of each prey item and do not usually kill their prey. Parasites also consume only part of each host, and also do not usually kill their host especially in the short term, but attack one or very few hosts in the course of their life, with which they therefore often form a relatively intimate association.

The subtleties of predation

Grazers and parasites, in particular, often exert their harm not by killing their prey immediately as do true predators, but by making the prey more vulnerable to some other form of mortality.

The effects of grazers and parasites on the organisms they attack are often *less* profound than they first seem because individual plants can compensate for the effects of herbivory and hosts may have defensive responses to attack by parasites.

The effects of predation on a population of prey are complex to predict because the surviving prey may experience reduced competition for a limiting resource, or they may produce more offspring, or other predators may take fewer of the prey.

Predator behavior

True predators and grazers typically *forage*, moving around within their habitat in search of their prey. Other predators "sit and wait" for their prey, though almost always in a selected location. With parasites and pathogens, there may be direct transmission between infectious and uninfected hosts, or contact between free-living stages of the parasite and uninfected hosts may be important.

Optimal foraging theory aims to understand why particular patterns of foraging behavior have been favored by natural selection (because they give rise to the highest net rate of energy intake).

Generalist predators spend relatively little time searching but include relative low-profitability items in their diet. Specialists only include high-profitability items in their diet but spend a relatively large amount of their time searching for them.

The population dynamics of predation

There is an underlying tendency for predators and prey to exhibit cycles in abundance, and cycles are observed in some predator–prey and host–parasite interactions. However, there are many important factors that can modify or override the tendency to cycle.

Crowding of either predator or prey is likely to have a damping effect on any predator–prey cycle.

Many populations of predators and prey exist as a metapopulation. In theory, and in practice, asynchrony in population dynamics in different patches and the process of dispersal tend to dampen any underlying population cycles.

Predation and community structure

There are many situations in which predation may hold down the densities of populations, so that resources are not limiting and individuals do not compete for them. When predation promotes the coexistence of species among which there would otherwise be competitive exclusion (because the densities of some or all of the species are reduced to levels at which competition is relatively unimportant) this is known as *predator-mediated coexistence.*

The effects of predation generally on a group of competing species depend on which species suffers most. If it is a subordinate species, then this may be driven to extinction and the total number of species in the community will decline. If it is the competitive dominants that suffer most, however, the result of heavy predation will usually be to free space and resources for other species, and species numbers may then increase.

Overall, the number of species in a community seems to be greatest at intermediate levels of predation.

Review Questions

⚠ = Challenge Question

1. With the aid of examples, explain the feeding characteristics of true predators, grazers, parasites, and parasitoids.

2. ⚠ True predators, grazers, and parasites can alter the outcome of competitive interactions that involve their prey populations. Discuss this assertion using one example for each category.

3. Discuss the various ways that plants may compensate for the effects of herbivory.

4. Predation is bad for the prey that get eaten. Explain why it may be good for those that do not get eaten.

5. ▲ Discuss the pros and cons, in energetic terms, of (a) being a generalist as opposed to a specialist predator and (b) being a sit-and-wait predator as opposed to an active forager.

6. In simple terms, explain why there is an underlying tendency for populations of predators and prey to cycle.

7. ▲ You have data that show cycles in nature among interacting populations of a true predator, a grazer and a plant. Describe an experimental protocol to determine whether this is a grazer–plant cycle or a predator–grazer cycle.

8. Define mutual interference and give examples for true predators and parasites. Explain how mutual interference may dampen inherent population cycles.

9. Discuss the evidence presented in this chapter that suggests environmental patchiness has an important influence on predator–prey population dynamics.

10. With the help of an example, explain why most prey species may be found in communities subject to an intermediate intensity of predation.

In previous chapters, we generally dealt with individual species or pairs of species in isolation, as ecologists often do. Ultimately, however, we must recognize that every population exists within a web of interactions with myriad other populations, across several trophic levels. Each population must be viewed in the context of the whole community, and we need to understand that populations occur in patchy and inconstant environments in which disturbance and local extinction may be common.

9

Population Processes—The Big Picture

Key Concepts

In this chapter you will

- appreciate the variety of interacting abiotic and biotic factors that account for the dynamics of populations.

- distinguish between the determination and the regulation of population abundance.

- understand how patchiness and dispersal between patches influence the dynamics of both populations and communities.

- recognize the influence of disturbance on community patterns and understand the nature of community succession.

- appreciate the importance of direct and indirect effects and distinguish between bottom-up and top-down control of food webs.

- understand the relationship between the structure and stability of food webs.

9.1 ▶ Introduction

Single-species populations have been the focus for many of the questions posed in previous chapters. In attempting to answer the most fundamental ecological question of all—what determines a species' abundance and distribution—we have chosen to ask separately about the role of conditions and resources, of migration, of competition (both intra- and interspecific), of mutualism, and of predation and parasitism. In reality, the dynamics of every population depend on a combination of these factors, though the importance of each varies from species to species. We need now, therefore, to view the population in the context of the whole community, since each population exists within a web of predator–prey and competitive interactions (Figure 9.1), and each responds differently to the prevailing abiotic conditions. In Section 9.2 we consider how abiotic and biotic factors combine to determine the dynamics of species populations.

Then, in Section 9.3, we revisit one of the major themes of this book—the importance of patchiness and dispersal between patches in ecological dynamics—and discuss especially the importance of the concept of the metapopulation.

Disturbances, such as forest fires and the storm battering of seashores, also play an important role in the dynamics of many populations and the composition of most communities. After each disturbance there is a pattern of reestablishment of species that is played out against a background of changing conditions, resources, and population interactions. We deal with temporal patterns in community composition, including community succession, in Section 9.4.

Finally, in Section 9.5 we broaden our view further to examine food webs, like the one illustrated in Figure 9.1, with usually at least three trophic levels (plant–herbivore–predator), emphazing the importance not only of direct but also of indirect effects that a species may have on others on the same trophic level or on levels below or above it.

9.2 ▶ Multiple Determinants of the Dynamics of Populations

fluctuations in abundance are caused by a wide variety of biotic and abiotic factors

Why are some species rare and others common? Why does a species occur at low population densities in some places and at high densities in others? What factors cause fluctuations in a species' abundance? These are crucial questions when we wish to conserve rare species, or control pests, or manage natural, living resources,

Figure 9.1

Community matrix illustrating how each species may interact with several others in competitive (among plant species 1, 2, and 3; or between grazers 4 and 5; or between predators 6 and 7) and predator-prey interactions (such as between 6 and 4 or 5 and 2).

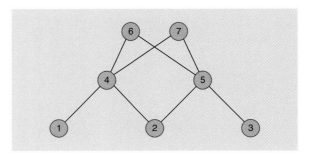

or when we wish simply to understand the patterns and dynamics of the natural world. To provide complete answers for even a single species in a single location, we need to know the physicochemical conditions, the level of resources available, the organism's life cycle, and the influence of competitors, predators, parasites, and so on—and how all these factors influence abundance through effects on birth, death, dispersal, and migration. We now bring these factors together and consider how we might discover which actually matter in particular examples.

The raw material for the study of abundance is usually some estimate of the numbers of individuals in a population. However, a record of numbers alone can hide vital information. Picture three human populations, shown to contain identical numbers of individuals. One is an old people's residential area, the second is a population of young children, and the third is a population of mixed age and sex. In the absence of information beyond mere numbers, it would not be clear that the first population was doomed to extinction (unless maintained by immigration), the second would grow fast but only after a delay, and the third would continue to grow steadily. The most satisfactory studies, therefore, estimate not only the numbers of individuals (and their parts, in the case of modular organisms) but also those of different age, sex, and size.

what total numbers can and cannot tell us

The data that accumulate from estimates of abundance may be used to establish correlations with external factors like weather. Correlations may be used to suggest causal relationships, but they cannot prove them. For example, low populations of a pest may regularly follow cold winters. Such a correlation might suggest that cold winters kill the overwintering stage of the pest, but it might equally mean that cold winters favor its enemies or deplete part of its food resources. Such data are also often used to suggest that internal factors are responsible for fluctuations in size of a population—for instance, that it grew fastest when numbers were low. Again, a correlation may be demonstrated—this time, say, between the size of a population and its growth rate. The correlation may hint that it is the size of the population itself that causes it to change, but it does not prove that this is the cause. It may be that when the population is large many individuals starve to death, or fail to reproduce, or become aggressive and drive out the weaker members. It is only by observing what is happening to the individuals that we can discover why a population changes in size.

what correlations can and cannot tell us

9.2.1 Fluctuation or stability?

Some populations appear to change very little in size. One population study that covers an extended time span—though it is not necessarily the most scientific!—has examined the swifts (*Micropus apus*) in the village of Selborne in southern England. In one of the earliest published works on ecology, Gilbert White, who lived in the village, wrote in 1778:

many populations are very stable

> I am now confirmed in the opinion that we have every year the same number of pairs invariably. . . . The number that I constantly find are eight pairs, about half of which reside in the church, and the rest in some of the lowest and meanest thatched cottages. Now, as these eight pairs—allowance being made for accidents— breed yearly eight pairs more, what becomes annually of this increase?

More than 200 years later, Lawton and May (1984) visited the village and, not surprisingly, found major changes. Swifts are unlikely to have nested in the church for 50 years, and the thatched cottages have disappeared or their thatched roofs have been covered with wire. Yet the number of breeding pairs of swifts regularly to be found in the village is now 12. In view of the many changes that have taken place in the intervening centuries, this number is remarkably close to the 8 pairs so consistently found by White.

but stability need not mean "nothing changes"

But the stability of a population may conceal complex underlying dynamics. In a study of creeping buttercup (*Ranunculus repens*) in an old permanent pasture in North Wales, detailed maps of the distribution of plants and seedlings in three separate sites within the pasture allowed the fate of every individual to be followed (an approach that is rarely possible for mobile animals). Overall, over the two years of the study the number of buttercups in the pasture (as estimated from the three sites) fell slightly—from 650 to 518. However, at site A the population halved over the two years, at site B it scarcely changed, and at site C it increased by 50 percent (Figure 9.2). Moreover, 1054 new individuals were gained by the population and 1189 individuals were lost from it. This illustrates clearly that different patches in the same "population" may show quite different dynamics, and that even when there is relatively little overall change in a population, very rapid fluxes of births and deaths may occur.

9.2.2 Theories of species abundance

Is the comparison between 8 and 12 pairs of swifts over 200 years, or between 650 and 518 buttercups over 2 years, an indication of consistency or of change? Is the

Figure 9.2

Changes in population size of creeping buttercup (*Ranunculus repens*) at site C (see text). Open circles, cumulative gains from seed germination and clonal growth; closed circles, cumulative losses; open squares, net population size. (After Sarukhán & Harper, 1973.)

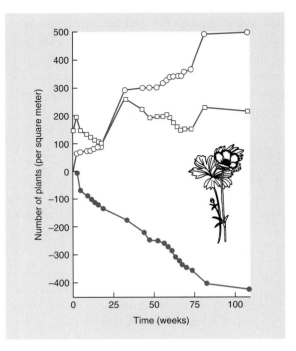

similarity between 8 and 12 of more interest—or is the difference between them? Some investigators have emphasized the apparent constancy of populations; others have emphasized the fluctuations. Indeed, the same data have been used by some to indicate that populations remain within narrow limits, and by others to stress that they change dramatically.

Those who have emphasized constancy argue that we need to look for stabilizing forces within populations (so-called density-dependent forces, for instance competition between crowded individuals for limited resources) to explain why the populations do not exhibit unfettered increase or a decline to extinction. Those who have emphasized fluctuations often look to external factors, for example weather, to explain the changes. The disagreements dominated much of ecology in the middle third of the twentieth century. By considering some of these arguments, it will be easier to appreciate the details of the modern consensus.

First, however, it is important to distinguish clearly between questions about the ways in which the abundance of individuals is *determined* and questions about the way in which the abundance is *regulated*. *Regulation* is the tendency of a population to decrease in size when it is above a particular level, but to increase in size when below that level. In other words, regulation of a population can, by definition, occur only as a result of one or more density-dependent processes (Chapters 3 and 5) that act on rates of birth and/or death and/or movement (Figure 9.3a). Various potentially density-dependent processes have been discussed in earlier chapters on competition, predation, and parasitism. We must look at regulation, therefore, to understand how it is that a population tends to remain within defined upper and lower limits.

On the other hand, the precise abundance of individuals will be determined by the combined effects of all the factors and all the processes that affect a population, whether they are dependent or independent of density (Figure 9.3b). We must look at the determination of abundance, therefore, to understand how it is that a particular population exhibits a particular abundance at a particular time, and not some other abundance.

The Australian A. J. Nicholson (1954), a theoretical and laboratory animal ecologist, is usually credited as the major proponent of the view that density-dependent, biotic interactions play the main role in determining population size, holding populations in a state of balance in their environments, where, in his words, "The mechanism . . . is almost always intraspecific competition, either amongst the animals for a critically important requisite, or amongst natural enemies for which the animals concerned are requisites." He recognized that "factors which are uninfluenced by density may produce profound effects upon density" (see Figure 9.3b), but considered that density dependence "is merely relaxed from time to time and subsequently resumed, and it remains the influence which adjusts population densities in relation to environmental favorability."

Two other Australian ecologists, Andrewartha and Birch (1954), also made a major contribution to the debate. Their research was concerned mainly with the control of insect pests in the wild, and it is likely, therefore, that their views were conditioned by the needs of being able to predict abundance, and predict, especially, the timing and intensity of pest outbreaks. It may also have influenced their opinion

the distinction between determination and regulation of abundance

the view that abundance is most influenced by the effects of crowding

the view that abundance is most influenced by the length of periods of population growth

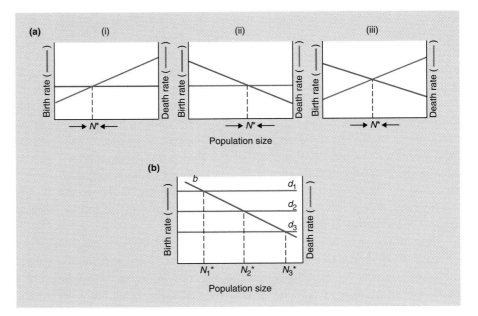

Figure 9.3

(a) Population regulation with (i) density-independent birth and density-dependent death; (ii) density-dependent birth and density-independent death; and (iii) density-dependent birth and death. Population size increases when birth rate exceeds death rate and decreases when death rate exceeds birth rate. N^* is therefore a stable equilibrium population size. The actual value of the equilibrium population size is seen to depend on both the magnitude of the density-independent rate and the magnitude and slope of any density-dependent process. (b) Population regulation with density-dependent birth, b, and density-independent death, d. Death rates are determined by physical conditions which differ in three sites (death rates d_1, d_2, and d_3). Equilibrium population size varies as a result (N_1^*, N_2^*, N_3^*).

that probably the most important factor limiting the numbers of organisms in natural populations was the shortage of time when the rate of increase in the population was positive. In other words, populations could be viewed as passing through a repeated sequence of setbacks (rate of "increase" negative) and recovery—a view that can certainly be applied to many insect pests that are sensitive to unfavorable environmental conditions but are able to bounce back rapidly. They also rejected any subdivision of the environment into physical and biotic "factors" or into density-dependent and density independent "factors," preferring instead to see populations as sitting at the center of an ecological web, where the essence was that various factors and processes interacted in their effects on the population.

a study of thrips   Among many other studies, Andrewartha and Birch attached considerable weight to work on the appleblossom thrips (*Thrips imaginis*), a small insect found in the flowers of fruit trees, garden plants, and weeds in southern Australia. The number of thrips was counted in 20 roses picked at random from a long rose hedge every day, except Sundays and certain holidays, for 81 consecutive months. Like many insects the thrips underwent very large fluctuations in abundance (Figure 9.4), but these fluctuations remained within bounds. The study was remarkable for both its

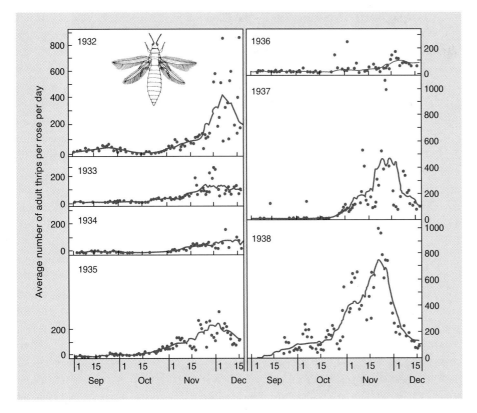

Figure 9.4
The numbers of *Thrips imaginis* per rose during the spring each year for 7 consecutive years. The points represent daily records and the curve is a 15-point moving average. (After Davidson & Andrewartha, 1948a.)

intensity and its duration. Some 6 million thrips had been recorded by the end. The work also provides an excellent example of identifying correlations: Andrewartha and Birch attached special significance to an analysis that revealed that a full 78 percent of the population's variation from year to year could be accounted for by correlations with concurrent variations in patterns of temperature and rainfall. Clearly, weather played a crucial role in determining the peak numbers of thrips each year—but Andrewartha and Birch went further and concluded that there was "no room" for a density-dependent factor as a determinant of peak thrips numbers.

The same data were then used by others, however, to suggest through different analytical techniques that the thrips populations were regulated by some factor internal to the populations (Smith, 1961). In fact, Davidson and Andrewartha themselves (1948b) felt that weather acted as a density-dependent component of the environment during the winter, by killing the proportion of the population inhabiting less favorable "situations." (If the number of safe sites is limited and remains roughly constant from year to year, then the proportion of individuals outside these sites that is killed by the weather will increase with the size of the whole population.) They also felt, however, that this did not fit the general density-dependence theory "since Nicholson clearly excludes climate from the list of possible 'density-dependent factors.'"

What conclusions can we draw from this? First, there seems to have been no great need for any argument: Andrewartha and Birch were simply more interested

some conclusions

in what determines abundance and Nicholson more interested in what regulates abundance—both are perfectly valid interests. Second, there seems little point in questioning whether climate (or anything else) *can* be a density-dependent factor—it is more profitable to focus on potentially density-dependent *processes*, which often arise from interaction of several factors. Next, it is indisputable that no population can be absolutely free of regulation—long-term unrestrained population growth is unknown, and unrestrained declines to extinction are rare. Furthermore, any suggestion that density-dependent processes are rare or generally of only minor importance would be wrong. A very large number of studies have been made of various kinds of animals, especially of insects. Density dependence has by no means always been detected but is commonly seen when studies are continued for many generations. For instance, density dependence was detected in 80 percent or more of studies of insects that lasted more than 10 years (Hassell et al., 1989; Woiwod & Hanski, 1992).

On the other hand, the weather did account for 78 percent of the variation in thrips numbers—if we wished to predict thrips abundance, weather would undoubtedly be the major determinant and other factors of relatively minor importance. Hence, it would be wrong to give regulation or density dependence some kind of preeminence. It may be occurring only infrequently or intermittently. And even when regulation is occurring, it may be drawing abundance toward a level that is itself changing in response to changing levels of resources. It is likely that no natural population is ever truly at equilibrium. Rather, it seems reasonable to expect to find some populations in nature that are almost always recovering from the last disaster (Figure 9.5a), others that are usually limited by an abundant resource (Figure 9.5b) or by a scarce resource (Figure 9.5c), and others that are usually in decline after sudden episodes of colonization (Figure 9.5d).

9.2.3 Key-factor analysis

All fluctuations in the abundance of a population, or differences between populations in their abundance, must be explicable in terms of births, deaths, immigrants, and emigrants. By assessing the contributions of each of these to the differences or changes that occur, we can focus research more precisely on the stages in the life cycle that are of greatest significance. In particular, the more frequently that numbers are estimated, and the more completely the details of a life cycle are exposed, the more likely we are to discover the crucial phases that determine or regulate population size. One approach, known as *key-factor analysis*, was developed for this purpose. It has been applied to many insects and some other animals and plants and is based on calculating what are known as *k*-values for each phase of the life cycle. Details are described in Box 9.1, but the approach can be understood simply by appreciating that these *k*-values measure the amount of mortality: the higher the *k*-value, the greater the mortality (*k* stands for "killing power").

the Colorado potato beetle   In fact, *key-factor analysis* is poorly named—what it actually does is identify key *phases* in the life of a study organism. Data are compiled in the form of a life table, such as that for a Canadian population of the Colorado potato beetle (*Leptinotarsa decemlineata*) in Box 9.1. The sampling program in that case provided estimates of the population at seven stages: eggs, early larvae, late larvae, pupal cells, summer

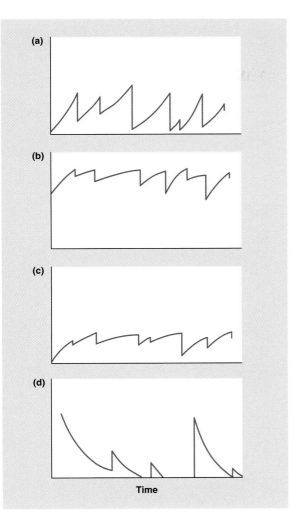

Figure 9.5
Idealized diagrams of population dynamics: (a) dynamics dominated by phases of population growth after disasters; (b) dynamics dominated by limitations on environmental carrying capacity—carrying capacity high; (c) same as (b) but carrying capacity low; (d) dynamics within a habitable site dominated by population decay after more or less sudden episodes of colonization or recruitment, for example, from the seed bank in the soil.

adults, hibernating adults, and spring adults. One further category was included, females × 2, to take account of any unequal sex ratios among the summer adults.

The first question we can ask is, How much of the total mortality tends to occur in each of the phases? The question can be answered by calculating the mean k-values, in this case determined over 10 seasons (that is, from 10 tables like the one in Box 9.1). These are presented in the first column of Table 9.2. Thus, the loss of summer adults (mostly through emigration) has by far the greatest effect (k_6 = 1.543); those of older larvae (starvation), of hibernating adults (frost-induced mortality), of young larvae (rainfall), and of eggs (cannibalization and nondeposition) all play substantial roles.

It is usually more valuable, however, not to know how much of the total mortality occurs in each of the phases, but to know the relative importance of these phases as determinants of year-to-year *fluctuations* in mortality, and hence of year-to-year fluctuations in abundance. This is rather different. For instance, a phase might repeatedly witness a significant toll being taken from a population (high mean k-value), but if that toll is always roughly the same, it will play little part in determining

when does most mortality occur?

the phases that determine abundance

Box 9.1

Determining *k*-Values for Key-Factor Analysis

Table 9.1 sets out a typical set of life table data, collected by Harcourt (1971) for the Colorado potato beetle, *Leptinotarsa decemlineata*, in Canada. The first column lists the various phases of the life cycle. *Spring adults* emerge from hibernation around the middle of June, when potato plants are breaking through the ground. Within 3 or 4 days egg laying begins, and it continues for about 1 month. The eggs are laid in clusters (approximately 34 eggs) on the lower leaf surface, and the larvae crawl to the top of the plant, where they feed throughout their development, passing through four stages. When mature, they drop to the ground and form pupal cells in the soil. *Summer adults* emerge in early August, feed, and then reenter the soil at the beginning of September to hibernate and become the next season's spring adults.

The next column lists the estimated numbers (per 96 potato hills) at the start of each phase, and the third column then lists the numbers dying in each phase, before the start of the next. This is followed, in the fourth column, by what were believed to be the main causes of deaths in each stage of the life cycle. The fifth and sixth columns then show how *k*-values are calculated. In the fifth column, the logarithms of the numbers at the start of each phase are listed. The *k*-values in the sixth column are then simply the differences between successive values in column 5. Thus, each value refers to deaths in one of the phases, and, similarly to column 3, the total of the column refers to the total death throughout the life cycle. Moreover, each *k*-value measures the rate or intensity of mortality in its own phase, whereas this in not true

for the values in column 3—there, values tend to be higher earlier in the life cycle simply because there are more individuals "available" to die. These useful characteristics of *k*-values are put to use in *key-factor analysis*.

An adult Colorado potato beetle (*Leptinotarsa decemlineata*) taking off from its host plant. Emigration by summer adults represents the key phase in the population dynamics of potato beetles.

the particular rate of mortality (and thus, the particular population size) in any particular year. In other words, this second question is much more concerned with discovering what *determines* particular abundances at particular times, and it can be addressed in the following way.

Mortality during a phase that is important in determining population change—referred to as a *key phase*—will vary very much in line with total mortality in terms of both size and direction. It is a key phase in the sense that when mortality during it is high, total mortality tends to be high and the population declines—whereas when phase mortality is low, total mortality tends to be low and the population tends to remain large, and so on. By contrast, a phase with a *k*-value that varies

Table 9.1

Life-table data for the Canadian Colorado potato beetle

AGE INTERVAL	NUMBERS PER 96 POTATO HILLS	NUMBERS DYING	MORTALITY FACTOR	FACTOR Log_{10} N	k-VALUE	
Eggs	11,799	2531	Not deposited	4.072	0.105	(k_{1a})
	9268	445	Infertile	3.967	0.021	(k_{1b})
	8823	408	Rainfall	3.946	0.021	(k_{1c})
	8415	1147	Cannibalism	3.925	0.064	(k_{1d})
	7268	376	Predators	3.861	0.024	(k_{1e})
Early larvae	6892	0	Rainfall	3.838	0	(k_2)
Late larvae	6892	3722	Starvation	3.838	0.337	(k_3)
Pupal cells	3170	16	Parasitism	3.501	0.002	(k_4)
Summer adults	3154	−126	Sex (52% ♀)	3.499	−0.017	(k_5)
♀ × 2	3280	3264	Emigration	3.516	2.312	(k_6)
Hibernating adults	16	2	Frost	1.204	0.058	(k_7)
Spring adults	14			1.146		
					2.926	(k_{total})

quite randomly with respect to total k will, by definition, have little influence on changes in mortality and hence little influence on population size. We need therefore to measure the relationship between phase mortality and total mortality and this is achieved by the *regression coefficient* of the former on the latter. The largest regression coefficient will be associated with the key phase causing population change, whereas phase mortality that varies at random with total mortality will generate a regression coefficient close to zero.

In the present example (Table 9.2), the summer adults, with a regression coefficient of 0.906, are the key phase. Other phases (with the possible exception of older larvae) have a negligible effect on the changes in generation mortality.

Table 9.2

Summary of the life-table analysis for Canadian Colorado beetle populations, showing k-values (see Box 9.1) and regression coefficients of phase k on total k

		MEAN	COEFFICIENT OF REGRESSION ON k_{total}
Eggs not deposited	k_{1a}	0.095	−0.020
Eggs infertile	k_{1b}	0.026	−0.005
Rainfall on eggs	k_{1c}	0.006	0.000
Eggs cannibalized	k_{1d}	0.090	−0.002
Eggs predation	k_{1e}	0.036	−0.011
Larvae 1 (rainfall)	k_2	0.091	0.010
Larvae 2 (starvation)	k_3	0.185	0.136
Pupae (parasitism)	k_4	0.033	−0.029
Unequal sex ratio	k_5	−0.012	0.004
Emigration	k_6	1.543	0.906
Frost	k_7	0.170	0.010
	$k_{total} =$	2.263	

(After Harcourt, 1971.)

and the factors that regulate abundance

What, though, about the possible role of these phases in the regulation of the Colorado beetle population? In other words, which, if any, act in a density-dependent way? This can be answered most easily by plotting k-values for each phase against the numbers present at the start of the phase. For density dependence, the k-value should be highest (that is, mortality greatest) when density is highest. For the beetle population, two phases are notable in this respect: for both summer adults (the key phase) and older larvae there is evidence that losses are density-dependent (Figure 9.6) and thus a possible role of those losses in regulating the size of the beetle population. In this case, therefore, the phases with the largest role in determining abundance are also those that seem likely to play the largest part in regulating abundance. But as we see next, this is by no means a general rule.

two further examples of key-factor analysis

Key-factor analysis has been applied to a great many insect populations, but to far fewer vertebrate or plant populations. Examples of these, though, are shown in Figure 9.7a and b, which summarizes the changes through time of total k and of the k-values of the various phases. The figure also displays the relationship between the k-values and densities for any phase in which there is evidence suggesting that mortality acts in a density-dependent manner.

In the population dynamics of the African buffalo (*Syncerus caffer*) in the Serengeti region of East Africa (Figure 9.7a), the juvenile period is the obvious key

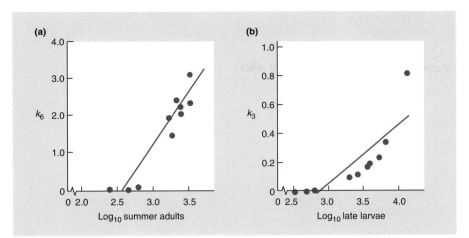

Figure 9.6
(a) Density-dependent emigration by Colorado beetle "summer" adults (slope 2.65). (b) Density-dependent starvation of larvae (slope = 0.37). (After Harcourt, 1971.)

phase, but in this case it does not operate in a density-dependent way. Juveniles suffered more than adults from a variety of endemic diseases and parasites, but this mortality, although heavy, was random with respect to density. On the other hand, adult mortality was found to be density-dependent (Figure 9.7a, right). Undernutrition appears to have been a primary agent.

The key phase in the life of a Polish population of the sand dune annual plant *Androsace septentrionalis* (Figure 9.7b) was found to be the seeds in the soil. Once again, however, mortality there did not operate in a density-dependent manner, whereas mortality of seedlings, which were not the key phase, was found to be density-dependent. Seedlings that emerge first in the season stand a much greater chance of surviving; this suggests that competition for resources may be intense (and density-dependent).

Overall, therefore, key-factor analysis (its rather misleading name apart) is useful in identifying important phases in the life cycles of study organisms, and useful too in distinguishing the variety of ways in which phases may be important: in contributing significantly to the overall sum of mortality; in contributing significantly to variations in mortality, and hence in *determining* particular population sizes; and in contributing significantly to the *regulation* of population size by virtue of the density dependence of the mortality. Box 9.2 presents an account of a topical problem, an understanding of which could benefit from key-factor analysis.

9.3 ▷ Dispersal, Patches, and Metapopulation Dynamics

In many studies of abundance, the assumption has been made that the major events all occur within the study area, and that immigrants and emigrants can be safely ignored. But migration can be a vital factor in determining and/or regulating abundance. We have already seen that emigration was the predominant reason for loss of summer adults of the Colorado potato beetle, which was both the key phase in determining population fluctuations and one in which loss was strongly density-dependent. We have also discussed previously a detailed study of the effect of seed

dispersal is ignored at the ecologist's peril

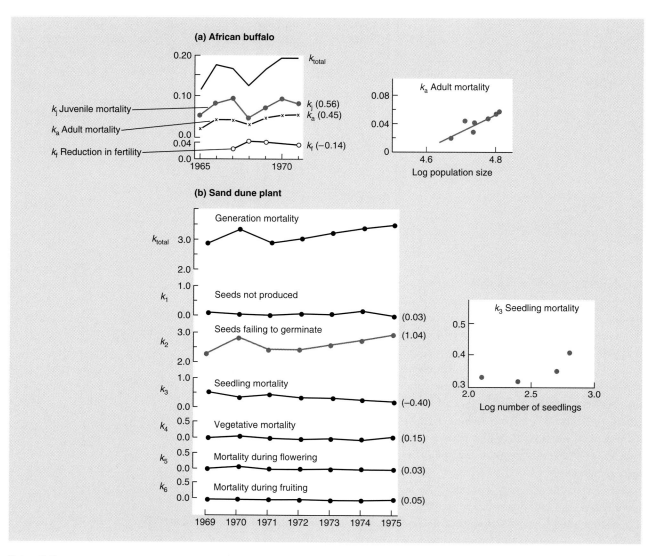

Figure 9.7

Key-factor analysis of two contrasting populations. In both cases, a graph of total generation mortality (k_{total}) and of various k-factors is presented. The values of regression coefficient of each individual k-value on k_{total} are given in brackets. The largest regression coefficient signifies the key phase and is shown as a colored line. Alongside are shown k-values that vary in a density-dependent manner. (a) African buffalo, *Syncerus caffer*. (After Sinclair, 1973.) (b) The sand dune annual plant *Androsace septentrionalis*. (After Symonides, 1979; analysis in Silvertown, 1982.)

populations may show a patchwork of dynamics

dispersal on the population dynamics of a Canadian annual plant (*Cakile edentula*— see Section 5.4.1).

Chapter 5 emphasized how dispersal plays a critical role in determining the abundance of species because most populations are fragmented and patchy and may include patches that have quite different dynamics. This was also clear in the study of buttercups in a pasture summarized in Figure 9.2. On a different scale, the aphids on one leaf may be crowded, but on other leaves on the same plant new colonists

Box 9.2   Topical ECOncerns

Acorns, Mice, Ticks, Deer, and Human Disease—Complex Population Interactions

Ecologists have been trying to uncover the complex interactions among acorn production, populations of mice and deer, parasitic ticks, and ultimately a bacterial pathogen carried by the ticks that can affect people. It is clear that a thorough understanding of the abiotic factors that determine the size of the acorn crop and of the various population interactions can enable scientists to predict years when the risk of human disease is high. This is the topic of the following newspaper article.

CONTRA COSTA TIMES Friday, February 13, 1998, by Paul Recer

More Acorns May Mean a Rise in Lyme Disease

A big acorn crop last fall could mean a major outbreak of Lyme disease next year, according to a study that linked acorns, mice and deer to the number of ticks that carry the Lyme disease parasite.

Based on the study, researchers at the Institute of Ecosystem Studies in Millbrook, New York, say that 1999 may see a dramatic upswing in the number of Lyme disease cases among people who visit the oak forests of the Northeast.

"We had a bumper crop of acorns this year, so in 1999, two years after the event, we should also have a bumper year for Lyme disease," said Clive G. Jones, a researcher at the Institute of Ecosystem Studies; "1999 should be a year of high risk for Lyme disease."

Lyme disease is caused by a bacterium carried by ticks. The ticks normally live on mice and deer, but they can bite humans. Lyme disease first causes a mild rash, but left untreated can damage the heart and nervous system and cause a type of arthritis.

Jones, along with researchers at the University of Connecticut, Storrs, and Oregon State University, Corvallis, found that the number of mice, the number of ticks, the deer population and even the number of gypsy moths are linked directly to the production of acorns in the oak forest.

Jones said that in years following a big acorn crop, the number of tick larvae is eight times greater than in years following a poor acorn crop. Additionally, he said, there are about 40 percent more ticks on each mouse.

The researchers tested the effect of acorns by manipulating the population of mice and the availability of acorns in forest plots along the Hudson River. Jones said the work, extended over several seasons, proved the theory that mice and tick populations rise and fall based on the availability of acorns.

How could a key-factor analysis be used to pinpoint the phases of importance in determining risk of human disease?

may be in the early phases of unimpeded population growth. On other, older leaves, the aphids may be becoming extinct as the leaf dies. Yet other leaves may still be unoccupied: "habitable sites" that dispersers have not yet reached.

patchiness is everywhere

Patchiness is the essence of many ecological processes. For example, plant and animal hosts are patches from the point of view of a parasite or pathogen. A population of pathogens persists if there are patches (i.e., hosts) some of which are successfully occupied (infected), remain occupied long enough to become infective, and are close enough to allow dispersal to (infection of) uncolonized hosts (see Chapter 7). Similarly, to understand the interactions between predator and prey populations, it is important to recognize that the prey are inevitably patchily distributed and that this distribution interacts profoundly with the searching behavior of predators (see Chapter 8).

habitable sites and dispersal distance

The processes determining the abundance of organisms can be linked through the two concepts of habitable site and dispersal distance. Then (following Gadgil, 1971) it can be suggested that a population may be small because

1. there are few patches (habitable sites) that provide the conditions and resources that it requires.
2. the habitable sites support few individuals.
3. the habitable sites remain habitable for only a short time.
4. the dispersal distance between habitable sites is great relative to the dispersibility of the species.

Habitable site, though, is a concept that may not be easy to put into practice. Habitable sites might remain uninhabited because individuals fail to disperse into them. If we are to discover the limitations that the number of habitable sites places on abundance, it is necessary to identify habitable sites that are not inhabited. One method involves identifying characteristics of habitat patches to which a species is restricted and then determining the distribution and abundance of similar, unoccupied patches. The water vole (*Arvicola terrestris*) lives in river banks, and in a survey of 39 sections of river bank in northern England, 10 contained breeding colonies of voles (core sites), 15 were visited by voles but they did not breed there (peripheral sites), and 14 were apparently never used or visited. The *core sites* were carefully characterized, and on this basis a further 12 unoccupied or peripheral sites that should have been suitable for breeding voles (i.e., habitable sites) were identified. Overall, about 30 percent of habitable sites were uninhabited by voles because they were too isolated to be colonized or in some cases suffered high levels of predation by mink (Lawton & Woodroffe, 1991).

Habitable sites can also be identified for a number of rare butterflies, because the larvae feed on only one or a few patchily distributed plant species. Thomas et al. (1992), for example, found that the silver-studded blue butterfly *Plebejus argus* was able to colonize virtually all habitable sites less than 1 km from existing populations, but those isolated from sources of dispersal remained uninhabited. What is more, the habitability of some of the isolated sites was established when the butterfly was successfully introduced there (Thomas & Harrison, 1992). This is presumably the crucial test of whether an uninhabited site is really habitable or not!

A radical change in the way ecologists think about populations has involved combining patchiness, dispersal, and within-patch dynamics in the concept of a *metapopulation*, the origins of which are described in Box 9.3. A population can be described as a metapopulation if it can be seen to comprise a collection of subpopulations, each one of which has a realistic chance of going extinct. The essence, then, is that the metapopulation persists, stably, as a result of the balance between random extinctions and recolonizations, even though none of the local populations is stable in its own right. An example of this is shown in Figure 9.8, where within a persistent, highly fragmented metapopulation of the Glanville fritillary butterfly (*Melitaea cinxia*) in Finland, even the largest subpopulations had a high probability of declining to extinction within 2 years.

metapopulations

Box 9.3

The Genesis of Metapopulation Theory

A classic book, *The Theory of Island Biogeography*, written by MacArthur and Wilson and published in 1967, was an important catalyst in radically changing ecological theory. They showed how the distribution of species on islands could be interpreted as a balance between the opposing forces of extinctions and colonizations (see Chapter 10) and focused attention especially on situations in which those species were all available for repeated colonization of individual islands from a common source—the mainland. They developed their ideas in the context of the floras and faunas of real (i.e., oceanic) islands, but their thinking has been rapidly assimilated into much wider contexts with the realization that patches everywhere have many of the properties of true islands—ponds as islands of water in a sea of land, trees as islands in a sea of grass, and so on.

At about the same time as MacArthur and Wilson's book was published, a simple model of metapopulation dynamics was proposed by Levins (1969). The concept of *metapopulation* was introduced to refer to a subdivided and patchy population in which the population dynamics operate at two levels:

1. The dynamics of individuals within patches (determined by the usual demographic forces of birth, death, and movement)

2. The dynamics of occupied patches, or subpopulations, within the overall metapopulation (determined by the rates of colonization of empty patches and of extinction within occupied patches)

Although both this and MacArthur and Wilson's theory embraced the idea of patchiness, and both focus on colonization and extinction rather than the details of local dynamics, MacArthur and Wilson's theory was based on a vision of mainlands as rich sources of colonists for whole archipelagos of islands, whereas in a metapopulation there is a collection of patches but no such dominating mainland.

Levins introduced the variable $p(t)$, the fraction of habitat patches occupied at time t. Note that the use of this single variable carries the profound notion that not all habitable patches are always inhabited. The rate of change in $p(t)$ depends on the rate of local extinction of patches and the rate of colonization of empty patches. It is not necessary to go into the details of Levin's model; suffice to say that as long as the intrinsic rate of colonization exceeds the intrinsic rate of extinction within patches, the total metapopulation will reach a stable, equilibrium fraction of occupied patches, even though none of the local populations is stable in its own right.

Perhaps because of the powerful influence on ecology of MacArthur and Wilson's theory, the whole idea of metapopulations was largely neglected during the 20 years after Levins's initial work. The 1990s, however, saw a great flowering of interest, both in underlying theory and in populations in nature that might conform to the metapopulation concept (Hanski, 1994).

Figure 9.8
Comparison of the subpopulation sizes in June 1991 (adults) and August 1993 (larvae) of the Glanville fritillary butterfly (*Melitaea cinxia*) on Åland Island in Finland. Multiple data points are indicated by numbers. Many 1991 populations, including many of the largest, had become extinct by 1993. (After Hanski et al., 1995.)

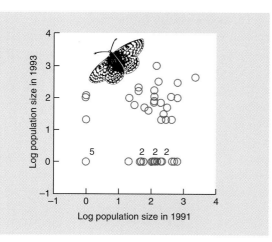

transient dynamics may be as important as equilibria

On the other hand, although stable equilibria can readily be generated in simple metapopulation models (Box 9.3), the observable dynamics of a species may often have more to do with the "transient" metapopulation behavior, far from equilibrium. For example, the silver-spotted skipper butterfly (*Hesperia comma*) declined steadily in Great Britain from a widespread distribution over most calcareous (limestone) hills in 1900, to 46 or fewer refuge localities in 10 regions by the early 1960s (Thomas & Jones, 1993). The probable reasons were changes in land use—increased plowing of grasslands, reduced stocking with grazing animals, and virtual elimination of rabbits by myxomatosis with its consequent profound vegetational changes. Throughout this nonequilibrium period, rates of local extinction generally exceeded those of recolonization. In the 1970s and 1980s, however, reintroduction of livestock and recovery of the rabbits led to increased grazing and suitable habitats increased again. Recolonization exceeded local extinction—but the spread of the skipper remained slow, especially into localities isolated from the 1960s refuges. Even in southeast England, where the density of refuges was greatest, it is predicted that the abundance of the butterfly will increase only slowly—and remain far from equilibrium—for at least 100 years.

a continuum of metapopulation types

In reality, moreover, there is likely to be a continuum of types of metapopulation: from collections of nearly identical local populations, all equally prone to extinction, to metapopulations in which there is great inequality between local populations, some of which are effectively stable in their own right. This contrast is illustrated in Figure 9.9 for the silver-studded blue butterfly (*Plejebus argus*) in North Wales.

9.4 Temporal Patterns in Community Composition

9.4.1 Founder-controlled and dominance-controlled communities

disturbances and the patch dynamics concept of community organization

The metapopulation concept is important when population dynamics are considered in the context of environmental patchiness, but when communities are the focus of attention we usually refer to the *patch dynamics* concept of community organization.

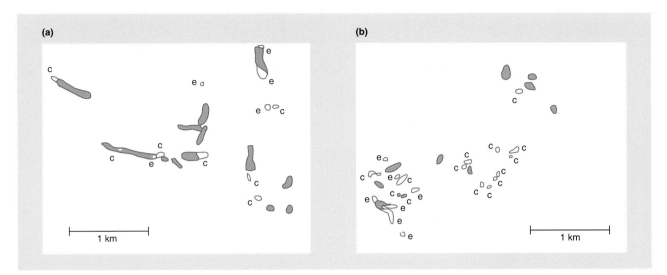

Figure 9.9
Two metapopulations of the silver-studded blue butterfly (*Plejebus argus*) in North Wales: (a) in a limestone habitat, where there was a large number of persistent (often larger) local populations among smaller, much more ephemeral local populations; (b) in a heathland habitat, where the proportion of smaller and ephemeral populations was much greater. Filled outlines, present in both 1983 and 1990; open outlines, not present at both times; e, present only in 1983 (presumed extinction); c, present only in 1990 (presumed colonization). (After Thomas & Harrison, 1992.)

The concepts are very closely related. Both accept that populations and communities are usually open systems, with dispersal between patches and differential dynamics within patches—and that a combination of patchiness and movement between patches can give rise to dynamics quite different from those that would be observed if a population or community consisted of just one, homogeneous patch.

Disturbances that open up gaps (patches) are common in all kinds of community. In forests, they may be caused by high winds, lightning, earthquakes, elephants, lumberjacks, or simply death of a tree through disease or old age. Agents of disturbance in grassland include frost, burrowing animals, and teeth, feet, or dung of grazers. On rocky shores or coral reefs, gaps in algal or sessile animal communities may be formed as a result of severe wave action during hurricanes, tidal waves, battering by logs or moored boats, fins of careless scuba divers, or action of predators.

From the viewpoint of the effect of disturbance and gap creation and reoccupancy, two fundamentally different kinds of community organization can be recognized according to the type of competitive relationships exhibited by the component species (Yodzis, 1986). Situations in which all species are good colonists and essentially equal competitors are described as *founder-controlled*, whereas those in which some species are strongly competitively superior can be described as *dominance-controlled*. The dynamics of these two situations are quite different, and we deal with them in turn.

In founder-controlled communities, species are approximately equivalent in their ability to invade gaps, are equally tolerant of the abiotic conditions, and can hold the gaps against all comers during their lifetime. Hence, the probability of competitive

founder-controlled communities—competitive lotteries

The Great Barrier Reef

**dominance-controlled communities
and community succession**

exclusion in the community as a whole may be much reduced where gaps are appearing continually and randomly. This can be referred to as a "competitive lottery." Figure 9.10 illustrates how the occupancy of a series of gaps is likely to change through time. On each occasion that an organism dies (or is killed) the gap is reopened for invasion. All conceivable replacements are possible and species richness can be expected to be maintained at a high level in the system as a whole.

Some tropical reef communities of fish may conform to this model (Sale & Douglas, 1984). They are extremely rich in species. For example, the number of species of fish on the Great Barrier Reef off the east coast of Australia ranges from 900 in the south to 1500 in the north, and more than 50 resident species may be recorded on a single patch of reef 3 m in diameter. Only a proportion of this species richness is likely to be attributable to resource partitioning of food and space—indeed the diets of many of the coexisting species are very similar. In this community, vacant living space seems to be a crucial limiting factor, and it is generated unpredictably in space and time when a resident dies or is killed. The life-styles of the species match this state of affairs. They breed often, sometimes year-round, and produce numerous clutches of dispersive eggs or larvae. It can be argued that the species compete in a lottery for living space in which larvae are the tickets, and the first arrival at the vacant space wins the site, matures quickly, and holds the space for its lifetime.

In dominance-controlled communities, by contrast, some species are competitively superior to others, and an initial colonizer of a patch cannot necessarily maintain its presence there. Dispersal between patches, or growth of an individual within a patch, will bring about a reshuffle and species may be competitively excluded locally. In these cases, disturbances that open up gaps lead to reasonably predictable species sequences because different species have different strategies for exploiting resources—early species are good colonizers and fast growers, whereas later species can tolerate lower resource levels and grow to maturity in the presence of early species, eventually outcompeting them. Such sequences are examples of community successions. The effect of the disturbance is to knock the community back to an earlier stage of succession (Figure 9.11). The open space is colonized by one or more of a group of opportunistic, early-successional species (p_1, p_2, etc., in Figure

Figure 9.10

Founder-controlled community dynamics. Occupancy of gaps that periodically become available: each of species *A* to *E* is equally likely to fill a gap, regardless of the identity of its previous occupant (this is illustrated in the inset). Species richness remains high and relatively constant. (Compare with Figure 9.11.)

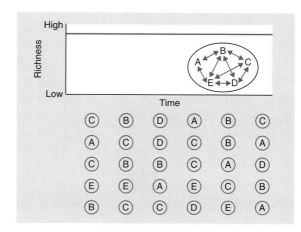

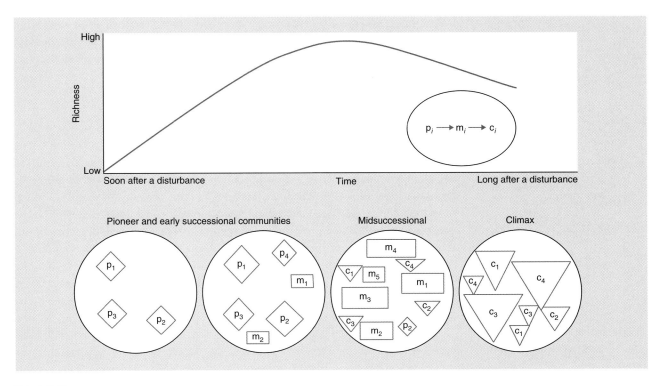

Figure 9.11

Hypothetical succession in a gap—an example of dominance control. The occupancy of gaps is reasonably predictable. Richness begins at a low level as a few pioneer (p_i) species arrive; reaches a maximum in midsuccession when a mixture of pioneer, midsuccessional (m_i) and climax (c_i) species occur together; and drops again as competitive exclusion by climax species takes place (compare with Figure 9.10).

9.11). As time passes, more species invade, often those with poorer powers of dispersal. These eventually reach maturity, dominating midsuccession (m_1, m_2, etc.), and many or all of the pioneer species are driven to extinction. Later still, the community regains a *climax* stage when the most efficient competitors (c_1, c_2, etc.) oust their neighbors. In this sequence, if it runs its full course, the number of species first increases (because of colonization) then decreases (because of competition).

One study that has provided support for this pattern was carried out on the rocky shores of southern California (Sousa, 1979a,b). In an intertidal algal community associated with boulders of various sizes, wave action disturbs small boulders more often than large. Hence, the successional sequence that follows the clearing of space (but in the absence of further disturbance) could be established by studying both large boulders that had been experimentally cleared and concrete blocks that had been implanted. Within the first month the surfaces were colonized by a mat of an ephemeral green alga, *Ulva* sp. Then, in the autumn and winter of the first year, several species of perennial red alga became established, including *Gelidium coulteri*, *Gigartina leptorhynchos*, *Rhodoglossum affine*, and *Gigartina canaliculata*. However, the last of these gradually came to dominate the community, holding 60–90 percent of the surface area after 2 to 3 years. Thus, as described in Figure 9.11, the number

illustrated by rocky shore data

of species on a boulder ("in a patch") increased during early stages of succession through a process of colonization but declined again subsequently because of competitive exclusion by *Gigartina canaliculata*. The same succession occurred on small boulders that had been made artificially stable.

9.4.2 Community succession

the importance of phasing

Some disturbances are synchronized, or phased, over extensive areas. A forest fire may destroy a huge tract of a climax community. The whole area then proceeds through a more or less synchronous succession, with diversity increasing through the early colonization phase and falling again through competitive exclusion as the climax is approached. Other disturbances are much smaller and produce a patchwork of habitats. If these disturbances are unphased, the resulting community comprises a mosaic of patches at different stages of succession. This is true of the California shore, described previously, where different boulders are disturbed at different times. For a more detailed description of a succession, however, we turn next to a much more synchronized disturbance.

community succession on fields abandoned by farmers

Successions on old fields have been studied primarily in the eastern United States, where many farms were abandoned by farmers who moved west after the frontier was opened up in the nineteenth century. Indeed, in many places, a series of sites are available that have been abandoned for different, recorded periods of time. Most of the precolonial mixed conifer–hardwood forest had been destroyed, but regeneration was swift after the "disturbance" caused by farmers came to an end. The early pioneers of the American West left behind exposed land that was colonized by pioneers of a very different kind!

The typical sequence of dominant vegetation is

annual weeds → herbaceous perennials → shrubs → early
successional trees → late successional trees

The pioneer species are those that can establish themselves quickly in the disturbed habitat of a recently cultivated field, either by rapid dispersal into the site or from propagules that are already present. Perhaps the most common annual of old field succession is *Ambrosia artemisiifolia*, which has seeds that survive for many years in the soil and germinate when disturbance brings them to the surface, where they experience unfiltered light, reduced CO_2 concentration, and fluctuating temperatures—all conditions that induce germination in many annuals. However, summer annuals such as *Ambrosia* sp. are often quickly eclipsed in importance by winter annuals that have small seeds with little or no dormancy but disperse over long distances. Their seeds germinate soon after they land, usually in late summer or autumn. Next spring they have a head start over summer annuals and thus they preempt the resources of light, water, space, and nutrients.

Early successional plants have a fugitive life-style. Their continued survival depends on dispersal to other disturbed sites. They cannot persist in competition with later species, and thus they must grow and consume the available resources rapidly. High growth and photosynthetic rates are crucial properties of the fugitive. Those of later successional plants are much lower (Table 9.3).

In contrast to the pioneer annuals, seeds of later successional plants can germinate in the shade—for example, beneath a forest canopy. They can continue to grow at these low light intensites, too—quite slowly but faster than the species they replace (Figure 9.12).

The trees that take part in the later stages of an old field succession can themselves be grouped into early and late successional classes. Many early successional trees have multilayered foliage. Leaves extend deep into the canopy, where they still receive enough light to add more to the plant (in photosynthate) than they take from it. Species such as eastern red cedar (*Juniperus virginiana*) are thus able to use the abundant light that is available early in the tree stage of the succession. In contrast, sugar maple (*Acer saccharum*) and American beech (*Fagus grandifolia*) are monolayered species. They possess a single layer of leaves in a shell around the tree and

early and late successional species have different properties

Table 9.3

Some representative photosynthetic rates (mg CO_2 dm^{-2} h^{-1}) of plants in a successional sequence. Late successional trees are arranged according to their relative successional position.

PLANT	RATE	PLANT	RATE
Summer annuals		Early successional trees	
Abutilon theophrasti	24	*Diospyros virginiana*	17
Amaranthus retroflexus	26	*Juniperus virginiana*	10
Ambrosia artemisiifolia	35	*Populus deltoides*	26
Ambrosia trifida	28	*Sassafras albidum*	11
Chenopodium album	18	*Ulmus alata*	15
Polygonum pensylvanicum	18		
Setaria faberii	38	Late successional trees	
		Liriodendron tulipifera	18
Winter annuals		*Quercus velutina*	12
Capsella bursa-pastoris	22	*Fraxinus americana*	9
Erigeron annuus	22	*Quercus alba*	4
Erigeron canadensis	20	*Quercus rubra*	7
Lactuca scariola	20	*Aesculus glabra*	8
		Fagus grandifolia	7
Herbaceous perennials		*Acer saccharum*	6
Aster pilosus	20		
(After Bazzaz, 1979.)			

Figure 9.12

Idealized light saturation curves for early, mid-, and late successional plants. (After Bazzaz, 1979.)

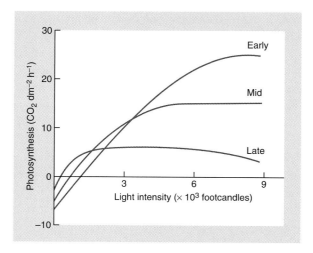

are more efficient in the crowded canopy of late succession. If a monolayered and a multilayered tree simultaneously colonize an open site, the multilayered tree will usually grow faster and dominate until it is crowded by its neighbors, when the more slowly growing monolayered tree emerges to dominance.

The early colonists among the trees usually have efficient seed dispersal; this in itself makes them likely to be early on the scene. They are usually precocious reproducers and are soon ready to leave descendants in new sites elsewhere. The late colonists are those with larger seeds, poorer dispersal, and long juvenile phases. The contrast is between the life-styles of the "quickly come, quickly gone" and "what I have, I hold."

experiments unravel the multiple causes of an old field succession

A particularly detailed study of old field succession has been performed at the Cedar Creek Natural History Area in Minnesota on well-drained and nutrient-poor soil (refer back to Chapter 1, Section 1.3.2, where this study is discussed in detail). One consequence of the poor growing conditions is that species replacement is slow: annual plants persist as dominants for up to 40 years after field abandonment and woody plants (mostly vines and shrubs) still provide only 13 percent of ground cover in 60-year-old fields. In other respects, however, the pattern of species replacement is similar to what has been described elsewhere (Figure 1.14).

animals are often affected by, but may also affect, plant successions

The fact that plants dominate most of the structure and succession of communities does not mean that the animals always follow the communities that plants dictate. This will often be the case, of course, because the plants provide the starting point for all food webs and determine much of the character of the physical environment in which animals live. But it is also sometimes the animals that determine the nature of the plant community, for example, through heavy grazing or trampling (see Box 9.4). More often, though, animals are passive followers of successions among the plants. This is the case for bird species in an old field succession (see Figure 4.10).

the concept of the climax

Do successions reach a climax? In fact, it is generally very difficult to identify a stable "climax" community in the field. Usually we can do no more than point out that the rate of change of succession slows to the point where any change is imperceptible to us. The succession of seaweeds on overturned boulders (Sousa,

Box 9.4   *Topical ECOncerns*

Conservation Sometimes Requires Manipulation of a Succession

Some endangered animal species are associated with particular stages of a succession. Their conservation then depends on a full understanding of the successional sequence, and intervention may be required to maintain their habitat at an appropriate successional stage.

An intriguing example is provided by a giant New Zealand insect, the weta *Deinacrida mahoenuiensis* (Orthoptera; Anostostomatidae). This species, which is believed to have been formerly widespread in forest habitat, was discovered in the 1970s in an isolated patch of gorse (*Ulex europaeus*). Ironically, in New Zealand gorse is an introduced weed that farmers spend much time and effort attempting to control. Its dense, prickly sward provides a refuge for the giant weta against other introduced pests, particularly rats but also hedgehogs, stoats, and possums, which could readily capture weta in their original forest home. Mammalian predation is believed to be responsible for weta extinction elsewhere.

New Zealand's Department of Conservation purchased this important patch of gorse from the landowner, who insisted that his cattle should still be permitted to overwinter in the reserve. Conservationists were unhappy about this, but the cattle subsequently proved to be part of the weta's salvation. By opening up paths through the gorse, cattle provided entry for feral goats that browse the gorse, producing a dense hedge-like sward and preventing the habitat from succeeding to a stage inappropriate to the wetas. This story involves a single endangered endemic insect together with a whole suite of introduced pests (gorse, rats, goats, etc.) and introduced domestic animals (cattle). Before the arrival of people in New Zealand, the island's only land mammals were bats, and New Zealand's endemic fauna has proved to be extraordinarily vulnerable to the mammals that arrived with people. However, by maintaining gorse succession at an early stage, the grazing goats provide a habitat in which the weta can escape the attentions of the rats and other predators.

Because of its economic cost to farmers, ecologists have been trying to find an appropriate biological control agent for gorse, ideally one that would eradicate it. How would you weigh up the needs of a rare insect against the economic losses associated with gorse on farms?

A giant weta on a gorse branch. (Photograph by Greg Sherley, Department of Conservation, Wellington, New Zealand.)

1979) is unusual because it may reach a climax in only a few years. Old field successions, on the other hand, might take 100–300 years to reach a climax, but in that time the probabilities of fire or severe hurricanes, which occur every 70 or so years in New England, for example, are so high that a process of succession may never go to completion. Bearing in mind that forest communities in northern temperate regions, and probably also in the tropics, are still recovering from the last glaciation, it is questionable whether the idealized climax vegetation is often reached in nature.

Finally, we again return to issues of scale that feature in almost every chapter of this book. A forest, or a rangeland, that appears to have reached a stable community structure when studied on a scale of hectares will always be a mosaic of miniature successions. Every time a tree falls or a grass tussock dies an opening is created in which a new succession starts. The patch-dynamics pattern of many communities is a result of the dynamic processes of deaths, replacements, and microsuccessions that the broad view may conceal.

9.5 ▷ Food Webs

No predator–prey, parasite–host, or grazer–plant pair exists in isolation. Each is part of a complex web of interactions with other predators, parasites, food sources, and competitors within its community. Ultimately, it is these food webs that ecologists wish to understand. However, it has been useful to isolate groups of competitors as we did in Chapter 6, and mutualists, symbionts and parasite–host pairs as in Chapter 7, and predator–prey pairs as in Chapter 8, simply because we have little or no hope of understanding the whole unless we have some understanding of the component parts. Toward the end of Chapter 8 (Section 8.6), our field of view was expanded to include the effects of predators on groups of competitors and to show, for example, the importance of predator-mediated coexistence.

food webs—shifting the focus to systems with at least three trophic levels

We now take this approach a stage further to focus on systems with at least three trophic levels (plant–herbivore–predator), and consider not only direct but also indirect effects that a species may have on others on the same or on other trophic levels. The effects of a predator on both the individuals and the populations of its herbivorous prey, for example, are direct and relatively straightforward. But these effects may also be felt by any plant population on which the herbivore feeds, by other predators and parasites of the herbivore, by other consumers of the plant, by competitors of the herbivore and the plant, and by the myriad species linked even more remotely in the food web.

9.5.1 Indirect and direct effects

The deliberate removal of a species from a community can be a powerful tool in unraveling the workings of a food web. We might expect such removal to lead to an increase in the abundance of a competitor, or, if the species removed is a predator, to an increase in the abundance of its prey. Sometimes, however, when a species is removed, a competitor may actually decrease in abundance, and removal of a predator can lead to a decrease in a prey species population. Such unexpected effects arise when direct effects are less important than effects that occur through indirect pathways. Thus, removal of a species might increase the density of one competitor, which in turn causes another competitor to decline. Or removal of a predator might increase the abundance of a prey species that is competitively superior to another, leading to a decrease in the density of the latter.

the direct and indirect effects of shorebirds on limpet populations

In one study, for example, predation by birds was experimentally manipulated over a 2-year period in an intertidal community on the Northwest coast of the United States to determine the consequences for three limpet species and their algal food.

Glaucous-winged gulls (*Larus glaucescens*) and oystercatchers (*Haematopus bachmani*) were excluded by means of wire cages from large areas (each 10 m²) in which limpets were common. It became evident that excluding the birds increased the overall abundance of one of the limpet species, *Lottia digitalis*, as might have been expected, but a second limpet species (*L. strigatella*) became rarer, and the third, *L. pelta*, which was the one most frequently consumed by the birds, did not vary in abundance. The reasons are complex and go well beyond the direct effects of birds' eating limpets (Figure 9.13).

   L. digitalis, a light-colored limpet, tends to occur on light-colored goose barnacles (*Pollicipes polymerus*), where it is camouflaged, whereas the dark *L. pelta* occurs primarily on dark Californian mussels (*Mytilus californianus*). Predation by birds normally reduces the area covered by goose barnacles, and so excluding the birds increased barnacle abundance and increased the abundance, too, of *L. digitalis* (Figure 9.13). Increasing barnacle abundance through bird-exclusion also led to a

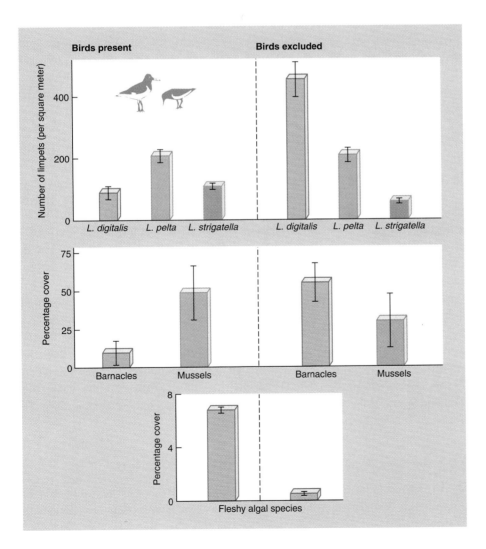

Figure 9.13

When birds are excluded from the intertidal community, barnacles increase in abundance at the expense of mussels, and three limpet species show marked changes in density, reflecting changes in the availability of cryptic habitat and competitive interactions as well as the easing of direct predation. Algal cover is much reduced in the absence of effects of birds on intertidal animals (means ± standard errors are shown). (After Wootton, 1992.)

decrease in the area covered by mussels, because they were now subject to more intense competition from the barnacles (Figure 9.13). This, one imagines, might have led to a decrease in the abundance of *L. pelta* living predominantly on those mussels. However, the third limpet species, *L. strigatella*, is competitively inferior to the others, and the increase in abundance of *L. digitalis* when birds were excluded therefore led to a decrease in the abundance of *L. strigatella*, which in turn released pressure on *L. pelta* such that overall its abundance remained effectively unchanged (Figure 9.13).

The effects of bird predation also cascade down to the plant trophic level, because by consuming limpets, the birds normally reduce the grazing pressure of the limpets on fleshy algae, and by consuming goose barnacles, the birds normally free up space for algal colonization. Hence, when the birds were excluded, algal cover decreased (Figure 9.13). It is apparent that short-term studies of pairwise interactions would have failed to reveal the rich array of direct and indirect interactions in this food web.

keystones in food web architecture

Some species are more intimately and tightly woven into the fabric of the food web than others. A species whose removal would produce a significant effect (extinction or a large change in density) in at least one other species may be thought of as a strong interactor. Removal of some strong interactors would lead to significant changes spreading throughout the food web—we refer to these as *keystone species*. In building construction, a keystone is the wedge-shaped block at the highest point of an arch that locks the other pieces together. Removal of the keystone species, just like removal of the keystone in an arch, leads to collapse of the structure. More precisely, it leads to extinction or large changes in abundance of several species, producing a community with a very different species composition and, to our eyes, an obviously different physical appearance.

Although the term was originally applied only to predators, it is now widely accepted that *keystone species* can occur at any trophic level. For example, lesser snow geese (*Chen caerulescens caerulescens*) are herbivores that breed in large colonies in coastal brackish and freshwater marshes along the west coast of Hudson Bay in Canada. At their nesting sites in spring, before growth of aboveground foliage begins, adult geese grub for roots and rhizomes of plants in dry areas and eat the swollen bases of shoots of sedges in wet areas. Their activity creates bare areas (1–5 m²) of peat and sediment. Few pioneer plant species are able to recolonize these patches, and recovery is very slow. Furthermore, in areas of intense summer grazing, "lawns" of *Carex* and *Puccinellia* sp. have become established. Here, therefore, high densities of grazing geese are essential to maintain the species composition of the vegetation and its aboveground production (Kerbes et al., 1990). The lesser snow goose is a keystone species—the whole structure and composition of these communities are drastically altered by its presence.

9.5.2 Top-down or bottom-up control of food webs?

One of the most fundamental questions about food webs is whether they are controlled from the top down or from the bottom up. *Top-down control* refers to situations in which the structure (abundance, species number) of lower trophic levels depends on the effects of consumers from higher trophic levels—"predators" controlling "prey." *Bottom-up control* occurs as dependence of community structure on factors, such as nutrient concentration and prey availability, that influence a trophic level from below; in this case, populations within a trophic level are affected predominantly by competition not predation.

It is helpful to dissect the question of whether predators or resources will dominate the dynamics of populations by considering first a hypothetical community with just one (plant) trophic level and then successively adding extra trophic levels, one at a time.

It is not easy to envisage real communities in nature that consist of only one trophic level, but if one existed, predation would be absent by definition, control would be bottom-up, and competition would be the predominant population interaction (Figure 9.14).

Two-trophic-level systems can be identified if we allow ourselves to isolate limited but significant components of real systems. For example, grazing by the giant tortoise, on the most remote island on Earth, Aldabra, crops the turf down to less than 5 mm over extensive areas (Strong, 1992). When the tortoises are excluded by fences, many species of trees, shrubs, and grasses, normally excluded by grazing, grow and dominate the plant community. No predators of the giant tortoises, or their eggs or young, have arrived on the island. In this two-trophic-level system, then, there are clear top-down control of the bottom trophic level, where the effects of predation predominate, and bottom-up control of the herbivores (Figure 9.14).

The saline Great Salt Lake of Utah provides a case of a three-trophic-level system, when the usual zooplankton–phytoplankton system is augmented by a third trophic level (a predatory insect, *Trichocorixa verticalis*) in unusually wet years when salinity declines. Normally, the zooplankton keep phytoplankton biomass at a low level. But when salinity declined from above 100 to 50 g/liter in 1985, *Trichocorixa* invaded and reduced zooplankton biomass from 720 to 2 mg/cubic meter, leading to a 20-fold increase in phytoplankton concentration. In this three-trophic-level community at least, the plants are subject to bottom-up control, having been released from heavy grazing pressure by the effects of carnivores on the herbivores. The herbivores are thus subject to top-down and the carnivores to bottom-up control (Figure 9.15).

dependence on the number of trophic levels

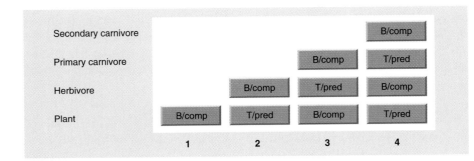

Figure 9.14
Diagrammatic representation of communities with one, two, three, or four trophic levels, illustrating for each trophic level whether control is predicted to be bottom-up (B) or top-down (T), and whether population dynamics are determined primarily by competition (comp) or predation (pred).

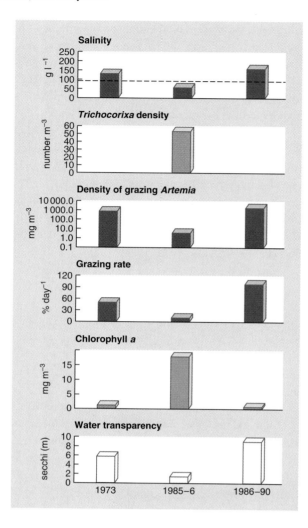

Figure 9.15

Variation in the open-water ecosystem of the Great Salt Lake during three periods that differed in salinity. (After Wurtsbaugh, 1992.)

UNANSWERED QUESTION:

how green is our world?

UNANSWERED QUESTION:

**is the world green or prickly and bad-
tasting?**

The top-down view was first introduced in a famous paper that proposed that "the world is green." It argued that green plant biomass accumulates because carnivores keep herbivores in check (Hairston et al., 1960). The Great Salt Lake study is one of several that tend to support this argument. An alternative view, however, which has been described as "the world is prickly and tastes bad" (Pimm, 1991), emphasizes that many plants have evolved physical and chemical defenses that make life difficult for herbivores (see Section 3.4.2). Thus, the world may be green—where it is green—because plants are wholly or partly inedible and not because herbivores are kept in check by their predators. Which of these views has most support, and under what circumstances, remains to be seen.

If the pattern developing thus far is continued (Figure 9.14), we expect that in a four-trophic-level system, the plants and the primary carnivores will be limited top-down, whereas the herbivores and secondary carnivores will be limited bottom-up. This is precisely what was found in a study of the food web in Eel River, northern California. Large fish reduced the abundance of fish fry and invertebrate predators, allowing their prey, tuft-weaving midge larvae, to attain high density and to exert intense grazing pressure on filamentous algae, whose biomass was thus kept low (Figure 9.16a).

However, in a four-trophic-level terrestrial community in the Bahamas, consisting of sea grape shrubs, which were fed upon by herbivorous arthropods, and then web spiders (primary carnivores) and lizards (secondary or "top" carnivores), the results of experimental manipulations indicated a strong effect of the lizards on the herbivores but a weaker effect of the lizards on the spiders. Consequently, the net effect of top predators on plants was positive and there was less leaf damage in the presence of lizards. In essence, this four-trophic-level community functions as if it has only three levels (compare Figure 9.16b with Figure 9.16a).

Overall, therefore, it is apparent, first, that although the food webs of the real world may be disassembled into series of simple predator–prey links for our convenience of study, these isolated pairs can never tell the whole story. It is also apparent, however, that not even the first line of that story could be written unless it were based on a solid foundation of understanding of predator–prey dynamics.

9.5.3 Community stability and food web structure

Of all the imaginable food webs in nature, are there particular types that we tend to observe repeatedly? Do actual (as opposed to imaginable) food webs have particular properties? Are some food web structures more stable than others? (We discuss what *stable* means in Box 9.5.) These are important practical questions. We require answers if we are to determine whether some communities are more fragile (and

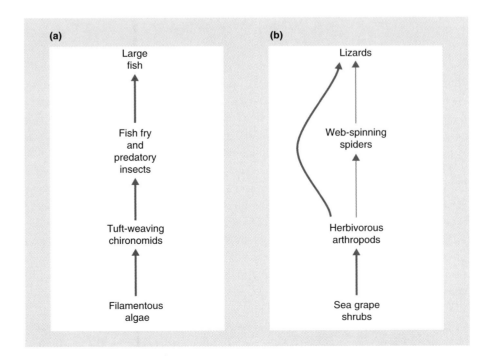

Figure 9.16

Two examples of food webs with four trophic levels. (a) The absence of omnivory (feeding at more than one trophic level) in this North American stream community means it functions, as predicted in Figure 9.9, as a four-trophic-level system. On the other hand, web (b) from a terrestrial Bahamanian community functions as a three-trophic-level web. This is because of the strong direct effects of omnivorous top predators on herbivores and their less influential effects on intermediate predators. (After Power, 1990; Spiller & Schoener, 1994, respectively.)

Box 9.5

What Do We Mean by "Community Stability"?

Among several, there are two important qualifications that can be made when we come to decide what we mean by stability. The first is the distinction between the resilience of a community and its resistance. A *resilient community* is one that returns rapidly to something like its former structure after that structure has been altered. A *resistant community* is one that undergoes relatively little change in its structure in the face of a disturbance.

The second distinction is between *fragile* and *robust stability*. A community has only *fragile stability* if it remains essentially unchanged in the face of a small disturbance but alters utterly when subjected to a larger disturbance, whereas one that stays roughly the same in the face of much larger disturbances is said to have stability that is *dynamically robust*.

To illustrate these distinctions by analogy, consider the following:

- a pool or billiard ball balanced carefully on the end of a cue

- the same ball resting on the table

- the ball sitting snugly in its pocket

The ball on the cue is stable in the narrow sense that it will stay there forever as long as it is not disturbed—but its stability is fragile, and both its resistance and its resilience are low: the slightest touch will send the ball to the ground, far from its former state (low resistance), and it has not the slightest tendency to return to its former position (low resilience).

The same ball resting on the table has a similar resilience: it has no tendency to return to exactly its former state (assuming the table is level), but its resistance is far higher: pushing or hitting it moves it relatively little. And its stability is also relatively robust: it remains "a ball on the table" in the face of all sorts and all strengths of assault with the cue.

The ball in the pocket, finally, is not only resistant but resilient too—it moves little and then returns—and its stability is highly robust: it will remain where it is in the face of almost everything other than a hand that carefully plucks it away.

more in need of conservation) than others; or whether there are certain "natural" structures that we should aim for when we construct communities ourselves; or whether communities that have been restored are likely to stay "restored." Progress in answering these questions, though, depends critically on the quality of data that are gathered from natural communities. This is not always as good as it needs to be. There are many unresolved questions.

a long-standing belief that complexity leads to stability

The aspect of structure that has received most attention from the point of view of stability has been community complexity. During the 1950s and 1960s, the conventional wisdom in ecology was that increased complexity within a community leads to increased stability (MacArthur, 1955; Elton, 1958). For example, MacArthur (1955) argued that the more possible pathways there were by which energy passed through a community, the less likely it was that the density of one of the constituent species would change in response to altered density in one of the other species. In other words, the greater the complexity (more pathways), the greater the resistance (less numerical change, Box 9.5) in the face of a perturbation.

that is not supported by mathematical models

However, that conventional wisdom has by no means always received support from more recent work and has been undermined particularly by the analysis of mathematical models of food webs (reviewed by May, 1981c). Briefly, these model food webs can be characterized by (1) the number of species they contain, (2) the *connectance* of the web (the fraction of all possible pairs of species that interact directly with one another—as competitors, mutalists or predators and prey), and (3) the average interaction strength between pairs of species. Most models lead to similar conclusions: increases in the number of species, increases in connectance, and increases in average interaction strength all tend to decrease the resilience aspect of instability. Yet each of these represents an increase in complexity. Thus these models suggest that complexity leads to *instability*. This runs counter to the conventional wisdom of Elton and MacArthur, and it certainly indicates that there is no necessary, unavoidable connection linking stability to complexity.

What is the evidence from real communities? A number of studies have sought to build on the models by examining the relationships among number of species, connectance, and interaction strength. The argument runs as follows. If the only communities we observe are those that are stable enough to exist, then those of them with more species can only be sufficiently stable if there are compensatory decreases in connectance and/or interaction strength. But data on interaction strengths for whole communities are unavailable. It is therefore usually assumed, for simplicity, that average interaction strength is constant, and hence communities with more species will only retain stability if there is an associated decline in average connectance.

A group of 40 food webs, including terrestrial, freshwater, and marine examples, was gleaned from the literature by Briand (1983). For each community, a single value for connectance was calculated as the total number of identified interspecies links as a proportion of the total possible links. As predicted, connectance decreased with species number (Figure 9.17a).

This compilation suffered from a very serious drawback, however—the data on which it was based were not collected for the purpose of quantitative study of food web properties, and, what is more, the accuracy of identification varied substantially from web to web—even in the same web, different taxa may have been grouped at the level of kingdom (plants), family (Diptera), and species (polar bear) (see review by Hall & Raffaelli, 1993). More recent studies, then, in which food webs were more rigorously documented, indicate that connectance may decrease with species number (as predicted) (Figure 9.17b), or may be independent of species number (Figure 9.17c), or may even increase with species number (Figure 9.17d). Thus, the stability argument does not receive consistent support from food web analyses.

The relationship between complexity and stability has also been investigated by manipulating two sets of plant communities. In the first, the perturbation consisted of adding plant nutrients to the soil; in the second it involved introducing African buffalo to graze the plants. In both, the effect was monitored in both species-rich and species-poor plant communities, and in both the perturbation significantly reduced the diversity of the species-rich but not the species-poor community (Table 9.4), suggesting that more complex communities are less likely to resist change in the face of a perturbation.

In contrast, in a "natural experiment" involving a set of eight varied grassland communities, those that were more diverse in species were more stable in community

complexity and stability in practice

a prediction supported by some studies but not others

a complex community that is less stable when disturbed

and another that is more stable

Figure 9.17
The relationships between connectance and species richness: (a) For a compilation from the literature of 40 food webs from terrestrial, freshwater, and marine environments. (After Briand, 1983.) (b) For a compilation of 95 insect-dominated webs from various habitats. (Schoenly et al., 1991.) (c) For seasonal versions of a food web for a large pond in northern England, varying in species richness from 12 to 32. (Warren, 1989.) (d) For food webs from swamps and streams in Costa Rica and Venezuela. (Winemiller, 1990; after Hall & Raffaelli, 1993.)

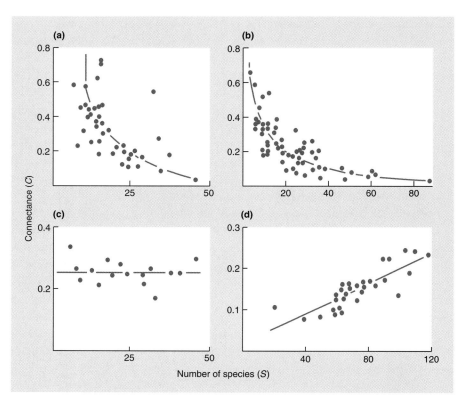

Figure 9.18

Relationship between resistance in grassland community composition and species diversity (Shannon's index, *H*) for a number of grassland areas in Yellowstone National Park. The resistance measure (*R*) is designed to be inversely related to the cumulative differences in species abundances at sites between 1988 (a year of severe drought) and 1989 (a year of normal rainfall). Thus a high value for *R* indicates that relative abundances changed little in the face of the drought, whereas a low value means they changed considerably. (After Frank & McNaughton, 1991.)

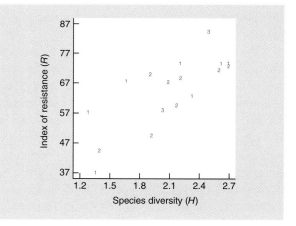

composition in the face of a severe summer drought in Yellowstone National Park (Figure 9.18). These experimental studies, too, therefore provide conflicting results in relation to the predicted complexity/stability relationship.

The conflicting results among both models and experiments, and the variety of types of stability that they have used, suggest, at least, that no single relationship between complexity and stability will be appropriate in all communities. It would be wrong to replace one sweeping generalization with another. Even if complexity

Table 9.4

The influence of (a) nutrient addition on species richness and diversity (Shannon's index, H; see Box 10.1) in two fields and (b) grazing by African buffalo on species diversity in two areas of vegetation

	CONTROL PLOTS	EXPERIMENTAL PLOTS	STATISTICAL SIGNIFICANCE
(a) *Nutrient addition*			
Species richness per 0.5 m² plot			
Species-poor plot	20.8	22.5	n.s.
Species-rich plot	31.0	30.8	n.s.
Equitability			
Species-poor plot	0.660	0.615	n.s.
Species-rich plot	0.793	0.740	$P < 0.05$
Diversity			
Species-poor plot	2.001	1.915	n.s.
Species-rich plot	2.722	2.532	$P < 0.05$
(b) *Grazing*			
Species diversity			
Species-poor plot	1.069	1.357	n.s.
Species-rich plot	1.783	1.302	$P < 0.005$

n.s., not significant.
(After McNaughton, 1977.)

and instability were connected in models, moreover, this would not necessarily mean that we should expect to see an association between complexity and instability in real communities. Unstable communities will fail to persist when they experience environmental conditions that reveal their instability. But the range and predictability of environmental conditions will vary from place to place. In a stable and predictable environment, a community will experience only a limited range of conditions, and a community that is dynamically fragile may nevertheless persist. But in a variable and unpredictable environment, only a community that is dynamically robust will be able to persist. Hence, we might expect to see (1) complex and fragile communities in stable and predictable environments, with simple and robust communities in variable and unpredictable environments; and (2) approximately the same recorded stability (in terms of population fluctuations, and so forth) in all communities, since this will depend on the inherent stability of the community combined with the variability of the environment.

This line of argument, moreover, carries a further, very important implication for the likely effects of unnatural perturbations caused by humans on communities.

Figure 9.19
Web size and average number of
feeding links per animal species
(number of prey species in diet) in
relation to intensity of flow-related
disturbances to the streambed. (After
Townsend et al., 1998.)

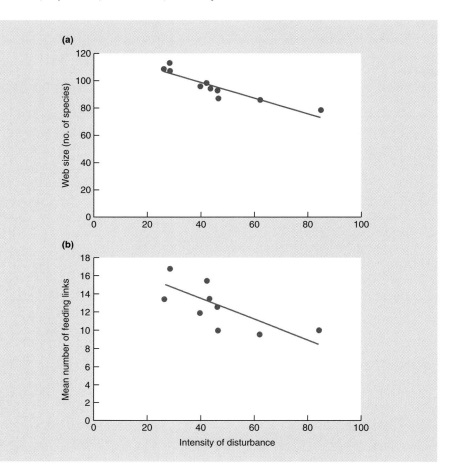

We might expect these to have their most profound effects on the dynamically fragile, complex communities of stable environments, which are relatively unaccustomed to perturbations, and least effect on the simple, robust communities of variable environments, which have previously been subjected to repeated (albeit natural) perturbations.

One study addressing the question of food web complexity and stability and environmental stability investigated 10 small streams in New Zealand that differ in the intensity and frequency of flow-related disturbances to their beds (Figure 9.19). Food webs in the more disturbed streams are characterized by smaller numbers of species and fewer links between each species and others in the community. This conforms to the prediction that simpler food webs should be found in more variable environments because they possess a structure that confers stability. These results are therefore intriguing, but they are tentative. There is a pressing requirement for more rigorous descriptions of food webs involving a standard methodology, and for analyses from related sets of habitats that differ in fundamental aspects. Many more empirical and experimental studies are needed before our understanding of food webs approaches maturity.

UNANSWERED QUESTION:

so far more questions than answers have arisen in the important field of food web theory

Summary

Multiple determinants of the dynamics of populations

To understand the factors responsible for the population dynamics of even a single species in a single location, it is necessary to have a knowledge of physicochemical conditions, available resources, the organism's life cycle, and the influence of competitors, predators, and parasites on rates of birth, death, immigration, and emigration.

There are contrasting theories to explain the abundance of populations. At one extreme, researchers emphasize the apparent stability of populations and point to the importance of forces that stabilize (density-dependent factors). At the other extreme, those who place more emphasis on density fluctuations may look at external (often density-independent) factors to explain the changes. Key-factor analysis is a technique that can be applied to life table studies to throw light both on determination and on regulation of abundance.

Dispersal, patches, and metapopulation dynamics

Dispersal can be a vital factor in determining and/or regulating abundance. A radical change in the way ecologists think about populations has involved combining ideas about patchiness, dispersal, and population dynamics within the patches in *metapopulation theory*.

Temporal patterns in community composition

Disturbances that open up gaps (patches) are common in all kinds of community. *Founder-controlled* communities are those in which all species are approximately equivalent in their ability to invade gaps and are equal competitors that can hold the gaps against all comers during their lifetime. On the other hand, *dominance-controlled* communities are those in which some species are competitively superior to others so that an initial colonizer of a patch cannot necessarily maintain its presence there.

The phenomenon of dominance control is responsible for many examples of community succession. Successions on old fields, for example, typically involve a vegetation sequence in which pioneer species are those that can establish themselves quickly in the disturbed habitat of a recently cultivated field, to be replaced later by species that are slower to arrive and establish but eventually outcompete the pioneers. It can be very difficult to identify when a succession reaches a stable climax community, since this may take centuries to achieve and in the meantime further disturbances are likely to occur.

Food webs

No predator–prey, parasite–host, or grazer–plant pair exists in isolation. Each is part of a complex food web involving other predators, parasites, food sources, and competitors within the various trophic levels of a community.

The effect of one species (say, a predator) on another (its herbivorous prey) may be *direct* and straightforward. But *indirect* effects may also be felt by any plant population on which the herbivore feeds, by other predators and parasites of the herbivore, by other consumers of the plant, by competitors of the herbivore and the plant, and by the myriad species linked even more remotely in the food web.

Some species are more tightly woven into the food web than others. A species whose removal would produce a significant effect (extinction or a large change in density) in at least one other species may be thought of as a strong interactor. Removal of some strong interactors would lead to significant changes that would spread throughout the food web. We refer to these as *keystone species*.

Top-down control of a food web occurs in situations in which the structure (abundance, species number) of lower trophic levels depends on the effects of consumers from higher trophic levels. *Bottom-up control*, on the other hand, occurs in a dependence of community structure on factors, such as nutrient concentration and prey availability, that influence a trophic level from below. The relative importance of these forces varies according to the trophic level under investigation and the number of trophic levels present.

The old conventional wisdom in ecology was that increased food web complexity leads to increased stability (though care is needed in deciding what is meant by stability). This view has been undermined by the analysis of mathematical models that have often concluded that complexity leads to instability. The evidence from real communities is conflicting, but many of the data gathered so far are unreliable. More rigorous studies will be needed before ecologists fully understand the causes and consequences of food web patterns.

Review Questions

▲ = Challenge Question

1. ▲ Construct a flow diagram (boxes and arrows) with a named population at its center to illustrate the wide range of abiotic and biotic factors that influence its pattern of abundance.

2. Population census data can be used to establish correlations between abundance and external factors such as weather. Why can such correlations not be used to prove a causal relationship that explains the dynamics of the population?

3. ◭ Consider the information given in Section 9.2.1 about the apparent constancy in numbers of swifts in the village of Selbourne. Now imagine first that you are H. G. Andrewartha and then A. J. Nicholson and construct an argument that you think each would have made to account for the population dynamics of swifts.

4. Distinguish between the *determination* and *regulation* of population abundance.

5. Discuss the pros and cons of key-factor analysis as a tool to unravel the dynamics of a population.

6. What is meant by a *metapopulation*?

7. Define founder control and dominance control as they apply to community organization. Why are communities in each category so different?

8. What factors are responsible for changes in species composition during an old field succession?

9. What are meant by bottom up and top down control of the structure of trophic levels? How does the importance of each of these control mechanisms vary with the number of trophic levels in a community?

10. Discuss the relationship between complexity and stability of food webs.

Recognizing and conserving the world's biological resources are becoming increasingly important. To conserve biodiversity we must understand why species richness varies widely across the face of the Earth. Why do some communities contain more species than others? Are there patterns or gradients in this biodiversity? If so, what are the reasons for these patterns?

Patterns in Species Richness

Chapter Contents

Key Concepts

In this chapter you will

- understand the meanings of *species richness*, *diversity indices*, and *rank–abundance diagrams*.

- appreciate that species richness is limited by available resources, the average portion of the resources used by each species (niche breadth), and the degree of overlap in resource use.

- recognize that species richness may be highest at intermediate levels of productivity, predation intensity, and disturbance, but tends to increase with spatial heterogeneity.

- appreciate the importance of habitat area and remoteness in determining richness, especially with reference to the *equilibrium theory of island biogeography*.

- understand richness gradients with latitude, altitude and depth, and during community succession, and the difficulties of explaining them.

- appreciate how theories of species richness can also be applied to the fossil record.

10.1 ▷ Introduction

The questions of why the number of species varies from place to place and from time to time present themselves not only to ecologists but to anybody who observes and ponders the natural world. They are interesting questions in their own right—but they are also questions of practical importance. The number of species in a community is a crucial aspect of that community's biodiversity. The meaning of *biodiversity* is discussed in Chapter 14 (Conservation), but for now it is clear that if we wish to conserve or restore biodiversity, then we must understand how species numbers are determined and how it comes about that they vary. We will see that there are plausible answers to the questions we ask, but these are by no means conclusive. Yet this is not so much a disappointment as a challenge to ecologists of the future. Much of the fascination of ecology lies in the fact that many of the problems are blatant, whereas the solutions are not. We will see that a full understanding of patterns in species richness must draw on our knowledge of all the areas of ecology discussed so far in this book.

determining species richness

The number of species in a community is referred to as its species richness. Counting or listing the species present in a community may sound a straightforward procedure, but in practice it is often surprisingly difficult, partly because of taxonomic problems, but also because only a subsample of the organisms in an area can usually be counted. The number of species recorded then depends on the number of samples that have been taken, or on the volume of the habitat that has been explored. The most common species are likely to be represented in the first few samples, and as more samples are taken, rarer species will be added to the list. At what point does one cease to take further samples? Ideally, the investigator should continue to sample until the number of species reaches a plateau. At the very least, the species richness of different communities should be compared only if they are based on the same sample sizes (in terms of area of habitat explored, time devoted to sampling, or, best of all, number of individuals included in the samples).

diversity indices and rank–abundance diagrams

An important aspect of the structure of a community is completely ignored, though, when its composition is described simply in terms of the number of species present—namely, that some species are rare and others common. Intuitively, a community of 10 species with equal numbers in each seems more diverse than another, again consisting of 10 species, with 91 percent of the individuals belonging to the most common species and just 1 percent in each of the other 9. Yet, each community has the same species richness. *Diversity indices* are designed to combine both species richness and the evenness or equitability of the distribution of individuals among those species (Box 10.1). Moreover, attempts to describe a complex community structure by one single attribute, such as richness or diversity, can still be criticized because so much valuable information is lost. A more complete picture of the distribution of species abundance in a community is therefore sometimes provided in a *rank–abundance diagram* (Box 10.1).

Nonetheless, for many purposes, the simplest measure, species richness, suffices. In the following sections, therefore, we examine the relationships between species richness and a variety of factors that may, in theory, influence richness in ecological communities. It will become clear that it is often extremely difficult to come up with unambiguous predictions and clean tests of hypotheses when dealing with something as complex as a community.

10.2 ▷ A Simple Model of Species Richness

To try to understand the determinants of species richness, it will be useful to begin with a simple model (Figure 10.1). Assume, for simplicity, that the resources available to a community can be depicted as a one-dimensional continuum, R units long (see Figure 10.1). Each species uses only a portion of this resource continuum, and these portions define the *niche breadths* (n) of the various species: the average niche breadth within the community is $\bar{n}$. Some of these niches overlap, and the overlap between adjacent species can be measured by a value o. The average niche overlap within the community is then $\bar{o}$. With this simple background, it is possible to consider why some communities should contain more species than others.

First, for given values of $\bar{n}$ and $\bar{o}$, a community will contain more species the larger the value of R, i.e., the greater the range of resources (Figure 10.1a).

Second, for a given range of resources, more species will be accommodated if $\bar{n}$ is smaller, i.e., if the species are more specialized in their use of resources (Figure 10.1b).

Alternatively, if species overlap to a greater extent in their use of resources (greater $\bar{o}$), then more may coexist along the same resource continuum (Figure 10.1c).

Finally, a community will contain more species the more fully saturated it is; conversely, it will contain fewer species when more of the resource continuum is unexploited (Figure 10.1d).

We can now consider the relationship between this model and two important kinds of species interactions described in previous chapters—interspecific competition and predation. If a community is dominated by interspecific competition (Chapter 6), the resources are likely to be fully exploited. Species richness will then depend on the range of available resources, the extent to which species are specialists, and the permitted extent of niche overlap (Figure 10.1a–c). We will examine a range of influences on each of these three.

competition and predation may influence species richness

Predation, on the other hand, is capable of exerting contrasting effects (Chapter 8). First, we know that predators can exclude certain prey species; in the absence of these species the community may then be less than fully saturated, in the sense that some available resources may be unexploited (Figure 10.1d). Second, though, predation may tend to keep species below their carrying capacities for much of the time, reducing the intensity and importance of direct interspecific competition for resources and permitting much more niche overlap and a greater richness of species than in a community dominated by competition (Figure 10.1c).

The next two sections examine a variety of determinants of species richness. To organize these, Section 10.3 focuses on factors that usually vary from place to place (though they can vary from time to time): productivity, predation intensity, spatial heterogeneity, and harshness; Section 10.4 focuses on factors that usually

Figure 10.1

A simple model of species richness. Each species utilizes a portion n of the available resources (R), overlapping with adjacent species by an amount o. More species may occur in one community than in another (a) because a greater range of resources is present (larger R), (b) because each species is more specialized (smaller average n), (c) because each species overlaps more with its neighbors (larger average o), (d) because the resource dimension is more fully exploited. (After MacArthur 1972.)

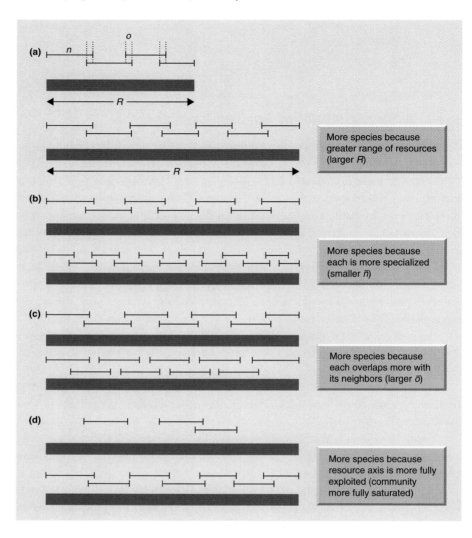

vary from time to time (though they can vary from place to place): climatic variation, disturbance, and evolutionary age.

10.3 > Spatially Varying Factors That Influence Species Richness

10.3.1 Productivity and resource richness

For plants, the productivity of the environment can depend on whichever nutrient or condition is most limiting to growth (dealt with in detail in Chapter 11). Broadly speaking, the productivity of the environment for animals follows the same trends as for plants, both as a result of the changes in resource levels at the base of the food chain, and as a result of the changes in critical conditions such as temperature.

If higher productivity is correlated with a wider *range* of available resources, then this is likely to lead to an increase in species richness (Figure 10.1a). However, a more productive environment may have a higher rate of supply of resources but not a greater variety of resources. This might lead to more individuals per species rather than more species. Alternatively again, it is possible, even if the overall variety of resources is unaffected, that rare resources in an unproductive environment may become abundant enough in a productive environment for extra species to be added, because more specialized species can be accommodated (Figure 10.1b).

increased productivity might be expected to lead to increased richness

In general then, we might expect species richness to increase with the richness of available resources and productivity, a contention that is supported by an analysis of the species richness of trees in North America in relation to climatic variables. Two measures of available environmental energy were used. The first, *potential evapotranspiration* (PET), is the amount of water that would evaporate or be transpired from a saturated surface, and can be taken as a crude integrated measure of available energy. But the actual amount of water evaporated or transpired depends on how much precipitation there has been, and thus how much water is available (surfaces may not be saturated); this is measured by the *actual evapotranspiration* (AET), which reflects the joint availability of energy and water. Tree species richness was most strongly correlated with AET (Figure 10.2), which can be regarded as a good correlate of the energy that can be captured by plants and is therefore an index of their productivity.

and often does

When this work was extended to four vertebrate groups, species richness was found to be correlated to some extent with tree species richness itself. However, the best correlations were consistently with PET (Figure 10.3). This raises the question of why animal species richness should be positively correlated with crude atmospheric energy. Perhaps the effect of extra atmospheric warmth for an ectotherm, such as a reptile, would be to enhance the intake and utilization of food resources, and for an endotherm, such as a bird, to require less expenditure of resources in maintaining body temperature. This could lead to faster individual and population growth and thus to larger populations. Warmer environments might therefore allow species with narrower niches to persist and therefore support more species in total (Figure 10.1b; Turner et al., 1996).

UNANSWERED QUESTION:

why should there be more animal species in regions that receive more energy?

Sometimes there seems to be a direct relationship between animal species richness and plant productivity. Thus, there are strong positive correlations between species richness and precipitation for both seed-eating ants and seed-eating rodents in the southwestern deserts of the United States (Figure 10.6a). In such arid regions, it is well established that mean annual precipitation is closely related to plant productivity and thus to the amount of seed resource available. It is particularly noteworthy that in the species-rich sites, the communities contain more species of very large ants (which consume large seeds) and more species of very small ants (which take small seeds) (Davidson, 1977). It seems that either the range of sizes of seeds is greater in the more productive environments (Figure 10.1a) or the abundance of seeds becomes sufficient to support extra consumer species (Figure 10.1b). There is also a positive relationship between the species richness of lizards in the deserts of the southwestern United States and the length of the growing season—an important aspect of productivity in desert environments (Figure 10.6b).

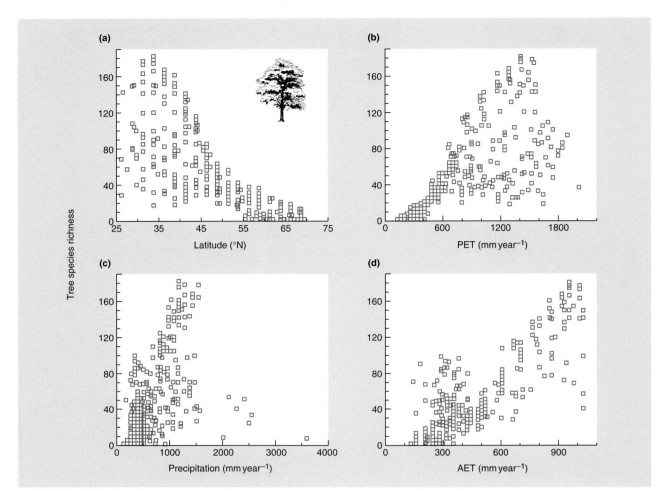

Figure 10.2
Species richness of trees in North America north of the Mexican border (in which the continent has been divided into 336 quadrats following lines of latitude and longitude) in relation to (a) latitude, (b) potential evapotranspiration (PET), (c) precipitation, (d) actual evapotranspiration (AET). The relationship with AET is strongest. (After Currie & Paquin, 1987; Currie, 1991.)

other evidence shows richness declining with productivity

On the other hand, an increase in diversity with productivity is by no means universal, as shown for example, by the unique Parkgrass experiment which started in 1856 at Rothamsted in England (see Box 10.1). An 8-acre pasture was divided into 20 plots, 2 serving as controls and the others receiving a fertilizer treatment once a year. While the unfertilized areas remained essentially unchanged, the fertilized areas showed a progressive decline in species richness (and diversity). Rosenzweig (1971) referred to this decline as "the paradox of enrichment." One possible solution to the paradox is that high productivity leads to high rates of population growth, bringing about the extinction of some of the species present because of a speedy conclusion to any potential competitive exclusion (Section 6.2.7). At lower productivity, the environment is more likely to have changed before competitive exclusion is

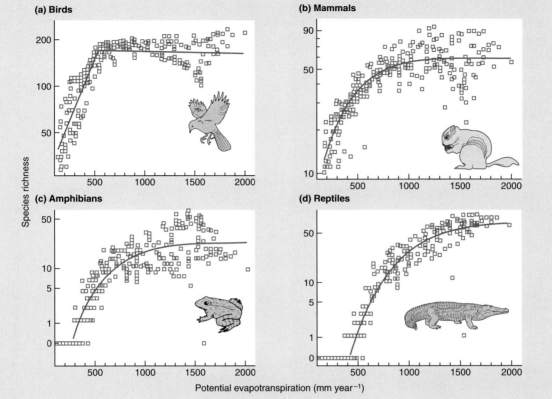

Figure 10.3
Species richness of birds, mammals, amphibians, and reptiles in North America in relation to potential evapotranspiration (PET). (After Currie, 1991.)

achieved. An association between high productivity and low species richness has been found in several other studies of plant communities. It can be seen when the cultural eutrophication (that is, an increased input of plant resources like nitrates and phosphates) of lakes, rivers, estuaries, and coastal marine regions consistently leads to a decrease in species richness of phytoplankton (despite an increase in their productivity), and it is paralleled by the fact that two of the most species-rich plant communities in the world occur on very nutrient-poor soils (the Fynbos of South Africa and the heath scrublands of Australia), whereas nearby communities on more nutrient-rich soils have much lower plant richness (all reviewed by Tilman, 1986).

It is perhaps not surprising, then, that several studies have demonstrated both an increase and a decrease in richness with increasing productivity—that is, that species richness may be highest at intermediate levels of productivity. For instance, there are humped curves when the number of Malaysian rain forest woody species is plotted against an index of phosphorus and potassium concentration, expressing the resource richness of the soil (Figure 10.6c), and when the species richness of

and further evidence suggests a "humped" relationship

Diversity Indices and Rank-Abundance Diagrams

The measure of the character of a community that is probably most commonly used to take into account both abundance patterns and species richness is known as the *Shannon* or the *Shannon-Weaver diversity index* (denoted by *H*). This is calculated by determining, for each species, the proportion of individuals or biomass (i.e., P_i for the *i*th species) that it contributes to the total in the sample. Then, if *S* is the total number of species in the community (i.e., the richness),

$$\text{diversity, } H = -\sum_{i=1}^{s} P_i \ln P_i$$

As required, the value of the index depends on both the species richness and the evenness (equitability) with which individuals are distributed among the species. Thus, for a given richness, *H* increases with equitability, and for a given equitability, *H* increases with richness.

An example of this kind of analysis is provided by the unusually long-term study that commenced in 1856 in an area of pasture at Rothamsted in England. Experimental plots received a fertilizer treatment once every year, and control plots did not. Figure 10.1 shows how species diversity (*H*) of grass species changed between 1856 and 1949. While the unfertilized area remained essentially unchanged, the fertilized

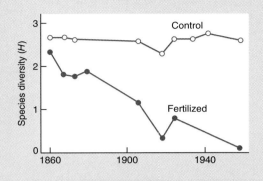

Figure 10.4

Species diversity (*H*) of a control plot and a fertilized plot in the Rothamstead Parkgrass experiment. (After Tilman, 1982.)

area showed a progressive decline. This is discussed in Section 10.3.1.

Rank–abundance diagrams make use of the full array of P_i values by plotting P_i against rank. Thus, the P_i for the most abundant species is plotted first, then the next most common, and so on, until the array is completed by the rarest species of all. A variety of models may then be fitted to the series of points so generated—

desert rodents is plotted against precipitation (and thus productivity) along a gradient in Israel (Figure 10.6d).

10.3.2 Predation intensity

The possible effects of predation on the species richness of a community were examined in Chapter 8: predation may increase richness by allowing otherwise competitively inferior species to coexist with their superiors (*predator-mediated coexistence*); but intense predation may reduce richness by driving prey species (whether or not they are strong competitors) to extinction. Overall, therefore, there may even be a humped relationship between predation intensity and species richness in a community, with greatest richness at intermediate intensities, such as that shown by the effects of rabbit grazing (refer to Figure 8.24, Section 8.6).

some statistical, some biological in origin. Rank–abundance diagrams, like indices of richness and diversity, should be viewed as abstractions of the highly complex structure of communities that may nonetheless sometimes be useful when making comparisons. For example, assuming that a particular model, the *geometrical series*, can be appropriately applied in the Rothamsted experiment, Figure 10.5 shows how the dominance of some species steadily increased (steeper slope) while species richness decreased over time.

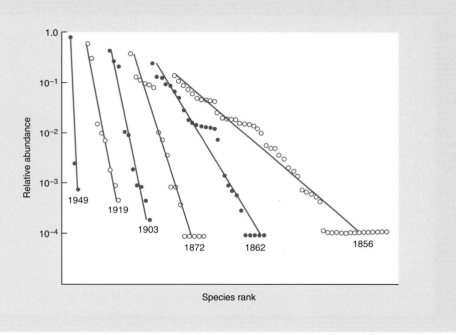

Figure 10.5

Change in the rank-abundance pattern of plant species in the Parkgrass fertilized plot from 1856 to 1949. (After Tokeshi, 1993.)

A further example of predator-mediated coexistence is provided by the pioneering work of Paine (1966) on the influence of a top carnivore on community structure on a rocky shore (Figure 10.7). The starfish *Pisaster ochraceus* preys on sessile filter-feeding barnacles and mussels, and also on browsing limpets and chitons and a small carnivorous whelk. These species, together with a sponge and four macroscopic algae, form predictable associations on rocky shores of the Pacific coast of North America. Paine removed all starfish from a typical stretch of shoreline about 8 m long and 2 m deep and continued to exclude them for several years. The structure of the community in nearby control areas remained unchanged during the study, but removal of *Pisaster* had dramatic consequences. Within a few months, the barnacle *Balanus glandula* settled successfully. Later mussels (*Mytilus californicus*) crowded it out, and eventually the site became dominated by these. All but one of the species of alga disappeared, apparently through lack of space, and the browsers tended to move away, partly because space was limited and partly because there

predator-mediated coexistence by starfish on a rocky shore

Figure 10.6

(a) Patterns of species richness of seed-eating rodents (triangles) and ants (circles) inhabiting sandy soils in a geographic gradient of precipitation and productivity. (After Brown & Davidson, 1977.) (b) Species richness of lizards at 11 sites in the southwestern United States plotted against length of the growing season. (After Pianka, 1967.) (c) Species richness of woody species in several Malaysian rain forests plotted against an index of phosphorus (P) and potassium (K) concentration. (After Tilman, 1982.) (d) Species richness of desert rodents in Israel plotted against rainfall for both rocky and sandy habitats. (After Abramsky & Rosenzweig, 1983.)

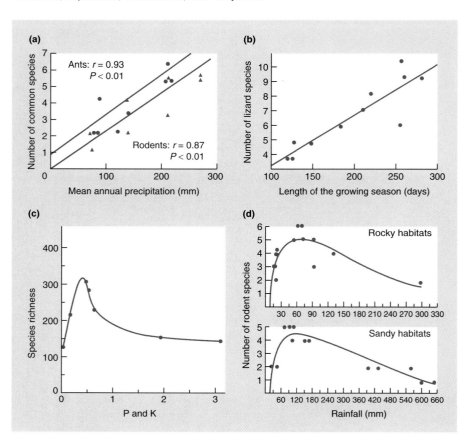

was a lack of suitable food. The main influence of the starfish *Pisaster* appears to be to make space available for competitively subordinate species. It cuts a swathe free of barnacles and, most importantly, free of the dominant mussels that would otherwise outcompete other invertebrates and algae for space. Overall, there is usually predator- (starfish-)mediated coexistence, but the removal of starfish led to a reduction in number of species from 15 to 8.

parasites, disease, and richness   Parasites and pathogens, like other types of predator, may also have a profound influence on species richness in an area. Thus, the extinction of nearly half the endemic bird fauna of the Hawaiian Islands has been attributed in part to the introduction of bird pathogens such as malaria and bird pox. Probably the largest single change wrought in the structure of communities by a parasite has been the destruction of the chestnut (*Castanea dentata*) in North American forests, where it had been a dominant tree over large areas until the introduction of the fungal pathogen *Endothia parasitica*, probably from China.

10.3.3 Spatial heterogeneity

Environments that are more spatially heterogeneous can be expected to accommodate extra species because they provide a greater variety of microhabitats, a greater range

Figure 10.7
Paine's rocky shore community.
(Modified from Paine, 1966.)

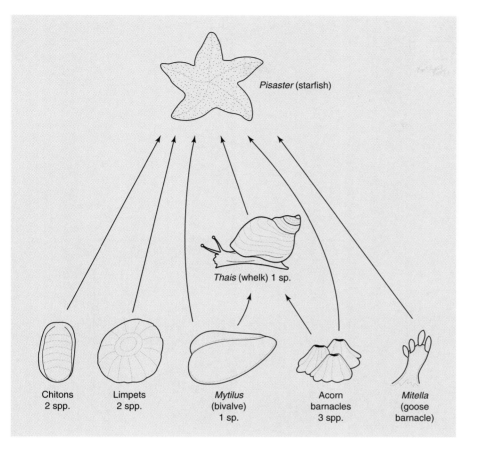

of microclimates, more types of places to hide from predators, and so on. In effect, the extent of the resource spectrum is increased (Figure 10.1a).

In some cases, it has been possible to relate species richness to the spatial heterogeneity of the abiotic environment. For instance, a study of plant species growing in 51 plots alongside the Hood River, Canada, revealed a positive relationship between species richness and an index of spatial heterogeneity (based, among other things, on the number of categories of substrate, slope, drainage regimes, and soil pH present) (Figure 10.8).

richness and the heterogeneity of the abiotic environment

Most studies of spatial heterogeneity have related the species richness of animals to the structural diversity of the plants in their environment (Figure 10.9). However, whether spatial heterogeneity arises intrinsically from the abiotic environment or is provided by other biological components of the community, it is capable of promoting an increase in species richness.

animal richness and plant spatial heterogeneity

10.3.4 Environmental harshness

Environments dominated by an extreme abiotic factor—often called harsh environments—are more difficult to recognize than might be immediately apparent.

An anthropocentric view might describe as extreme both very cold and very hot habitats, unusually alkaline lakes, and grossly polluted rivers. However, species

Figure 10.8

Relationship between number of plants per 300-m^2 plot beside the Hood River, Northwest Territories, Canada, and an index (ranging from 0 to 1) of spatial heterogeneity in abiotic factors associated with topography and soil. (After Gould and Walker, 1997.)

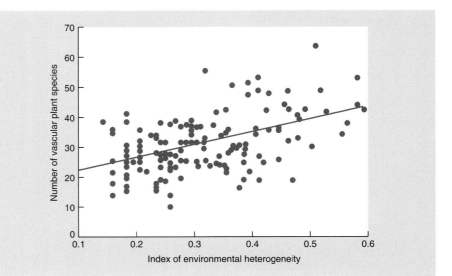

Figure 10.8

Relationship between number of plants per 300-m^2 plot beside the Hood River, Northwest Territories, Canada, and an index (ranging from 0 to 1) of spatial heterogeneity in abiotic factors associated with topography and soil. (After Gould and Walker, 1997.)

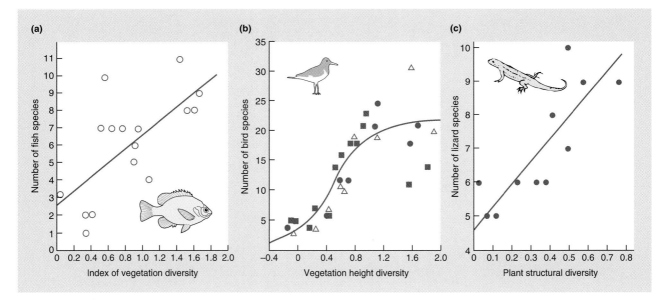

Figure 10.9

Relationships between animal species richness and an index of structural diversity of the vegetation: (a) For freshwater fish in 18 Wisconsin lakes. (After Tonn & Magnuson, 1982.) (b) For birds in Mediterranean-type habitats in California (triangles), central Chile (circles) and southwest Africa (squares). (After Cody, 1975.) (c) For lizards in desert sites in the southwestern United States. (After Pianka, 1967.)

have evolved and live in all such environments, and what is very cold and extreme for us must seem benign and unremarkable to a penguin.

We might try to get around the problem of defining environmental harshness by "letting the organism decide." An environment may be classified as *extreme* if organisms, by their failure to live there, show it to be so. But if the claim is to be made—as it often is—that species richness is lower in extreme environments, then this definition is circular, and it is designed to prove the very claim we wish to test.

Perhaps the most reasonable definition of an extreme condition is one that requires, of any organism tolerating it, a morphological structure or biochemical mechanism that is not found in most related species and is costly, either in energetic terms or in terms of the compensatory changes in the biological processes of the organism that are needed to accommodate it. For example, plants living in highly acidic soils (low pH) may be affected directly through injury by hydrogen ions or indirectly via deficiencies in the availability and uptake of important resources such as phosphorus, magnesium, and calcium. In addition, aluminum, manganese, and heavy metals may have their solubility increased to toxic levels, and mycorrhizal activity and nitrogen fixation may be impaired. Plants can only tolerate low pH if they have specific structures or mechanisms allowing them to avoid or counteract these effects.

Environments that experience low pHs can be considered harsh, and the mean number of plant species recorded per square meter in unmanaged grasslands in northern England was lowest in soils of low pH (Figure 10.10a). Similarly, the species richness of benthic stream invertebrates in the Ashdown Forest (southern England) was markedly lower in the more acidic streams (Figure 10.10b). Further examples of extreme environments that are associated with low species richness include hot springs, caves, and highly saline water bodies such as the Dead Sea. The problem with these examples, however, is that they are also characterized by other features associated with low species richness such as low productivity and low spatial heterogeneity. In addition, many occupy small areas (caves, hot springs), or areas that are rare compared to other types of habitat (only a small proportion of the streams in southern England are acidic). Hence extreme environments can often be seen as small and isolated islands. We will see in Section 10.5.1 that these features, too, are usually associated with low species richness. Although it appears

UNANSWERED QUESTION:

are harsh environments species-poor?

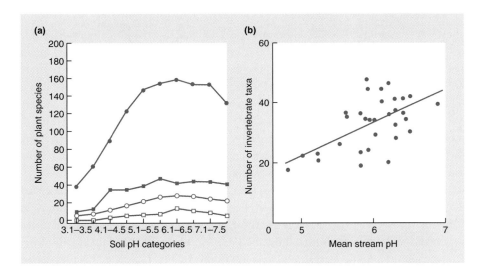

Figure 10.10

(a) Mean (open circles), maximum (filled squares), and minimum (open squares) plant species richness per square meter, and total number (filled circles) of species in categories of surface soil pH in unmanaged grasslands in northern England. (After Grime, 1973.) (b) Number of taxa of invertebrates in streams in Ashdown Forest, southern England, plotted against pH of the stream water. (After Townsend et al., 1983.)

reasonable that intrinsically extreme environments should as a consequence support few species, this has proved an extremely difficult proposition to establish.

10.4 ▷ Temporally Varying Factors That Influence Species Richness

Temporal variation in conditions and resources may be predictable or unpredictable and operate on time scales from minutes through to centuries and millennia. All may influence species richness in profound ways.

10.4.1 Climatic variation

temporal niche differentiation in seasonal environments

The effects of climatic variation on species richness depend on whether the variation is predictable or unpredictable (measured on time scales that matter to the organisms involved). In a predictable, seasonally changing environment, different species may be suited to conditions at different times of the year. More species might therefore be expected to coexist in a seasonal environment than in a completely constant one (Figure 10.1a). Different annual plants in temperate regions, for instance, germinate, grow, flower, and produce seeds at different times during a seasonal cycle; while phytoplankton and zooplankton pass through a seasonal succession in large, temperate lakes with a variety of species dominating in turn as changing conditions and resources become suitable for each.

specialization in nonseasonal environments

On the other hand, there are opportunities for specialization in a nonseasonal environment that do not exist in a seasonal environment. For example, it would be difficult for a long-lived obligate fruit-eater to exist in a seasonal environment when fruit is available for only a very limited portion of the year. But such specialization is found repeatedly in nonseasonal, tropical environments where fruit of one type or another is available continuously.

UNANSWERED QUESTION: does climatic instability increase or decrease richness?

Unpredictable climatic variation (climatic instability) could have a number of effects on species richness. On the one hand, (1) stable environments may be able to support specialized species that would be unlikely to persist where conditions or resources fluctuated dramatically (Figure 10.1b), (2) stable environments are more likely to be saturated with species (Figure 10.1d); and (3) theory suggests that a higher degree of niche overlap will be found in more stable environments (Figure 10.1c). All these processes could increase species richness. On the other hand, populations in a stable environment are more likely to reach their carrying capacities, the community is more likely to be dominated by competition, and species are therefore more likely to be excluded by competition ($\bar{o}$ smaller, Figure 10.1c).

Some studies have seemed to support the notion that species richness increases as climatic variation decreases. For example, there is a significant negative relationship between species richness and the range of monthly mean temperatures for birds, mammals, and gastropods that inhabit the West Coast of North America (from Panama in the south to Alaska in the north) (Figure 10.11). However, this correlation does not prove causation, since there are many other things that change between

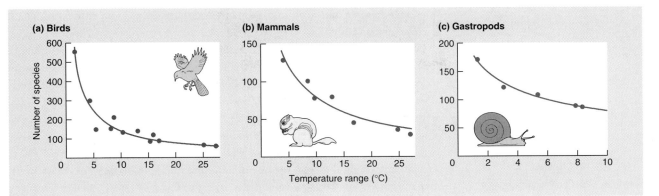

Figure 10.11
Relationships between species richness and the range of monthly mean temperature at sites along the west coast of North America for (a) birds, (b) mammals, and (c) gastropods. (After MacArthur, 1975.)

Panama and Alaska. There is no established relationship between climatic instability and species richness.

10.4.2 Disturbance

Previously, in Section 9.4, the influence of disturbance on community structure was examined, and it was demonstrated that when a disturbance opens up a gap, and the community is *dominance controlled* (strong competitors can replace residents), there tends in a community succession to be an initial increase in richness as a result of colonization, but a subsequent decline in richness as a result of competitive exclusion.

If the frequency of disturbance is now superimposed on this picture, it seems likely that very frequent disturbances will keep most patches in the early stages of succession (where there are few species) but also that very rare disturbances will allow most patches to become dominated by the best competitors (where there are also few species). This suggests an *intermediate disturbance hypothesis*, in which communities are expected to contain most species when the frequency of disturbance is neither too high nor too low (Connell, 1978). The intermediate disturbance hypothesis was originally proposed to account for patterns of richness in tropical rain forests and coral reefs. It has occupied a central place in the development of ecological theory because all communities are subject to disturbances that exhibit different frequencies and intensities.

Among a number of studies that have provided support for this hypothesis, we turn again to the study of green and red algae on different sized boulders on the rocky shores of southern California, discussed in Section 9.4 (Sousa, 1979a,b). Small boulders had a monthly probability of movement of 42 percent, intermediate-sized boulders a probability of 9 percent, and large boulders a probability of only 0.1 percent. The successional sequence that occurs in the absence of disturbance was described in Section 9.4: first *Ulva*, the ephemeral green alga; then several

the intermediate disturbance hypothesis

supported by more rocky shore data

species of perennial red alga; and then predominantly the competitive dominant, *Gigartina canaliculata.*

The percentage of bare space decreased from small to large boulders (Figure 10.12), and mean species richness was lowest on the regularly disturbed small boulders. These were dominated most commonly by *Ulva* sp. The highest levels of species richness were consistently recorded on the intermediate boulder class. Most held mixtures of three to five abundant species from all successional stages. The largest boulders had a lower mean species richness than the intermediate class, though a *G. canaliculata* monoculture was achieved on only a few boulders. These results, therefore, offer strong support for the intermediate disturbance hypothesis.

and from a study of small streams

Disturbances in small streams often take the form of bed movements during periods of high discharge. Because of differences in flow regimes and in the substrata of streambeds, some stream communities are disturbed more frequently and to a larger extent than others. This variation was assessed in 54 stream sites in the Taieri

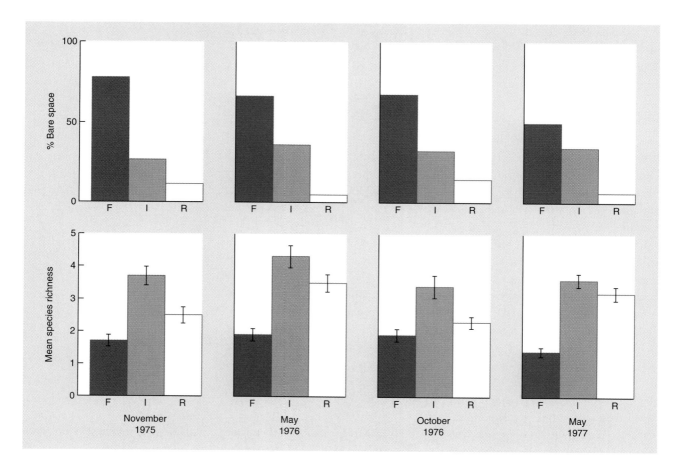

Figure 10.12

Seasonal patterns in bare space and species number ($\pm$ standard error) on boulders in each of three classes categorized according to the force (in newtons, N) required to move them. The three classes are frequently disturbed (F: requires less than 49 N to move), disturbed at an intermediate rate (I: requires 50–294 N to move), and rarely disturbed (R: requires more than 294 N to move). (After Sousa, 1979b.)

River in New Zealand by recording the average percentage of the bed that moved during a series of high-discharge events (assessed on five occasions during a year using painted particles of sizes characteristic of the bed of the stream in question). The pattern of richness of macroinvertebrate species conformed with the intermediate disturbance hypothesis (Figure 10.13).

10.4.3 Environmental age: evolutionary time

It has often been suggested that communities that are "disturbed" only on very extended timescales may nonetheless lack species because they have yet to reach an ecological or an evolutionary equilibrium. Thus communities may differ in species richness because some are closer to equilibrium and are therefore more saturated than others (Figure 10.1d).

For example, many have argued that the tropics are richer in species than are more temperate regions at least in part because the tropics have existed over long and uninterrupted periods of evolutionary time, whereas the temperate regions are still recovering from the Pleistocene glaciations. It seems, however, that the long-term stability of the tropics has in the past been greatly exaggerated by ecologists. Whereas the climatic and biotic zones of the temperate region moved toward the equator during the glaciations, the tropical forest appears to have contracted to a limited number of small refuges surrounded by grasslands. A simplistic contrast between the unchanging tropics and the disturbed and recovering temperate regions is therefore untenable. If the lower species richness of communities nearer the poles is to be attributed partly to their being well below an evolutionary equilibrium, then some complex (and unproven) argument must be constructed. (Perhaps movements of temperate zones to a different latitude caused many more extinctions than the contractions of the tropics into smaller areas at the same latitude.) Thus, although it seems likely that some communities are further below equilibrium than others, it is impossible at present to pinpoint those communities with certainty or even confidence.

UNANSWERED QUESTION:

are temperate communities less saturated with species?

10.5 ▷ **Gradients of Species Richness**

Sections 10.3 and 10.4 have demonstrated how difficult explanations for variations in species richness are to formulate and test. It is easier to describe patterns, especially

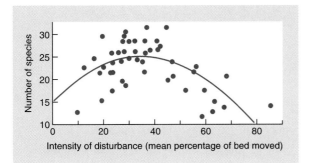

Figure 10.13

Relationship between insect species richness and intensity of disturbance: average percentage of the bed that moved (polynomial regression fitted, relationship significant at $P < 0.001$) assessed at 54 stream sites in the Taieri River. (After Townsend et al., 1997.)

gradients, in species richness. These are discussed next. Explanations for these, too, however, are often very uncertain.

10.5.1 Habitat area and remoteness—island biogeography

species–area relationships on oceanic islands

habitat islands and areas of mainland

It is well established that the number of species on islands decreases as island area decreases. Such a *species–area* relationship is shown in Figure 10.14a for amphibians and reptiles on oceanic islands of varying sizes.

"Islands," however, need not be islands of land in a sea of water. Lakes are islands in a "sea" of land, mountaintops are high-altitude islands in a low-altitude ocean, gaps in a forest canopy where a tree has fallen are islands in a sea of trees,

Figure 10.14
Species–area relationships: (a) For amphibians and reptiles on West Indian islands. (After MacArthur & Wilson, 1967.) (b) For birds inhabiting lakes in Florida. (After Hoyer and Canfield, 1994.) (c) For invertebrates inhabiting different sized clumps of intertidal mussels (*Brachidontes rostratus*) off the south coast of Victoria, Australia. (After Peake & Quinn, 1993.) (d) For fish living in Australian desert springs having source pools of different sizes. (After Kodric-Brown & Brown, 1993.) (e) For flowering plants found in different-sized sample areas of England. (After Williams, 1964. From Gorman, 1979.)

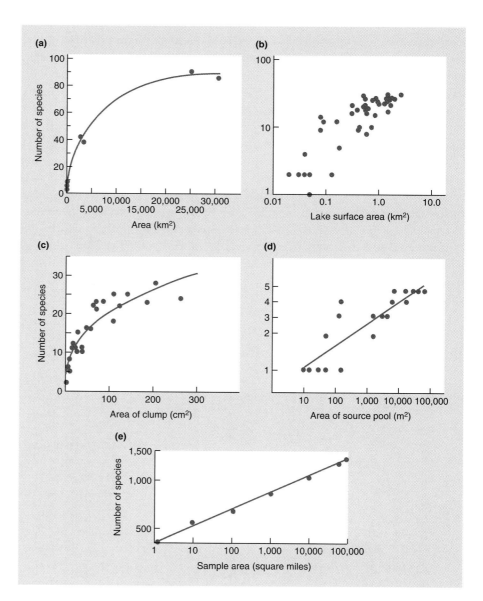

and there can be islands of particular geological types, soil types or vegetation types surrounded by dissimilar types of rock, soil, or vegetation. Species–area relationships can be equally apparent for these types of islands (Figure 10.14b–d). In fact, species–area relationships are not restricted to islands at all. They can also be seen when the numbers of species occupying different-sized arbitrary areas of the same geographical region are compared (Figure 10.14e).

The relationship between species richness and habitat area is one of the most consistent of all ecological patterns. However, the pattern raises an important question: Is the impoverishment of species on islands more than would be expected in comparably small areas of mainland? In other words, does the characteristic isolation of islands contribute to their impoverishment of species? These are important questions for an understanding of community structure since there are many oceanic islands, many lakes, many mountaintops, many woodlands surrounded by fields, many isolated trees, and so on.

"island effects" and community structure

Probably the most obvious reason why larger areas should contain more species is that larger areas typically encompass more different types of habitat. However, MacArthur and Wilson (1967) believed this explanation to be too simple. In their *equilibrium theory of island biogeography*, they argued that the number of species on an island is determined by a balance between immigration and extinction, and that this balance is dynamic, with species continually going extinct and being replaced (through immigration) by the same or by different species (Box 10.2).

MacArthur and Wilson's theory makes several predictions:

1. The number of species on an island should eventually become roughly constant through time.
2. This should be a result of a continual *turnover* of species, with some becoming extinct and others immigrating.
3. Large islands should support more species than small islands.
4. Species number should decline with the increasing remoteness of an island.

On the other hand, a higher richness on larger islands would be expected simply as a consequence of larger islands having more habitat types. Does richness increase with area at a rate greater than could be accounted for by increases in habitat diversity alone? Some studies have attempted to partition species–area variation on islands into that which can be entirely accounted for in terms of habitat heterogeneity, and that which remains and must be accounted for by island area in its own right. In the case of species richness of beetles on the Canary Islands, the relationship with plant species richness (an important component of habitat diversity for beetles) is much stronger than that with island area, and this is particularly marked for the herbivorous beetles, presumably because of their particular food-plant requirements (Figure 10.16a). At the opposite extreme, there was a positive correlation between bird species richness and the area of islands off the coast of Western Australia, but no relationship between species richness and habitat diversity (Figure 10.16b).

partitioning variation between habitat diversity and area itself

Another way of distinguishing a separate effect of island area is to compare species–area graphs for islands with those for arbitrarily defined areas of mainland. The relationships in the latter should be due to habitat diversity alone, because what would be an extinction on an island is soon reversed by the exchange of

Box 10.2

MacArthur and Wilson's Equilibrium Theory of Island Biogeography

Taking immigration first, imagine an island that as yet contains no species at all. The rate of immigration of *species* will be high, because any colonizing individual represents a species new to that island. However, as the number of resident species rises, the rate of immigration of new, unrepresented species diminishes. The immigration rate reaches zero when all species from the source pool (i.e., from the mainland or from other nearby islands) are present on the island in question (Figure 10.15a).

The immigration graph is drawn as a curve, because immigration rate is likely to be particularly high when there are low numbers of residents and many of the species with the greatest powers of dispersal are yet to arrive. In fact, the curve should really be a blur rather than a single line, since the precise curve will depend on the exact sequence in which species arrive, and this will vary by chance. In this sense, the immigration curve can be thought of as the *most probable curve*.

The exact immigration curve will depend on the degree of remoteness of the island from its pool of potential colonizers (Figure 10.15a). The curve will always reach zero at the same point (when all members of the pool are resident), but it will generally have higher values on

islands close to the source of immigration than on more remote islands, since colonizers have a greater chance of reaching an island the closer it is to the source. It is also likely that immigration rates will generally be higher on a large island than on a small island, since the larger island represents a larger target for the colonizers (Figure 10.15a).

The rate of species extinction on an island (Figure 10.15b) is bound to be zero when there are no species there, and it will generally be low when there are few species. However, as the number of resident species rises, the extinction rate is assumed by the theory to increase, probably at a more than proportionate rate. This is thought to occur because with more species, competitive exclusion becomes more likely, and the population size of each species is on average smaller, making it more vulnerable to chance extinction. Similar reasoning suggests that extinction rates should be higher on small than on large islands—population sizes will typically be smaller on small islands (Figure 10.15b). As with immigration, the extinction curves are best seen as most probable curves.

In order to see the net effect of immigration and extinction, their two curves can be superimposed (Figure 10.15c). The number of species where the curves cross

individuals between local areas. Thus, if there *is* a separate island effect, an arbitrarily defined area of mainland should contain more species than an otherwise equivalent island, and that means that the slopes of species–area graphs for islands should be steeper than those for mainland areas, since the effect of island isolation should be most marked on small islands, where extinctions are most likely. Despite a certain amount of variation, the island graphs *do* typically have steeper slopes.

bird species richness on Pacific islands decreases with remoteness

An example of species impoverishment on more remote islands can be seen in Figure 10.17 for nonmarine, lowland birds on tropical islands in the southwest Pacific. With increasing distance from the large source island of Papua New Guinea, there is a decline in the number of species, expressed as a percentage of the number present on an island of similar area but close to Papua New Guinea.

species missing because of insufficient time for colonization

A more transient but nonetheless important reason for the species impoverishment of islands, especially remote islands, is the fact that many lack species that they could potentially support, simply because there has been insufficient time for the species to colonize. An example is the island of Surtsey, which emerged in 1963

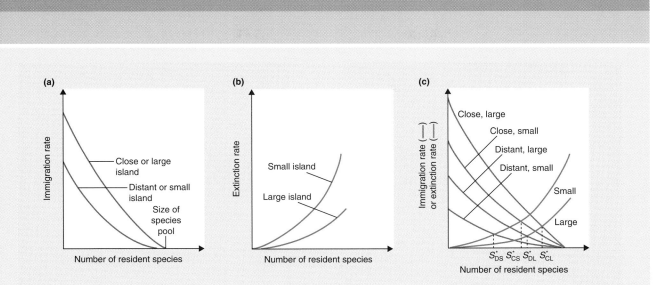

Figure 10.15

MacArthur and Wilson's (1976) equilibrium theory of island biogeography. (a) The rate of species immigration on to an island, plotted against the number of resident species on the island, for large and small islands and for close and distant islands. (b) The rate of species extinction on an island, plotted against the number of resident species on the island, for large and small islands. (c) The balance between immigration and extinction on small and large and on close and distant islands. In each case, S^* is the equilibrium species richness (S = small, L = large, D = distant, C = close).

(S^*) is a dynamic equilibrium and should be the characteristic species richness for the island in question. Below S^*, richness increases (immigration rate exceeds extinction rate); above S^*, richness decreases (extinction exceeds immigration). The theory, then, makes a number of predictions, described in the text.

as a result of a volcanic eruption (Fridriksson, 1975). The new island, 40 km southwest of Iceland, was reached by bacteria and fungi, some seabirds, a fly, and seeds of several beach plants within 6 months of the start of the eruption. Its first established vascular plant was recorded in 1965, and the first moss colony in 1967. By 1973, 13 species of vascular plant and more than 66 mosses had become established (Figure 10.18).

Finally, it is important to reiterate that no aspect of ecology can be fully understood without reference to evolutionary processes (Chapter 2), and this is particularly true for an understanding of island communities. On isolated islands, the rate at which new species evolve may be comparable to or even faster than the rate at which they arrive as new colonists. Clearly, the communities of these islands will be incompletely understood by reference only to ecological processes. Take the remarkable numbers of *Drosophila* species (fruitflies) found on the remote volcanic islands of Hawaii. There are probably about 1500 *Drosophila* species worldwide but at least 500 of these are found on the Hawaiian Islands; they have evolved, almost

evolution rates on islands may be faster than colonization rates

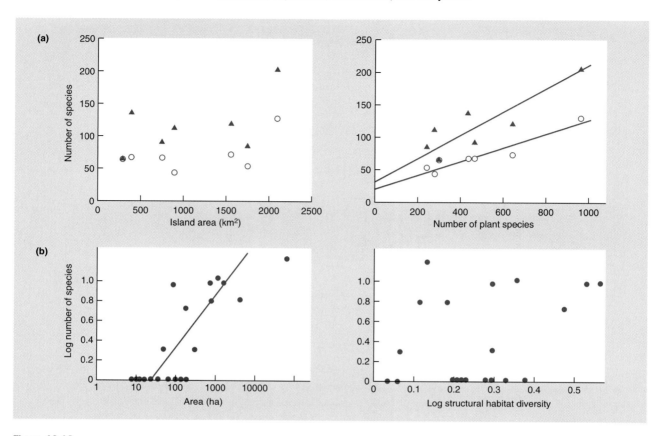

Figure 10.16

(a) The relationships between species richness of herbivorous (circles) and carnivorous (triangles) beetles of the Canary Islands and both island area and plant species richness. (After Becker, 1992.) (b) The relationship between species richness of birds on islands off the coast of Western Australia and both island area and structural habitat diversity (on log scales). Regression lines are included where $P < 0.05$. (After Abbott, 1978.)

Figure 10.17

The number of resident, nonmarine, lowland bird species on islands more than 500 km from the larger source island of Papua New Guinea expressed as a proportion of the number of species on an island of equivalent area but close to Papua New Guinea and plotted as a function of island distance from Papua New Guinea. (After Diamond, 1972.)

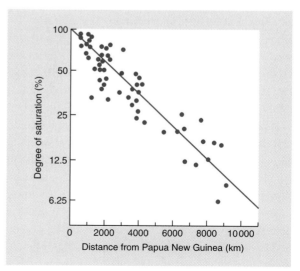

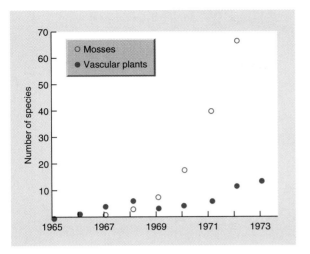

Figure 10.18
The number of species of mosses and vascular plants recorded on the new island of Surtsey from 1965 to 1973. (After Fridriksson, 1975.)

entirely, on the islands themselves. The communities of which they are a part are clearly much more strongly affected by local evolution and speciation than by the processes of invasion and extinction.

10.5.2 Latitudinal gradients

One of the most widely recognized patterns in species richness is the increase that occurs from the poles to the tropics. This can be seen in a wide variety of groups, including trees (Figure 10.2a), marine fish and invertebrates, ants, lizards, and birds (Figure 10.19). The pattern can be seen, moreover, in terrestrial, marine and freshwater habitats.

A number of explanations have been put forward for the general latitudinal trend in species richness, but not one of these is without problems. In the first place, the richness of tropical communities has been attributed to a greater intensity of predation and to more specialized predators. More intense predation could reduce the importance of competition, permitting greater niche overlap and promoting higher richness (Figure 10.1c), but predation cannot readily be forwarded as the root cause of tropical richness, since this begs the question of what gives rise to the richness of the predators themselves.

Second, increasing species richness may be related to an increase in productivity as one moves from the poles to the equator. Certainly, on average there are both more heat and more light energy in more tropical regions, and as discussed in Section 10.3.1, both of these have tended to be associated with greater species richness, though increased productivity in at least some cases has been associated with reduced richness.

productivity as an explanation?

Moreover, light and heat are not the only determinants of plant productivity. Tropical soils tend, on average, to have lower concentrations of plant nutrients than temperate soils. The species-rich tropics might therefore be seen, in this sense, as reflecting their *low* productivity. In fact, tropical soils are poor in nutrients because most of the nutrients are locked up in the large tropical biomass. A productivity

Figure 10.19
Latitudinal patterns in species richness: (a) Of marine bivalves. (After Flessa & Jablonski, 1995.) (b) Of swallowtail butterflies. (After Sutton and Collins, 1991.) (c) Of lizards in the United States. (After Pianka, 1983.) (d) Of breeding birds in North and Central America. (After Dobzhansky, 1950.)

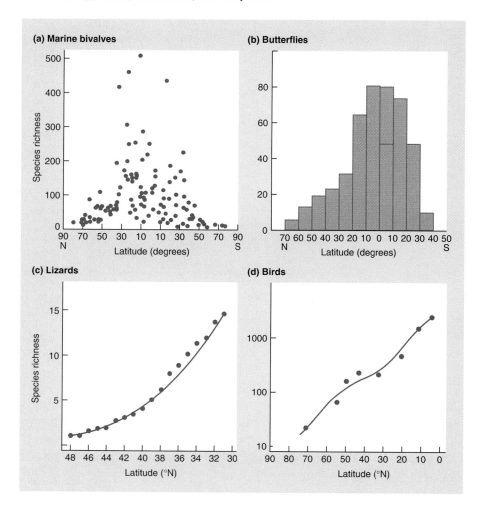

Figure 10.19
Latitudinal patterns in species richness: (a) Of marine bivalves. (After Flessa & Jablonski, 1995.) (b) Of swallowtail butterflies. (After Sutton and Collins, 1991.) (c) Of lizards in the United States. (After Pianka, 1983.) (d) Of breeding birds in North and Central America. (After Dobzhansky, 1950.)

argument might therefore have to run as follows. The light, temperature, and water regimes of the tropics lead to high biomass communities but not necessarily to diverse communities. This, though, leads to nutrient-poor soils and perhaps a wide range of light regimes from the forest floor to canopy far above. These in turn lead to high plant species richness and thus to high animal species richness. There is certainly no *simple* "productivity explanation" for the latitudinal trend in richness.

climatic variation as an explanation? Some ecologists have invoked the climate of low latitudes as a reason for their high species richness. Specifically, equatorial regions are generally less seasonal than temperate regions, and this may allow species to be more specialized (i.e., have narrower niches, Figure 10.1b). This suggestion has now received a number of tests. For example, a comparison of bird communities in temperate Illinois and tropical Panama showed that both tropical shrub and tropical forest habitats contained very many more breeding species than their temperate counterparts, and up to a half of the increase in richness consisted of specialist fruit-eaters in the tropical habitats—a specialized way of life that could not be sustained year-round nearer the poles. In contrast, there are also more species of bark and ambrosia beetles in the tropics than in temperate regions, but in this case the tropical species are less specialized (Figure 10.20).

The greater evolutionary "age" of the tropics has also been proposed as a reason for their greater species richness, and another line of argument suggests that the repeated fragmentation and coalescence of tropical forest refugia promoted genetic differentiation and speciation, accounting for much of the high richness in tropical regions. These ideas, too, are plausible but very far from proven.

Overall, therefore, the latitudinal gradient lacks an unambiguous explanation. This is hardly surprising. The components of a possible explanation—trends with productivity, climatic stability, and so on—are themselves understood only in an incomplete and rudimentary way, and the latitudinal gradient intertwines these components with one another, and with other, often opposing forces: isolation, harshness, and so on.

10.5.3 Gradients with altitude and depth

In terrestrial environments, a decrease in species richness with altitude is a phenomenon almost as widespread as a decrease with latitude. Examples are given in Figure 10.21 for birds, mammals, and vascular plants in the Himalayan Mountains of Nepal.

At least some of the factors instrumental in the latitudinal trend in richness are also likely to be important as explanations for this altitudinal trend (though the

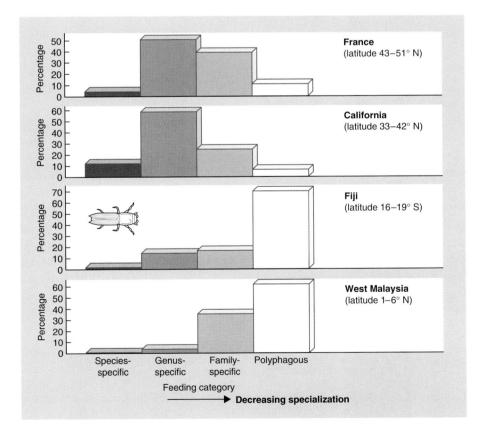

Figure 10.20

Number of species of ambrosia beetles and bark beetles (combined) in various feeding categories. These species are generally more specialized in their diets at high latitudes. (After data in Beaver, 1979.)

Figure 10.21
Relationship between species richness
and altitude in the Nepalese Himalayas:
(a) for breeding bird species, (b) for
mammals (After Hunter & Yonzon,
1992.), and (c) for vascular plants.
(After Whittaker, 1977, from data of K.
Yoda.)

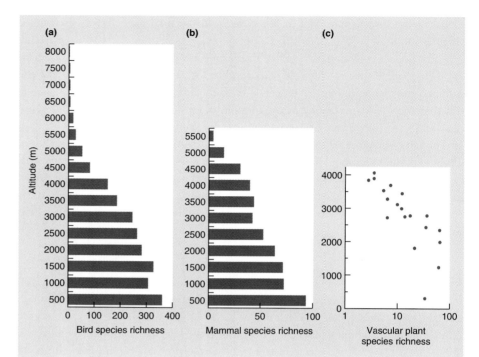

problems in explaining the latitudinal trend apply equally to altitude). In addition, however, high-altitude communities almost invariably occupy smaller areas than lowlands at equivalent latitudes, and they will usually be more isolated from similar communities than lowland sites. The effects of area and isolation are certain to contribute to the decrease in species richness with altitude.

In aquatic environments, the change in species richness with depth shows strong similarities to the terrestrial gradient with altitude. In larger lakes, the cold, dark, oxygen-poor abyssal depths contain fewer species than the shallow surface waters. Likewise, in marine habitats, plants are confined to the photic zone (where they can photosynthesize), which rarely extends below 30 m. In the open ocean, therefore, there is a rapid decrease in richness with depth, reversed only by the variety of often bizarre animals living on the ocean floor. Interestingly, however, in coastal regions the effect of depth on the species richness of benthic (bottom-dwelling) animals is to produce not a single gradient, but a peak of richness at about 1000 m, possibly reflecting higher environmental predictability there (Figure 10.22). At greater depths, beyond the continental slope, species richness declines again, probably because of the extreme paucity of food resources in abyssal regions.

10.5.4 Gradients during community succession

Section 9.4 described how, in community successions, if they run their full course, the number of species first increases (because of colonization) but eventually decreases

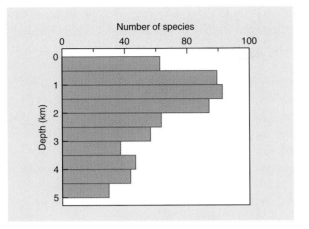

Figure 10.22
Depth gradient in species richness of the megabenthos (fish, decapods, holothurians, and asteroids) in the ocean southwest of Ireland. (After Angel 1994.)

(because of competition). This is most firmly established for plants, but the few studies that have been carried out on animals in successions indicate, at least, a parallel increase in species richness in the early stages of succession. Figure 10.23 illustrates this for birds and insects associated with different old-field successions.

To a certain extent, the successional gradient is a necessary consequence of the gradual colonization of an area by species from surrounding communities that are at later successional stages; that is, later stages are more fully saturated with species (Figure 10.1d). However, this is a small part of the story, since succession involves a process of replacement of species and not just the mere addition of new ones.

Indeed, as with the other gradients in species richness, there is something of a cascade effect with succession: one process that increases richness kick-starts a

a cascade effect?

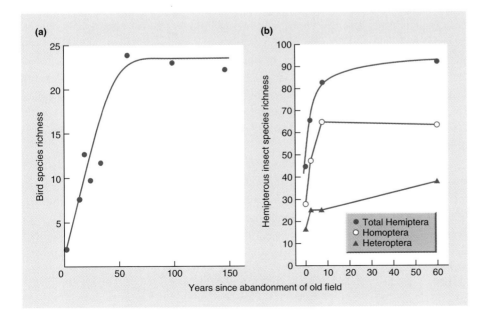

Figure 10.23
The increase in species richness during old-field successions. (a) Birds. (After Johnston & Odum, 1956.) (b) Hemipterous insects. (After Brown & Southwood, 1983.)

second, which feeds into a third, and so on. The earliest species will be those that are the best colonizers and the best competitors for open space. They immediately provide resources (and introduce heterogeneity) that were not previously present. For example, the earliest plants generate resource-depletion zones in the soil that inevitably increase the spatial heterogeneity of plant nutrients. The plants themselves provide a new variety of microhabitats, and for the animals that might feed on them they provide a much greater range of food resources (Figure 10.1b). The increase in herbivory and predation may then feed back to promote further increases in species richness (predator-mediated coexistence: Figure 10.1c), which provides further resources and more heterogeneity, and so on. In addition, temperature, humidity, and wind speed are much less variable within a forest than in an exposed early successional stage, and the enhanced constancy of the environment may provide a stability of conditions and resources that permits specialist species to build up populations and persist (Figure 10.1b). As with the other gradients, the interaction of many factors makes it difficult to disentangle cause from effect. But with the successional gradient of richness, the tangled web of cause and effect appears to be of the essence.

10.6 Patterns in Taxon Richness in the Fossil Record

Finally, it is of interest to take the processes that are believed to be instrumental in generating present-day gradients in richness and apply them to trends occurring over much longer time spans. The imperfection of the fossil record has always been the greatest impediment to the palaeontological study of evolution. Nevertheless, some general patterns have emerged, and our knowledge of six important groups of organisms is summarized in Figure 10.24.

the Cambrian explosion—exploiter-mediated coexistence?

Until about 600 million years ago, the world was populated virtually only by bacteria and algae, but then almost all the phyla of marine invertebrates entered the fossil record within the space of only a few million years (Figure 10.24a). Given that the introduction of a higher trophic level can increase richness at a lower level, it can be argued that the first single-celled herbivorous protist was probably instrumental in the Cambrian explosion in species richness. The opening up of space by cropping of the algal monoculture, coupled with the availability of recently evolved eukaryotic cells, may have caused the biggest burst of evolutionary diversification in Earth's history.

the Permian decline—a species-area relationship?

In contrast, the equally dramatic decline in the number of families of shallow-water invertebrates at the end of the Permian (Figure 10.24a) could have been a result of the coalescence of the Earth's continents to produce the single supercontinent of Pangaea; the joining of continents produced a marked reduction in the area occupied by shallow seas (which occur around the periphery of continents) and thus a marked decline in the area of habitat available to shallow-water invertebrates. Moreover, at this time the world was subject to a prolonged period of global cooling in which huge quantities of water were locked up in enlarged polar caps and

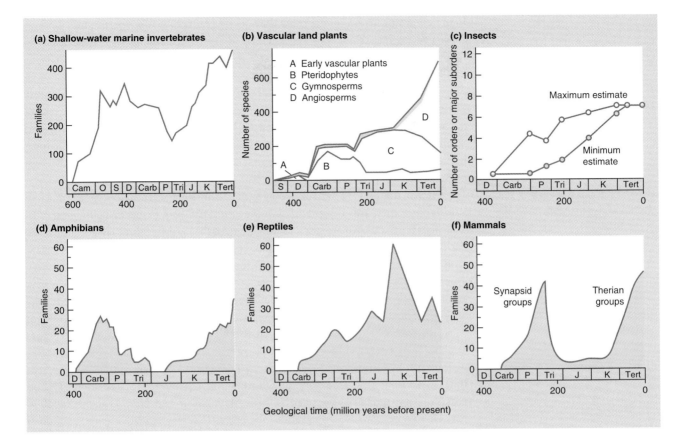

Figure 10.24
Curves showing patterns in taxon richness through the fossil record. (a) Families of shallow-water invertebrates. (After Valentine, 1970.) (b) Species of vascular land plants in four groups—early vascular plants, pteridophytes, gymnosperms, and angiosperms. (After Niklas et al., 1983.) (c) Major orders and suborders of insects. The minimum values are derived from definite fossil records; the maximum values include "possible" records. (From Strong et al., 1984.) (d), (e), and (f) Vertebrate families of amphibians, reptiles and mammals, respectively. (After Webb, 1987.) Key to geological periods: Cam, Cambrian; O, Ordovician; S, Silurian; D, Devonian; Carb, Carboniferous; P, Permian; Tri, Triassic; J, Jurassic; K, Cretaceous; Tert, Tertiary.

glaciers, causing a widespread reduction of warm shallow sea environments. Thus, a species–area relationship may be invoked to account for the reduction in richness of this fauna.

The analysis of fossil remains of vascular land plants (Figure 10.24b) reveals four distinct evolutionary phases: (1) a Silurian–mid-Devonian proliferation of early vascular plants, (2) a subsequent late-Devonian–Carboniferous radiation of fern-like lineages, (3) the appearance of seed plants in the late Devonian and the adaptive radiation to a gymnosperm-dominated flora, and (4) the appearance and rise of flowering plants (angiosperms) in the Cretaceous and Tertiary. It seems that after initial invasion of the land, made possible by the appearance of roots, the diversification of each plant group coincided with a decline in species numbers of the previously

competitive displacement among the major plant groups?

dominant group. In two of the transitions (early plants to gymnosperms, and gymnosperms to angiosperms), this pattern may reflect the competitive displacement of older, less specialized taxa by newer and presumably more specialized taxa.

The first undoubtedly phytophagous insects are known from the Carboniferous. Thereafter, modern orders appeared steadily (Figure 10.24c) with the Lepidoptera (butterflies and moths) arriving last on the scene, at the same time as the rise of the angiosperms. Reciprocal evolution and counterevolution between plants and herbivorous insects has almost certainly been, and still is, an important mechanism driving the increase in richness observed in both land plants and insects through their evolution.

Toward the end of the last ice age, the continents were much richer in large animals than they are today. For example, Australia was home to many genera of giant marsupials; North America had its mammoths, giant ground sloths, and more than 70 other genera of large mammals; and New Zealand and Madagascar were home to giant flightless birds, the moas (Dinorthidae) and elephant bird (*Aepyornis*), respectively. Over the past 30,000 years or so, a major loss of this biotic diversity has occurred over much of the globe. The extinctions particularly affected large terrestrial animals (Figure 10.25a); they were more pronounced in some parts of the world than others; and they occurred at different times in different places (Figure 10.25b). The extinctions mirror patterns of human migration. Thus, the arrival in Australia of ancestral aborigines occurred between 40,000 and 30,000 years ago; stone spear points became abundant throughout the United States about 11,500 years ago; and humans have been in both Madagascar and New Zealand for 1000 years. It can be convincingly argued, therefore, that the arrival of efficient human hunters led to the rapid overexploitation of vulnerable and profitable large prey. Africa, where humans originated, shows much less evidence of loss, perhaps because coevolution of large animals alongside early humans provided ample time for them to develop effective defenses (Owen-Smith, 1987).

The Pleistocene extinctions herald the modern age, in which the influence upon natural communities of human activities has been increasing dramatically.

extinctions of large animals in the Pleistocene—prehistoric overkill?

10.7 ▷ Appraisal of Patterns in Species Richness

There are many generalizations that can be made about the species richness of communities. We have seen how richness may peak at an intermediate level of disturbance frequency and how richness declines with a reduction in island area or an increase in island remoteness. We find also that species richness decreases with increasing latitude, altitude, and (after an initial rise) depth in the ocean. In more productive environments, a wider range of resources may promote species richness, whereas just more of the same can lead to a reduction in richness. A positive correlation between species richness and temperature has also been documented. Species richness increases with an increase in spatial heterogeneity but may decrease with an increase in temporal heterogeneity (increased climatic variation). It increases,

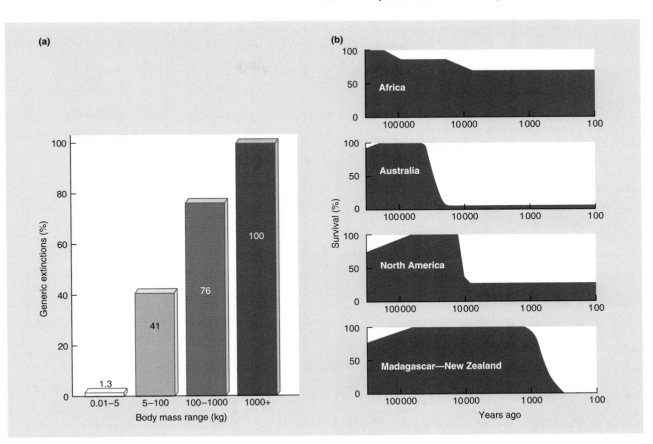

Figure 10.25
(a) The percentage of genera of large mammalian herbivores that have gone extinct in the last 130,000 years is strongly size-dependent (data from North and South America, Europe, and Australia combined). (After Owen-Smith, 1987.) (b) Percentage survival of large animals on three continents and two large islands (New Zealand and Madagascar). (After Martin, 1984.)

at least initially, during the course of succession and with the passage of evolutionary time. However, for many of these generalizations important exceptions can be found, and for most of them the current explanations are not entirely adequate.

It also needs to be recognized that global patterns of species richness have been disrupted in dramatic ways by human activities, such as land-use development, pollution and the introduction of exotic species (Box 10.3).

Unraveling richness patterns is one of the most difficult and challenging areas of modern ecology. Clear, unambiguous predictions and tests of ideas are often very difficult to devise and will require great ingenuity of future generations of ecologists. Because of the increasing importance of recognizing and conserving the world's biodiversity, though, it is crucial that we come to understand thoroughly these patterns in species richness. We will assess the adverse effects of human activities, and how they may be remedied, in Chapters 12–14.

Box 10.3   Topical ECOncerns

The Flood of Exotic Species

Throughout the history of the world, species have invaded new geographical areas, as a result of chance colonizations (e.g., dispersed to remote areas by wind or to remote islands on floating debris; see Section 10.5.1) or during the slow northwards spread of forest trees in the centuries since the last ice age (see Section 4.3). However, human activities have increased this historical trickle to a flood, disrupting global patterns of species richness.

Some human-caused introductions are an accidental consequence of human transport. Other species have been introduced intentionally, perhaps to bring a pest under control (Section 12.5), to produce a new agricultural product or to provide new recreational opportunities. Many invaders become part of natural communities without obvious consequences. But some have been responsible for driving native species extinct or changing natural communities in significant ways (Section 14.2.3).

The alien plants of the British Isles illustrate a number of general points about invaders. Species inhabiting areas where people live and work are more likely to be transported to new regions, where they will tend to be deposited in habitats like those where they originated. As a result, more alien species are found in disturbed habitats close to human transport centers (docks, railways, and cities) and fewer in remote mountain areas (see figure [a]). Moreover, more invaders to the British Isles are likely to arrive from nearby geographic locations (e.g., Europe) or from remote locations whose climate matches that of Britain (e.g., New Zealand) (see figure [b]). Note the small number of alien plants from tropical environments; these species usually lack the frost-hardiness required to survive the British winter.

Review the options available to governments to prevent (or reduce the likelihood) of invasions of undesirable alien species.

The alien flora of the British Isles (a) according to community type (note the large number of aliens in open, disturbed habitats close to human settlements) and (b) by geographic origin (reflecting proximity, trade, and climatic similarity). (After Godfray & Crawley, 1998.)

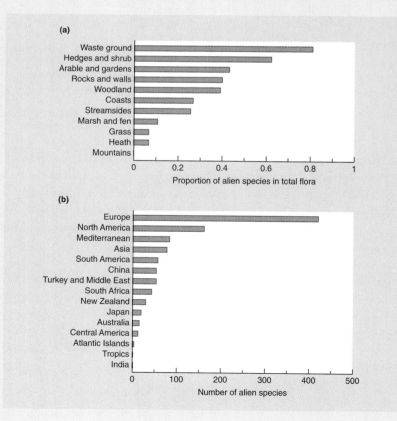

Summary

Richness and diversity

The number of species in a community is referred to as its *species richness*. Richness, though, ignores the fact that some species are rare and others common. *Diversity indices* are designed to combine species richness and the evenness of the distribution of individuals among those species. Attempts to describe a complex community structure by one single attribute, such as richness or diversity, can still be criticized because so much valuable information is lost. A more complete picture is therefore sometimes provided in a *rank–abundance diagram*.

A simple model of species richness

A simple model can help us understand the determinants of species richness. Within it, a community will contain more species the greater the range of resources, if the species are more specialized in their use of resources, if species overlap to a greater extent in their use of resources, or if the community is more fully saturated.

Productivity and resource richness

If higher productivity is correlated with a wider *range* of available resources, then this is likely to lead to an increase in species richness, but more of the same might lead to more individuals per species rather than more species. In general, though, species richness often increases with the richness of available resources and productivity, although in some cases the reverse has been observed—the paradox of enrichment—and others have found species richness to be highest at intermediate levels of productivity.

Predation intensity

Predation can exclude certain prey species and reduce richness or permit more niche overlap and a greater richness (predator-mediated coexistence). Overall, therefore, there may be a humped relationship between predation intensity and species richness in a community, with greatest richness at intermediate intensities.

Spatial heterogeneity

Environments that are more spatially heterogeneous often accommodate extra species because they provide a greater variety of microhabitats, a greater range of microclimates, more types of places to hide from predators, and so on—the resource spectrum is increased.

Environmental harshness

Environments dominated by an extreme abiotic factor—often called harsh environments—are more difficult to recognize than might be immediately apparent. Some apparently harsh environments do support few species, but any overall association has proved extremely difficult to establish.

Climatic variation

In a predictable, seasonally changing environment, different species may be suited to conditions at different times of the year. More species might therefore be expected to coexist than in a completely constant environment. On the other hand, there are opportunities for specialization (e.g., obligate fruit-eating) in a nonseasonal environment that do not exist in a seasonal environment. Unpredictable climatic variation (climatic instability) could decrease richness by denying species the chance to specialize or increase richness by preventing competitive exclusion. There is no established relationship between climatic instability and species richness.

Disturbance

The *intermediate disturbance hypothesis* suggests that very frequent disturbances keep most patches in the early stages of succession (where there are few species), but very rare disturbances allow most patches to become dominated by the best competitors (where there are also few species). Originally proposed to account for patterns of richness in tropical rain forests and coral reefs, the hypothesis has occupied a central place in the development of ecological theory.

Environmental age: evolutionary time

It has often been suggested that communities may differ in richness because some are closer to equilibrium and therefore more saturated than others, and that the tropics are rich in species in part because the tropics have existed over long and uninterrupted periods of evolutionary time. A simplistic contrast between the unchanging tropics and the disturbed and recovering temperate regions, however, is untenable.

Habitat area and remoteness—island biogeography

Islands need not be islands of land in a sea of water. Lakes are islands in a sea of land; mountaintops are high-altitude islands in a low-altitude ocean. The number of species on islands decreases as island area decreases, in part because larger areas typically encompass more different types of habitat. However,

MacArthur and Wilson's *equilibrium theory of island biogeography* argues for a separate island effect based on a balance between immigration and extinction, and the theory has received much support. In addition, on isolated islands especially, the rate at which new species evolve may be comparable to or even faster than the rate at which they arrive as new colonists.

Gradients in species richness

Richness increases from the poles to the tropics. Predation, productivity, climatic variation, and the greater evolutionary age of the tropics have been put forward as explanations, but not one of these is without problems.

In terrestrial environments, richness decreases with altitude. Factors instrumental in the latitudinal trend are also likely to be important in this, but so are area and isolation. In aquatic environments, richness usually decreases with depth for similar reasons.

In successions, if they run their full course, richness first increases (because of colonization) but eventually decreases (because of competition). There may also be a *cascade effect*: one process that increases richness kick-starts a second, which feeds into a third, and so on.

Patterns in taxon richness in the fossil record

The Cambrian explosion of taxa may have been an example of exploiter-mediated coexistence. The Permian decline may have been a species–area relationship when the Earth's continents coalesced into Pangaea. The changing pattern of plant taxa may reflect the competitive displacement of older, less specialized taxa by newer, more specialized ones. The extinctions of many large animals in the Pleistocene may reflect the hand of human predation and hold lessons for the present day.

Review Questions

▲ = Challenge Question

1. Explain species richness, diversity index and rank–abundance diagram and compare what each measures.

2. What is the paradox of enrichment, and how can the paradox be resolved?

3. Explain, with examples, the contrasting effects that predation can have on species richness.

4. ▲ Researchers have reported a variety of hump-shaped patterns in species richness, with peaks of richness occurring at intermediate levels of productivity, predation pressure, disturbance, and depth in the ocean. Review the evidence and consider whether these patterns have any underlying mechanisms in common.

5. Why is it so difficult to identify harsh environments?

6. Explain the intermediate disturbance hypothesis.

7. Islands need not be islands of land on an ocean of water. Compile a list of other types of habitat islands over as wide a range of spatial scales as possible.

8. ▲ An experiment was carried out to try to separate the effects of habitat diversity and area on arthropod species richness on some small mangrove islands in the Bay of Florida. These consist of pure stands of the mangrove species *Rhizophora mangle,* which support communities of insects, spiders, scorpions, and isopods. After a preliminary faunal survey, some islands were reduced in size by

means of a power saw and brute force! Habitat diversity was not affected, but arthropod species richness on three islands nonetheless diminished over a period of 2 years (Figure 10.26). A control island, the size of which was

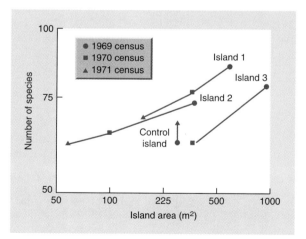

Figure 10.26

The effect on the number of arthropod species of artificially reducing the size of mangrove islands. Islands 1 and 2 were reduced in size after both the 1969 and 1970 censuses. Island 3 was reduced only after the 1969 census. The control island was not reduced, and the change in its species richness was attributable to random fluctuations. (After Simberloff, 1976.)

unchanged, showed a slight increase in richness over the same period. Which of the predictions of island biogeography theory are supported by the results in the figure? What further data would you require to test the other predictions? How would you account for the slight increase in species richness on the control island?

9. ◢ A cascade effect is sometimes proposed to explain the increase in species richness during a community succes-sion. How might this cascade concept apply to the commonly observed gradient of species richness with latitude?

10. Describe how theories of species richness that have been derived on ecological time scales can also be applied to patterns observed in the fossil record.

Like all biological entities, ecological communities, at scales from the local to the global, require matter for their construction and energy for their activities. A healthy community, like a healthy body, needs to be adequately fed and watered. We need to understand the routes by which matter and energy enter and leave ecological communities, and the pathways they follow, and the processes that affect them while they are there.

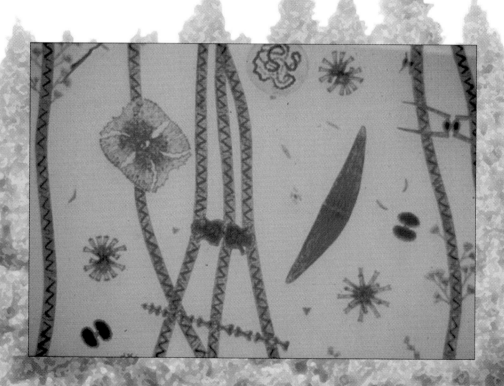

The Flux of Energy and Matter through Ecosystems

Key Concepts

In this chapter you will

- recognize that communities are intimately linked with the abiotic environment by fluxes of energy and matter.

- understand that net primary productivity is not evenly spread across the Earth.

- appreciate that transfer of energy between trophic levels is always inefficient—secondary productivity by herbivores is approximately an order of magnitude less than the primary productivity on which it is based.

- recognize that the decomposer system processes much more of a community's energy and matter than the live-consumer system.

- appreciate that the result of decomposition is that complex energy-rich molecules are broken down by their consumers (decomposers and detritivores) into carbon dioxide, water, and inorganic nutrients.

- understand that in global geochemical cycles, nutrients are moved over vast distances by winds in the atmosphere and in the moving waters of streams and ocean currents.

11.1 ▷ Introduction

When we focus on the processes driving the fluxes of energy and matter between and within communities, we are concerned more with the ways in which areas of land or water receive and process incident radiation and inorganic nutrients, and with the integrated performance of whole communities and their abiotic environment, and less concerned with the behavior of individuals, species, and populations. At this level of organization, we use the term *ecosystem* to denote the biological community together with the abiotic environment in which it is set. Lindeman (1942) laid the foundations for ecological energetics, a science with profound implications, both for an understanding of ecosystem processes and for human food production (Box 11.1).

standing crop and primary and secondary productivity

In order to examine ecosystem processes, it is important to understand some key terms.

■ The bodies of the living organisms within a unit area constitute a *standing crop* of biomass.

■ By *biomass* we mean the mass of organisms per unit area of ground (or water) and this is usually expressed in units of energy (e.g., joules per square meter) or dry organic matter (e.g., metric tons per hectare). In practice we include in biomass all those parts, living or dead, that are attached to the living organism. Thus, it is conventional to regard the whole body of a tree as biomass, despite the fact that most of the wood is dead. Organisms (or their parts) cease to be regarded

---

BOX 11.1

Ecological Energetics and the Biological Basis of Productivity and Human Welfare

A classic paper by Lindeman (1942) laid the foundations of a science of ecological energetics. He attempted to quantify the concept of food chains and food webs by considering the efficiency of transfer between trophic levels—from incident radiation received by a community through its capture by green plants in photosynthesis to its subsequent use by bacteria, fungi, and animals.

Lindeman's paper was a major catalyst that stimulated the International Biological Programme (IBP; Worthington 1975). The subject of the IBP was the biological basis of productivity and human welfare. Given the problem of a rapidly increasing human population, it was recognized that scientific knowledge would be required for rational resource management. Cooperative international research programs focused on the ecological energetics of areas of land, fresh waters, and the seas. The IBP provided the first occasion on which biologists throughout the world were challenged to work together to a common end.

More recently, another pressing issue has galvanized the ecological community into action. Deforestation, the burning of fossil fuels, and other human influences are expected to cause dramatic changes to global climate and atmospheric composition and can be expected in turn to influence patterns of productivity and the composition of vegetation on a global scale. The International Geosphere–Biosphere Programme (IGBP) has established a core project, Global Change and Terrestrial Ecosystems (GCTE), with the prime objective of predicting the effects of changes in climate and atmospheric composition on terrestrial ecosystems, including agricultural and production forest systems (Steffen et al., 1992).

as biomass when they die (or are shed) and become components of dead organic matter.

- The *primary productivity* of a community is the rate at which biomass is produced *per unit area* by plants, the primary producers. It can be expressed either in units of energy (e.g., joules per square meter per day) or of dry organic matter (e.g., kilograms per hectare per year).
- The total fixation of energy by photosynthesis is referred to as *gross primary productivity* (GPP). A proportion of this, however, is respired away by the plants themselves and is lost from the community as respiratory heat (R).
- The difference between GPP and R is known as *net primary productivity* (NPP); it represents the actual rate of production of new biomass that is available for consumption by heterotrophic organisms (bacteria, fungi, and animals).
- The rate of production of biomass by heterotrophs is called *secondary productivity*.

A proportion of primary production is consumed by herbivores, which, in turn, are consumed by carnivores. These are sometimes said to constitute "the grazer system," but we saw in Section 8.1 that not all such consumption is done by "grazing." Hence, we refer here to a *live-consumer system*. The fraction of NPP that is not eaten by herbivores is then said to pass through the *decomposer system*. We distinguish two groups of organisms responsible for the decomposition of dead organic matter (detritus): bacteria and fungi are called *decomposers;* animals that consume dead matter are known as *detritivores*.

live-consumer systems and decomposer systems

11.2 > Primary Productivity

11.2.1 Geographic patterns in primary productivity

The functioning of the biota of the Earth, and of the communities across its surface, depends crucially on the levels of productivity that the plants are able to achieve. If we wish to understand and protect these communities, and to be able to detect when their functioning is faulty or under threat, then we must understand the underlying patterns of productivity.

Global terrestrial net primary productivity is estimated to be about 115×10^9 metric tons dry weight per year, and in the sea 55×10^9 metric tons per year. The fact that productivity is not evenly spread across the Earth is emphasized in Table 11.1. A large proportion of the globe produces less than 400 g per square meter per year. This includes over 30 percent of the land surface and 90 percent of the ocean. At the other extreme, the most productive systems are found among swamps and marshlands, estuaries, algal beds and reefs, and cultivated land.

In the forest biomes of the world, there is a general latitudinal trend of increasing productivity from northern (boreal), through temperate, to tropical conditions (Table 11.1). The same trend may exist in the productivity of tundra and grassland communities (Figure 11.1a) and in various cultivated crops (Figure 11.1b), but there is much variation. In aquatic communities this trend is clear in lakes (Figure 11.1c) but not

the productivity of forests, grasslands, crops, and lakes follows a latitudinal pattern

Table 11.1

Net annual primary productivity and standing crop biomass estimates for contrasting communities

ECOSYSTEM TYPE	AREA (10^6 km²)	NPP, PER UNIT AREA (g m⁻² or t km⁻²) NORMAL RANGE	WORLD NPP (10^9 t)	BIOMASS PER UNIT AREA (kg m⁻²) NORMAL RANGE	WORLD BIOMASS (10^9 t)
Tropical forest	24.5	1000–3500	49.4	6–80	1025
Temperate forest	12.0	600–2500	14.9	6–200	385
Northern coniferous forest	12.0	400–2000	9.6	6–40	240
Woodland and shrubland	8.5	250–1200	6.0	2–20	50
Savannah	15.0	200–2000	13.5	0.2–15	60
Temperate grassland	9.0	200–1500	5.4	0.2–5	14
Tundra and alpine	8.0	10–400	1.1	0.1–3	5
Desert and semidesert shrub	18.0	10–250	1.6	0.1–4	13
Extreme desert, rock, sand, and ice	24.0	0–10	0.07	0–0.2	0.5
Cultivated land	14.0	100–3500	9.1	0.4–12	14
Swamp and marsh	2.0	800–3500	4.0	3–50	30
Lake and stream	2.0	100–1500	0.5	0–0.1	0.05
Total continental	149		115		1837
Open ocean	332.0	2–400	41.5	0–0.005	1.0
Upwelling zones	0.4	400–1000	0.2	0.005–0.1	0.008
Continental shelf	26.6	200–600	9.6	0.001–0.04	0.27
Algal beds and reefs	0.6	500–4000	1.6	0.04–4	1.2
Estuaries	1.4	200–3500	2.1	0.01–6	1.4
Total marine	361		55.0		3.9
Full total	510		170		1841

(After Whittaker, 1975.)

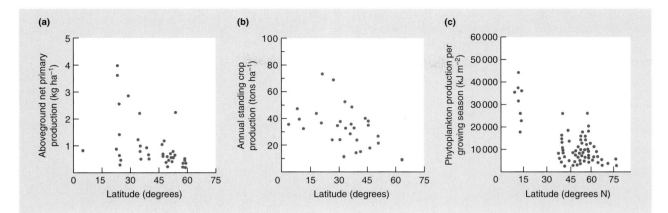

Figure 11.1
Despite much variation, there is a general trend of increasing net primary productivity from the poles (latitude 90 degrees) to the equator (latitude 0 degrees). (a) For grassland and tundra ecosystems. (After Cooper, 1975.) (b) For cultivated crops. (After Cooper, 1975.) (c) For lakes. (After Brylinsky & Mann, 1973.)

in oceans. The trend with latitude suggests that radiation (a resource) and temperature (a condition) may be the factors usually limiting the productivity of communities. Other factors can, however, constrain productivity within even narrower limits. In the sea, productivity is more often limited by a shortage of nutrients.

11.2.2 Factors limiting primary productivity

What, then, limits primary productivity? In terrestrial communities, solar radiation, carbon dioxide, water, and soil nutrients are the resources required for primary production; temperature, a condition, has a strong influence on the rate of photosynthesis. CO_2 is normally present at about 0.03 percent of atmospheric gases and seems to play no significant role in determining differences between the productivities of different communities (although global increases in CO_2 concentration may cause big changes). On the other hand, the intensity of radiation, availability of water and nutrients, and temperature all vary dramatically from place to place. They are all candidates for the role of limiting factor. Which of them actually sets the limit to primary productivity?

Depending on location, something between 0 and 5 joules of solar energy strikes each square meter of the Earth's surface every minute. If all this were converted by photosynthesis to plant biomass (that is, if photosynthetic efficiency were 100 percent), there would be a prodigious generation of plant material, 10 to 100 times greater than recorded values. However, only about 44 percent of incident short-wave radiation occurs at wavelengths suitable for photosynthesis. Yet, even when this is taken into account, productivity still falls well below the maximum possible. For example, the conifer communities shown in Figure 11.2 had the highest net photosythetic efficiencies, but these were only between 1 percent and 3 percent. For a similar level of incoming radiation, deciduous forests achieved 0.5 percent to 1 percent, and, despite their greater energy income, deserts managed only 0.01

terrestrial communities use radiation inefficiently

Figure 11.2

Photosynthetic efficiency (percentage of incoming photosynthetically active radiation converted to aboveground net primary production) for three sets of terrestrial communities in the United States. (After Webb et al., 1983.)

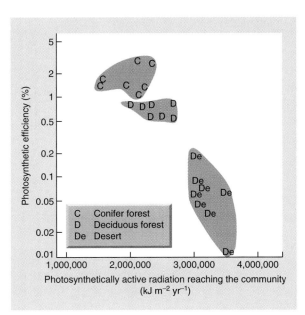

percent to 0.2 percent. These can be compared with short-term peak efficiencies achieved by crop plants under ideal conditions, when values from 3 percent to 10 percent can be achieved.

water and temperature as critical factors

There is no doubt that available radiation would be used more efficiently if other resources were in abundant supply. The much higher values of community productivity from agricultural systems bear witness to this. Shortage of water—an essential resource both as a constituent of cells and in photosynthesis—is often the critical factor. It is not surprising, therefore, that the rainfall of a region is quite closely correlated with its productivity (Figure 11.3a). There is also a clear relationship between the aboveground NPP and mean annual temperature (Figure 11.3b), but note that higher temperatures are associated with rapid transpiration, and thus they increase the rate at which water shortage may become important. Water shortage has direct effects on the rate of plant growth, but it also leads to less dense vegetation.

Figure 11.3

Relationships between forest net primary productivity and (a) annual precipitation and (b) temperature. (After Reichle, 1970.)

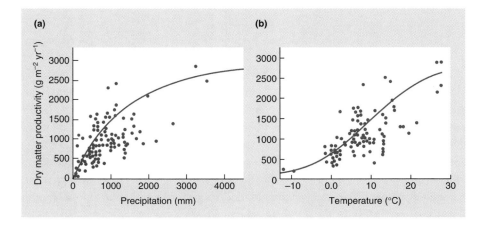

Vegetation that is sparse intercepts less radiation (much of which falls on bare ground). This accounts for much of the difference in productivity noted earlier between desert vegetation and forest.

The productivity of a community can be sustained only for that period of the year when the plants bear photosynthetically active foliage. Deciduous trees have a self-imposed limit on the period of the year during which they bear foliage, whereas evergreen trees hold a canopy throughout the year (though in some seasons they may barely photosynthesize at all).

NPP increases with the length of the growing season

No matter how brightly the sun shines and how often the rain falls, no matter how equable is the temperature, productivity must be low if there is no soil in a terrestrial community, or if the soil is deficient in essential mineral nutrients. Of all the mineral nutrients, the one with the strongest influence on community productivity is fixed nitrogen. There is probably no agricultural or forestry system that does not respond to applied nitrogen by increased primary productivity, and this may well be true of natural vegetation as well. The deficiency of other elements, particularly phosphorus, can also hold the productivity of a community far below what is theoretically possible.

NPP may be low because appropriate mineral resources are deficient

In fact, in the course of a year, the productivity of a terrestrial community may be limited by a succession of factors. The primary productivity of grasslands may be far below the theoretical maximum because the winters are too cold and the intensity of radiation is low, the summers are too dry, the rate of nitrogen mobilization is too slow, or especially because overgrazing reduces the standing crop of photosynthetic leaves and much of the incident radiation falls on bare ground.

a succession of factors may limit primary productivity through the year

In aquatic communities, the factors that most frequently limit primary productivity are the availability of nutrients (particularly nitrate and phosphate) and the intensity of solar radiation that penetrates the column of water. Productive aquatic communities occur where, for one reason or another, nutrient concentrations are unusually high (e.g., Figure 11.4). Lakes receive nutrients by weathering of rocks

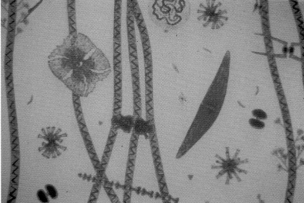

Figure 11.4
Relationship between the net primary productivity of phytoplankton (microscopic plants) in the open water of lakes throughout the world and phosphorus concentration. (After Schindler, 1978.) Photo: pond water algae.

and soils in their catchment areas, in rainfall, and as a result of human activity (fertilizer and sewage input; see Chapter 13); they vary considerably in nutrient availability.

productive aquatic communities occur where nutrient concentrations are high

In the oceans, locally high levels of primary productivity are associated with high nutrient inputs from two sources. First, nutrients may flow continuously into coastal shelf regions from estuaries (Figure 11.5). Productivity in the inner shelf region is particularly high because nutrient concentrations are high and the relatively clear water provides a reasonable depth within which net photosynthesis is positive (the *euphotic zone*). Closer to land, the water is richer in nutrients but highly turbid and its productivity is lower. The least productive zones are on the outer shelf and in the open ocean, where, although the water is clear and the euphotic zone is deep, there are generally extremely low concentrations of nutrients. Local regions of high productivity occur in the open ocean only where there are upwellings from deep nutrient-rich water.

11.3 ▷ The Fate of Primary Productivity

Fungi, animals, and most bacteria are heterotrophs: they derive their matter and energy either directly by consuming plant material or indirectly from plants by eating other heterotrophs. Plants, the primary producers, constitute the first trophic level in a community; primary consumers occur at the second trophic level; secondary consumers (carnivores) at the third; and so on.

11.3.1 The relationship between primary and secondary productivity

Since secondary productivity depends on primary productivity, we should expect a positive relationship between the two variables in communities. Figure 11.6 illustrates this general relationship in aquatic and terrestrial examples. Secondary productivity

Figure 11.5
Variation in phytoplankton net primary productivity, nutrient concentration, and euphotic depth along a transect from the coast of Georgia to the edge of the continental shelf. Primary productivity is high where there is a reasonable supply of nutrients and the water is clear enough for the zone of photosynthesis (euphotic zone) to be reasonably deep. Lower productivity occurs both where turbid water causes the euphotic zone to be excessively shallow close to land and out to sea, where nutrient concentrations are very low. (After Haines, 1979.)

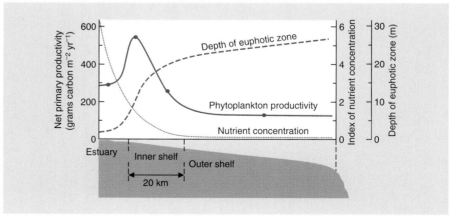

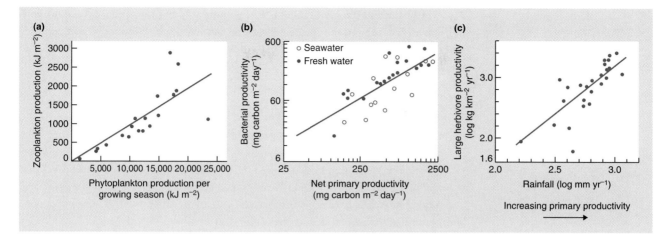

Figure 11.6
The relationship between primary and secondary productivity: (a) For zooplankton in lakes. (After Brylinsky & Mann, 1973.) (b) For bacteria in fresh water and seawater. (After Cole et al., 1988.) (c) For large mammalian herbivores in African game parks—primary productivity in these communities is directly related to annual rainfall. (After Coe et al., 1976.)

by zooplankton (small animals in the open water), which principally consume phytoplankton cells, is positively related to phytoplankton productivity in a range of lakes in different parts of the world (Figure 11.6a). The productivity of heterotrophic bacteria in lakes and oceans also parallels that of phytoplankton (Figure 11.6b); they metabolize dissolved organic matter released from intact phytoplankton cells or produced as a result of "messy feeding" by grazing animals. Figure 11.6c shows how in a series of African savanna game parks, secondary productivity by large mammalian herbivores is positively related to estimates of the primary productivity of plants.

In both aquatic and terrestrial communities, secondary productivity by herbivores is approximately a tenth of the primary productivity on which it is based. This results in a pyramidal structure in which the productivity of plants provides a broad base upon which a smaller productivity of primary consumers depends, with a still smaller productivity of secondary consumers above that. Trophic levels may also have a pyramidal structure when expressed in terms of density or biomass. But food chains that depend on phytoplankton often have inverted pyramids of biomass: a highly productive but small biomass of short-lived algal cells maintains a larger biomass of longer-lived zooplankton.

The productivity of herbivores is invariably less than that of the plants on which they feed. Where has the missing energy gone? First, not all of the plant biomass produced is consumed alive by herbivores. Much dies without being grazed and supports a community of decomposers (bacteria, fungi, and detritivorous animals). Second, not all the plant biomass that is eaten by herbivores (nor herbivore biomass eaten by carnivores) is assimilated and available for incorporation into consumer biomass. Some is lost in feces, and this also passes to the decomposers. Third, not all the energy that has been assimilated is actually converted to biomass. A proportion is lost as respiratory heat. This occurs both because no energy conversion process

there is a general positive relationship between primary and secondary productivity

is ever 100 percent efficient (some is lost as unusable random heat, consistent with the second law of thermodynamics) and also because animals do work that requires energy, again released as heat. These three energy pathways occur at all trophic levels and are illustrated in Figure 11.7.

11.3.2 The fundamental importance of energy transfer efficiencies

Figure 11.8 provides a more complete description of the trophic structure of a community. It has two additional elements of realism. First, it adds a *decomposer system*; second, it recognizes that there are subcomponents of each trophic level that operate in different ways. Thus distinctions are made between invertebrate and vertebrate categories, between microbes and detritivores that use dead organic matter, and between consumers of microbes (microbivores) and detritivores.

possible pathways of a joule of energy through a community

A unit of energy (a joule) may be consumed and assimilated by an invertebrate herbivore that uses it to do work and loses it as respiratory heat. Or it might be consumed by a vertebrate herbivore and later be assimilated by a carnivore that dies and enters the dead organic matter compartment. Here, what remains of the joule

Figure 11.7
The pattern of energy flow through a trophic compartment (represented as the red box).

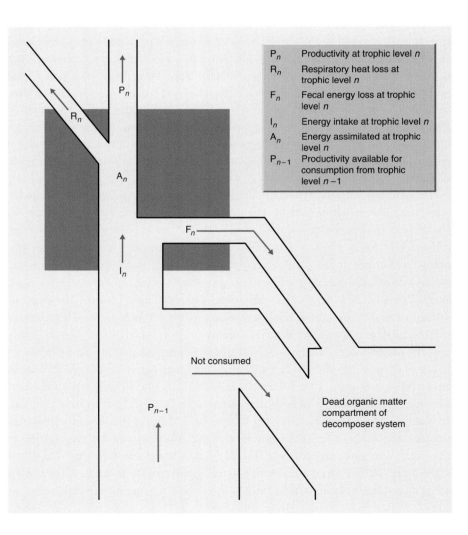

P_n	Productivity at trophic level n
R_n	Respiratory heat loss at trophic level n
F_n	Fecal energy loss at trophic level n
I_n	Energy intake at trophic level n
A_n	Energy assimilated at trophic level n
P_{n-1}	Productivity available for consumption from trophic level $n-1$

Not consumed

Dead organic matter compartment of decomposer system

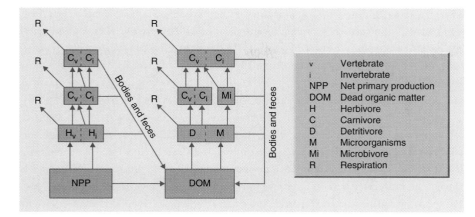

Figure 11.8
A generalized model of trophic structure and energy flow for a terrestrial community. (After Heal & MacLean, 1975.)

may be assimilated by a fungal hypha and consumed by a soil mite, which uses it to do work, dissipating a further part of the joule as heat. At each consumption step, what remains of the joule may fail to be assimilated and pass in the feces to the dead organic matter; or it may be assimilated and respired, or assimilated and incorporated into growth of body tissue (or the production of offspring). The body may die and what remains of the joule enter the dead organic matter compartment, or it may be captured alive by a consumer in the next trophic level, where it meets a further set of possible branching pathways. Ultimately, each joule will have found its way out of the community, dissipated as respiratory heat at one or more of the transitions in its path along the food chain. Whereas a molecule or ion may cycle endlessly through the food chains of a community, energy passes through just once.

The possible pathways in the herbivore–carnivore (live-consumer) and decomposer systems are the same, with one critical exception—feces and dead bodies are lost to the live-consumer system (and enter the decomposer system), but those from the decomposer system are simply sent back to the dead organic matter compartment at its base. Thus, the energy available as dead organic matter may finally be completely metabolized—and all the energy lost as respiratory heat—even if this requires several circuits through the decomposer system. The exceptions to this are situations (1) in which matter is exported out of the local environment to be metabolized elsewhere, for example, when detritus is washed out of a stream, and (2) those in which local abiotic conditions have inhibited decomposition and left pockets of incompletely metabolized high-energy matter, otherwise known as oil, coal, and peat.

The proportions of net primary production that flow along each of the possible energy pathways depend on *transfer efficiencies* from one step to the next. A knowledge of the values of just three categories of transfer efficiency is all that is required to predict the pattern of energy flow. These are *consumption efficiency* (CE), *assimilation efficiency* (AE), and *production efficiency* (PE).

Consumption efficiency is the percentage of total productivity available at one trophic level that is consumed ("ingested") by a trophic level above. For primary consumers, CE is the percentage of joules produced per unit time as net primary productivity that finds its way into the guts of herbivores. In the case of secondary consumers, it is the percentage of herbivore productivity eaten by carnivores. The remainder dies without being eaten and enters the decomposer system. Reasonable

consumption, assimilation, and production efficiencies determine the relative importance of energy pathways

average figures for consumption efficiency by herbivores are approximately 5 percent in forests, 25 percent in grasslands, and 50 percent in phytoplankton-dominated aquatic communities. We know much less about the consumption efficiencies of carnivores; vertebrate predators may consume 50–100 percent of production from vertebrate prey but perhaps only 5 percent from invertebrate prey, whereas invertebrate predators consume perhaps 25 percent of available invertebrate prey production.

Assimilation efficiency is the percentage of food energy taken into the gut of consumers in a trophic compartment that is assimilated across the gut wall and becomes available for incorporation into growth or is used to do work. The remainder is lost as feces and enters the decomposer system. An "assimilation efficiency" is much less easily ascribed to microorganisms. Food does not pass through a "gut" and feces are not produced. Bacteria and fungi digest dead organic matter externally and, between them, typically absorb almost all the product: they are often said to have an "assimilation efficiency" of 100 percent. Assimilation efficiencies are typically low for herbivores, detritivores, and microbivores (20–50 percent) and high for carnivores (around 80 percent). The way that plants allocate production to roots, wood, leaves, seeds, and fruits also influences their usefulness to herbivores. Seeds and fruits may be assimilated with an efficiency as high as 60–70 percent, and leaves with about 50 percent efficiency; the assimilation efficiency for wood may be as low as 15 percent.

Production efficiency is the percentage of assimilated energy that is incorporated into new biomass—the remainder is entirely lost to the community as respiratory heat. Production efficiency varies mainly according to the taxonomic class of the organisms concerned. Invertebrates in general have high efficiencies (30–40 percent), losing relatively little energy in respiratory heat. Among the vertebrates, ectotherms (whose body temperature varies according to environmental temperature; Section 3.2.6) have intermediate values for PE (around 10 percent), and endotherms, with their high energy expenditure associated with maintaining a constant temperature, convert only 1–2 percent of assimilated energy into production. Microorganisms, including protozoa, tend to have very high production efficiencies.

The overall *trophic transfer efficiency* from one trophic level to the next is simply CE × AE × PE. In the period after Lindeman's (1942) pioneering work (Box 11.1), it was generally assumed that trophic transfer efficiencies were around 10 percent; indeed, some ecologists referred to a "10 percent law." However, there is certainly no law of nature that precisely one tenth of the energy that enters a trophic level transfers to the next. For example, a recent compilation of trophic studies from a wide range of freshwater and marine environments revealed that trophic level transfer efficiencies varied between about 2 and 24 percent—although the mean *was* 10.13 percent (Figure 11.9).

11.3.3 The relative roles of the live-consumer and decomposer systems

Given specified values for net primary productivity at a site, and CE, AE, and PE for the various trophic groupings shown in the model in Figure 11.8, it is possible to map out the relative importance of different pathways.

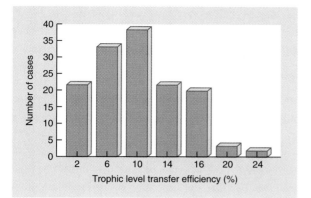

Figure 11.9
The frequency distribution of trophic level transfer efficiencies in 40 trophic studies of aquatic communities. There is considerable variation among studies and among trophic levels. The mean is 10.13 percent (SE = 0.49). (After Pauly & Christensen, 1995.)

There have been only a very few studies in which all the community compartments have been studied together. However, some generalizations are possible if we compare the gross features of contrasting systems. Figure 11.10 illustrates the patterns of energy flow in a forest, a grassland, a plankton community (of the ocean or a large lake), and the community of a small stream or pond. The decomposer system

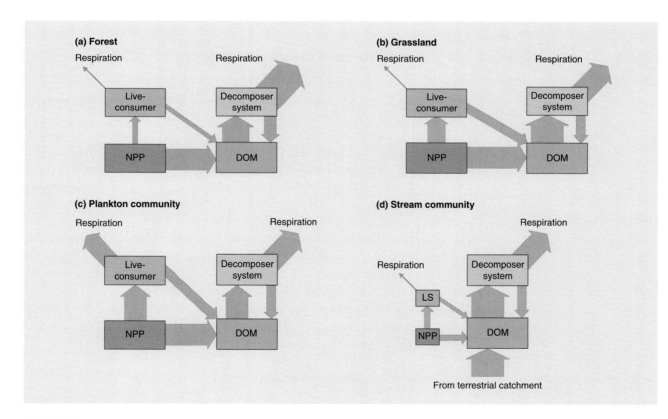

Figure 11.10
General patterns of energy flow for (a) a forest, (b) a grassland, (c) a plankton community, (d) the community of a stream or small pond. Relative sizes of boxes and arrows are proportional to relative magnitudes of compartments and flows. NPP, net primary production; DOM, dead organic matter; LS, live-consumer system.

is probably responsible for the majority of secondary production, and therefore respiratory heat loss, in every community in the world. The live-consumers have their greatest role in plankton communities, where a large proportion of net primary productivity is consumed alive and assimilated at quite high efficiency. Even here, though, very high densities of heterotrophic bacteria in the plankton community subsist on dissolved organic molecules excreted by phytoplankton cells, perhaps consuming more than 50 percent of primary productivity as "dead" organic matter in this way.

The live-consumer system holds little sway in terrestrial communities because of low herbivore consumption and assimilation efficiencies, and it is almost nonexistent in many small streams and ponds simply because primary productivity is so low. The latter often depend for their energy base on dead organic matter produced in the terrestrial environment, which falls or is washed or blown into the water. The deep-ocean benthic community has a trophic structure very similar to that of streams and ponds. In this case, the community lives in water too deep for photosynthesis to be appreciable or even to take place at all, but it derives its energy base from dead phytoplankton, bacteria, animals, and feces that sink from the autotrophic community in the euphotic zone above. From a different perspective, the ocean bed is equivalent to a forest floor beneath an impenetrable forest canopy.

11.4 ▶ The Process of Decomposition

Given the profound importance of the decomposer system, and thus of decomposers (bacteria and fungi) and detritivores, it is important to appreciate the range of organisms and processes involved in decomposition.

decomposition defined

Immobilization is what occurs when an inorganic nutrient element is incorporated into organic form, primarily during the growth of green plants: for example, when carbon dioxide becomes incorporated into a plant's carbohydrates. Energy (in the case of plants, from the sun) is required for this. Conversely, decomposition involves the release of energy and the *mineralization* of chemical nutrients—conversion of elements from organic back to inorganic form. *Decomposition* is the gradual disintegration of dead organic matter (i.e., dead bodies, shed parts of bodies, and feces) which is brought about by both physical and biological agencies. It culminates in the breakdown of complex energy-rich molecules by their consumers (decomposers and detritivores) into carbon dioxide, water, and inorganic nutrients. Ultimately, the incorporation of solar energy in photosynthesis, and the immobilization of inorganic nutrients into biomass, is balanced by the loss of heat energy and organic nutrients when the organic matter is mineralized.

11.4.1 The decomposers: bacteria and fungi

bacteria and fungi are early colonists of newly dead material

If a scavenging animal, a vulture or a burying beetle perhaps, does not take a dead resource immediately, the process of decomposition usually starts with colonization by bacteria and fungi. Bacteria and fungal spores are always present in the air and the water and are usually present on (and often in) dead material before it is dead. The early colonists tend to use soluble materials, mainly amino acids and sugars, that are freely diffusible. The residual resources, though, are not diffusible and are more resistant to attack. Subsequent decomposition therefore proceeds more slowly and involves micro-bial specialists that can use structural, complex carbohydrates like celluloses and lignins and break down the more complex proteins, suberin (cork) and cuticle.

11.4.2 The detritivores and specialist microbivores

The microbivores are a group of animals that operate alongside the detritivores and that can be difficult to distinguish from detritivores. The name *microbivore* is reserved for the minute animals that specialize in feeding on bacteria or fungi but are able to exclude detritus from their gut. In fact, though, the majority of the detritivorous animals involved in the decomposition of dead organic matter are generalist consum-ers, of both the detritus itself and the associated bacterial and fungal populations. The invertebrates that take part in the decomposition of dead plant and animal materials are a taxonomically diverse group. In terrestrial environments they are usually classified according to their size (Figure 11.11). This is not an arbitrary basis for classification, because size is an important feature for organisms that reach their resources by burrowing or crawling among cracks and crevices of litter or soil.

> most detritivores consume both detritus and its associated bacteria and fungi

In freshwater ecology, on the other hand, the study of detritivores has been concerned less with the size of the organisms than with the ways in which they obtain their food (refer to Figure 4.17). For example, *shredders* are detritivores that feed on coarse particulate organic matter (often tree leaves fallen into the stream); they serve to fragment the material. *Collectors*, on the other hand, feed on fine particulate organic matter (< 2 mm), and *grazer–scrapers* have mouthparts appropriate for scraping off and consuming the algae growing on rocks.

> aquatic detritivores are usually classified according to their feeding mode

11.4.3 Consumption of plant detritus

Two of the major organic components of dead leaves and wood are cellulose and lignin. These pose considerable digestive problems for animal consumers. Digesting cellulose requires *cellulase* enzymes, and, surprisingly, cellulases of animal origin have been definitely identified in only one or two species. The majority of detritivores, lacking their own cellulases, rely on the production of cellulases by associated bacteria or fungi or, in some cases, protozoa. The interactions are of a range of types: (1) *obligate mutualisms* between a detritivore and a specific and permanent gut microflora (e.g., bacteria) or microfauna (e.g., termites); (2) *facultative mutualisms,* in which the animals make use of cellulases produced by a microflora ingested with detritus as it passes through an unspecialized gut (for example, wood lice); or (3) "external rumens," in which animals simply assimilate the products of the cellulase-producing microflora associated with decomposing plant remains or feces (for example, spring-tails—Collembola).

> UNANSWERED QUESTION:
>
> given the abundance of cellulose as a resource, why do so few animals possess their own cellulase enzymes?

The decomposition of dead material is not simply due to the sum of the activities of decomposers and detritivores; it is largely the result of interaction between the two (Lussenhop, 1992). This can be illustrated by taking an imaginary journey with a leaf fragment through the process of decomposition, focusing attention on a part of the wall of a single cell. Initially, when the leaf falls to the ground, the piece of cell wall is protected from microbial attack because it lies within the plant tissue. The leaf is now chewed and the fragment enters the gut of a wood louse. Here it

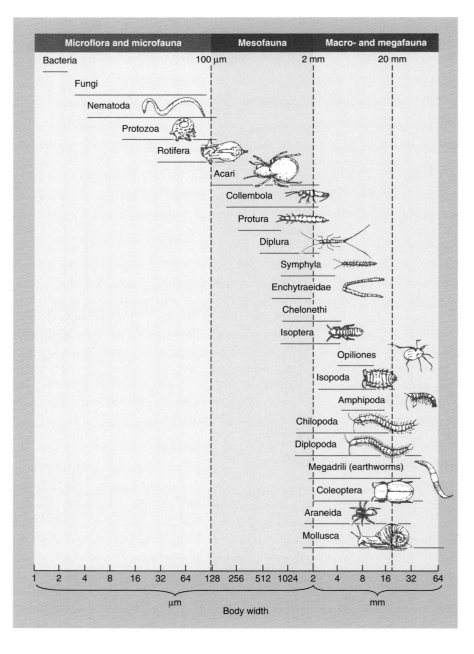

meets a new microbial flora in the gut and is acted on by the digestive enzymes of the wood louse. The fragment emerges, changed by passage through the gut. It is now part of the wood louse's feces and is much more easily attacked by microorganisms, because it has been fragmented and partially digested. While microorganisms are colonizing the feces, it may again be eaten, perhaps by a springtail, and pass through the new environment of the springtail's gut. Incompletely digested fragments may again appear, this time in springtail feces, yet more easily accessible to microorganisms. The fragment may pass through several other guts in its progress from being a piece of dead tissue to its inevitable fate of becoming carbon dioxide and minerals.

11.4.4 Consumption of feces and carrion

The dung of carnivorous vertebrates is relatively poor quality stuff. Carnivores assimilate their food with high efficiency (usually 80 percent or more is digested) and their feces retain only the least digestible components; their decomposition is probably effected almost entirely by bacteria and fungi. In contrast, herbivore dung still contains an abundance of organic matter and is sufficiently thickly spread in the environment to support its own characteristic fauna, consisting of many occasional visitors but with several specific dung feeders. A good example is provided by elephant dung (Kingston, 1977). Within a few minutes of dung deposition the area is alive with beetles. The adult dung beetles feed on the dung, but they also bury large quantities along with their eggs to provide food for developing larvae (Figure 11.12).

All animals defecate and die, yet feces and dead bodies are not generally very obvious in the environment. This is because of the efficiency of the specialist

Figure 11.12

(a) African dung beetle rolling a ball of dung. (Photograph courtesy of Heather Angel.) (b) The larva of the dung beetle *Heliocopris dilloni* excavates a hollow as it feeds within the dung ball. (After Kingston & Coe, 1977.)

consumers of these dead organic products. On the other hand, where consumers of feces are absent, a dramatic buildup of dung may occur (Box 11.2).

When considering the decomposition of dead bodies, it is helpful to distinguish three categories of organisms that attack carcasses. As before, decomposers (bacteria and fungi) and invertebrate detritivores have roles to play, but, in addition, scavenging vertebrates are often of considerable importance. Many carcasses of a size to make a single meal for one or a few of these scavenging detritivores will be removed completely within a very short time of death, leaving nothing for bacteria, fungi, or invertebrates. This role is played, for example, by arctic foxes and skuas in polar regions; by crows, wolverines, and badgers in temperate areas; and by a wide variety of birds and mammals, including kites, jackals, and hyenas, in the tropics.

11.5 ▷ The Flux of Matter through Ecosystems

Chemical elements and compounds are vital for the processes of life. When living organisms expend energy (as they all do, continually), they do so, essentially, in order to extract chemicals from their environment and to hold on to them and use them for a period before they lose them again. Thus, the activities of organisms profoundly influence the patterns of flux of chemical matter.

BOX 11.2   *Topical ECOncerns*

An Australian Dilemma: Bovine Dung but no Bovine Dung Beetles

Bovine dung has provided an extraordinary and economically very important problem in Australia. During the past 200 years the cow population has risen from just 7 individuals (brought over by the first English colonists in 1788) to about 30 million. These produce some 300 million cow pats per day, covering as much as 6 million acres per year with dung. Deposition of bovine dung poses no particular problem elsewhere in the world, where bovines have existed for millions of years and have an associated fauna that exploits the fecal resources. However, the largest herbivorous animals in Australia until European colonization were marsupials such as kangaroos. The native detritivores that deal with the dry, fibrous dung pellets that they leave cannot cope with cow dung, and the loss of pasture under dung has imposed a huge economic burden on Australian agriculture. The decision was therefore made in 1963 to establish in Australia beetles of African origin, able to dispose of bovine dung in the most important places and under the most prevalent conditions where cattle are raised; so far 21 species have been introduced (Doube et al., 1991).

Adding to the problem, Australia is plagued by native bushflies (*Musca vetustissima*) and buffalo flies (*Haematobia irritans exigua*), which deposit eggs on dung pats. The larvae do not survive in dung that has been buried by beetles, and the presence of beetles has been shown to be effective at reducing fly abundance (Ridsdill-Smith, 1991). Success depends on dung being buried within about 6 days of production, the time it takes for the fly egg (laid on fresh dung) to hatch and develop to the pupal stage. Edwards and Aschenborn (1987) surveyed the nesting behavior in southern Africa of 12 species of dung beetles in the genus *Onitis*. They concluded that *Onitis uncinatus* was a prime candidate for introduction to Australia for fly-control purposes, since substantial amounts of dung were buried on the first night after dung colonization. The least suitable species, *O. viridualus*, spent several days constructing a tunnel and did not commence burying until 6–9 days had elapsed.

Introduced species sometimes become undesirable aliens, invading and disrupting native communities. What steps can be taken to minimize the risk that a purposeful introduction will become a disruptive invader?

The great bulk of living matter in any community is water. The rest is made up mainly of carbon compounds, and this is the form in which energy is accumulated and stored. Carbon enters the trophic structure of a community when a simple molecule, CO_2, is taken up in photosynthesis. If it becomes incorporated in net primary productivity, it is available for consumption as part of a sugar, a fat, a protein, or, very often, a cellulose molecule. It follows exactly the same route as energy, being successively consumed and either defecated, assimilated, or used in metabolism, during which the energy of its molecule is dissipated as heat while the carbon is released again to the atmosphere as CO_2. Here, though, the tight link between energy and carbon ends.

Once energy is transformed into heat, it can no longer be used by living organisms to do work or to fuel the synthesis of biomass. The heat is eventually lost to the atmosphere and can never be recycled: life on Earth is only possible because a fresh supply of solar energy is made available every day. In contrast, the carbon in CO_2 can be used again in photosynthesis. Carbon and all other nutrient elements (e.g., nitrogen and phosphorus, etc.) are available to plants as simple inorganic molecules or ions in the atmosphere (CO_2) or as dissolved ions in water (e.g., nitrate, phosphate, potassium). Each can be incorporated into complex carbon compounds in biomass. Ultimately, however, when the carbon compounds are metabolized to CO_2, the mineral nutrients are released again in simple inorganic form. Another plant may then absorb them, and so an individual atom of a nutrient element may pass repeatedly through one food chain after another.

Unlike the energy of solar radiation, moreover, nutrients are not in unalterable supply. The process of locking up some in living biomass reduces the supply remaining to the rest of the community. If plants, and their consumers, were not eventually decomposed, the supply of nutrients would become exhausted and life on Earth would cease.

energy cannot be cycled and reused—matter can

We can conceive of pools of chemical elements existing in compartments. Some compartments occur in the *atmosphere* (carbon in carbon dioxide, nitrogen as gaseous nitrogen), some in the rocks of the *lithosphere* (calcium as a constituent of calcium carbonate, potassium in feldspar), and others in the water of soil, stream, lake, or ocean—the *hydrosphere* (nitrogen in dissolved nitrate, phosphorus in phosphate, carbon in carbonic acid). In all these cases the elements exist in inorganic form. In contrast, living organisms (the biota) and dead and decaying bodies can be viewed as compartments containing elements in organic form (carbon in cellulose or fat, nitrogen in protein, phosphorus in adenosine triphospate ATP). Studies of the chemical processes occurring within these compartments and, more particularly, of the fluxes of elements between them, constitute the science of biogeochemistry.

Nutrients are gained and lost by communities in a variety of ways (Figure 11.13). We can construct a nutrient budget if we can identify and measure all the processes on the credit and debit sides of the equation.

11.5.1 Nutrient budgets in terrestrial ecosystems

Weathering of parent bedrock and soil, by both physical and chemical processes, is the dominant natural and ultimate source of nutrients such as calcium, iron,

nutrient inputs

Figure 11.13

Components of the nutrient budgets of a terrestrial and an aquatic system. Note how the two communities are linked by stream flow, which is a major output from the terrestrial system but a major input to the aquatic one. Inputs are shown in red and outputs in black.

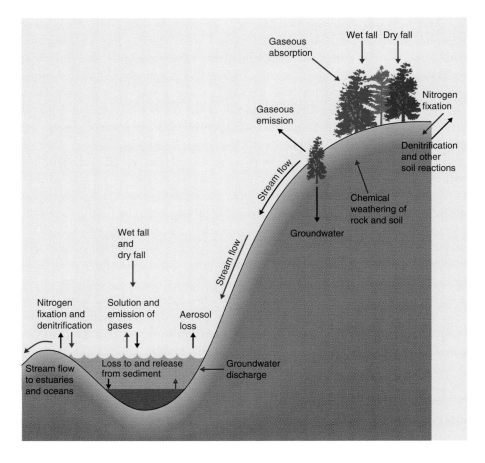

magnesium, phosphorus, and potassium, which may then be taken up via the roots of plants.

Atmospheric carbon dioxide is the source of the carbon content of terrestrial communities. Similarly, gaseous nitrogen from the atmosphere provides most of the nitrogen content of communities. Several types of bacteria and blue–green algae possess the enzyme nitrogenase, which converts gaseous nitrogen to ammonium ions (NH_4^+) which can then be taken up through the roots and used by plants. All terrestrial ecosystems receive some available nitrogen through the activity of free-living nitrogen-fixing bacteria, but communities containing plants such as legumes and alder trees (*Alnus* sp.), with their root nodules containing symbiotic nitrogen-fixing bacteria (Section 7.2.3), may receive a very substantial proportion of their nitrogen in this way.

Other nutrients from the atmosphere become available to communities in *dry fall* (settling of particles during periods without rain) or *wet fall* (in rain, snow, and fog). Rain is not pure water but contains chemicals derived from a number of sources: (1) trace gases, such as oxides of sulfur and nitrogen; (2) aerosols (produced when tiny water droplets from the oceans evaporate in the atmosphere and leave behind particles rich in sodium, magnesium, chloride, and sulfate); and (3) dust particles from fires, volcanoes, and windstorms, often rich in calcium, potassium, and sulfate. Nutrients dissolved in precipitation mainly become available to plants when the water reaches the soil and can be taken up by plant roots. However, some are absorbed by leaves directly.

Nutrients may circulate within the community for many years. Alternatively, the atom may pass through the system in a matter of minutes, perhaps without interacting with the biota at all. Whatever the case, the atom will eventually be lost through one of the variety of processes that remove nutrients from the system (Figure 11.13). These processes constitute the debit side of the nutrient budget equation.

Release to the atmosphere is one pathway of nutrient loss. In many communities there is an approximate annual balance in the carbon budget; the carbon fixed by photosynthesizing plants is balanced by the carbon released to the atmosphere as carbon dioxide from the respiration of plants, microorganisms, and animals. Plants themselves may be direct sources of gaseous and particulate release. For example, forest canopies produce volatile hydrocarbons (e.g., terpenes) and tropical forest trees appear to emit aerosols containing phosphorus, potassium, and sulfur (Waring & Schlesinger, 1985). Finally, ammonia gas is released during the decomposition of vertebrate excreta. Other pathways of nutrient loss are important in particular instances. For example, fire (either natural, or that, for instance, caused by agricultural practices that include the burning of stubble) can turn a very large proportion of a community's carbon into carbon dioxide in a very short time, and the loss of nitrogen, as volatile gas, can be equally dramatic.

For many elements, the most substantial pathway of loss is in stream flow. The water that drains from the soil of a terrestrial community into a stream carries a load of nutrients that is partly dissolved and partly particulate. With the exception of iron and phosphorus, which are not mobile in soils, the loss of plant nutrients is predominantly in solution. Particulate matter in stream flow occurs both as dead organic matter (mainly tree leaves) and as inorganic particles.

It is the movement of water, under the force of gravity, that links the nutrient budgets of terrestrial and aquatic communities (Figure 11.13). Terrestrial systems lose dissolved and particulate nutrients into streams and groundwaters; aquatic systems (including the stream communities themselves, and ultimately the oceans) gain nutrients from stream flow and groundwater discharge. Refer to Section 1.3.3 for discussion of a study (at Hubbard Brook) that explored the chemical linkages at the land–water interface.

11.5.2 Nutrient budgets in aquatic communities

Aquatic systems receive the bulk of their supply of nutrients from stream inflow. In stream and river communities, and also in lakes with a stream outflow, export in outgoing stream water is a major factor. By contrast, in lakes without an outflow (or where outflow is low relative to lake volume), and also in oceans, nutrient accumulation in permanent sediments is often the major export pathway.

Many lakes in arid regions, lacking a stream outflow, lose water only by evaporation. The waters of these *endorheic* lakes (those having internal flow) are thus more concentrated than their freshwater counterparts, being particularly rich in sodium but also in other nutrients such as phosphorus. Saline lakes should not be considered as oddities; globally, they are just as abundant in terms of numbers and volume as freshwater lakes (Williams, 1988).

The largest of all endorheic "lakes" is the ocean—a huge basin of water supplied by the world's rivers and losing water only by evaporation. Its great size, in comparison to the input from rain and rivers, leads to its remarkably constant chemical composition. Nutrients in the surface water come from two sources—river inputs and water welling up from the deep (Figure 11.14). A given phosphorus atom, for example, arriving in the surface waters, is taken up by a phytoplankton cell or bacterium, or by microscopic picoplankton, and is passed along the food chain (very much as described for a lake). Detrital particles are continuously sinking to the deep waters, where the majority are decomposed, releasing soluble phosphorus that eventually finds its way back to the surface water by upwelling. Only about 1 percent of detrital phosphorus is lost to the sediment in each oceanic mixing cycle.

11.6 ▷ Global Biogeochemical Cycles

Nutrients are moved over vast distances by winds in the atmosphere and by the moving waters of streams and ocean currents. There are no boundaries, either natural or political. It is appropriate, therefore, to conclude this chapter by moving to an even larger spatial scale to examine global biogeochemical cycles.

11.6.1 The hydrological cycle

The principal source of water is the oceans; radiant energy makes water evaporate into the atmosphere, winds distribute it over the surface of the globe, and precipitation

Figure 11.14

A simple representation of the major fluxes of phosphorus (P) in the ocean.

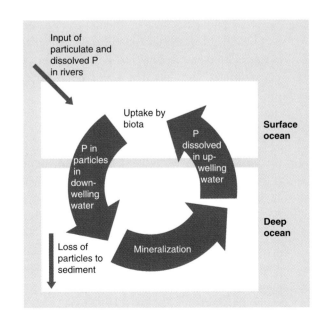

brings it down to Earth (with a net movement of atmospheric water from oceans to continents), where it may be stored temporarily in soils, lakes, and ice fields (Figure 11.15). Loss occurs from the land through evaporation and transpiration or as liquid flow through stream channels and groundwater aquifers, eventually to return to the sea. The major pools of water occur in the oceans (97.3 percent of the total for the biosphere; Berner & Berner, 1987), the ice of polar icecaps and glaciers (2.06 percent), deep in the groundwater (0.67 percent), and in rivers and lakes (0.01 percent). The proportion that is in transit at any time is very small—water draining through the soil, flowing along rivers, and present as clouds and vapor in the atmosphere constitutes only about 0.08 percent of the total. However, this small percentage plays a crucial role, both by supplying the requirements for survival of living organisms and for community productivity and by transporting many chemical nutrients as it moves.

The hydrological cycle would proceed whether or not a biota were present. However, terrestrial vegetation can modify to a significant extent the fluxes that occur. Vegetation can intercept water at two points on this journey, preventing some from reaching the stream and causing it to move back into the atmosphere: (1) by catching some in foliage from which it may evaporate and (2) by preventing some from draining from the soil water by taking it up in the transpiration stream. We have seen earlier how cutting down the forest in a catchment in Hubbard Brook (Section 1.3.3) increased the throughput to streams of water together with its load of dissolved and particulate matter. It is small wonder that large-scale deforestation around the globe, usually to create new agricultural land, can lead to loss of topsoil, nutrient impoverishment, and increased severity of flooding.

Water is a very valuable commodity, and this is reflected in the difficult political exercise of dealing with competing demands of diverting river water for hydroelectric power generation or irrigation for agriculture as opposed to maintaining the intrinsic values of an unmanipulated river (Box 11.3).

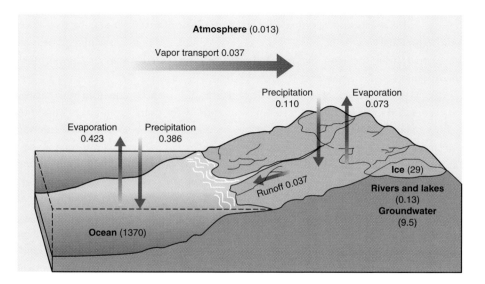

Figure 11.15

The hydrological cycle showing fluxes and sizes of reservoirs (x 10^6 km³). (After Berner & Berner, 1987.)

Box 11.3   Topical ECOncerns

Conflicting Opinions on the Value of a River

The Snowy River in Australia, famous for its mountain scenery, deep gorges, and sheer bluffs, was once an inspiration for poets and scriptwriters. But in 1949 the Snowy Mountains Hydro Power Act was passed; by the time the scheme was finished in the late 1960s fully 99 percent of the Snowy River's discharge had been diverted to neighboring rivers (the Murray and Murrumbidgee) to be used for hydroelectric power generation and irrigation farming. The Snowy has virtually disappeared in many parts of its upper reaches and is profoundly impacted throughout its length.

Poignant testimony comes from people in the Snowy River community. Charlie Robertson (born in 1919), who lived within a kilometer of the river for most of his life, says, "We could always hear the Snowy singing from home. That is how I used to describe the sound of the river. Now we don't hear it at all."

In 1998 a large-scale inquiry was set up to determine whether the allocation of Snowy River water for power, irrigation, and intrinsic value should change. Members of the local community and conservationists believe that at least 28 percent of original flows should be reinstated to restore part of the river's original ecological and scenic character. On the other hand, the owners of the hydro scheme warn against anything but minimal reallocation of the valuable water. A decision is expected to be reached before the new millennium. Watch http:/www.snowywaterenquiry.org.au for further developments.

If you were heading this enquiry, how would you weigh up the costs in loss of ecological character and recreational amenity against the benefits of "clean" power generation (which does not pump greenhouse gases into the atmosphere) and the agricultural production and employment associated with irrigation?

The world's major abiotic reservoirs for nutrients are illustrated in Figure 11.16. We consider these cycles in turn.

11.6.2 The phosphorus cycle

The principal stocks of phosphorus occur in the water of the soil, rivers, lakes, and oceans and in rocks and ocean sediments. The phosphorus cycle may be described as a sedimentary cycle because of the general tendency for mineral phosphorus to be carried from the land inexorably to the oceans, where ultimately it becomes incorporated in the sediments (Figure 11.16a).

the life history of a phosphorus atom

A "typical" phosphorus atom, released from the rock by chemical weathering, may enter and cycle within the terrestrial community for years, decades, or centuries before it is carried via the groundwater into a stream. Within a short time of entering the stream (weeks, months, or years), the atom is carried to the ocean. It then makes, on average, about 100 round trips between surface and deep waters, each lasting perhaps 1000 years. During each trip, it is taken up by surface-dwelling organisms, before eventually settling into the deep again. On average, on its hundredth descent (after 10 million years in the ocean) it fails to be released as soluble phosphorus, but instead enters the bottom sediment in particulate form. Perhaps 100 million years later, the ocean floor is lifted up by geological activity to become dry land. Thus, our phosphorus atom will eventually find its way back via a river to the sea,

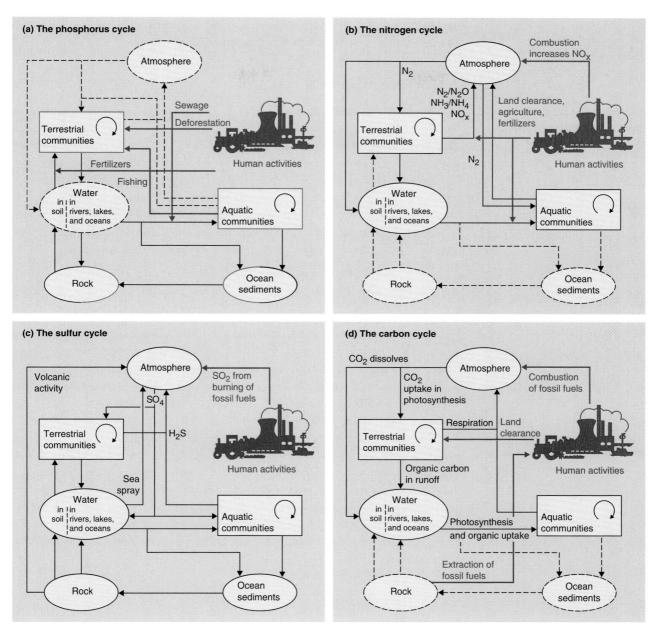

Figure 11.16

The major global pathways of nutrients between the abiotic "reservoirs" of atmosphere, water (hydrosphere), and rock and sediments (lithosphere) and the biotic "reservoirs" constituted by terrestrial and aquatic communities. Human activities (in red) affect nutrient fluxes through terrestrial and aquatic communities both directly and indirectly, via their effects on global biogeochemical cycling through the release of extra nutrients into the atmosphere and water. Cycles are presented for four important nutrient elements: (a) phosphorus, (b) nitrogen, (c) sulfur, and (d) carbon. Insignificant compartments and fluxes are represented by dashed lines.

and to its existence of cycle (biotic uptake and decomposition) within cycle (ocean mixing) within cycle (continental uplift and erosion).

11.6.3 The nitrogen cycle

the nitrogen cycle has an atmospheric phase of overwhelming importance

The atmospheric phase is predominant in the global nitrogen cycle, in which nitrogen fixation and denitrification by microbial organisms are of particular importance (Figure 11.16b). Atmospheric nitrogen is also fixed by lightning discharges during storms and reaches the ground as nitric acid dissolved in rainwater, but only about 3–4 percent of fixed nitrogen derives from this pathway. The magnitude of the flux in stream flow from terrestrial to aquatic communities is relatively small, but it is by no means insignificant for the aquatic systems involved. This is because nitrogen is one of the two elements (along with phosphorus) that most often limit plant growth. Finally, there is a small annual loss of nitrogen to ocean sediments.

11.6.4 The sulfur cycle

Three natural biogeochemical processes release sulfur to the atmosphere: the formation of sea-spray aerosols, anaerobic respiration by sulfate-reducing bacteria, and volcanic activity (relatively minor) (Figure 11.16c). Sulfur bacteria release reduced sulfur compounds, particularly H_2S, from waterlogged bog and marsh communities and from marine communities associated with tidal flats. A reverse flow from the atmosphere involves oxidation of sulfur compounds to sulfate, which returns to Earth as both wet fall and dry fall.

the sulfur cycle has an atmospheric phase and a lithospheric phase of similar magnitude

The weathering of rocks provides about half the sulfur draining off land into rivers and lakes, the remainder deriving from atmospheric sources. On its way to the ocean, a proportion of the available sulfur (mainly dissolved sulfate) is taken up by plants, is passed along food chains and, via decomposition processes, becomes available again to plants. However, in comparison to that of phosphorus and nitrogen, a much smaller fraction of the flux of sulfur is involved in internal recycling in terrestrial and aquatic communities. Finally, there is a continuous loss of sulfur to ocean sediments.

11.6.5 The carbon cycle

the opposing forces of photosynthesis and respiration drive the global carbon cycle

Photosynthesis and respiration are the two opposing processes that drive the global carbon cycle. It is predominantly a gaseous cycle, with CO_2 as the main vehicle of flux among atmosphere, hydrosphere, and biota. Historically, the lithosphere played only a minor role; fossil fuels lay as dormant reservoirs of carbon until human intervention in recent centuries (Figure 11.16d).

Terrestrial plants use atmospheric CO_2 as their carbon source for photosynthesis, whereas aquatic plants use dissolved carbonates (i.e., carbon, from the hydrosphere). The two subcycles are linked by exchanges of CO_2 between atmosphere and oceans. In addition, carbon finds its way into inland waters and oceans as bicarbonate resulting from weathering (carbonation) of calcium-rich rocks such as limestone and chalk. Respiration by plants, animals, and microorganisms releases the carbon locked in photosynthetic products back to the atmospheric and hydrospheric carbon compartments.

11.6.6 Human impacts on biogeochemical cycles

It almost goes without saying that human activities contribute significant inputs of nutrients to ecosystems and disrupt local and global biogeochemical cycles. For example, the amounts of carbon dioxide and oxides of nitrogen and sulfur in the atmosphere have been increased by the burning of fossil fuels and by the exhausts of cars, and the concentrations of nitrate and phosphate in stream water have been raised by agricultural practices and sewage disposal. These changes have far-reaching consequences, which will be discussed in Chapter 13.

Summary

Patterns in primary productivity

Primary production on land is limited by a variety of factors—the quality and quantity of solar radiation; the availability of water, nitrogen, and other key nutrients; and physical conditions, particularly temperature. Productive aquatic communities occur where, for one reason or another, nutrient concentrations are unusually high and the intensity of radiation is not limiting.

The fate of primary productivity

Secondary productivity by herbivores is approximately an order of magnitude less than that of the primary productivity on which it is based. Energy is lost at each feeding step because consumption efficiencies, assimilation efficiencies, and production efficiencies are all less than 100 percent. The *decomposer system* processes much more of a community's energy and matter than the *live-consumer system*. The energy pathways in the live-consumer and decomposer systems are the same, with one critical exception—feces and dead bodies are lost to the live-consumer system (and enter the decomposer system), but feces and dead bodies from the decomposer system are simply sent back to the dead organic matter compartment at its base.

The process of decomposition

The result of decomposition is that complex energy-rich molecules are broken down by their consumers (decomposers and detritivores) into carbon dioxide, water, and inorganic nutrients. Ultimately, the incorporation of solar energy in photosynthesis and the immobilization of inorganic nutrients into biomass are balanced by the loss of heat energy and organic nutrients when the organic matter is decomposed. This is brought about partly by physical processes, but mainly by the activities of decomposers (bacteria and fungi) and detritivores (animals that feed on dead organic matter).

The flux of matter through ecosystems

Nutrients are gained and lost by communities in a variety of ways. Weathering of parent bedrock and soil, by both physical and chemical processes, is the dominant source of nutrients such as calcium, iron, magnesium, phosphorus, and potassium, which may then be taken up via the roots of plants. Atmospheric carbon dioxide and gaseous nitrogen are the principal sources of the carbon and nitrogen content of terrestrial communities, whereas other nutrients from the atmosphere become available as *dry fall* or in rain, snow, and fog. Nutrients are lost again through release to the atmosphere or in the water that feeds into streams and rivers. Aquatic systems (including the stream communities themselves, and ultimately the oceans) gain nutrients from stream flow and groundwater discharge and from the atmosphere by diffusion across their surfaces.

Global biogeochemical cycles

The principal source of water in the hydrological cycle is the oceans; radiant energy makes water evaporate into the atmosphere, winds distribute it over the surface of the globe, and precipitation brings it down to Earth. Phosphorus derives mainly from the weathering of rocks (lithosphere); its cycle may be described as sedimentary because of the general tendency for mineral phosphorus to be carried from the land to the oceans, where ultimately it becomes incorporated in the sediments. The sulfur cycle has an atmospheric phase and a lithospheric phase of similar magnitude. The atmospheric phase is predominant in both the global carbon and nitrogen cycles. Photosynthesis and respiration are the two opposing processes that drive the global carbon cycle; nitrogen fixation and denitrification by microbial organisms are of particular importance in the nitrogen cycle. Human activities contribute significant inputs of nutrients to ecosystems and disrupt local and global biogeochemical cycles.

Review Questions

▲ = Challenge Question

1. A large proportion of the surface of the ocean has primary production of less than 400 g per square meter per year. The open ocean is, in effect, a marine desert. Why?

2. ▲ Describe the general latitudinal trends in net primary productivity. Suggest reasons why such a latitudinal trend does not occur in the oceans.

3. ▲ Table 11.2 presents the results of a study that contrasted the productivity of a deciduous beech (*Fagus sylvatica*) with that of a nearby evergreen spruce (*Picea abies*) forest. The beech leaves photosynthesized at a greater rate (per gram dry weight) than those of spruce and beech "invested" a considerably greater amount of biomass in its leaves each year. Why were these differences not reflected in overall primary productivity? If these species were grown together, which would you expect to come to dominate the forest? What factors other than productivity might influence the relative competitive status of the two species?

4. What evidence suggests that the productivity of many terrestrial and aquatic communities is limited by nutrients?

5. ▲ In both aquatic and terrestrial communities, secondary productivity by herbivores is approximately a tenth of the primary productivity upon which it is based. This has led some to suggest the operation of a 10 percent law. Do you subscribe to this view?

6. Account for the observation that in most communities much more energy is processed through the decomposer system than through the live-consumer system.

7. Outline the role played by bacteria and fungi (decomposers) in the flux of energy and matter through a named ecosystem. Imagine what would happen if bacteria and fungi were magically removed—describe the resulting scenario in detail.

8. Energy cannot be cycled and reused—matter can. Discuss this assertion and its significance for ecosystem functioning.

9. Is the ocean simply a large lake in terms of patterns of flux of energy and matter?

10. The hydrological cycle would proceed whether or not a biota were present. Discuss how the presence of vegetation modifies the flow of water through an ecosystem.

Table 11.2

Characteristics of representative trees of two contrasting species growing within 1 km of each other on the Solling Plateau, Germany

	BEECH	NORWAY SPRUCE
Age (years)	100	89
Height (m)	27	25.6
Leaf shape	Broad	Needle
Annual production of leaves	Higher	Lower
Photosynthetic capacity per unit dry weight of leaf	Higher	Lower
Length of growing season (days)	176	260
Primary productivity (metric tons of carbon per hectare per year)	8.6	14.9
(After Schulze, 1970; and Schulze et al., 1977a,b.)		

Web Research Questions

1. Many of the world's forests have been cut down for timber or to open up land for agriculture. Discuss the consequences of deforestation for the communities that live in rivers and the human communities that live in their flood-plains.

2. Consider a current topic in the news whose understanding depends on complex population interactions (e.g., prediction of the incidence of Lyme disease based on acorn crop and mouse population increase). Review reports on the topic from a number of commentators and evaluate the variation in their science content. What is needed for a readable AND well-informed piece on the topic?

3. To an extent that is often underestimated, pathogens (causing infectious disease) may affect populations of their hosts. Consider human popula-tions, and select one or more human pathogens. How many deaths does each pathogen cause per year? Who, exactly, dies? To what extent is the pathogen the primary cause of death—or does it interact with other factors? Which factors?

4. Many pests are pests because they are strong interspecific competitors. This is especially true of weeds competing against crop plants. Consider one or more of the world's important crops—perhaps those growing close to your home. What are the main problem weeds of those crops? What is the range of ways in which those weeds are controlled? To what extent may the different methods be seen as attempts to shift the competitive balance between crop and weed?

5. Many birds undergo migrations, whose study poses considerable problems. Discuss the role of satellite tracking and other technologies in studying bird migration behavior. Contrast the ecological problems of migratory birds whose populations are so large that they may be considered pests in part of their range (e.g., snowgeese) with those whose populations are so small that they are in danger of extinction (e.g., whooping crane).

Applied Issues in Ecology

The sustainability of human activities, and of the size and distribution of the human population, has increasingly become a preoccupation of the general public and of the politicians who represent them. But attaining or even approaching sustainability requires more than a will to do so—it requires ecological understanding, carefully acquired and even more carefully applied.

Sustainability—Exploitation and Agriculture

Key Concepts

In this chapter you will

- appreciate the underlying dynamics of human population growth and its relationship to the sustainable (or unsustainable) use of resources.

- understand the biological basis of sustainable harvesting of wild populations—particularly in fisheries.

- recognize the benefits and costs of farming monocultures.

- understand that much agricultural practice has not been sustainable because of associated loss and degradation of soil.

- appreciate that accessible supplies of water may be the least sustainable of global resources.

- recognize the benefits and costs of different methods of pest control and the importance of devising integrated management practices.

12.1 ▷ Introduction

what is "sustainability"?

To call an activity sustainable means that it can be continued or repeated for the foreseeable future. Concern has arisen, therefore, precisely because so much human activity is clearly unsustainable. We cannot go on increasing the size of the global human population; we cannot (if we wish to have fish to eat in future) continue to remove fish from the sea faster than the remaining fish can replace their lost companions; we cannot continue to harvest agricultural crops or forests if the quality and quantity of the soil deteriorate or water resources become inadequate; we cannot continue to use the same pesticides if increasing numbers of pests become resistant to them; we cannot maintain the diversity of nature if we continue to drive other species to extinction. Sustainability has thus become one of the core concepts—perhaps *the* core concept—in an ever-broadening concern for the fate of the Earth and the ecological communities that occupy it.

In defining sustainability we used the words *foreseeable future*. We did so because, when an activity is described as sustainable, it is characterized on the basis of what is known at the time. But many factors remain unknown or unpredictable. Conditions may take a turn for the worse (as when adverse oceanographic conditions damage a fishery already threatened with overexploitation), or some unforeseen additional problem may be discovered (resistance may appear to some previously irresistible pesticide). On the other hand, technological advances may allow an activity that previously seemed unsustainable to be sustained (new types of pesticide may be discovered that are more finely targeted on the pest itself rather than "innocent bystander" species). However, there is a real danger that we observe the many technological and scientific advances that have been made in the past and act on the faith that there will always be a technological fix coming along to solve our present problems, too. We cannot accept unsustainable practices on the basis of faith that future advances will make them sustainable after all.

sustainability comes of age

The recognition of sustainability's importance as a unifying idea in applied ecology has grown gradually, and although it is inevitably misleading to claim that particular ideas were "born" at a particular time, there is something to be said for the claim that sustainability came of age in 1991. In the first place, the primary body of ecological scientists in the United States, the Ecological Society of America, published, in the scientific journal *Ecology*, the article "The sustainable biosphere initiative: an ecological research agenda," as "a call-to-arms for all ecologists" with a list of 16 coauthors (Lubchenko et al., 1991); in the same year, three international organizations more concerned with environmental protection, the World Conservation Union, the United Nations Environment Programme and the World Wide Fund for Nature, jointly published *Caring for the Earth: A Strategy for Sustainable Living* (IUCN/UNEP/WWF, 1991). Then, in 1997, the secretary general of the United Nations reported the findings of the UN Commission on Sustainable Development; these were published in 1998 as *Global Change and Sustainable Development: Critical Trends*. The detailed contents and proposals of these documents are less important than their existence. They indicate a growing preoccupation with sustainability, shared by scientists and pressure groups, and a recognition that much of what we do is not sustainable.

12.2 ▷ The Human Population "Problem"

12.2.1 Introduction

The root of most, if not all, of the environmental problems facing us is the population problem: the effects of a large and growing population of humans. More people means an increased requirement for energy, a greater drain on nonrenewable resources like oil and minerals, more pressure on renewable resources like fish and forests (Section 12.3), more need for food production through agriculture (Section 12.4), and so on. The issue is undoubtedly one of sustainability: matters cannot go on the way they are. Yet it is not clear exactly what the problem is (Box 12.1). Here, therefore, we examine first the size and growth rate of the global human population and how we reached our current state, then the success we can expect in projecting

what is the human population "problem"?

Box 12.1   *Topical ECOncerns*

The Human Population Problem

What is the human population problem? This is not an easy question to answer, but what follow are some possible versions of it (Cohen, 1995). The real problem, of course, may be a combination of these—or of these and others. There is little doubt, though, that there *is* a problem—and that the problem is ours, collectively.

What role or responsibility does the individual, as opposed to government, have in responding to this problem? Which of the following variants of the problem pose particular questions of the relationship between the developed and the developing parts of the world or between the haves and the have-nots?

- **The present size of the global human population is unsustainably high.** Around A.D. 200, when there were about 0.25 billion people on Earth, Quintus Septimus Florens Tertullianus wrote "we are burdensome to the world, the resources are scarcely adequate to us." By 1990 the total had risen to an estimated 5.3 billion.

- **The present rate of growth in size of the global human population is unsustainably high.** Prior to the widespread agricultural revolution of the 18th century, the human population, very roughly, had taken 1000

years to double in size. Recently, the total has been doubling about every 43 years.

- **It is not the size but the distribution over the Earth of the human population that is unsustainable.** The fraction of the population living, highly concentrated, in an urban environment rose from around 3 percent in 1800 to 29 percent in 1950 to 45 percent in 1995.

- **It is not the size but the age distribution of the global human population that is unsustainable.** In the developed regions of the world, the percentage of the population that was elderly (over 65) rose from 7.6 percent in 1950 to 12.1 percent in 1990. This proportion will jump dramatically after 2010, when the large cohorts born after World War II pass age 65.

- **It is not the size but the uneven distribution of resources within the global population that is unsustainable.** In 1992, the 830 million people of the world's richest countries enjoyed an average income equivalent to $22,000 per annum. The 2.6 billion people in the middle-income countries received $1600. But the 2 billion in the poorest countries got just $400. These averages themselves hide enormous inequalities.

forward into the future, before finally addressing the problem more directly by asking the question, How many people can the Earth support?

12.2.2 Population growth up to the present

When the finger is pointed at human population *growth* as the key issue, it is often said that what is wrong is that the global population has been growing exponentially. But in an *exponentially* growing population (Chapter 5) the rate of increase per individual is constant. The population as a whole grows at an accelerating rate (a plot of numbers against time sweeps upward), because the population growth rate is a product of the individual rate (constant) and the accelerating number of individuals. In Chapter 5, such exponential growth was contrasted with a population limited by intraspecific competition (such as one described by the logistic equation), in which the rate of increase per individual *decreases* as population size increases. In the case of the global human population, however (Box 12.2; Cohen, 1995) the rate of increase per individual (and also therefore the *annual percentage increase* in size: the rate of increase per 100 individuals) has certainly not been decreasing—but neither has it remained constant. Rather, the individual rate has itself been accelerating. Even exponential growth would be unsustainable, but the more-than-exponential growth that we have been witnessing would, if continued, become unsustainable even sooner.

12.2.3 Predicting the future

It is interesting to see what has happened to the total human population in the past—and to do so alerts us to the scale of the problem we face—but the major practical importance of such a survey lies in the opportunity it might provide to predict future population sizes and rates of growth. There is an enormous difference, however, between projection and prediction. Simply to project forward would be to make the almost certainly false assumption that conditions in the future will be the same as they have been in the past.

Prediction, by contrast, requires an *understanding* of what has happened in the past, as well as how the present differs from the past, and finally how these differences might translate into future patterns of population growth. In particular, it is essential to recognize that the global population of humans is a collection of smaller populations, each with its own often very different characteristics. Like all ecological populations, the human population is heterogeneous.

One common way in which subpopulations have been distinguished has been in terms of the demographic transition. The term *demographic transition* refers both to an idealized pattern of changing birth and death rates as they have occurred during the history of populations, and to a hypothesis proposed to explain this pattern (Cohen 1995). Three groups of nations can be recognized: those that passed through the demographic transition early (pre-1945), late (since then), or not yet (pretransition countries). The pattern is as follows. Initially, both the birth rate and the death rate are high, but the former is only slightly greater than the latter, so the overall rate of population increase is only moderate or small. (This is presumed to have been the case in all human populations at some time in their relatively undeveloped past.) Next, the death rate declines while the birth rate remains high, so the population growth rate increases. Subsequently, however, the birth rate also declines until it is similar to or perhaps even lower than the death rate. The population growth rate therefore declines again and may even become negative—though at a population size far greater than that before the transition began (Figure 12.1).

The hypothesis commonly proposed to explain this transition, put simply, is that it is an inevitable consequence of industrialization, education, and general modernization leading, first, through medical advances, to the drop in death rates, and then, through the choices people make (delaying having children, and so on), to the drop in birth rates. The question remains, though: Is the demographic transition a real repeatable phenomenon or just an academic ideal? This is an important question. If we can understand the demographic transition where it has occurred, we might predict how it will occur in those countries that have yet to pass through it, and hence predict future rates of growth much more accurately than by extrapolation. Future patterns may even be amenable to manipulation by appropriately enlightened policies.

There is some support for the idealized pattern over the past 150 years if the early-transition countries of Europe are taken as a whole (Figure 12.2) and this apparent demographic transition was, as we have seen, associated with a steady decline in population growth rate in the early-transition countries (Figure 12.1). But there has been enormous variation in the speed with which rates have declined, in the length of the delay between the drops in death rates and birth rates, in the present rates themselves, and so on. It is clear that neither the pattern nor the

UNANSWERED QUESTION:

can demographic transitions be understood?—and manipulated?

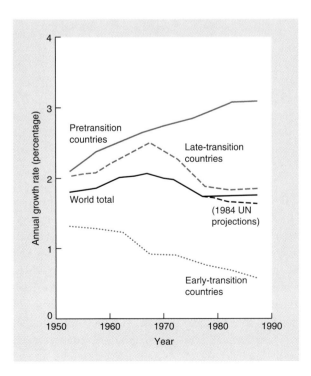

Figure 12.1

The annual rates of population growth, 1950–1990, according to United Nations estimates, for early-transition, late-transition, and pretransition countries and for the global total. (After Cohen, 1995.)

Box 12.2

The Growth of Human Populations

Figure (a) shows estimates, from a variety of sources (indicated by different symbols) of the size of the global human population from 2000 years ago to the present. Apart from the occasional hesitation and even rarer downturn (such as that caused by the ravages of the Black Death toward the end of the 14th century) the overall picture is clearly one of ever more rapid population growth: the slope of the curve gets steeper and steeper.

But is this exponential growth? The answer is a conclusive no. Figure (b) shows this same graph (with the individual estimates omitted), but also shows (1) what an exponentially growing population that started at the same point 2000 years ago and finished at the present population size would have looked like; and (2) for the

sake of contrast, a population anchored at the same start and finish points but growing according to the logistic equation.

Disregarding the logistic equation as utterly unrealistic (we have not approached a global carrying capacity), it is also clear that exponential growth is much more gradual than what has actually been observed. The crux of the differences among these three graphs is shown in Figure (c), which uses the same information, but this time plots the changing individual rate of increase against time (the percentage increases would look the same except that all figures would be 100 times greater). For the logistic equation, under the influence of increasingly intense intraspecific competition, this declines in a straight line down

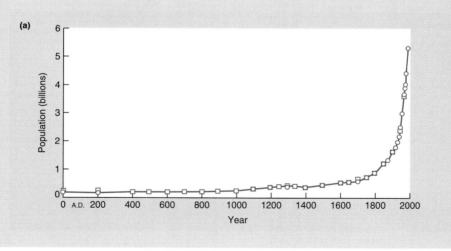

(a)

underlying mechanism is either simple or always the same. To take just one example, fertility rate only began to fall in England and Wales once literacy levels were high and a mere 15 percent of men were employed in agriculture, whereas fertility began to fall in Bulgaria, and long before that in France, when literacy levels were low and the populations were still largely agricultural (van de Walle & Knodel, 1980; Cohen, 1995). Comparable variations exist among countries that are just entering or have just started demographic transitions—the result, perhaps, of differences in government policy, religion, and so on. There is still much to be learned about demographic transitions before the future can be predicted with any degree of certainty, much less manipulated.

to zero—as it always does for the logistic equation. For exponential growth, the rate is constant—again by definition. But the actual growth curve gives rise to an individual rate that not only increases with time as the global population has increased but increases more than linearly: it accelerates.

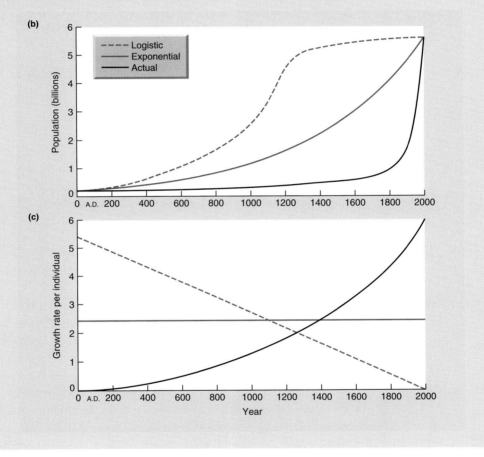

12.2.4 Two future inevitabilities

If it were possible to effect some kind of demographic transition in all countries of the world, so that the birth rates were no higher than the death rates, would the population problem be solved? The answer, sadly, is no, for at least two important reasons. First, there is a big difference in age structure between a population with equal birth and death rates when both those rates are high and one when both are low. When life tables were described in Chapter 5, we made the point that the net reproductive rate of a population was a reflection of the age-related patterns of survival and birth. A given net reproductive rate, though, can be arrived at through

unsustainable age structures?

Figure 12.2
The decline in the annual rate of
population growth in Europe since
1850 has been associated with a
decline in the death rate, followed by
a decline in the birth rate, and an
overall narrowing of the gap between
the two. (After Cohen, 1995.)

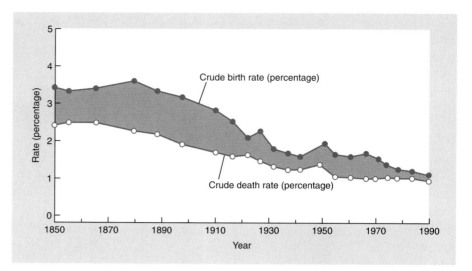

a literally infinite number of different birth and death patterns, and these different combinations themselves give rise to different age structures within the population. If birth rates are high but survival rates low (pretransition), then there will be many young, and relatively few old individuals in the population. But if birth rates are low but survival rates high—the ideal to which we might aspire posttransition—then relatively few young, productive individuals will be called upon to support many who are old, unproductive, and dependent (Box 12.1). The size and growth rates of the human population are not the only problems; the age structure of a population is another.

the momentum of population growth

Moreover, suppose that our understanding were so sophisticated, and our power so complete, that we could establish equal birth and death rates tomorrow. Would the human population stop growing? The answer, once again, is no. Population growth has its own momentum which would still have to be contended with. Even with a birth rate matched to the death rate, there would be many years before a stable age structure was established, and in the meantime there would be considerable further population growth before numbers leveled off. One estimate, for example, suggests that even under these circumstances, the global population of 5.3 billion in 1990 would not level off until around 2150, at a population size of 8.4 billion. The reason, simply, is that there are, for example, many more babies now than there were 25 years ago, and so even if birth rate per capita drops considerably now, there will still be many more births in 25 years, when these babies grow up, than at present, and these children, in turn, will continue the momentum effect before an approximately stable age structure is eventually established.

12.2.5 A global carrying capacity?

The current rate of increase in the size of the global population is unsustainable even though it is lower now than it has been: in a finite space and with finite resources, no population can continue to grow forever. What is an appropriate response to this? To suggest an answer, it is necessary to have some sense of a target, and thus it is interesting, and may be important, to know how large a population of humans could be sustained on Earth. What is the global carrying capacity?

Many estimates have been proposed over the last 300 or so years, a number of which are surveyed in Figure 12.3. There is astonishing variation in these estimates—we are no nearer an agreed carrying capacity now than we have ever been—and there is as much variation among estimates now as at any time in the past: even the estimates in the figure since 1970 span three orders of magnitude, from 1 billion to 1000 billion. A few examples of how estimates have been arrived at, however, described in Box 12.3 (Cohen, 1995), will help to make sense of the apparent confusion reflected in Figure 12.3.

In the first place, it is clear from Box 12.3 that there is a difference between the number the Earth can support and the number that can be supported with an acceptable standard of living. The higher estimates come closer to the concept of a carrying capacity we normally apply to other organisms (Chapter 5)—a number imposed by the limiting resources of the environment. But it is unlikely that many of us would choose to live crushed up against an environmental ceiling or wish it on our descendants.

It is also clear, therefore, that any suggestions we make about a global carrying capacity depend on choices we make both for ourselves and for others. The first are relatively easy: most of us would choose to live at least as well as we do at present. The latter are much more difficult: Can the global population afford to choose for the whole world to live at least as well as those in developed countries do now? What Figure 12.3 and Box 12.3 illustrate more than anything else, therefore,

UNANSWERED QUESTION:

what do we mean by the global carrying capacity?

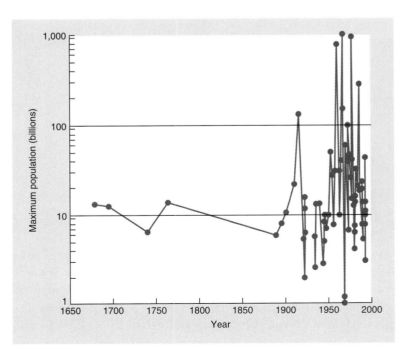

Figure 12.3
Some estimates of how many people the Earth can support plotted against the year the estimates was made. Note the logarithmic scale. Where a range of estimates was provided by an author, the highest has been given. (After Cohen, 1995.)

Box 12.3

Some Estimates of the "Global Carrying Capacity"

In 1967, C. T. De Wit, a plant population ecologist, asked the question, How many people can live on Earth if photosynthesis is the limiting process? The answer he arrived at was roughly 1000 billion. He built into his calculation the fact that the length of the potential growing season varied with latitude, but assumed, among other things, that neither water nor minerals were limiting. He acknowledged, "This is how many people could live *from* the Earth; not *on* the Earth." In other words, if people wanted to eat meat, or wanted what most of us consider a reasonable amount of living space, and so on, then the estimate would be much lower.

By contrast, H. R. Hulett in 1970 assumed that United States levels of affluence and consumption were "optimal" for the whole world, and that not only food but requirements for renewable resources like wood and nonrenewable resources like steel and aluminum had to be included in the calculations. The figure he came up with was no more than 1 billion (less than one-fifth of the 1990 total). He further assumed, however, that the global supply of the various limiting resources would remain the same as the number of producers changed (in his scenario, declined) and the distribution of consumption among them became equal. This seems unlikely.

This last assumption was also made by R. W. Kates and others in a series of reports in 1988, although they worked from global rather than United States averages for consumption rates. With these they estimated support for 5.9 billion people on a basic diet (principally vegetarian), 3.9 billion on an "improved" diet (about 15 percent of calories from animal products), and 2.9 billion on a diet with 25 percent of calories from animal products.

In any case, it is a big step to assume that the human population is limited from below by its resources rather than from above by its natural enemies. Infectious disease, in particular, which not long ago was considered to be an enemy largely vanquished, is now once again perceived, for example by the World Health Organization, as a major threat to human welfare. In a growing global tuberculosis epidemic, for example, 2.7 million people died in 1993, but an estimated 1.7 billion were infected but had not yet manifested the disease (Dobson & Carper, 1996). We saw in Chapter 8 that many infectious diseases thrive best in the densest populations.

is that the answer to a question depends on what is meant by that question. What do we mean by the global carrying capacity?

12.3 ▷ Harvesting Living Resources from the Wild

A major limit to the number of people the Earth can support is the food that can be obtained. Populations of many species living freely in the wild are exploited for food by humans, who cull or harvest a proportion of the population, leaving some individuals behind to grow and reproduce to yield material for future harvests. Primitive human societies obtained all their resources in this way, by hunting and gathering from nature, and we continue to garner some resources in this way. The resources may be fish from the sea, deer from a moorland, or timber from a forest. There is an important difference between resources obtained in this way and those that are farmed (Sections 12.4 and 12.5). Farmed resources are obtained by taking chosen species of plant or animal, domesticating them (often changing them genetically), and growing or rearing them in more or less controlled monocultures. These resources tend to be owned and managed by a farmer or organization. In contrast, most of the oceans and forests that are fished and hunted have at one time been common property, open to free-for-all unsustainable looting by allcomers. Recently, though, fishing and hunting have also come under increasing national and international regulation and national claims to ownership. Many of our examples in this section are of fish or fisheries, but the principles apply to the harvesting of any natural resource.

12.3.1 Fisheries—maximum sustainable yields

Whenever a natural population is exploited there is a risk of overexploitation. Too many individuals are removed and the population is driven into biological jeopardy or economic insignificance—perhaps even to extinction. Global catches of marine fish rose five-fold between 1950 and 1989. However, many of the world's fish stocks are now at or near the point of overexploitation (Figure 12.4). But harvesters also want to avoid underexploitation. If fewer individuals are removed than the population can bear, the harvested crop is smaller than necessary, potential consumers are deprived, and those who do the harvesting are underemployed. It is not easy to tread the narrow path between under- and overexploitation. It is asking a great deal of a policy that it combine the well-being of the exploited species, the profitability of the harvesting enterprise, continuing employment for the workforce and the maintenance of traditional life-styles, social customs, and natural biodiversity.

aiming for the narrow path between over- and underexploitation

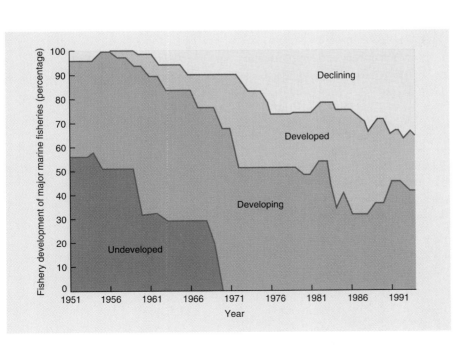

Figure 12.4

Changes in the contribution to global marine fish production made by fisheries in phases of their development. Note that there have been no significant unexploited fisheries remaining since 1970. (UN, 1998.)

population dynamics in the absence of exploitation—humped net recruitment curves

The most fundamental aspects of ecology that we need to understand here were introduced in Chapter 5 in the discussion of the effects of intraspecific competition on populations. To determine the best way to exploit a population, it is necessary to know what the consequences of different exploitation strategies will be. But in order to know these consequences, we first need some understanding of the dynamics of the population in the absence of, or prior to, exploitation. It is usual to assume that, before it is exploited, a harvestable population is crowded and intraspecific competition is intense. Summarizing from Chapter 5, and remembering that these are broad generalizations:

1. Populations in the absence of exploitation can be expected to settle around their carrying capacity, but exploitation will reduce numbers to less than this.
2. Exploitation, by reducing the intensity of competition, moves the population leftwards along the humped net recruitment curve, increasing the net number of recruits to the population per unit time (Figure 12.5).

MSY—the narrow path?

In fact, we can go further with Figure 12.5, since it is clear from the shape of the curve that there must be an intermediate population size at which that rate of net recruitment is highest. Consider a time scale of years. The peak of the curve might be 10 million new fish each year. This is then also the highest number of new fish that could be removed from the population each year that the population itself could replenish. It is known as the *maximum sustainable yield* (MSY): the largest harvest that can be removed from the population regularly and indefinitely. It appears that a fishery could tread the narrow path between under- and overexploitation if those doing the fishing could find a way to achieve this maximum sustainable yield.

the MSY concept has shortcomings

The MSY concept has been the guiding principle in resource management for many years in fishery, forestry, and wildlife exploitation, but it is very far from being the perfect answer for a variety of reasons:

1. By treating the population as a number of similar individuals, it ignores all aspects of population structure such as size or age classes and their differential rates of growth, survival, and reproduction.
2. By being based on a single recruitment curve, it treats the environment as unvarying.

Figure 12.5

The humped relationship between the net recruitment into a population (births - deaths) and the size of that population, resulting from the effects of intraspecific competition (discussed in Chapter 5). Population size increases from left to right, but increasing rates of exploitation take the population from right to left.

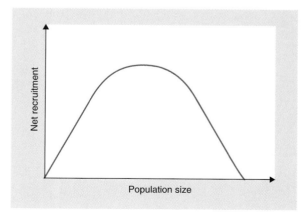

3. In practice, it may be impossible to obtain a reliable estimate of the MSY.
4. Achieving an MSY is by no means the only, nor necessarily the best, criterion by which success in the management of a harvesting operation should be judged. (It may, for example, be more important to provide stable long-term employment for the workforce.)

12.3.2 Obtaining MSYs through fixed quotas

There are two simple ways of obtaining an MSY on a regular basis: through a fixed quota and through a fixed effort. In *fixed quota MSY harvesting* (Figure 12.6), the same amount, the MSY, is removed from the population every year. If (and it is a big if) the population remained at exactly the peak of its net recruitment curve, then this system could work: each year the members of the population, through their own growth and reproduction, would add exactly what the harvesting removed. But if by chance numbers fell even slightly below those at which the curve peaked, then the numbers harvested would exceed those recruited. Population size would then decline to below the peak of the curve, and if a fixed quota at the MSY level were maintained the population would continue to decline until it was extinct (Figure 12.6). Furthermore, if the MSY were even slightly overestimated (and reliable estimates are hard to come by), then harvesting rate would always exceed the recruitment rate and extinction would again follow. In short, a fixed quota at the MSY level might be desirable and reasonable in a wholly predictable world about which we had perfect knowledge, but in the real world of fluctuating environments and imperfect data sets, these fixed quotas are open invitations to disaster (Clark, 1981).

Nevertheless, a fixed-quota strategy has frequently been used, where a management agency formulates an estimate of the MSY, which is then adopted as the annual quota. On a specified day in the year, the fishery is opened and the accumulated

the fragility of fixed quota harvesting

borne out in practice

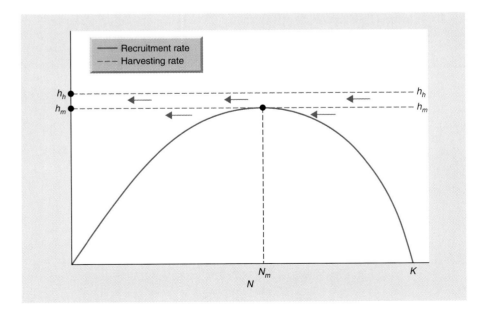

Figure 12.6

Fixed-quota harvesting. The figure shows a single recruitment curve (—) and two fixed-quota harvesting curves (- - -): high quota (h_h) and MSY quota (h_m). Arrows in the figure refer to changes to be expected in abundance under the influence of the harvesting rate to which the arrows are closest. •, = equilibria. At h_h the only "equilibrium" occurs when the population is driven to extinction. The MSY is obtained at h_m because it just touches the peak of the recruitment curve (at a density N_m): populations greater than N_m are reduced to N_m, but populations smaller than N_m are driven to extinction.

catch is logged. A fairly typical example is provided by the Peruvian anchoveta (*Engraulis ringens*) fishery (Figure 12.7). From 1960 to 1972 this was the world's single largest fishery, and it constituted a major sector of the Peruvian economy. Fisheries experts advised that the MSY was around 10 million metric tons annually, and catches were limited accordingly. But the fishing capacity of the fleet expanded, and in 1972 the catch crashed. Overfishing seems at least to have been a major cause of the collapse, although its effects were compounded by the influences of profound environmental fluctuations, discussed later. A moratorium on fishing might have allowed the stocks to recover, but this was not politically feasible: twenty thousand people were dependent on the anchoveta industry for employment. The Peruvian government has therefore allowed fishing to continue. The catches have never recovered.

12.3.3 Obtaining MSYs through fixed effort

the relative robustness of fixed-effort harvesting

An alternative to trying to maintain a constant harvest is to maintain a constant *harvesting effort* (e.g., the number of trawler days in a fishery or the number of gun days with a hunted population). With such a regime the amount harvested should increase with the size of the population being harvested (Figure 12.8). Now, in contrast to Figure 12.6, if density drops below the peak, new recruitment exceeds the amount harvested and the population recovers. The risk of extinction is much reduced. The disadvantages, however, are that, first, because there is a fixed effort, the yield varies with population size (there are good, but, more to the point, bad years); second, steps need to be taken to ensure that nobody makes a greater effort than he or she is supposed to. Nonetheless, there are many examples of harvests being managed by legislative regulation of effort. Harvesting of the important Pacific halibut, for example, is limited by seasonal closures and sanctuary zones, though heavy investment in fisheries protection vessels is needed to control lawbreakers.

12.3.4 Beyond MSY: environmental fluctuations and population structure

There is no doubt that fishing pressure often exerts a great strain on populations. But the collapse of fish stocks in one year rather than another is often the result of

Figure 12.7

Catch history of the Peruvian anchoveta fishery. (After Hilborn & Walters, 1992.)

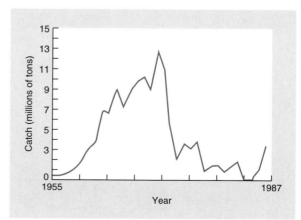

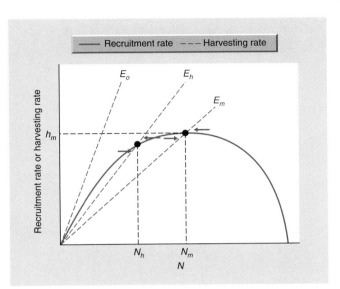

Figure 12.8
Fixed-effort harvesting. Curves, arrows, and dots as in Figure 12.6. The MSY is obtained with an effort of E_m, leading to a stable equilibrium at a density of N_m with a yield of h_m. At a somewhat higher effort (E_h), the equilibrium density and the yield are both lower than with E_m but the equilibrium is still stable. Only at a much higher effort (E_o) is the population driven to extinction.

the occurrence of unusually unfavorable environmental conditions, rather than simple overfishing.

Harvests of the Peruvian anchoveta (Figure 12.7) collapsed from 1972 to 1973, but a previous steady rise in catches had already dipped in the mid-1960s as a result of an *El Niño event*: this event occurs when warm tropical water from the north reduces the upwelling, and hence the productivity, of the nutrient-rich cold Peruvian current coming from the south. By 1973, however, commercial fishing had so greatly increased that the subsequent El Niño event had even more severe consequences. There were some signs of recovery from 1973 to 1982, but a further collapse occurred in 1983, associated with yet another El Niño event. It is unlikely that the El Niño events would have had such severe effects if the anchoveta had only been lightly fished. It is equally clear, though, that the history of the Peruvian anchoveta fishery cannot be explained simply in terms of overfishing.

So far, this account has ignored population structure of the exploited species. This is a serious fault for two reasons. First, most harvesting practices are primarily interested in only a portion of the harvested population (mature trees, fish that are large enough to be saleable, and so on). Second, "recruitment" is, in practice, a complex process incorporating adult survival, adult fecundity, juvenile survival, juvenile growth, and so on, each of which may respond in its own way to changes in density and harvesting strategy. An example of a model that takes some of these variables into account was that developed for the Arcto-Norwegian cod fishery, the most northerly fish stock in the Atlantic ocean. The number of fish in different age classes was known for the late 1960s and this information was used to predict the tonnage of fish likely to be caught with different intensities of harvesting and with different net mesh sizes. The model predicted that the long-term prospects for the fishery were best ensured with a low intensity of fishing (less than 30 percent) and a large mesh size. These gave the fish more opportunity to grow and reproduce before they were caught (Figure 12.9). The recommendations derived from the model were ignored, and, as predicted, the stocks of cod fell disastrously.

the anchoveta and El Niño

population structure and the Arctic cod

Figure 12.9
Predictions for the stock of Arctic cod under three intensities of fishing and three different sizes of mesh in nets. (After Pitcher & Hart, 1982.)

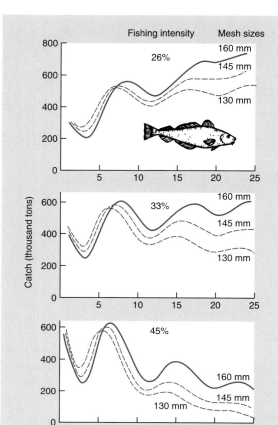

12.4 ▷ The Farming of Monocultures

Globally there is abundant food. Between 1961 and 1994 the per capita food supply in developing countries has increased by 32 percent (Figure 12.10) and the proportion of the world's population that is undernourished has fallen from 35 to 21 percent though this is very unevenly distributed (Figure 12.11). Yet 840 million people remain hungry worldwide, and the rate of increase in per capita food production is falling.

Fishing and hunting (Section 12.3) have been human activities since our early history as hunter–gatherers. But the harvest that can be taken from nature was totally inadequate to support the main phases of growth of human populations. Increasingly, both animals and plants were domesticated and managed in ways that allowed much greater rates of production. The great bulk of the human food resource is now farmed—produced usually as dense populations of single species (*monocultures*). This system allows them to be managed in specialized ways that can maximize their productivity—whether as immense monocultures of rice, wheat, or corn, or as livestock factories producing beef, pork, or poultry. Fish, indeed, are increasingly managed in the same way (*aquaculture*), reared in enclosures, fed with controlled

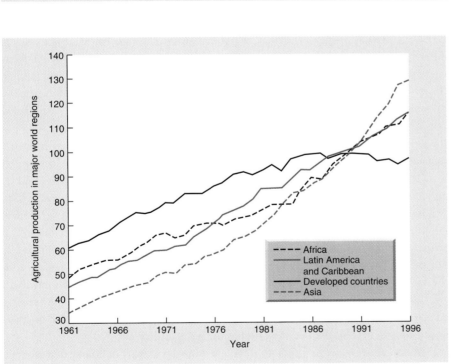

Figure 12.10
Changes in world food production and food produced per capita over the period 1961 to 1996. The values in 1989–1991 are adjusted to 100 to make comparisons easier. (From UN, 1998.)

Figure 12.11
Changes in agricultural food production in the major regions of the world over the period from 1961 to 1996. The values in 1989–1991 are adjusted to 100 to make comparisons easier. Developed countries including the former Soviet Union, Africa, Latin America and the Caribbean, and Asia. (From UN, 1998.)

diets, and harvested in mass production. Nearly a quarter of the fish supply in Asia is already produced in this way.

Only monoculture can maximize the rate of food production. This is because it allows the farmer to control and optimize with high precision the density of the

monoculture—and beyond

Agricultural monoculture: wheat as far as the eye can see.

population (livestock or crop plants), the quantity and quality of their resources (food supplied to livestock, and fertilizer and water to crops) and often even the physical conditions of temperature and humidity. With many animals, the monocultures extend to segregating livestock or poultry into narrow age bands or age classes. There need be none of the uneconomic mixing of calves with cows or chickens with hens; fish eggs and fry can be segregated from potentially cannibalistic adults; the grossly uneconomic equality of the sex ratio that is common in nature can be distorted by culling, to yield efficient all-female dairy herds of cattle, or all-hen populations in batteries for egg production. This is a far cry from the ecology of the primitive human hunter–gatherers, who subsisted on their gleanings from the tangled bank of wild nature!

but disease spreads in monocultures

To what extent, though, are modern farming methods sustainable? There is abundant evidence that a high price has to be paid to sustain the high rates of food production achieved by farmed monocultures. For example, they offer the ideal conditions for the epidemic spread of diseases such as mastitis, brucellosis, and swine fever among livestock and coccidiosis among poultry. Farmed animals are normally kept at densities far higher than their species would meet in nature with the result that transmission rates are magnified (see Chapter 8). In addition, high rates of transmission between herds occur as animals are sold from one farming enterprise to another, and it is easy for the farmers themselves, with mud on their boots and their vehicles, to act as vectors of pests and disease.

Monocultures of crop plants, too, can illustrate the fragility of human dependence on monocultures. The potato, for example, was not introduced across the Atlantic to Europe until the second half of the 16th century, but three centuries later other foods had given way to it, and it had become the almost exclusive food crop of the poorer half of the population of Ireland. Dense monoculture, though, provided ideal conditions for the devastating spread of late blight (the fungal pathogen *Phytophthora infestans*) when it also crossed the Atlantic in the 1840s. The disease spread rapidly, dramatically reducing potato yields and also decomposing the tubers in storage. Of the Irish population of about 8 million, 1.1 million died in the resulting famine and another 1.5 million emigrated to Britain and the United States.

In more modern history, an outbreak of southern corn leaf blight (caused again by a fungus, *Helminthosporium maydis*) developed in the southeastern United States in the late 1960s and spread rapidly after 1970. Most of the corn grown in the area had been derived from the same stock and was genetically almost uniform. This extreme monoculture allowed one specialized race of the pathogen to have devastating consequences. The damage was estimated as at least $1 billion in the United States and had repercussions on grain prices worldwide

12.4.1 The degradation and erosion of soil

A United Nations report (1998) stated, "Agricultural intensification in recent decades has taken a heavy toll on the environment. Poor cultivation and irrigation techniques and excessive use of pesticides and herbicides have led to widespread soil degradation and water contamination." About 300 million hectares is now severely degraded around the world and a further 1.2 billion hectares—10 percent of the Earth's vegetated surface—can be described as moderately degraded. Clearly much of agricultural practice has not been sustainable.

Land without soil can support only very small primitive plants such as lichens and mosses that can cling onto a rock surface. The rest of the world's terrestrial vegetation has to be rooted in soil that gives it physical support. The soil also serves as a store of essential mineral nutrients and water that are extracted by the roots during plant growth. Soil develops by the accumulation of finely divided mineral products of rock weathering and decomposing organic residues from previous vegetation. The characteristics of the soil under natural vegetation in any particular climatic region and on any particular rock type depend on the balance between these processes of accumulation and forces that degrade and remove the soil.

agriculture and forestry require soil

The formation and persistence of soil in a region depend on local natural checks and balances. Soil may be lost by being washed away by rain or blown away to sea or to be redeposited as an accumulation of fine textured loess somewhere else. Soil is best protected when it contains organic matter, is always wholly covered with vegetation, is finely interwoven with roots and rootlets, and is on horizontal ground. Natural soil systems are probably *always* too fragile to be fully sustained when land is brought into cultivation. Dramatic evidence of unsustainable land use was the Dust Bowl disaster in the Great Plains of the United States (Box 12.4).

soil forms—and is lost

In an ideal sustainable world, new soil would be formed as fast as the old was lost. In Britain about 0.2 metric ton of new soil is produced naturally per hectare per year, and it has been suggested that a tolerable (although not sustainable) rate of soil erosion might be about 2.0 metric tons per hectare per year. However, recorded rates of erosion have varied from 0.1 to 47.8 metric tons per hectare per year (Morgan, 1985)!

soil maintenance

Almost all (perhaps all) agricultural land will support higher yields if artificial fertilizers are applied to supplement the nitrogen, phosphorus, and potassium supplied naturally by the soil. Fertilizers are cheap, are easy to handle, have a guaranteed

BOX 12.4

Soil Erosion and the Dust Bowl

Large areas of southeastern Colorado, and southwestern Kansas and parts of Texas, Oklahoma, and northeastern New Mexico had been used to support rangeland management of livestock. The vegetation consisted largely of native perennial grasses and had been neither plowed nor sown with seed.

At the time of the First World War, much of the land was plowed and annual crops of wheat were grown. In the early 1930s there were poor crops due to severe drought and the top soil was exposed and carried away by the wind. Black blizzards of windblown soil blocked out the sun and piled the dirt in drifts. Occasionally the dust storms swept completely across the country to the East Coast. Thousands of families were forced to leave the region at the height of the Great Depression in the early and mid-1930s. The wind erosion was gradually halted with federal aid: windbreaks were planted and much

of the grassland was restored. By the early 1940s the area had largely recovered (Encyclopaedia Brittanica, 1994–1998).

Dust bowl field and abandoned farm.

composition, and allow even and accurate application and higher and more predictable yields. When there is an overwhelming reliance on them, however, maintaining the organic matter capital of the soil tends to be neglected, and this capital has declined worldwide.

The degradation of soil by agriculture can be prevented, or at least slowed, by (1) incorporating farmyard manure, crop residues, and animal wastes; (2) alternating years under cultivation with years of fallow; or (3) returning the land to grazed pasture or rangeland. Such practices conserve soil quality in technologically sophisticated agricultures in temperate regions.

contour plowing and terracing—Agenda 21

But soil degradation is most serious and less easily prevented in developing countries. The problems are greatest in high-rainfall areas and on steeply sloping ground in the tropics, where organic matter in the soil also decomposes more rapidly. The soil conservation strategy of Agenda 21, as formulated at UNCED (Rio de Janeiro, 1992), is now coordinated by the United Nations Commission on Sustainable Development and implemented through national and local authorities. Agenda 21 recommended measures to prevent soil erosion and promote erosion control in all sectors, in particular minimizing soil runoff and sedimentation. It is important to keep the ground covered by a crop or by a mulch of dead organic matter to protect it against the direct impact of raindrops. Trees can be used to provide this form of shelter and also to act as windbreaks.

The most cost-effective technology used in reducing soil erosion, however, is considered to be contour-based cultivation (Figure 12.12). In India, contour ditches

Figure 12.12
Terracing of hill and mountainous land.

have helped to quadruple the survival chances of tree seedlings and quintuple their early growth in height. Deeply rooted, hedge-forming vetiver grass, planted in contour strips across hill slopes, slows water runoff dramatically, reduces erosion, and increases the moisture available for crop growth. Currently 90 percent of soil conservation efforts in India are based on such biological systems. Simple technologies involving rock *bunds* (embankments) constructed along contour lines for soil and water conservation have succeeded. OXFAM has promoted this technique among farmers to improve water harvesting in Burkina Faso (in West Africa). Bunded fields yielded an average of 10 percent more than traditional fields in a normal year and, in drier years, almost 50 percent more (UN, 1998). Such terracing provides a very high quality of soil conservation but is possible only where labor is cheap. On less steep slopes, by plowing and cultivating in strips along the contours, runoff of soil can be significantly reduced.

Agricultural land is also highly susceptible to degradation in arid and semiarid regions. Both overgrazing and excessive cultivation expose the soil directly to erosion by the wind and to rare but fierce rainstorms. In the process of *desertization*, land that is arid or semiarid but has supported subsistence or nomadic agriculture gives way to desert. The process has often been slowed for a time by irrigating the land. This gives a temporary remission but lowers the water table, and salts accumulate in the topsoil (*salinization*). Once salts have started to accumulate, the process of salinization tends to spread and leads to an expansion of sterile white salt deserts. This has been a particular hazard in irrigated areas of Pakistan.

desertization and salinization

Forests protect soil from erosion because the canopy absorbs the direct impact of the rain on the soil surface, the perennial root systems bind the soil, and leaf fall continually adds organic matter. But when forests are clear-felled and then replanted, there is an open window of opportunity for soil erosion until the forest canopy closes again. Cultivation and replanting along contours give some control over soil erosion during this danger period, but the best precaution is to avoid clear felling

forests protect, except when harvested by clear felling

and extract only a proportion of a forest stand at each harvest. This can often be technically difficult and expensive.

12.4.2 The sustainability of water as a resource

In the 1960s and 1970s, the main worry about the sustainability of global resources concerned energy supplies that were recognized to be finite and exhaustible. Although energy resources remain finite, concern has shifted because exploration has revealed much larger reserves of oil, gas, and even coal than had been entered into earlier environmental balance sheets. Water has now come into sharper focus. Fresh water, which is used in crop irrigation and domestic consumption, is of crucial importance (Figure 12.13). On a global scale, agriculture is the largest consumer of fresh water, taking about 70 percent of available supplies and more than 90 percent in parts of Africa, South America, and central Asia.

Figure 12.13
(a) Changes in global withdrawal of water for agriculture, industry, municipal (largely domestic), and other uses during the period from 1990 and projected to the year 2025. (b) The contrast between the extraction of water for agriculture, industry, and domestic use in developing and developed countries. (From UN, 1998.)

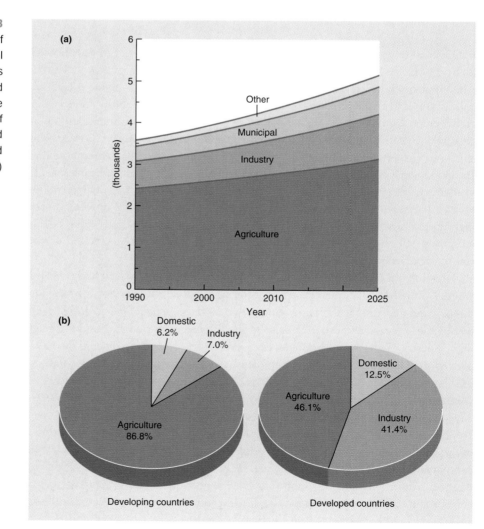

The soils around the Nile can support plant growth where they are irrigated, but where irrigation stops (note the background here) there is desert.

There is a fixed stock of water on the globe and it is continually recycled as it evaporates from vegetation, land, and sea and is then condensed and redistributed as precipitation. The human species now uses, directly or indirectly, more than half of the world's accessible water supply. The fresh water available per capita worldwide fell from 17,000 cubic meters in 1950 to 7300 cubic meters in 1995. Many assessments of the problems of water supply suggest that countries with less than 1000 cubic meters per person per year experience chronic scarcity. Water is widely thought to be the resource that future wars will be fought over. Even at a national level, the allocation of water resources can cause political problems, as occur, for example, in conflicts in California between urban and agricultural demands for water from the Colorado River. At the international level, conflict arises between countries that are upstream of their neighbors and are in a position to dam and divert water supplies. There are bitter cross-border disputes in South America, Africa, and the Middle East between nations that share river basins.

water—the resource that future wars will be fought over?

One response to chronic scarcity of water is to pump it from underground aquifers—but this often happens faster than the aquifers can be recharged. Such activity is clearly profligate and unsustainable. It is also the main cause of the loss of land from agriculture due to salinization. The demand for accessible supplies of water for both agriculture and domestic use has led to the plumbing of the Earth's river systems on a vast scale. The number of river dams more than 15 meters high has increased from about 5000 in 1950 to 38,000 in the 1990s.

In Chapter 13, we discuss the pollution of water by excreta, and by the pesticides and fertilizers applied in agriculture. Water that is uncontaminated by disease, nitrates, or pesticides is an especially valuable commodity, but contamination is easy and removing contaminants (e.g., nitrates) is very expensive. Major dams built to control and conserve water in North and West Africa create large bodies of open water in which contamination spreads easily. One consequence is that schistosomiasis (a flatworm disease of humans) has spread rapidly along rivers and infection rates have risen from 1–10 percent to 98–100 percent.

contamination and conservation

Maintaining water supplies for human use also creates problems for the conservation of wildlife (Chapter 14). The water flow in many of the world's larger rivers is now very heavily controlled—in some cases little water now reaches the sea and wetlands have been lost or are at risk. Moreover, silt accumulates upriver instead of spreading into deltas and flood plains. The results may be catastrophic for wildlife areas and for human communities as well. For example, there is reason to believe that a decline in silt deposition in the Nile delta (together with rising sea levels) may cause Egypt to lose up to 19 percent of its habitable land and displace 16 percent of its population within 60 years.

12.5 ▷ Pest Control

what is a pest?

Pest control is another area in which the sustainability of agricultural practice may be threatened. A pest species is one that humans consider undesirable. Estimates suggest that there are around 67,000 species of pests that attack agricultural crops worldwide: 8000 weeds that compete with crops, and 9000 insects and mites and 50,000 plant pathogens that prey on them (Pimentel, 1993). Here we consider the sustainability of insect pest control in agriculture to illustrate the types of problems that arise in managed monocultures. We could equally well have chosen the control of weeds, mollusks, eelworms, or the pests and diseases of farmed livestock, poultry, or fish.

12.5.1 Aims of pest control: economic injury levels and action thresholds

EILs for pests, nonpests, and potential pests

Economics and sustainabilty are intimately tied together. Market forces ensure that uneconomic practices are not sustainable. One might imagine that the aim of pest control is total eradication of the pest, but this is not the general rule. Rather, the aim is to reduce the pest population to a level at which it does not pay to achieve yet more control (the *economic injury level* [EIL]). The EIL for a hypothetical pest is illustrated in Figure 12.14a. It is greater than zero (eradication is not profitable) but it is also below the typical, average abundance of the species—it is precisely this that makes the species a pest. If the species were naturally self-limited to a density below the EIL, then it would never make economic sense to apply control measures, and the species could not be a pest (Figure 12.14b). There are other species, though, that have a carrying capacity (Chapter 5) in excess of their EIL but have a typical abundance that is kept below their EIL by their natural enemies (Figure 12.14c). These are potential pests. They can become actual pests if their enemies are removed.

When a pest population has reached a density at which it is causing economic injury, however, it is generally too late to start controlling it. More important, then, is the *control action threshold* (CAT): the density of the pest at which action should be taken to prevent it from reaching the EIL. CATs are predictions based on detailed studies of past outbreaks or sometimes on correlations with climatic records. They may take into account the numbers not only of the pest itself but also of its natural enemies. As an example, in order to control the spotted alfalfa aphid (*Therioaphis*

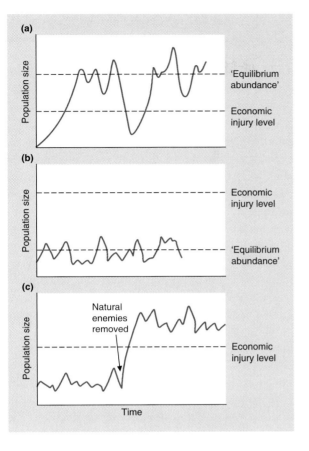

Figure 12.14
(a) The population fluctuations of a hypothetical pest. Abundance fluctuates around an "equilibrium abundance" set by the pest's interactions with its food, predators, and so on. It makes economic sense to control the pest when its abundance exceeds the economic injury level (EIL). Because it is a pest, its abundance exceeds the EIL most of the time (assuming it is not being controlled). (b) By contrast, a species that cannot be a pest fluctuates always below its EIL. (c) Potential pests fluctuate normally below their EIL but rise above it in the absence of one or more of their natural enemies.

trifolii) on hay alfalfa in California, control measures have to be taken at the times and under the circumstances specified:

1. In spring when the aphid population reaches 40 aphids per stem.
2. In summer and autumn when the population reaches 20 aphids per stem, but the first three cuttings of hay are not treated if the ratio of ladybugs (beetle predators of the aphids) to aphids is one adult per 5–10 aphids or three larvae per 40 aphids on standing hay or one larva per 50 aphids on stubble.
3. During winter when there are 50–70 aphids per stem (Flint & van den Bosch, 1981).

12.5.2 The problems with chemical pesticides—and their virtues

Pesticides gain a bad name if, as is usually the case, they kill more species than the one at which they are aimed. They may then become pollutants (Chapter 13). However, in the context of the sustainability of agriculture, pesticides especially justify their bad name if they kill the pests' natural enemies and so contribute to undoing what they were employed to do. Thus, the numbers of a pest sometimes increase rapidly some time after application of pesticide. This is known as *target pest resurgence* and occurs when treatment kills both large numbers of the pest *and*

target pest resurgence and secondary pest outbreaks

large numbers of their natural enemies. Pest individuals that survive the pesticide or that migrate into the area later find themselves with a plentiful food resource but few if any natural enemies. The pest population may then explode.

The after-effects of applying a pesticide may involve even more subtle reactions. When a pesticide is applied, it may not be only the target pest that resurges. Alongside the target are likely to be a number of potential pest species, which had been kept in check by their natural enemies (Figure 12.14). If the pesticide destroys these, the potential pests become real ones: *secondary pests*. A dramatic example are the insect pests of cotton in Central America. In 1950, when mass dissemination of organic insecticides began, there were two primary pests: the boll weevil and the Alabama leafworm. Organochlorine and organophosphate insecticides were applied fewer than five times a year and initially had apparently miraculous results—yields soared. By 1955, however, three secondary pests had emerged: the cotton bollworm, the cotton aphid, and the false pink bollworm. The pesticide applications rose to 8 to 10 per year. This reduced the problem of the aphid and the false pink bollworm, but led to the emergence of five further secondary pests. By the 1960s, the original two pest species had become eight and there were, on average, 28 applications of insecticide per year (Flint & van den Bosch, 1981). Clearly, such a rate of pesticide application is not sustainable.

evolved resistance

Chemical pesticides lose their role in sustainable agriculture if the pests evolve resistance. The evolution of pesticide resistance is simply natural selection in action (see Chapter 2). It is almost certain to occur when vast numbers of a genetically variable population are killed. One or a few individuals may be unusually resistant (perhaps because they possess an enzyme that can detoxify the pesticide). If the pesticide is applied repeatedly, each successive generation of the pest will contain a larger proportion of resistant individuals. Pests typically have a high intrinsic rate of reproduction, and so a few individuals in one generation may give rise to hundreds or thousands in the next, and resistance spreads very rapidly in a population.

This problem was often ignored in the past, even though the first case of DDT resistance was reported as early as 1946 (houseflies in Sweden). Now, the scale of the problem is illustrated in Figure 12.15, which shows the exponential increases in the number of insect species resistant to insecticides. The housefly has developed resistance worldwide to virtually every chemical that has been employed against it. The evolution of pesticide resistance can be slowed, though, by changing from one pesticide to another, in a repeated sequence that is rapid enough for resistance not to have time to emerge (Roush & McKenzie, 1987).

pesticides work

If chemical pesticides produced nothing but problems, however—if their use was intrinsically and acutely unsustainable—then they would already have fallen out of widespread use. This has not happened. Instead, their rate of production has increased rapidly (Figure 12.16). The ratio of cost to benefit for the individual producer has remained in favor of pesticide use: they do what is asked of them. In the United States, insecticides are estimated to benefit the agricultural producer to the tune of around 5 dollars for every 1 dollar spent (Pimentel et al., 1978).

Moreover, in many poorer countries, the prospect of imminent mass starvation or of an epidemic disease are so frightening that the social and health costs of using pesticides have to be ignored. In general the use of pesticides is justified by objective

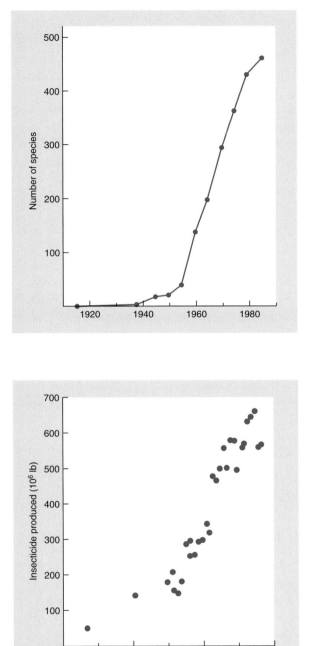

Figure 12.15
The increase in the number of insect species reported to be resistant to at least one insecticide. (After Metcalf, 1980; Miller, 1988.)

Figure 12.16
Various estimates for the quantity of insecticide produced in the United States, showing the rapid increase over time. (After Metcalf, 1980.)

measures such as lives saved, economic efficiency of food production, and total food produced. In these very fundamental senses, their use may be described as sustainable. In practice, sustainability depends on continually developing new pesticides that keep at least one step ahead of the pests, that are less persistent, that are biodegradable, and that are more accurately targeted at the pests.

12.5.3 Biological control

Outbreaks of pests occur repeatedly and so does the need to apply pesticides. But biologists have other tools that can sometimes replace chemicals, that do the same job and cost a great deal less. These involve the manipulation of the natural enemies of pests (biological control) and the controlled evolution of host resistance.

Most crops have been introduced to the areas in which they are grown and some of their pests have accompanied them or arrived later. However, the pest's own natural enemies have often been left behind. The pest populations have then exploded with disastrous consequences. A search of the area where the crop originated might reveal predators or diseases of the pest that can be introduced to control it. There is then a hope that the introduced enemy will persist and keep the pest population below an economically damaging threshold indefinitely.

three types of biological control

There are three main types of biological control. The natural enemy of a pest may be *imported* from another geographical area—very often the area in which the pest originated. Probably the best example of this has become a classic: the control of the cottony cushion scale (Box 12.5). Sometimes the natural enemy is unable to persist throughout the year but is repeatedly reintroduced (*inoculated*). Finally, a crop may be deliberately *inundated* with large numbers of a natural enemy of the pest, for example, a bacterium or virus, that does not persist, but acts, like a chemical pesticide, to kill those pests that are present at the time.

control by inoculation

Inoculation is widely practiced in the biological control of pests in greenhouses, as crops are removed, along with the pests and their natural enemies, at the end of the growing season. The two species of natural enemy that are used most widely

---

Box 12.5

The Cottony Cushion Scale: A Classic Case of Importation

The cottony cushion scale insect, *Icerya purchasi*, was first discovered as a pest of Californian citrus orchards in 1868. By 1886 it had brought the citrus industry close to the point of destruction. Would-be pest controllers initiated a worldwide correspondence to try and discover the natural home and natural enemies of the scale. In 1887 they received a report from Adelaide, Australia, of apparently natural populations of the scale along with a dipteran parasitoid, *Cryptochaetum* sp. Thus, in 1888, an expert on insect taxonomy, Koebele, was dispatched to collect parasitoids for importation into California. He made a widespread tour of Australia but both scale and parasitoid were very hard to find. Eventually he was successful—but he also found a ladybug beetle, the vedalia *Rodolia cardinalis*, feeding on the scale insect. Then, on his way home, he visited New Zealand and found the

vedalia doing very well on the scale insects. He sent approximately 12,000 *Cryptochaetum* sp. and 500 vedalia back to California.

Initially, the parasitoids seemed simply to have disappeared, but the predatory beetles underwent such a population explosion that all infestations of the scale insects in California were controlled by the end of 1890. In fact, although the beetles have usually taken most or all of the credit, the long-term outcome has been that the vedalia has been largely instrumental in keeping the scale in check inland, but *Cryptochaetum* sp. is the main agent of control on the coast (Flint & van den Bosch, 1981).

Biological control of the cottony cushion scale has subsequently been transferred to 50 other countries and savings have been incalculable: the original outlay, including salaries, was no more than $5000.

in this way are *Phytoseiulus persimilis*, a mite that preys on a spider mite pest of cucumbers and other vegetables, and *Encarsia formosa*, a parasitoid wasp that attacks a whitefly pest of tomatoes and cucumbers. Already by 1985, around 500 million individuals of each species were being produced for release each year in Western Europe.

Insects have been the main agents of biological control against both insect pests and weeds. Table 12.1 summarizes the extent to which they have been used and the proportion of cases in which the establishment of an agent has greatly reduced or eliminated the need for other control measures.

biological control: excellent—when it works

12.6 Integrated Farming Systems

The desire for sustainable agriculture has increasingly led to more ecologically oriented approaches to food production, which are often given the label "integrated farming systems." Part of this, and something that preceded it historically, is a similar approach to pest control: integrated pest management (IPM).

IPM is a practical philosophy of pest management. It combines physical control (for example, simply keeping pests away from crops), cultural control (for example, rotating the crop planted in a field so pests cannot build up their numbers over several years), biological and chemical control, and the use of resistant varieties. It has come of age as part of the reaction against the unthinking use of chemical pesticides in the 1940s and 1950s.

integrated pest management

IPM is ecologically based but uses all methods of control—including chemicals— where appropriate. It relies heavily on natural mortality factors, such as enemies and weather, and seeks to disrupt them as little as possible. It aims to control pests below an economically damaging level (EIL), and it depends on monitoring the abundance of pests and their natural enemies and using various control methods as complementary parts of an overall program. IPM therefore calls for specialist pest managers or advisers. Broad-spectrum pesticides in particular, although not excluded,

Table 12.1

The record of insects as biological control agents against insect pests and weeds

	INSECT PESTS	WEEDS
Control agent species	563	126
Pest species	292	70
Countries	168	55
Cases where agent has become established	1063	367
Substantial successes	421	113
Successes as a percentage of establishments	40	31
(After Waage & Greathead, 1988.)		

are used only very sparingly, and if chemicals are used at all it is in ways that minimize the costs and the quantities used. The essence of the IPM approach is to make the control measures fit the pest problem, and no two pest problems are the same—even in adjacent fields.

IPM of cotton pests in California

One example of IPM concerns crops of cotton in the San Joaquin Valley, California. The crops had been afflicted by target pest resurgence (Figure 12.17a), secondary pest outbreaks (Figure 12.17b and c), and the evolution of resistance to chemical insecticides (Figure 12.17d). The key pest was the lygus bug *Lygus hesperus*, which feeds primarily on fruiting cotton buds. Another major pest was the cotton bollworm (*Heliothis zea*). However, in the San Joaquin Valley this seems largely to have been a secondary pest induced by insecticides, as were two other lepidopterous pests, the beet army-worm (*Spodoptera exigua*) and the cabbage looper (*Trichoplusia ni*). An IPM program aimed to reduce insecticide usage to prevent secondary outbreaks, and also to avoid certain commonly used insecticides that decrease the yield of cotton plants.

Prior to the development of the program, a control action threshold (CAT) had been established for the lygus bug—but from little information. Hence, the threshold remained the same throughout the year—a serious error. Further research showed that the lygus bug could inflict serious damage only during the budding season (broadly, early June to mid-July). Thus, in the IPM program the threshold was set at 10 bugs per 50 sweep-net sweeps on two consecutive sampling dates between June 1 and July 15. After that time, insecticide application for lygus bug control was avoided. In fact, the threshold was subsequently refined to one based on a bug/bud ratio, which was more flexible and more accurate.

Another aspect of control within the system centered on manipulation of alfalfa, a favored host plant of the lygus bug. In particular, the interplanting of narrow alfalfa strips (5–10 m) in the cotton fields proved effective in attracting the bugs out of the cotton. In addition, attempts were made to augment the populations of natural enemies by providing them with supplementary nutrition.

The key to a successful IPM program is a good field monitoring system. Here, each field was sampled twice a week from the beginning of budding (roughly mid-May) to the end of August. Plant development, pest data, and natural enemy population data were all collected. Additional sampling for the cotton bollworm began around August 1 and continued until mid-September. Overall, the program represents a harmonious integration of naturally occurring (but augmented) biological controls and carefully timed insecticide treatments. The success of the program for cotton and also for citrus in the San Joaquin Valley is illustrated in the Table 12.2.

integrated farming systems—LISA, IFS, and LIFE

It has increasingly become apparent, in an agricultural context at least, that implicit in the philosophy of IPM is the idea that pest control cannot be isolated from other aspects of food production and is especially bound up with the means by which soil fertility is maintained and improved. Thus, a number of programs have been initiated to develop and put into practice sustainable food production methods that incorporate IPM, including not only integrated farming systems (IFSs) but also low input sustainable agriculture (LISA) in the United States and lower input farming and environment (LIFE) in Europe (International Organisation for

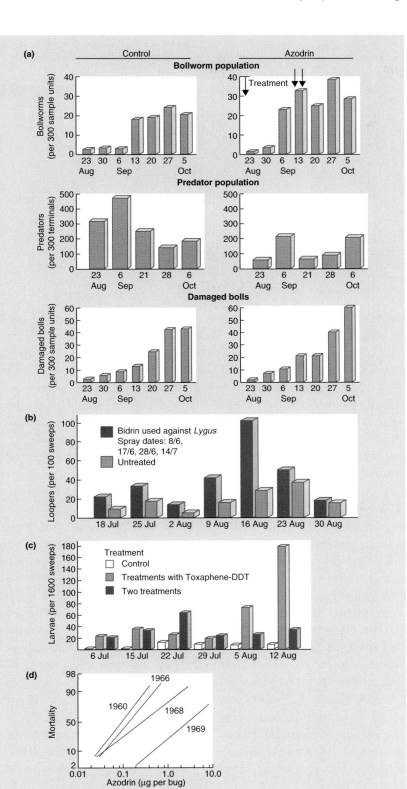

Figure 12.17

Pesticide problems among cotton pests in San Joaquin Valley, California. (a) Target pest resurgence: bollworms resurged because the abundance of their natural predators was reduced—the number of damaged bolls was higher. (b) An increase in cabbage loopers and (c) in beet army worm were seen when plots were sprayed against lygus bugs; both are examples of secondary pest outbreaks. (d) Increasing resistance of lygus bugs to Azodrin. (After van den Bosch et al., 1971.)

Table 12.2

The success of integrated pest management (IPM) in San Joaquin Valley, California 1970–71. With IPM the outlay on insecticides was cut drastically yet the financial yield was not reduced. (Figures in brackets are standard deviations.)

	IPM USERS	NON-USERS
Average insecticide cost per acre of cotton (US $)	4.9 (±3.9)	12.0 (±7.4)
Average yield per acre of cotton (US $)	270.2 (±27.5)	247.8 (±55.1)
Average insecticide cost per acre of citrus (US $)	20.5 (±15.0)	42.4 (±18.3)
Average yield per acre of citrus (US $)	515.8 (±260.6)	502.9 (±157.0)
(After Hall et al., 1975; Flint & van den Bosch, 1981.)		

Biological Control, 1989; National Research Council, 1990). All share a commitment to the development of sustainable agricultural systems.

These approaches have advantages in terms of reduced environmental hazard. Even so, it is unreasonable to suppose that they will be adopted widely unless they are also sound in economic terms. As we have already noted, in an industry such as agriculture, practices that are economically unsustainable are, ultimately, unsustainable overall. Mounting evidence, though, suggests that the ecologically sustainable practices described here are frequently economically sound too. For example, Table 12.2 gives details of the San Joaquin Valley cotton and citrus industries concerning (1) expenditure on insecticides and (2) economic yields for users and nonusers of IPM consultancies. IPM makes economic as well as environmental good sense—partly, but not entirely, as a result of the reduction in pesticide costs.

environmental and economic sustainability

More broadly, a number of studies in Europe and the United States support the economic claims of the "sustainable" approach. For example, in Germany the Lautenbach project has, since 1978, compared the performance of a sustainable farm with that of a matched, adjoining conventional farm, both growing cereals, sugar beet, and legumes (El Titi, 1989). The sustainable farm used significantly less mineral fertilizer and pesticide (36 percent less) and suffered significantly less soil erosion. Thus, it fared substantially better environmentally. In addition, however, it produced greater yields of oats, sugar beet, and beans (although less wheat); achieved an increased net profit on winter wheat, sugar beet, and beans (although not on spring wheat and oats); and managed to produce an increased net profit of 4 percent overall. Hence, the environmentally sustainable farm fared better than the conventional farm in narrow economic terms.

but many challenges for the future

The ecological advantages of IPM and sustainable agriculture generally are clear, then, and the potential economic advantages are apparent. Yet, certain disadvantages are apparent too: (1) the difficulty of training the necessary army of advisers, (2) the difficulty of developing a sufficiently detailed ecological understanding of the agricultural ecosystems, and (3) the low initial rate of return on investment during training and ecological investigations. When we add to this the commercial pressures exerted on producers to use simpler, chemical methods of control, and the more immediate benefits that this approach brings, it is easy to see why, in spite of widespread acceptance among ecologists that fully sustainable agricultural systems are the ideal objective, actual examples of fully implemented programs have accumulated only slowly. On the other hand, as we move into the new millennium, the number of examples seems certain to accelerate (e.g., Corey et al., 1993). This is to be welcomed. It is worth recalling that in the United States, before 1945 and the extensive use of synthetic pesticides, crop losses to insects averaged around 7 percent. By 1991, after the use of insecticides in agriculture had increased 10-fold, crop losses to insects averaged 13 percent (Pimentel, 1993).

❯ Summary

The human population problem

Resource use by humans is defined as *sustainable* if it can be continued for the forseeable future. The root of most environmental problems is the "population problem": a large human population that has been growing at a more-than-exponential rate.

Three groups of nations can be recognized: those that passed through the demographic transition early, late or not yet. Even if it were possible instantaneously to bring about the transition in all remaining countries of the world, the population problem would not be solved, partly because population growth has its own momentum.

The global carrying capacity is variously estimated at between 1 billion and 1000 billion, depending mainly on what is deemed to constitute an acceptable standard of living.

Harvesting living resources from the wild

Whenever a natural population is exploited by harvesting there is a risk of overexploitation. But harvesters also want to avoid underexploitation: potential consumers are deprived and those who do the harvesting are underemployed.

The concept of the maximum sustainable yield (MSY) has been a guiding principle in the exploitation of natural populations. There are two simple ways of obtaining an MSY on a regular basis: through a fixed quota and through a fixed effort. Limitations of the MSY approach are (1) that it treats populations as a number of similar individuals, and (2) that it treats the environment as unvarying. Improved harvesting strategies take both these factors into account.

The farming of monocultures

Increasingly, animals and plants have been domesticated and managed in ways that allowed much greater harvests— usually as monocultures. But a high price may be paid to sustain these high rates of food production. Monocultures offer ideal conditions for the epidemic spread of diseases and lead to widespread degradation of land.

The sustainability of soil and of water supplies

In an ideal sustainable world, new soil would be formed as fast as the old was lost, but in most agricultural systems this is not achieved. When there is an overwhelming reliance on artificial fertilizers, maintaining the organic matter capital of the soil tends to be neglected and this capital has declined worldwide.

Soil degradation can be slowed by incorporating manures and residues, alternating cultivation and fallow, or returning the land to grazed pasture. In tropical regions, terracing is widely practiced over hilly and mountainous terrain. In arid regions, overgrazing and excessive cultivation can lead to desertization and salinization.

Water is widely thought to be the resource that will cause future wars. On a global scale, agriculture is the largest consumer of fresh water. Pumping water from underground aquifers is the main cause of loss of agricultural land through salinization.

Pest control

The aim of pest control is to reduce the pest population to its economic injury level (EIL), but control action thresholds (CATs) may be of more immediate importance.

Pesticides may kill species other than their target and may give rise to target pest resurgence or secondary pest outbreaks. Pests may also evolve pesticide resistance.

Biologists may also manipulate the natural enemies of pests (biological control) through importation, inoculation, or inundation.

Integrated farming systems

Integrated pest management (IPM) is a practical philosophy of pest management that is ecologically based but uses all methods of control where appropriate. It relies heavily on natural mortality factors and calls for specialist pest managers or advisers.

Implicit in the philosophy of IPM is the idea that pest control cannot be isolated from other aspects of food production. A number of programs have been initiated to develop and put into practice sustainable food production methods that incorporate IPM. Evidence has been accumulating that this sustainable farming approach can yield improved economic returns too.

Review Questions

▲ = Challenge Question

1. ▲ What is sustainability? Is it possible to have sustainable population growth? Sustainable use of fossil fuels? Sustainable use of forest trees? Justify your answers.

2. Describe what is meant by the *demographic transition* in a human population. Explain why it might be important, for future management of human population growth, to discover whether the demographic transition is an academic ideal or a process through which all human populations necessarily pass.

3. ▲ The number of people that Earth can support depends on their standard of living. Argue the case either for or against developing nations having the right to expect standards of living those in the developed world take for granted.

4. Contrast the ways in which fixed-quota and fixed-effort harvesting strategies seek to extract maximum sustainable yields from natural populations.

5. Discuss the pros and cons of agricultural monocultures.

6. One of the main bodies regulating the production of organic food (food produced without synthetic fertilizers or pesticides) in the United Kingdom is the Soil Association. Explain why you think it has adopted this name.

7. Explain the meaning and importance of the terms *economic injury level* and *control action threshold*.

8. Weigh up the advantages and disadvantages of the chemical and biological control of pests.

9. Explain why methods of pest control and methods of soil fertility maintenance need to be considered together in integrated farming systems.

10. ▲ Hilborn and Walters (1992) have suggested that there are three attitudes that ecologists can take when they enter the public arena. The first is to claim that ecological interactions are too complex, and our understanding and our data too poor, to warrant making definite pronouncements (for fear of being wrong). The second possibility is for ecologists to concentrate exclusively on ecology and arrive at a recommendation designed to satisfy purely ecological criteria. The third is for ecologists to make ecological recommendations that are as accurate and realistic as possible, but to accept that these will be incorporated with a broader range of factors when management decisions are made—and may be rejected. Which of these courses do you favor, and why?

As populations grow, organisms become more crowded and both deplete their environment of the essentials for life and contaminate it with products that hinder their activities. The human species is unique in using fire (and recently nuclear fission) to provide the energy for doing work. This has given the species remarkable power to change the environment. When we don't like what has changed in our environment we call it pollution.

Pollution

Key Concepts

In this chapter you will

- recognize that *Homo sapiens* is just one species among many whose activities can reduce the quality of their environment.

- understand that social and urban living exaggerate the problems of waste disposal and that corpses, feces, and urine can be serious pollutants.

- appreciate that human use of fire and fuel makes our species a unique polluter and that there is a multitude of consequences in atmospheric pollution.

- understand that nuclear radiation caused by human activities adds to that already present.

- realize that quarrying and mining, smelting, and disposal of metals are seriously polluting.

- appreciate that most terrestrial pollutants ultimately pollute rivers, lakes, oceans, or the atmosphere.

- understand that applied ecology has a role in the restoration of polluted environments.

13.1 ▷ Introduction

The verb *pollute* derives from the Latin *polluere,* meaning "to soil" and has come to have meanings as diverse as "to render ceremonially or morally impure" (see Box 13.1) or, more appropriately in the present context, "to make physically impure, foul, filthy." Environmental pollution means, to most people, the contamination of the environment by human waste, especially sewage, and by the unwanted products of human activities, including gases from power plants, leaks of radioactive material, car exhaust fumes, and runoff from agricultural land of pesticides and fertilizers.

The human species is not unique in the way that individuals deplete their environment of the essentials for life and soil their environment with products that can hinder their activities. However, we are certainly unique in using fire, fossil fuels, and nuclear fission to provide the energy for doing work. This power generation has extremely far reaching consequences for the state of the land, aquatic ecosystems, and the atmosphere. Moreover, the generated energy has given *Homo sapiens* remarkable power to transform landscapes through activities such as urbanization, industrial agriculture, and forestry; to mine, smelt, and use metals (causing further pollution); to create new chemical diversity with attendant dangers; to add to the hazard of natural radiation; and to alter the atmosphere and even the nature of the climate. When we don't like what we have changed in our environment we call it pollution.

13.2 ▷ Urban Pollution

As human populations grew and became increasingly concentrated in cities, two sources of pollution became particularly important: feces and corpses. Plagues killed large proportions of local populations in a very short time, and disposal of the bodies caused appalling problems (as it still does after major disasters, such as genocide in Rwanda). A series of plagues in England included the Black Death of 1348–1349 when a third to a half of the population died in less than 2 years and ended in 1665 with the Great Plague of London, when 70,000 individuals (about a sixth of the population) died in 1 year.

Black Death and the Great Plague

Plague, caused by the bacterium *Yersinia pestis,* was spread by bites from fleas and rats and by direct human contact, especially with corpses. It was a good example of the simple rule that nothing pollutes a human environment so much as a lot more humans.

13.2.1 Burial and cremation

Both Romans and Jews had recognized that cemeteries were sources of pollution and disease and they placed their graveyards outside the city walls of Jerusalem and Rome. But in other areas bodies were buried within cities. In the Middle Ages in London coffins were placed tier upon tier until sometimes they were within a few centimeters of the surface; to make room for more burials, the gravediggers moved bones and partly decomposed remains to nearby pits. In Western societies the safe

BOX 13.1

Moral Codes

Historically almost all the functions of the human body have been considered by human societies to be pollution. These include urine, feces, saliva, sweat, blood, and processes such as menstruation, sexual intercourse, birth, illness, and death. Most societies have built moral codes into their religions in which one or more of these types of pollution becomes a type of sin and purity a virtue.

The Jews were pioneers in developing moral codes of personal hygiene. This is illustrated in three chapters of Leviticus that relate the Lord's extraordinarily detailed instructions to Moses for the hygienic management and quarantine of one of the worst pollutants of the time—patients with leprosy. Leprosy (caused by *Mycobacterium leprae*) had been a major cause of plague in the early history of Middle Eastern countries and the chapters set out the procedures for quarantine of infected individuals (pollutants), and for washing and cleaning their

clothes and scraping and replastering their houses to free them from the plague. This is all presented in a purely religious context, together with the instructions on how most properly to make religious sacrifice of two birds in order to cure the patient.

The detailed dietary instructions that often form part of religious laws and practices can also be seen as warnings and protection against the spread of disease. The prohibition against eating species of animal that associate with feces and garbage gives some protection against contamination and infection by intestinal parasites. The absolute Jewish and Islamic ban on pig meat as a food (or as a pollutant in other foods) has obvious hygienic significance in protecting against the tapeworm (*Taenia* sp.), which is easily transmitted in undercooked pork (see Section 7.2.2).

and hygienic disposal of corpses in religious rites has increasingly become a civic responsibility.

Cremation has become increasingly acceptable in Western societies, partly in reaction to the massive loss of countryside that is converted into cemeteries—itself a dramatic example of environmental pollution! In Hindu communities corpses are disposed of by burning on an open fire. At first sight this appears to be much less damaging to the environment than burial, but enormous volumes of wood are used in funeral pyres. More than 50 percent of human corpses (and those of many pets) are now cremated in England and nearly 100 percent in Japan, but in the United States the figure is only about 10 percent. Modern cremation contributes its small share to environmental pollution in the form of carbon dioxide (and mercury from tooth fillings!).

cremation

Perhaps the least polluting methods for disposing of corpses are burial at sea and direct exposure on land, as practiced by Parsees. The naked corpses are exposed in special isolated sites for vultures to strip off the flesh. This takes only 1 or 2 hours and the bared bones then dry in the sun and are swept away.

the use of scavengers to deal with corpses

Humans are not unique in regarding corpses as pollutants that must be removed from the community—the process has even been given a name, *necrophoresis*. Worker ants recognize corpses and remove them from the nest. If material from a dead ant is sprayed onto a live one, workers will quickly carry it out of the nest, as if it were a corpse.

disposal of corpses in nonhuman species

13.2.2 Feces, urine, and other urban waste

and the problems of feces in nonhuman species

Feces and urine are unlikely to cause problems in natural communities of nonsocial species—indeed both feces and urine are sometimes put to use to mark territories. However, they may act as pollutants in the environment of social animals. For example, cattle avoid grass near their dung for more than a month, apparently in response to certain volatile chemicals, and sheep choose to eat poor quality herbage rather than herbage of high quality that has been polluted by feces (Hutchings et al., 1999). Most mammals that live as social families in burrows avoid soiling them and deposit urine and feces outside, even in special latrines. Birds avoid fouling their nests and may carry away nestlings' feces in special fecal sacs.

All human body products, but most notably feces and urine, can be regarded as pollutants. The Greeks were probably the first to control the accumulation of pollution within towns, and a law of 320 B.C. forbade dumping of waste in the streets. The Romans were also very pollution conscious, dumping city waste in pits outside the city walls. When Roman and Greek civilizations perished, their quite sophisticated control of urban pollution collapsed. Medieval castles, for example, were often designed with latrines projecting from castle walls that simply dumped waste at the base of the walls (the accumulated wastes give archaeologists a direct record of historic diets and infestation with intestinal worms!). Until the 14th and 15th centuries the open streets again became the main, and often the only, destination for human and animal feces and urine, from which they were washed during rainstorms into rivers, lakes, and oceans. Horses, used for transport, and farm animals, brought into cities and kept alive for milk or until they were slaughtered for meat, contributed their share of feces to the burden of pollution. A special trade developed, that of the scavenger, who was paid to carry waste to dumps outside the cities; in 1714 every city in England had an official scavenger (the forerunner of the Environmental Protection Agency!). Intense city pollution developed later in America, but municipal collection of garbage had started in Boston before the end of the 18th century.

Even when water closets (invented by Thomas Crapper) began to be installed early in the 19th century, the underground reservoirs (cesspools) into which they emptied often overflowed and contaminated drinking water. Outbreaks of cholera in the middle of the 19th century were traced directly to this source of contamination, and this discovery led to the connection of household waste directly to sewers in both Britain and the United States.

At first glance, the easiest way to cope with accumulated feces and urine might appear to be to dilute them in large bodies of water. However, someone is likely to drink this water, and it is not easy to dispose of human waste and at the same time provide healthy drinking water. In addition to health issues there can also be profound ecological effects of disposing of sewage in water bodies.

when natural ecosystems cannot cope with the extra supply of organic matter and nutrients in human waste

All natural ecosystems have an inherent capacity to decompose feces (see Section 10.3.4) and up to a point natural decomposition processes in rivers, lakes, and oceans may cope with increased organic matter from human sewage without obvious changes to the nature of the biological communities they contain. However, problems arise when the rate of input of sewage exceeds this capacity. First, excessively high rates of decomposition of dead organic matter in rivers and lakes can lead to anaerobic conditions (causing the death of fish and invertebrates). This is because oxygen is used up by the decomposer microorganisms faster than it is replenished from photosynthesis by aquatic plants and diffusion from the air. Second, the supply of nutrients such as phosphate and nitrate that normally limit plant growth in water bodies may be increased to a level where the growth of algae is so great that it shades and kills other aquatic plants. This leads to high rates of algal decomposition and consequent oxygen-poor conditions (a process known as *cultural eutrophication*).

Modern sewage systems were developed as ecological devices for pollution management. They aim to capture pollutants from wastewater and to clean it, usually in a drainage system separate from that which carries the heavy flows of storm water. Ideally a sewage system cleans polluted water to a state suitable for drinking before discharging it back into rivers, lakes, and the sea. The full treatment of sewage has three stages, though in many places only the first or second stage are actually used before discharge into the environment. After paper, rags, and plastic have been removed by passing the sewage through screens, *primary treatment* is a physical process in which much of the solid organic sewage waste is allowed to settle to the bottom of settlement tanks, from where it is removed as sludge. *Secondary treatment*, on the other hand, is a biological process designed to mimic (and indeed enhance) natural decomposition. In its simplest version, the partly cleaned water is sprayed onto a layer of crushed rock within which microorganisms have been encouraged to grow. As the water trickles down through these *percolating* or *trickling filters*, natural decomposition mineralizes much of the remaining organic matter, releasing carbon dioxide to the atmosphere. A more sophisticated and efficient method of secondary treatment is the *activated sludge method*, in which the sewage is passed into aerated tanks containing sludge that is activated, or seeded with microorganisms. After secondary treatment the remaining solids are settled to yield more sludge. The wastewater now appears clean, but it still contains two types of "impurities," namely, disease organisms and high concentrations of mineral nutrients that may have health consequences (see Section 13.3.2) but also cause eutrophication if released into rivers and lakes. A final "polishing" stage usually includes chlorination, and sometimes irradiation with ultra violetlight to kill bacteria. Full *tertiary treatment* involves removing mineral nutrients, largely by artificial and expensive chemical processes (Figure 13.1).

Untreated sewage is obviously a pollutant, with adverse health and ecological consequences for water bodies into which it is discharged. However, discharge of sewage that has only been subject to primary treatment is still likely to cause eutrophication because it remains rich in organic matter and nutrients. Moreover, even secondary treatment removes only the organic matter, leaving wastewater rich in plant nutrients. The sludge that accumulates in settling tanks is itself a pollutant that has to be disposed of, usually by dumping at sea or burying in landfill sites. Buried sludge decomposes anaerobically, sometimes taking more than 20 years to mineralize completely, and it produces methane, which contributes to atmospheric pollution (Section 13.4.2). Sludge can be more appropriately used as a fertilizer, either dried or as a liquid sprayed onto the land; in this way the nutrient cycle

sewage treatment systems are needed

products of sewage treatment are themselves pollutants

Figure 13.1
The sequence of treatments commonly
applied to the sewage waste from a
modern urban community.

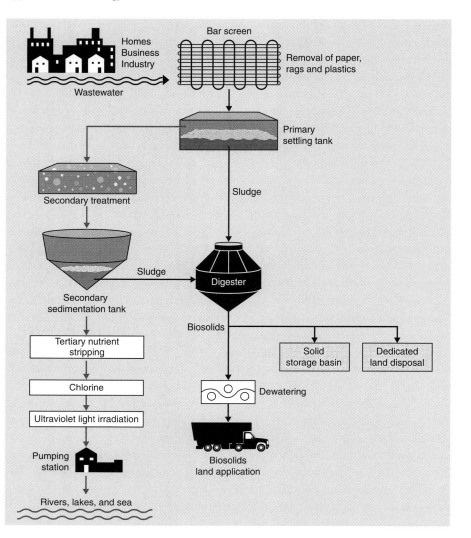

Figure 13.1
The sequence of treatments commonly
applied to the sewage waste from a
modern urban community.

can be reconstituted by returning nutrients, assimilated from crops by people, to
agricultural land to be taken up by future crops.

The range of products of sewage treatment (wastewater, nutrients, sludge, methane) illustrates very elegantly how each solution to a problem of environmental
pollution leads almost inevitably to a new type of pollution. Moreover, almost every
stage in pollution control is demanding of energy and therefore causes more pollution.
An environmentally attractive fate for sludge involves digesting it in a controlled
microbial process to produce methane that can then be burned to produce the
energy that powers the mechanical stages of sewage treatment.

other urban waste   Urban waste includes a very wide range of materials from widely different
sources—the chemical industry, the military, and household items such as plastics,
wood, and metal. Such waste is commonly deposited in landfill sites, which are
then capped with soil and passed to agriculture or urban development. There have
been many cases in which these have subsequently caused problems as toxic chemicals
start to leach into neighboring watercourses and noxious gases are released. Pollution

BOX 13.2

Love Canal: An Example of Industrial Urban Pollution

An attempt to build a canal joining the upper and lower Niagara River in 1892 was abandoned, leaving a section 60 feet wide and 3000 feet long. This was sold in 1920 and became used as a site for disposing of chemical waste. The Hooker Chemical Corporation, the city of Niagara, and the U.S. Army all deposited material at the site (including perhaps materials for chemical warfare). The canal was then filled with about 20,000 tons of toxic waste and was covered with soil and sold to the Niagara Board of Education (with a warning that chemical wastes had been deposited). A school for about 400 students was built on the land near the center of the landfill and completed in 1955. A community of 800 single family homes and 240 low-income apartments had been built around the site by 1978.

Residents complained of foul odors and smelly materials emerging from the soil and consultants were called in. They found storage drums just beneath the surface and high concentrations of polychlorinated biphenyls (PCBs) in the storm sewage system. Little was done except to place window fans in a few homes where the residues were especially high. The New York State Department of Health (NYSDOH) collected air and soil samples in 1978 and made a study of the health of the 239 families who lived close to the canal. They found high levels of pollution in both the soil and the atmosphere and reproductive problems among the women.

The NYSDOH issued a health order on August 2, 1978, recommending that the school be closed, that pregnant women and children below the age of 2 years be evacuated and that residents eat nothing from their gardens and avoid spending time in their basements. The state agreed to buy 239 homes closest to the canal.

In the period from January 1979 to February 1980 there were 22 pregnancies among women from Love Canal, 18 of which ended in miscarriage, stillbirth, or a child with birth defects.

This account of the saga of Love Canal is a précis of an article in Encyclopedia Britannica CD '98; a much more detailed calendar can be found in Gibbs (1998).

is not necessarily stopped by burying the pollutants out of sight. A classic example, Love Canal, is described in Box 13.2.

13.3 ▶ Agricultural Pollution

When intensive livestock production forces animals to live the equivalent of urban life, their waste is produced faster than natural decomposers and detritivores can handle it. All the problems of human urban overpopulation then affect domestic livestock. Intensive agriculture is also associated with increased amounts of nitrates that leach from the soil into rivers and lakes (and into drinking water) and problems associated with the use of insecticides and herbicides.

13.3.1 Intensive livestock management

Pigs, cattle, and poultry are the three major contributors to pollution in industrialized agriculture feedlots. The waste from factory farmed poultry is easily dried and forms a transportable, inoffensive, and valuable crop and garden fertilizer. In contrast, the excreta from cattle and pigs are 90 percent water and have an unpleasant smell

refuse from poultry farming is easy to
transport and use as fertilizer

but excreta from cattle and pigs are
much more bulky (and smelly)

(Table 13.1). A commercial unit for fattening 10,000 pigs produces as much pollution as a town of 18,000 inhabitants.

The law in the United States and Europe increasingly restricts the discharge of agricultural slurry into watercourses. The simplest and often the most economically sound practice returns the material to the land as semisolid manure or as sprayed slurry. This dilutes its concentration in the environment to what it might have been in a more primitive and sustainable type of agriculture and converts pollutant into fertilizer. In turn, of course, new pollution is created—in this case the smell of applied slurry can be offensive. Soil microorganisms decompose the organic components of sewage and slurry and release CO_2 into the atmosphere, while most of the mineral nutrients become immobilized in the soil, available to be absorbed as nutrients by the vegetation. Nitrogen is a special case: nitrate ions are not adsorbed in the soil and rainfall leaches them into drainage (and therefore potential drinking) water (Figure 13.2) to become a new pollutant.

13.3.2 Nitrate

Excess nitrate in drinking water is a health hazard—the Environmental Protection Agency in the United States recommends a maximum concentration of 10 mg per liter in drinking water. Nitrates may contribute to the formation of carcinogenic nitrosamines and in young children may reduce the oxygen-carrying capacity of the blood. Human and animal wastes are just two of the contributors to the nitrate pollution of water; agriculture and forestry are also major culprits.

nitrates in drinking water are a hazard
to health

Most of the fixed nitrogen in natural communities is present in the vegetation and in the organic fraction of the soil. As organisms die they contribute organic matter to the soil, and this decomposes to release CO_2 so that the ratio of carbon to nitrogen falls; when the ratio approaches 10:1, nitrogen begins to be released from the soil organic matter as ammonium ions. In aerated regions of the soil, the ammonium ions become oxidized to nitrite and then to nitrate ions, which are leached by rainfall down the soil profile, dissolved in the drainage waters and then enter aquifers or rivers en route to the sea.

Table 13.1

The average percentage nitrogen and dry matter content of manures and slurries

SOURCE	NITROGEN	DRY MATTER
Cattle farmyard manure	0.6	25
Pig farmyard manure	0.6	25
Poultry broiler litter	2.4	70
Cattle slurry	0.5	10
Pig slurry	0.6	10
(From Gostick, 1982.)		

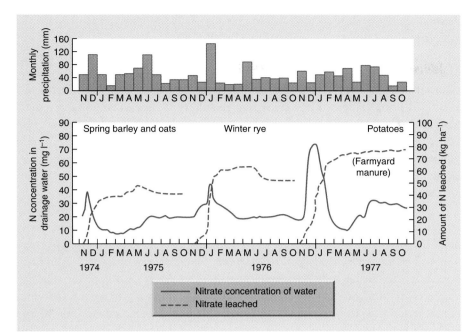

Figure 13.2
The changes in nitrate concentration (milligrams of nitrogen per liter) in water draining from an arable field at Hannover, Germany, and the amounts of nitrate leached (kilograms of nitrogen per hectare). The monthly rainfall during the period is shown above the graphs. Note how most leaching occurs during the winter months when crop growth is slow and before the concentration of nitrate has reached its high spring and summer levels. (After Duynisveld et al., 1988.)

Both the processes of organic matter decomposition and the formation of nitrates are usually fastest in the summer, when natural vegetation is growing most quickly. Nitrates are then absorbed by the growing vegetation as fast as they are formed. They are not present in the soil long enough for significant quantities to be leached out of the plants' rooting zone and lost to the community. Natural vegetation is a "nitrogen-tight" ecosystem.

In contrast, there are several reasons why nitrates leach more easily from agricultural land and managed forests than from natural vegetation.

1. For part of the year agricultural land carries little or no living vegetation to absorb nitrates (and for many years forest biomass is below its maximum).
2. Crops and managed forests are usually monocultures that can capture nitrates only from their own rooting zones, whereas natural vegetation often has a diversity of rooting systems and depths.
3. When straw and forestry waste are burned, the organic nitrogen within them is returned to the soil as nitrates.
4. When agricultural land is used for grazing animals their metabolism speeds up the rate at which carbon is respired, reduces the C:N ratio, and increases nitrate formation and leaching.
5. Nitrogen in agricultural fertilizer is usually applied only once or twice a year rather than being steadily released as it is during the growth of natural vegetation; it is therefore more readily leached and finds its way into drainage waters.

Because nitrogen is not efficiently recycled on agricultural land or managed forests, repeated cropping leads to losses of nitrogen from the ecosystem and thus to decreasing crop productivity. The growth of human populations places heavy

and leach readily from agricultural and forest soils

most agricultural crops depend on
fertilizer nitrogen

or nitrogen fixation by legumes

UNANSWERED QUESTION:

can intensive agriculture or forestry be
practiced without creating nitrate
pollution of the drainage water?

demands on agriculture to produce ever greater yields; to maintain crop yields the available nitrogen has to be supplemented with fertilizer nitrogen.

Some of the nitrogen used in agricultural fertilizers is obtained by mining potassium nitrate in Chile and Peru, but the majority comes from the energy-expensive industrial process of nitrogen fixation, in which nitrogen is catalytically combined with hydrogen under high pressure to form ammonia and, in turn, nitrate. Nitrogen fertilizers are applied in agriculture either as nitrates or as urea or ammonium compounds (which are oxidized to nitrates). However, it is wrong to regard artificial fertilization as the only practice that leads to nitrate pollution; nitrogen fixed by crops of legumes such as alfalfa, clover, peas, and beans also finds its way into nitrates that leach into drainage water.

The problem of nitrate pollution in drinking water is tackled mainly by encouraging wetland plants to use it in their growth or by modifying agricultural and forestry practices to ease the problem. This may involve maintaining ground cover of vegetation year-round, by practicing mixed cropping rather than monoculture, returning organic matter to the soil, maintaining low stocking levels, and applying nitrogen fertilizers only in the season when they are most rapidly absorbed by the crop. However, all these restrictions reduce crop yields and generate other costs. Nitrate pollution of drainage waters appears to be an inevitable consequence of intensive agriculture or forestry.

13.3.3 Pesticides

Many of the manufactured chemicals that are used to kill pests have become important environmental pollutants. The most widely polluting pesticides are those that are used to control the pests and weeds that damage crops in agriculture, horticulture, and forestry or to kill pests that transmit diseases of livestock and humans. All are sprayed or dusted onto the areas in which the pests live, and only a very small proportion hits the target—most lands on the (resistant) crop or on bare ground. Such pesticides are therefore used in much larger quantities than strictly necessary. Characteristics of the most widely used pesticides are described in Chapter 12.

In the early industrial development of pesticides, manufacturers were not much concerned with the specificity of their product: they could damage anything, provided it did not harm the crop, humans, or domestic animals. The potential for disaster is illustrated by the occasion when massive doses of the insecticide dieldrin were applied to large areas of Illinois farmland from 1954 to 1958 to "eradicate" a grassland pest, the Japanese beetle. Cattle and sheep on the farms were poisoned, 90 percent of the cats on the farms and a number of dogs were killed, and among the wildlife 12 species of mammals and 19 species of birds suffered losses (Luckman & Decker, 1960). Even sulfuric acid, a very general poison, was commonly used as a herbicide until the 1950s. Sprayed onto a cereal crop, sulfuric acid damaged almost everything that it wetted, but most of it ran off and did not wet the cereal leaves.

Chemical insecticides are generally intended to control particular target pests at particular places and times. Problems arise when they are toxic to many more species than just the target and particularly when they drift beyond the target

pesticides are most polluting when
they are unselective, persistent, and
accumulate in food chains

areas and persist in the environment beyond the target time. The organochlorine insecticides have caused particularly severe problems because they are *biomagnified*. Biomagnification happens when a pesticide is present in an organism that becomes the prey of another and the predator fails to excrete the pesticide. It then accumulates in the body of the predator. The predator may itself be eaten by a further predator, and the insecticide becomes more and more concentrated as it passes up the food chain. Top predators in freshwater and terrestrial food chains, which were never intended as targets, can then accumulate extraordinarily high doses (Figure 13.3 and Box 13.3).

but some species reabsorb and concentrate the pollutants to dangerous levels

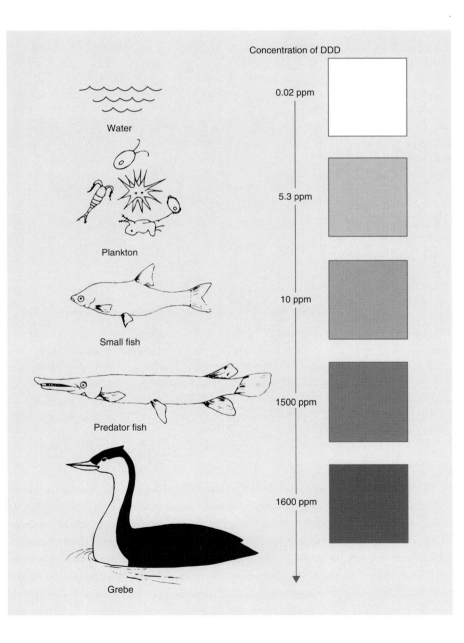

Figure 13.3
The concentration of the insecticide dichlorodiphenyl dichloroethane (DDD) in a food chain of organisms in Clear Lake, California. The pesticide had been added to the lake water to control gnats and had been deliberately chosen because it was less toxic to fish than DDT. Biomagnification by the food chain led to its quite unexpected accumulation in the grebe. (After Flint & Bosch, 1981.)

BOX 13.3   *Topical ECOncerns*

Pollution and the Thickness of Birds' Eggshells

The peregrine falcon (*Falco peregrinus*) is a particularly distinctive and beautiful bird of prey with an almost worldwide range. Until the 1940s about 500 pairs bred regularly in the eastern states of the United States and about 1000 pairs in the West and Mexico. In the late 1940s their numbers started a rapid decline, and by the mid-1970s the bird had disappeared from almost all the eastern states and its numbers had fallen by 80–90 percent in the West. Similar dramatic declines were occurring in Europe. Peregrine falcons were listed as *endangered species* (at risk of extinction). The decline occurred in many other birds of prey and was traced to failure to hatch normal broods. There was very high breakage of eggs in the nest. The cause was eventually identified as the accumulation of DDT in the parent birds. The pesticide had apparently contaminated seeds and insects which had then been eaten by small birds and accumulated in their tissues. In turn these had been caught and eaten by birds of prey and the pesticide interfered with their reproduction—in particular causing the eggs to have thin shells and be more likely to break. The use of DDT was banned in the United States in 1972. Programs were developed to breed peregrines in captivity and at least 4000 peregrines were bred and released to the wild. Peregrines are now breeding successfully over much of the United States and are no longer considered an endangered species. In Britain recovery has been so successful that the peregrine is becoming regarded as a pest by pigeon fanciers and lovers of the smaller songbirds.

It was possible to identify DDT pollution as a cause of eggshell thinning because egg shells had been collected as dated specimens in museums and private collections.

A measure of eggshell thickness (EI) in collections of eggs of the sparrow hawk (*Accipiter nisus*) showed a sudden stepwise fall of 17 percent in 1947, when DDT began to be used widely in agriculture, and has increased steadily since DDT was banned (see figure).

It was a surprise to ornithologists in Britain to find evidence of a decline in eggshell thickness of 2–10 percent in four species of thrush (*Turdus*) since the mid-19th century (Green, 1998). This started long before the development of organic pesticides and there was no sudden change when DDT was introduced. Snails are an

The sparrow hawk. A common small bird of prey that is resident in much of Europe but a migrant in parts of northern Europe and Asia. Like the peregrine falcon, the sparrow hawk has suffered from the characteristic eggshell thinning and nestling mortality that has followed the wide spread use of organochlorine insecticides.

most terrestrial pollution finds its way via rivers and estuaries to the sea—the ultimate dilution?

The oceans receive whatever drains and leaches from the land, and the seemingly ideal fate for a human pollutant is to be diluted and carried away in the vastness of the oceans. However, what has been diluted in the sea may subsequently be accumulated by biological processes, and, like their freshwater and terrestrial counterparts, predators at the top of marine food chains can accumulate damaging concentrations from what appear to be innocuously dilute sources. For example, persistent chlorinated biphenyls (CBs) can accumulate to concentrations 1000 times that found in invertebrates lower in the food chain in the liver of cetaceans (whales, dolphins, and

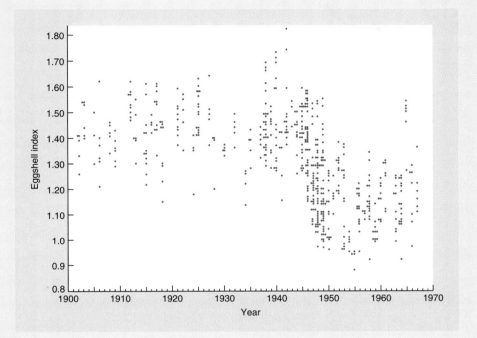

Graph showing change in sparrow hawk eggshell thickness. (From Ratcliffe 1970.)

important part of the diet of thrushes, and contribute much of the calcium for their egg shells. There is convincing evidence that acid rain (see Section 13.4.4) since the Industrial Revolution has acidified leaf litter and reduced its calcium content, leading to a reduction in snail populations and to the calcium content of their shells. The shells of wild birds' eggs have therefore recorded two of the major, but quite different forces of environmental pollution—pesticides and acid rain.

Is the thinness of an eggshell (which may be a tragedy for the individual bird) a problem in itself—or only when this is translated into a decline in the size of the bird population (as with the peregrine falcons)? Consider the value of museum collections as means of providing benchmarks against which present-day animals and plants and even soils may be matched.

porpoises), which are at particular risk because they cannot metabolize CBs efficiently. Furthermore, top predators may suffer not only from the direct effects of pesticide residues; these compounds may also damage the immune system and lay the organisms open to infection from pathogens.

Herbicides are used in very large amounts and on a worldwide scale. They are active against pest plants and when used at commercial rates appear to have few significant effects on animals. Herbicide pollution of the environment has not aroused the passions associated with insecticides. However, conservationists worry about the loss of many species of "weed," which are the food for larvae of butterflies and

herbicides and pollution

Box 13.4   *Topical ECOncerns*

Agent Orange and Problems with Dioxins

A mixture of the herbicides 2,4,5-T and 2,4-D known as Agent Orange was used as part of military strategy in the Vietnam War (see figure). Subsequently a large number of veterans of the war complained of various symptoms, including birth defects in their children and certain rare cancers, and it was found that low levels of 2,4,5-T caused high rates of birth defects in laboratory animals. The veterans were compensated in an out-of-court settlement by the manufacturers of Agent Orange, although no guilt was admitted. A dioxin, TCCD, a manufacturing impurity present at very low levels, was probably the culprit chemical rather than the herbicides themselves, though they are now known to be carcinogenic under some circumstances.

The Agent Orange affair has turned a lot of attention to pollution by dioxins and related furans. These have not had significant commercial use but arise as by-products of the manufacture of chlorine and its compounds, and also from the incineration of municipal waste, steel production and smelting, and burning of coal, wood, plastics, and used tires. In 1881, soil samples had been collected from a field at Rothamsted, England, and sealed in bottles. They were opened recently and found

to contain concentrations of dioxin of 0.7 nanogram per kilogram of soil—about half the concentration at the same site at the present time. Dioxins and furans can be detected in the sediment cores of lakes in the United States, which record their history of pollution. Levels remained low until 1920 and reached their peak and then declined after 1980. They seem not to be easily decomposed except by photochemical processes and accumulate in the environment, ultimately in lake and marine sediments. Pollution by dioxins and the related furans is now taking the blame for a wide range of human problems such as damage to the immune system with enhanced risk of disease, including cancer and mental disorders.

Who gains and who loses, do you think, from "out of court" settlements? Is the production and use of a toxin blameworthy even when its toxicity is only discovered subsequently, especially if all legal requirements have been met? Do producers have a responsibility to assume toxicity until a product is proved safe? Can any product ever be proved safe?

other insects and whose seeds form the main diet of many birds. A recent development has been the genetic modification of crops such as soybean to become resistant to the nonselective herbicide glyphosate. When these specialized crops are sown, the farmer can use glyphosate to kill all species of weed within the crop, a strategy that is particularly worrying to conservationists because it removes even minor species that would not have harmed crop yield.

In Europe, many of the most beautiful cornfield weeds can now be found only in the more primitive agricultures of Central and Eastern Europe while the glorious floristic diversity of Texan roadsides contrasts vividly with the crop monocultures at their side. It may be important to maintain roadsides, hedges, and field margins in an unpolluted condition if some residue of the natural diversity of fauna and flora is to persist.

(Left) Landscape near Khe Sanh, South Vietnam extensively defoliated by aerial application of herbicides (Agent Orange). Photographed in November 1968. (Right) Areas of South Vietnamese forest defoliated by Agent Orange are extremely slow to regenerate. This photograph was taken in 1995.

Between 1962 and 1970 herbicides were used in the Vietnam War to defoliate swamps and forests in South Vietnam. This was done partly to expose enemy troops concealed in the vegetation and partly to destroy food crops on which the enemy depended. The far-reaching consequences of its use are described in Box 13.4.

the use of herbicides in the Vietnam War

13.4 Atmospheric Pollution

In many of the ways in which humans pollute their environment they are just another social animal, facing essentially the same problems of accumulating waste as ants, bees, and termites. But humans have added a type of pollution of a quite different order—we are unique in the way we obtain and manage external sources of energy.

The rich biodiversity of agricultural weeds has been largely destroyed by the use of selective herbicides. This photograph shows a range of arable weeds that still persist in primitive peasant farms in Europe but are nearing extinction elsewhere.

In its simplest form, this is the use of fire to cook food (rendering a wide variety of plants, and particularly seeds, digestible). The development of cooking was also the first step toward humans' contributing more than just their own metabolic carbon dioxide to the atmosphere: this was now augmented by the carbon dioxide produced from burned fuel. To cooking was added in turn the burning of fuel to give warmth, bake clay, smelt ores, work metals, fix atmospheric nitrogen, drive machinery, and power transport. All these operations are peculiar to the human species and gave it unique powers to pollute the environment.

humans are unique among animals in using fuel to release energy for work

The history of humans as a unique polluter can be written in terms of their primary sources of energy as they have changed from using wood to coal to oil to nuclear fuel. In theory, wood is a sustainable fuel, but in practice this is only true if the population that uses it remains small. As the populations of peasant communities who use wood as a fuel grow, trees must be felled farther and farther from home. In parts of Ethiopia the women from a village may spend a day walking to the nearest source of wood and a second day carrying it back home. A major element of the Industrial Revolution was the switch from using wood to coal (and later oil) as a source of power. No fossil fuel is sustainable in the long term—like wood, both coal and oil are exhaustible resources as well as pollutants. Each type of energy source, including nuclear power, makes its own special contribution to pollution. Even hydroelectric, wind, and wave power, and photoelectric processes (which use photovoltaic cells to convert solar radiation directly into usable electric power) use machinery that has been made by polluting industries.

The use of fossil fuels has many adverse effects. For example, the mining and transport of oil have caused major problems for marine life (Box 13.5). However, the most profound and far-reaching consequences are those of atmospheric pollution. From the mid-19th to the mid-20th century the burning of fossil fuels (together with deforestation) contributed about 9×10^{10} metric tons of carbon dioxide to pollute the atmosphere and a further 9×10^{10} metric tons has been added since 1950.

Box 13.5 *Topical ECOncerns*

Oil Pollution—A Major Contaminant of Marine and Coastal Environments

More than 4,000,000 tons of oil enters the world's waterways every year; some as seepage from the ocean floor, some from industry (e.g., gasoline solvents), and more than 1,000,000 pour directly into the oceans from oil tankers or from wells drilled into the seabed. Major accidents around the world have released large volumes of oil close to shore and caused damage to wildlife and to seashores and beaches. Oil in and on the sea affects wildlife in many ways. It reduces the level of aeration of the water, and it reduces the light penetrating the surface. Damage to invertebrates can be widespread, affecting chitons, mussels, and crustaceans, and encrusting species of hydroids and bryozoans as well as seaweeds and kelps. Feathers become choked with oil so that seabirds cannot fly and fish gills become coated and cease to function.

The largest accident in the United States occurred on March 24, 1989, when the oil tanker *Exxon Valdez* ran aground in Prince William Sound, Alaska. It spilled nearly 50,000 tons of crude oil, which spread along the coast for nearly 1000 kilometers, contaminating the shores of a national forest, five state parks, four state critical habitat areas, and a state game sanctuary. The episode is believed to have killed 300 harbor seals, 2800 sea otters, 250,000 birds, and possibly 13 killer whales. Many commercial fisheries were closed for a year or more because of concern that fish caught in the area might find their way into the human food chain. By 1996, 28 species and resources were still listed as having failed to recover.

Are all accidents avoidable—given sufficient care and attention? Should our plans for possible future oil spillages be directed mostly at prevention or at cure?

13.4.1 Carbon dioxide—a major atmospheric pollutant

The concentration of carbon dioxide in the atmosphere has increased from about 280 parts per million (ppm) in 1750 to about 350 ppm today, and it is projected to continue to rise, to 700 ppm by the year 2100 unless there are rather profound changes in human behavior. A remarkable census of atmospheric carbon dioxide was started in 1958 at Mauna Loa Observatory in Hawaii and this detected the extraordinary pattern shown in Figure 13.4. The principal cause of this increase

since 1750 the concentration of carbon dioxide has risen from 280 to 350 ppm

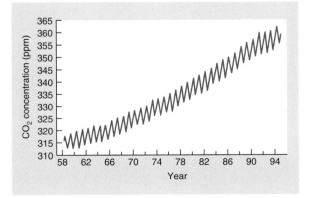

Figure 13.4

The concentration of atmospheric CO_2 measured at the Mauna Loa Observatory, Hawaii, showing the seasonal cycle (resulting from changes in photosynthetic rate) and the long-term increase that is due largely to the burning of fossil fuels. (After Keeling et al., 1995.)

due to the burning of fossil fuels and deforestation

has been the burning of fossil fuels, which in 1980 released about 5.2×10^9 metric tons of carbon into the atmosphere (Table 13.2).

The clearing and burning of tropical forest to make way for agriculture or timber production and the decay of the residues make a further contribution to atmospheric carbon dioxide (Table 13.2). Some of this is recaptured in photosynthesis by the succeeding vegetation, but this is least when forest is converted to grassland, which has a much lower biomass. In total about 1.0×10^9 metric tons per year is released through changes in tropical land use (Detwiler and Hall, 1988). This calculation was made for 1980, and the figure must now be significantly greater as a result of the uncontrollable spread of forest fires in Indonesia and in South America following the droughts associated with the El Niño phenomenon of 1997–1998.

Figure 13.5 shows the overwhelming contribution made to atmospheric carbon dioxide by the combustion of fossil fuel in northern industrial countries and the relatively smaller contribution from changes in land use in the tropics and subtropics.

It should be possible to draw up a balance sheet that shows how the carbon dioxide produced by human activities translates into the changes in concentration in the atmosphere. Table 13.2 is an attempt to do this, but it exposes a problem. Human activities release 5.1–7.5×10^9 metric tons of carbon (C) to the atmosphere every year. But the increase in atmospheric CO_2 (2.9×10^9 metric tons) accounts for only 60 percent of this, a percentage that has remained remarkably constant for 40 years (Hansen et al., 1998). The oceans absorb carbon dioxide from the atmosphere, and it is estimated that they may absorb 1.8 to 2.5×10^9 metric tons of the carbon released by human activities. The balance sheets in Table 13.2 do not

Table 13.2

Balancing the global carbon budget to account for increase in atmospheric carbon caused by human activities

	EXTREME ESTIMATE	MEDIAN ESTIMATE	EXTREME ESTIMATE
Release to atmosphere			
Fossil fuel combustion	4.7	5.2	5.7
Cement production	0.1	0.1	0.1
Tropical forest clearance	0.4	1.0	1.6
Nontropical forest clearance	−0.1	0.0	0.1
Total release	5.1	6.3	7.5
Accounted for			
Atmospheric increase	−2.9	−2.9	−2.9
Ocean uptake	−2.5	−2.2	−1.8
Missing?	−0.3	+1.2	+2.8

In the row labeled "Missing" the minus sign indicates the need to identify an unknown source of carbon of the size shown whereas the plus signs indicate the need for a sink. (After Detwiler & Hall, 1988.)

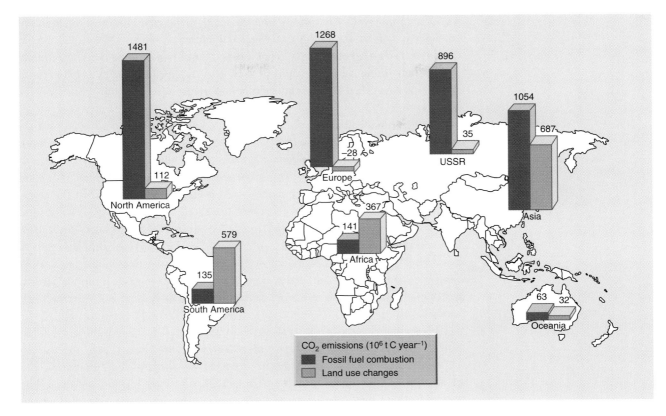

Figure 13.5
Emissions of CO_2 resulting from fossil fuel consumption and land use changes (mainly forest clearance) in 1980. (After UNEP, 1991.)

add up. It seems most likely that we have underestimated the buffering power of the oceans.

13.4.2 Other atmospheric pollutants

Carbon dioxide is not the only gas with which humans pollute the atmosphere. Figure 13.6 shows other changes in the atmosphere from 1850 to 1985 and changes predicted for 1985 to 2100.

The sources of the atmospheric pollution by methane are not all understood. The gas is produced when organic matter is fermented anaerobically. The anaerobic digestion that occurs in the stomach of ruminant animals releases methane: a cow produces about 40 liters of methane per day. The growth of the dairy and livestock fattening industries will also have contributed methane when slurry decomposes anaerobically. The major source of the increase in atmospheric methane may, however, have come from the increased areas of anaerobic soil in land flooded for rice cultivation. Nitrous oxide (N_2O) is produced under many of the same conditions as methane and is a product of denitrification when nitrates are used as a source of oxygen in microbial respiration.

Natural processes contribute carbon dioxide, methane and nitrous oxide to the atmosphere; humans have simply accelerated the process (see Figure 4.19). However, a further major group of atmospheric pollutants, the chlorofluorocarbons, are entirely the products of human activity and did not start to appear in the atmosphere until after 1950, when they became industrially important in freezers and propellants for aerosols.

Further contamination of the atmosphere comes from sulfur compounds. Some derives from sea spray (about 44×10^6 metric tons per year) and some from anaerobic respiration by sulfate-reducing bacteria (estimates are very vague—perhaps 33–230 metric tons per year). Anaerobic respiration releases sulfur as H_2S from waterlogged bog and marsh communities and from marine tidal flats. In the atmosphere H_2S oxidizes to SO_2 and joins that produced by the burning of fossil fuels (oil contains 2–3 percent sulfur and 1–5 percent coal). SO_2 is further oxidized in the atmosphere and converted into sulfuric acid to form aerosol droplets, mostly less than 1 μm in size. The amount of SO_2 released to the atmosphere naturally is about the same as that from human activity, but the latter releases up to 90 percent of the total near the major industrial centers in northern Europe and eastern North America.

The polluted atmosphere that forms in and above cities is commonly called "smog" (*smoke* + *fog*). It was first recognized as a human hazard early in the 20th century when more than 1000 deaths occurred in Glasgow and Edinburgh as a result of "sulfurous smog" produced by sulfur oxides mainly from coal fires. The famous London fogs ("peasoupers") were responsible for darkness and gloom in Victorian London and a high incidence of asthma and bronchitis. The problems continued until the mid-20th century, when control of smoke emission and laws allowing only smokeless fuels almost entirely removed the problem.

At the same time that sulfurous smog was beginning to be controlled, a second form of smog, "photochemical smog," was developing in cities with a high density of automobiles—notably Los Angeles. Automobile exhausts emit nitrogen oxides and hydrocarbons, and these interact in the presence of sunlight to produce the toxic gas ozone and some nitrogen dioxide. The atmosphere develops a brownish color—especially noticeable from the air—visibility is reduced, breathing becomes difficult, and eyes become sore. Catalytic converters force the more complete combustion of automobile fuel and greatly reduce the problem.

humans are not the only polluters of the atmosphere

The pollen of many wind pollinated flowering plants is a major problem for asthmatics and hayfever sufferers. In part, this problem has human causes and so can be described as pollution. Most of the plant species responsible are either arable weeds or cereals and pasture grasses that have increased in abundance enormously in agricultural environments. The feces of the house mite (*Dermatophagoides* sp.) are allergens and pollute the atmosphere for asthma sufferers. Again humans are indirectly the polluters, having provided abundant environments of centrally heated carpeted houses with pillows and cushions that provide a specialized niche for house mites.

major environmental consequences of atmospheric pollution

The major environmental consequences of atmospheric pollution are indirect. They account for three of the most profound changes in global ecology and provide direct challenges to the well-being and economy of modern industrial society. These three changes are the greenhouse effect, acid rain, and damage to the ozone layer. They have many features in common, but it is convenient to consider them separately.

13.4.3 The greenhouse effect

The atmosphere behaves as a greenhouse: during the day solar radiation warms up the Earth's surface, which then reradiates energy outward, principally as infrared radiation. Carbon dioxide, nitrous oxide, methane, ozone, and the chlorofluorocarbons (CFCs) absorb infrared radiation and, like the glass on a greenhouse, prevent the radiation from escaping and keep the temperature high. The concentrations of all these gases have increased rapidly since the Industrial Revolution; the time scale of these changes and predictions for the future are shown in Figure 13.6. It appears that the present air temperature at the land surface is $0.5 \pm 0.2°C$ warmer than in preindustrial times. Doubling of atmospheric CO_2 from its present level is predicted

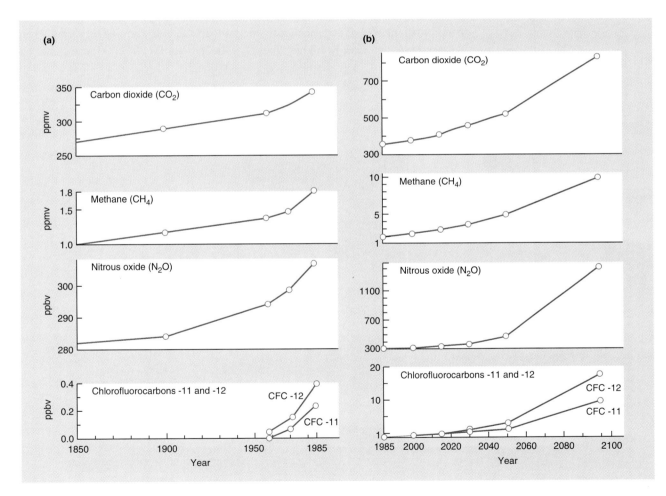

Figure 13.6

The increase in atmospheric concentration of the principal greenhouse gases (a) recorded from 1850 to 1985 and (b) predicted for 1985 to 2100. (After World Meteorological Organization, 1985.)

industrial gases accumulate in the
atmosphere—and this leads to global
warming

to lead to a further warming of 3.5–4.2°C. Such changes would (probably will) lead to a melting of the ice-caps, a consequent rising of sea level, and large changes in the pattern of global climates and the distribution of species. In addition, changes in temperature and global climate can be expected to influence world patterns of famine and disease.

Global warming is not evenly distributed over the surface of the Earth. Figure 13.7 shows the measured global change as the trends in surface temperature over the 46 years from 1951 to 1997. Areas in North America (Alaska) and Asia experienced rises of 1.5–2°C in that period, and these areas are predicted to continue the fastest warming in the first half of the next century. In some regions the temperature has apparently not changed during this period (New York, for example) and should not change greatly in the next 50 years. There are also some areas, notably Greenland and the northern Pacific Ocean, where surface temperatures have fallen.

vegetation has migrated following postglacial warming but has failed to keep pace with changing climates

Global temperatures have changed naturally in the past, most notably during the recurring ice ages. We are currently approaching the end of one of the warming periods that started 20,000 years ago and during which global temperatures have risen by about 8°C. The greenhouse effect adds to global warming at a time when temperatures are already higher than they have been for 400,000 years. Buried

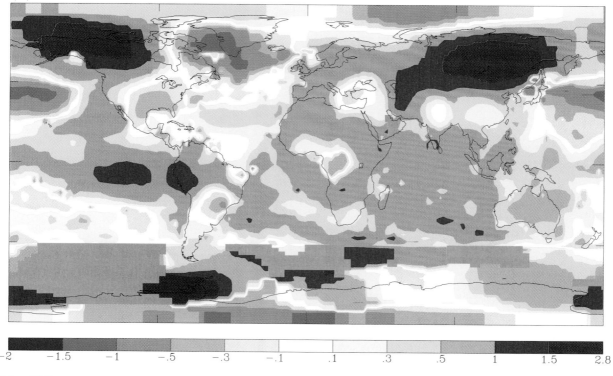

Figure 13.7
Change in the surface temperature of the globe expressed as the linear trend over the 46 years from 1951 to 1997. ΔT, degrees Celsius. (From Hansen et al., in press.)

pollen gives us evidence of past changes in vegetation and shows that North American forest boundaries have migrated north at rates of 100 to 500 meters per year since the last ice age. However this rate of advance has not been fast enough to keep pace with postglacial warming, and the present pattern of forest boundaries is still catching up. The rate of global warming forecast to result from the greenhouse effect is 50–100 times faster than that of postglacial warming!

Of all the types of environmental pollution caused by human activities none may have such profound effects. We must expect widespread extinctions as floras and faunas fail to track and keep up with the rate of change in global temperatures. Extinctions will almost certainly be much faster than those after glacial cycles because humans have now fragmented the landscapes and destroyed the continuity of migration routes.

The growth rate at which greenhouse gases are forcing climatic change may have reached its peak in the late 1970s. Since that time the damaging effects of CFCs have been recognized and substitutes have been found. The growth rate of methane pollution has also fallen quite sharply—perhaps because land under rice cultivation has not been increasing so fast. Moreover, the growth rate of carbon dioxide in the atmosphere has fallen slightly below prediction, perhaps because the oceans absorb more than had been estimated (Hansen et al., 1998). However, there is no reason for complacency—greenhouse levels are still increasing, though perhaps not quite so fast.

> the greenhouse effect warms 50–100 times faster than postglacial warming

13.4.4 Acid rain

Of the pollutants that humans release into the atmosphere most are returned to Earth, about half as gases or particles and half dissolved or suspended in rain, snow, and fog. They may be carried in the wind for hundreds of miles across state and national borders, and when they cause harm they can be the source of bitter international dispute. Atmospheric pollution by sulfur dioxide (SO_2) and oxides of nitrogen (NO_X) interacts with water and oxygen in the atmosphere to form dilute sulfuric and nitric acids, which fall as "acid rain."

Rain water has a pH of about 5.6, but pollutants lower it to below 5.0 and values as low as 2.4 have been recorded in Britain, 2.8 in Scandinavia, and even 2.1 in the United States. Acid rain acidifies the water in lakes and streams, especially when the composition of the underlying soil and rock does not help to neutralize the acidity. A high concentration of hydrogen ions may itself be toxic, but changes in the availability of nutrients and other toxins are usually more important. At pH below 4.0–4.5 the concentrations of aluminum (Al^{3+}), iron (Fe^{3+}), and manganese (Mn^{2+}) become toxic to most plants and to aquatic animals that expose delicate tissues directly to the water (such as the gills of fish). Acid rain is most damaging in water that is already naturally acidic: it may then lower the pH so far that it sterilizes the environment for many of the native species (e.g., Figure 13.8).

> oxides of sulfur and nitrogen form acids in the atmosphere . . . and fall as acid rain

Dry deposits of acidic pollutant may accumulate on vegetation and in snow and then be released suddenly ("episodic acidification") by rainstorms or snow melt. The pH of the water then falls rapidly and populations of fish can be killed on a massive scale. The U.S. Environmental Protection Agency estimates that in parts of

> dry acids washed off vegetation in rain cause lethal "episodic acidification"

Figure 13.8
The history of the diatom flora of a Scottish lake (Round Loch, Glenhead) can be traced by taking cores from the sediment at the bottom of the lake. The percentage of various diatom species at different depths reflects the flora present at various times in the past (four species are illustrated). The age of layers of sediment can be determined by the radioactive decay of lead-210. We know the pH tolerance of the diatom species from their present distribution, and this can be used to reconstruct what the pH of the lake has been in the past. Note how rapidly the waters acidified after 1872. The diatom *Cyclotella kutzingiana* became extinct in the past 200 years whereas the acid tolerant *Tabellaria quadriseptata* was not present until the 19th century. (After Flower & Battarbee, 1983.)

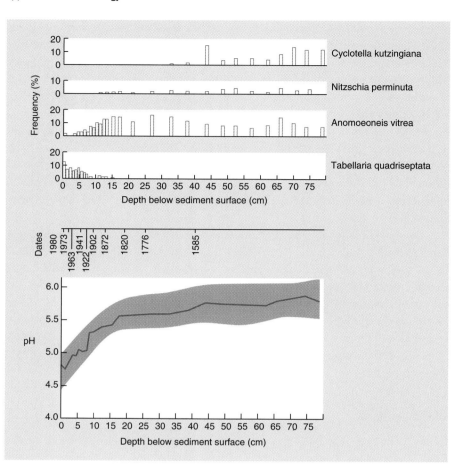

Figure 13.9
Acid rain damage to spruce forest.

the eastern United States, three times as many lakes are at risk from episodic acidification as are chronically acid. This means that a lake or river may appear to be reasonably healthy when it is monitored at intervals through the year—yet it is the occasional lethal episodes that are the real danger to the flora and fauna.

Many of the most visibly dramatic effects of acid rain have been observed in the forests of Central Europe where industry has depended on low-quality coal with a high sulfur content, and as a result there has been forest die-back on a massive scale. For example, in the Czech Republic nearly 60 percent of the forests have been damaged or destroyed. Even in the United States high-elevation spruce forests have been affected, including the Shenandoah and Great Smoky Mountain national parks (Figure 13.9).

There are two main responses to the acid rain problem: reduce its origin or soften its consequences. The burning of fossil fuel to generate electricity in the United States accounts for about 70 percent of the SO_2 and 30 percent of the NO_x emitted to the atmosphere. SO_2 emissions from the generation of electricity can be reduced by switching from coal (1–5 percent sulfur) and oil (2–3 percent sulfur) to gas, which has a lower sulfur content. Alternatively, when coal is burned the sulfur can be extracted before or after burning but in either case the extracted sulfur (e.g., as calcium sulfate) is itself a potential pollutant. The process uses limestone, and when this started to be used on a large scale to reduce sulfur acid emissions there were angry protests at the quarrying that damaged landscapes and conservation sites!

The Clean Air Act in the United States, which is designed to reduce annual emissions of SO_2 (and NO_X) by power plants, sets an interesting, peculiarly American, model for pollution control called *allowance trading*. Each power plant was allocated an allowance of units based on its past record of fuel consumption and emission. A unit allows 1 metric ton of SO_2 to be emitted in a year. The units can be bought, sold, or banked, and this gives a financial incentive to the power company to control its emissions. If it curtails them, there will be units of allowance that can be sold to another company with more lax controls. In effect, this allowance system confers market costs and values on environmental pollution. In most other parts of the world, when governments intervene to reduce emissions, they do so by taxing or fining the polluter.

The acidification of lakes can be reduced by applying lime, but like almost all measures that control pollution this creates environmental problems of its own. The necessary lime, which has to be quarried from limestone, requires a great deal of heat and contributes CO_2 to the atmosphere.

liming can reduce lake acidity—but causes new problems

13.4.5 Thinning of the ozone layer

Ozone is produced by the influence of sunlight on oxygen and during the oxidation of carbon monoxide and hydrocarbons such as methane. It has three very different roles in environmental pollution. The first two are negative, in the sense that undesirable "polluting" consequences occur as the concentration of ozone increases. First, in atmospheres polluted with methane, industrial hydrocarbons, NO_X gases, and carbon monoxide, ozone can reach concentrations that are toxic to plants and that contribute to smog. Second, as we have already described, ozone is also one of the greenhouse gases, though it is not particularly significant in this respect.

ozone can have adverse consequences locally

However, ozone also accumulates in the upper atmosphere; this "ozone layer" is beneficial. It absorbs most of the ultraviolet (UV) radiation (wavelength 200–300 nanometers) incident on the Earth's upper atmosphere and so makes the Earth habitable for plants and animals. The frequency of skin cancer among humans has focused attention on the damage caused by exposure to the sun and on the importance of stability of the ozone layer.

but in the upper atmosphere it shields the Earth from damaging UV radiation

Evidence that nitric oxide produced by supersonic aircraft might contribute massively to reduce atmospheric ozone led to the halting of their large-scale development. However, that has by no means been the end of the story. Chlorofluorocarbon compounds (CFCs) had been developed as aerosols and refrigerants and used on a very large international scale. It became clear that these posed the threat that their chlorine content could interact with and destroy atmospheric ozone. International agreements to phase them out have already had effects (Figure 13.6). Similarly, methyl bromide, a soil fumigant used on a large scale especially in southern Europe, was recognized to be an important pollutant because it destroys ozone in the upper atmosphere. The plan to ban its use after 2001 could make the largest single contribution to protecting the ozone layer.

chlorine compounds and other pollutants decompose ozone in the atmosphere and need to be phased out

UNANSWERED QUESTION:

is it possible to minimize ozone concentrations close to home but to stabilize them high in the atmosphere?

Ozone chemistry is very complicated, and methane, nitrous oxide, and carbon monoxide may all play a part in its decomposition. The upper atmosphere is not the easiest place in which to study the chemical characteristics of gases! But pollution of the upper atmosphere poses questions of the greatest significance for environmental scientists, especially since the discovery that the concentration of ozone in the atmosphere over Antarctica (Figure 13.10) had started to decline by 1978 and was doing so very rapidly after 1982. It is clearly in the interests of humans and probably most other organisms that ozone concentrations should remain low close to home (e.g., minimizing smog) but high in the upper atmosphere, and that we should find out how to ensure this.

13.5 > Nuclear Radiation

Fossil fuels are exhaustible, increasingly costly to extract, pollute the atmosphere, and contribute to global warming. Humans have been slow to develop the full potential for hydroelectric power or power harvested from nonpolluting wind, wave, and tidal energy (but note again that the engineering work needed to harvest such sources of energy is certainly not pollution free). We have been quicker to exploit energy from nuclear fission, a source which in the beginning was viewed as an almost ideal, long-term source of industrial, domestic, and military power. However, the view that the release of radiation could be readily controlled faded rapidly. Some leakage occurs from nuclear power reactors, and it is doubtful whether the reprocessing of waste nuclear fuel can ever be made completely clean. Moreover, the polluting power of radioactive wastes has a time scale that may be orders of magnitude greater than that of other human pollutants. For example, plutonium-239 has a half-life of about 25,000 years. Plutonium is separated and recovered from the spent fuel in nuclear reactors and stocks are expected to have risen to more

Figure 13.10
Thinning of the ozone layer over the Antarctic.

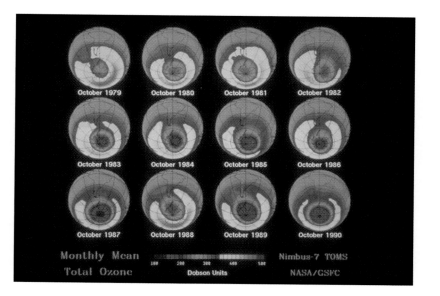

than 100 metric tons by 2010. Ways have to be found to protect against risks of leakage over this sort of time scale, perhaps by burial in deep mines after incorporation in glass. In the longer term it may become necessary to devise reactions that can burn it more efficiently or technology that can fire such pollutants into outer space or into the sun (the ultimate dilution!).

The radiation received by an organism derives from human activities (nuclear warfare, leakage from nuclear plants, accidents such as the Chernobyl meltdown, and medical use) together with a very similar sized contribution from "background radiation," mainly cosmic rays; the radioactive decay of materials such as radium and thorium in the Earth's crust; and radioactive isotopes that are present in food and that decay within the body (Table 13.3). It is a sobering thought that the total radiation given to a cancer tumor in a patient is many thousand times greater than the total normal exposure from the combined natural and artificial background radiation. Background radiation is not evenly distributed. Cosmic rays are more intense at high altitudes, and the radioactive gas radon, which is formed by the decay of radium in the soil, can be a significant source of radiation over certain rock formations, from where it may leak into households.

natural background radiation and that produced by human activities are of similar magnitude

Table 13.3

Estimates of the average annual dose from various sources of irradiation received per human per year by members of the United States population

SOURCE OF RADIATION	AVERAGE DOSE RATES (MICROSIEVERT/YEAR)	% OF ANNUAL TOTAL
Natural environmental		
Cosmic radiation	0.27–1.30	13.7
Terrestrial radiation	0.28–1.15	14.2
Internal radioactive isotopes (K^{40}, C^{14}, etc.)	0.36	18.3
Subtotal	0.91	46.2
Human-made environmental		
Technologically enhanced	0.04	2.0
Global fallout	0.04	2.0
Nuclear power	0.002	0.1
Medical diagnostic	0.79	39.6
Radiopharmaceutical	0.14	7.0
Occupational	0.01	0.5
Miscellaneous	0.05	2.5
Subtotal	1.06	53.7
Total	**1.97**	**100**

(From *Encyclopaedia Britannica.* Copyright © 1994–1998.)

different isotopes follow different pathways and become concentrated in different tissues

Some radioisotopes become concentrated in particular tissues. Iodine-131, a product of fallout from nuclear explosions that emits dangerous beta and gamma radiation, is preferentially absorbed and accumulated by the thyroid gland, where it reaches concentrations 100 times higher than in the rest of the body. It has a half-life of only 8 days, but the high local concentration can lead to thyroid cancer, particularly in children. In contrast, cesium-137 (37 years) and strontium-90 (28 years) have much longer half-lives. Released into the environment in nuclear warfare and by leakage from nuclear industry, they are similar to major plant nutrients (cesium similar to potassium and strontium to calcium) and are readily absorbed by plants from the soil. They then pass into meat and milk and so into the diet of humans, where they can cause damage. Strontium, for example, may be laid down in bone as if it were calcium. Radioactive isotopes that enter rivers, lakes, and oceans are also absorbed and concentrated up food chains and may ultimately find their way into human diet.

Chernobyl—the worst nuclear pollution disaster so far

A major accident in 1986 at a nuclear power station at Chernobyl in the Ukraine released 50–185 million curies of radionucleides into the atmosphere. Close to the explosion, 32 deaths occurred within a very short time. Farther away, individuals contracted radiation sickness and some died. Effects in the locality have continued to appear—livestock have been born deformed, and thousands of radiation-induced illnesses and deaths from cancer are expected in the longer term. Farther afield, wind-dispersed atmospheric pollution from Chernobyl was detected in Sweden 3 days after the accident. Fallout also reached the British Isles. Figure 13.11 shows the persistence of cesium-137 in the acid soils of the northwest of Britain, where it has been absorbed by plants and eaten by sheep. The sale of sheep for food was still banned more than 10 years after the accident because of persistence of the isotope at dangerous levels.

Chernobyl is the only really major nuclear disaster that has occurred to date, excepting atomic bombs dropped on Hiroshima and Nagasaki which released less than half as much radiation (though to far greater effect in terms of human life). However, there have been "near misses," most dramatically in 1979 at Three Mile Island, near Harrisburg, Pennsylvania, when a reactor core was partly melted. Large amounts of radioactive material were released—but into a special containment structure that had been specifically designed to deal with such a happening.

species differ in their sensitivity to nuclear radiation

The effects of environmental pollution are usually described in terms of the harm they cause humans. Taking a broader, ecological view, a benchmark experiment was performed in an oak–pine forest at the Brookhaven National Laboratory, on Long Island, New York (Woodwell, 1970; Figure 13.12). A 9500-curie radiation source of cesium-137 was exposed in the forest in the fall of 1961. Trees close to the source soon died, and over time a concentric pattern of vegetation developed around the source, reflecting different sensitivities to radiation damage. Some mosses colonized ground close to the source. Farther away annuals and then perennials persisted and shrubs and trees formed the most distant zone. It was very noticeable that the species most tolerant to radiation were also those that were quickest to establish after other natural hazards in the forest, such as fire and hurricane. Moreover, the order of tolerance to radiation was generally similar to the order of tolerance to drought, frost, and fire. The most tolerant organisms were usually small, rapidly

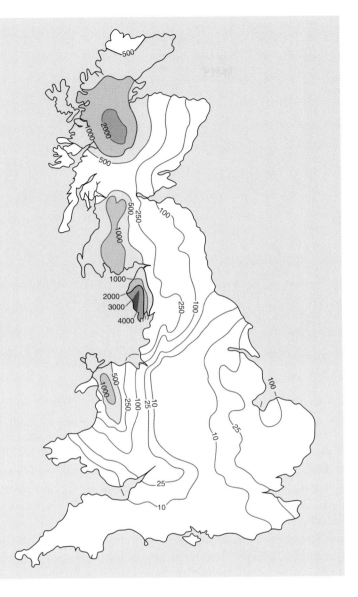

Figure 13.11
An example of long-distance environmental pollution. The distribution in Great Britain of fallout of cesium-137 from the Chernobyl nuclear accident in the Soviet Union in 1986. The contours show the persistence of cesium on acid upland soils, where it is recycled through soil, plants, and animals. On typical lowland soils cesium does not persist in food chains. (After NERC, 1990.)

reproducing species—the ones generally regarded as *r*-strategists (see the discussion in Section 5.6).

13.6 ❯ Mining and Quarrying

Many animals are ecological engineers: they dig, burrow, and build and can change the physical character of their environment (Jones et al., 1997). Human engineering is on a grander scale: quarrying, building, and even manipulating whole landscapes as in contour agriculture, the Great Wall of China, the Egyptian pyramids, and

Figure 13.12
Aerial view of the Gamma Forest at Brookhaven National Laboratory, New York, taken in September 1976, showing damage to the vegetation caused by exposure to gamma radiation from a 9500-curie source of cesium-137. The experiment began in 1961. The cesium source was housed in a metal pipe in the center of the forest and was lowered into the ground by remote control for 4 hours each day to allow safe entry by workers. The radiation field ranged from more than 1000 roentgens a day immediately adjacent to the source to 10 to 20 roentgens a day near the limit of this photograph. (Courtesy of George M. Woodwell.)

quarrying and mining are "ecological engineering" on a vast scale

Stonehenge. These examples of human engineering on a massive scale were powered, like the ecological engineering of other species, solely by metabolic energy. The work was done by the muscles of humans and their draft livestock using only energy derived from their food. It was, in a sense, no more polluting of the environment than the natural burrowing activities of gophers and moles or the architectural creations of termites. The ability of humans to use nonmetabolic energy sources is responsible for the unique scale of modern engineering. In particular the ability to control and use fire made it possible for humans to extract metals from their ores and so ultimately made the Industrial Revolution possible.

Metals were first used by humans in the late Stone Age, about 6500 years ago. Gold, silver, and copper were the first metals used; they are easy to extract because they exist in nature as the metals themselves rather than as chemical compounds. Nuggets of pure metallic gold were found in riverbeds and were beaten and molded for decoration. Once such metals were valued it was an obvious step to dig and mine for them, and from that point almost every phase in the extraction and industrial use of metals involves a sequence of phases of environmental pollution.

Each type of metal has its own peculiarities. In this chapter the mining and purification of copper are used to illustrate pollution by the extraction of metal. Copper is present in deposits either as the metal or as copper sulfide or oxide. Like most metal deposits it usually exists in a mixture with other metals, some of which may be worth saving (e.g., gold), whereas others are discarded in more or less hazardous waste.

The mining industry may pollute at every stage of extraction, purification, and disposal:

- Mining or quarrying exposes the metal and its ores. Many of the world's copper reserves are close to the surface and are easily extracted by open cut mining: the copper mines of Bougainville (Papua New Guinea) and of Utah are among the largest human scars on the Earth's surface (Figure 13.13).
- The ores are crushed and finely ground. This immediately exposes ores to the elements, and even after the best has been extracted the residues are copper-rich and the metal leaches as toxic waste into rivers and lakes. Waters close to copper mines are commonly brilliantly blue–green with copper salts and quite sterile.
- The finely ground ore is agitated in water, and the metal becomes concentrated in the froth and dried to a cake. The remainder, which is still rich in copper, may be further concentrated to recover more of the metal. Ultimately water and solid "tailings" have insufficient copper to be worth further extraction but contain sufficient copper to form a hazardous and polluting waste.
- The concentrate is then roasted to 1230–1300°C, polluting the atmosphere by the burning of the necessary fuel. The roasting drives off a host of pollutants such

Figure 13.13
Binyon Canyon Mine, Utah. A toxic and sterile environment created by the world's largest excavation.

as arsenic, mercury, and sulfur into the atmosphere. Further refining at high temperatures removes more of the sulfur (as sulfur dioxide) and yields metal copper of 99.5 percent purity.

■ The copper can now be purified by electrolysis, which leaves most of the other metals in a sludge that may be further purified (to remove gold, for example) but ultimately contributes yet more toxic waste.

There are many variations in the commercial process of copper extraction, smelting, and refining, and some have been designed to reduce pollution. For example, some of the SO_2 can be removed and converted to sulfuric acid that can then be used in the electrolytic phase.

Each type of metal presents its own peculiar problems to the mining and smelting engineer, and increasingly the levels of environmental pollution are controlled by law to ensure low rates of pollution and efficient cleanup. There is, however, a history of accumulating hazardous waste from before the time when communities became sensitive to environmental pollution. The Environmental Protection Agency (EPA) administers these historic problem sites in the United States (by means of the Superfund program) and funds pollution control through a tax on the chemical and petroleum industries.

The major role of some metals as environmental pollutants occurs after they have been purified and used industrially and are then released into the environment as industrial waste. Lead and mercury are particularly striking examples. Lead became an environmental pollutant from the moment that the Romans started to use it to make water pipes and so started to pollute their drinking water. Though lead is ranked by the EPA as number 1 in their list of 275 hazardous substances, 3.8 million children in the United States still drink lead-contaminated water. Lead poses a particular risk for the development of the nervous system in young children and in the fetus and is being phased out of many commercial uses. Where lead tetraethyl is still used as an additive to gasoline, lead pollutes the atmosphere, especially in cities.

It is not clear whether lead pollution has significant consequences for wildlife on the land or in aquatic environments, but it does not appear to become concentrated along food chains. This is a major contrast with mercury.

Mercury is used in a variety of specialized applications in industry and medicine, such as electric switches, batteries, fluorescent and mercury vapor lights, thermometers, barometers, and dental amalgams. The main culprits in releasing mercury to the atmosphere are, in order of importance, coal-fired power plants, medical waste incinerators, municipal waste incinerators, and industrial boilers. In the natural environment mercury can be converted by microbial activity to methylmercury, a form that is readily absorbed and accumulated up food chains, especially in lakes and estuaries. Fish, the top predators, may accumulate concentrations of mercury 10,000 to 100,000 times that in the surrounding water. Native peoples who hunt and eat wildlife can accumulate even higher concentrations. Mercury is a serious poison that can cause permanent damage to the human brain and kidneys, and particularly to the developing fetus. It may also damage the immune system.

the disposal of hazardous waste may be regulated by legislation

lead and mercury can be especially dangerous pollutants

13.7 > **Restoration Ecology**

Ecologists are concerned to understand the relationship between organisms and their environment, and as ecological science matures they are becoming better able to manage this relationship. *Restoration ecology* is a discipline directed at managing communities that have been damaged by pollution. Management may aim to restore communities to their prepollution state or to create new types of community that can tolerate the pollution.

After an oil spill, for example, the aim of the restoration ecologist is usually to remove the polluting oil and help the marine communities to recover to their original condition. Cleaning up after a major spill is peculiarly difficult. Physical removal is the preferred option, but it can be enormously costly. The oil can sometimes be skimmed off, but only in calm water. Straw and ash can be spread on the water to soak up the oil, but then there is the problem of dealing with the oil-soaked residue! In Europe, detergents have been sprayed onto oil slicks to disperse the oil, but the detergent probably does more harm than good to the biota of the polluted environment. Seaweeds and kelps and their fauna cannot be sown like a grass sward, and so recovery of the flora and fauna has to await dispersal from undamaged parts of the coastline.

Restoration ecologists have been particularly successful at reintroducing species that have been damaged by pollution (e.g., Box 13.3) and revegetating the open scars on the landscape caused by mining and quarrying. The latter may involve deliberately sowing species or varieties of plants that tolerate toxic materials (see the discussion in Chapter 2 of the evolution of genetic strains of grasses and other plants that tolerate high concentrations of copper and other heavy metals). Often, however, it is the sheer absence of soil and its nutrients (particularly nitrogen) that prevents mines and quarries from being naturally colonized by vegetation.

Ecologists and managers now appreciate the role played by wetlands and riparian (bankside) vegetation in reducing the transport from land to river of soil and organic particles and, more particularly, in absorbing nutrients from enriched agricultural, forestry, and urban runoff. Economists are beginning to make some tentative valuations of such natural water-purifying "ecosystem services," in multimillion dollar annual estimates. Innovative water managers are restoring marshy areas (drained long ago for agricultural or urban development) by building artificial wetlands consisting of a linked series of treatment and enhancement marshes, through which secondarily treated sewage is transformed by natural vegetation into a progressively cleaner form. In essence, these artificial wetlands are being used for tertiary treatment of sewage (Section 13.2.2), and in the process provide an environment for wildlife and recreation in an urban setting.

Prevention of pollution is, of course, better than cure. Most of the major disasters of environmental pollution have been caused by human decisions to take high risks and shortcuts. But it may be as unrealistic to imagine that such human behavior can change as to suppose that we will learn to prevent earthquakes, volcanoes, and hurricanes.

⟩ Summary

Waste from humans, their livestock, and other species

Almost all functions of the human body have been regarded historically as polluting the environment. Feces, urine, and corpses are primary pollutants in the crowded environments of most social animals where waste is produced faster than it can be dealt with by natural decomposition. Patterns of behavior have evolved that reduce pollution within the community, for example, fecal sacs of nesting birds, latrines of social mammals, removal of corpses and other waste from the nests of colonial insects. All of these activities have their parallels in human societies in rituals of burial and in the technology for disposing of feces and urine that contributes to community hygiene and helps to prevent the spread of disease.

Sewage systems use water to carry human excreta away from crowded communities and accelerate its decomposition by managed microbial populations. The resultant sludge is dumped at sea, buried, or used as fertilizer. The intensive production of livestock in factory farming is seriously polluting, and agricultural slurry may need to be thinly dispersed over extensive farmland to dilute it to a level that the natural microbial flora can decompose and then release as fertilizer.

Nitrates and pesticides

Nitrates released from human or agricultural waste may leach into potential drinking waters and create a health hazard: they are expensive to remove. Natural vegetation can absorb nitrates and maintain low concentrations in soil and water. But both agriculture and forestry make ecosystems "leaky" and allow nitrates to leach more readily into drainage (potential drinking) water.

Pesticides can be serious environmental pollutants, especially to wildlife. The greatest damage is caused by accidents and misuse and by those pesticides (like chlorinated biphenyls) that accumulate along food chains and reach very high concentrations in top predators. Contaminants of commercial pesticides with other toxins, such as dioxins and furans, can also pose serious risks.

Fossil fuels and atmospheric pollution

Humans are unique polluters of the environment because they burn fossil fuels to give warmth, cook, smelt metals, fix atmospheric nitrogen, drive machinery, and power transport.

Each of these activities pollutes the environment. Deforestation to provide wood, mining for coal, and drilling (and spillage) of oil all harm the environment and contribute in various ways to pollute the atmosphere with CO_2, SO_2, oxides of nitrogen (NO_X), and hydrocarbons.

Atmospheric pollutants contribute to the "greenhouse effect" by which the Earth's atmosphere hinders back-radiation from the Earth's surface and forces global warming. Predicted rates of warming are 50–100 times faster than that experienced during the past 20,000 years of global warming and are expected to lead to widespread migration and extinctions of species.

Atmospheric pollution by SO_2 and NO_X leads to the deposition of acid rain, which can be toxic to vegetation directly or by changing soil pH and in turn changing the toxicity of mineral ions. The Clean Air Act aims to control such pollution in the United States.

Atmospheric pollution by industrial products, especially the fumigant methyl bromide and chlorine compounds, such as chlorofluorocarbons, reduces the ozone concentration in the upper atmosphere. This allows increased carcinogenic UV radiation to reach the Earth's surface. Thinning of the ozone layer was first observed in 1978 and increased rapidly. Use of the more guilty chemicals is being phased out internationally.

Nuclear radiation

Fossil fuels are exhaustible: nuclear fission provides an alternative source of industrial energy as well as a powerful military weapon. Theoretically it produces little pollution, but in practice power stations leak and human failure has lead to major disasters (e.g., Chernobyl). Bioaccumulation of radioactive isotopes such as iodine, cesium, and strontium intensifies the radiation hazard above that present in the environment.

Mining, quarrying, and restoration

Mining and quarrying for metals and their extraction and purification result in severe environmental pollution at every stage of the process. Copper, lead, and mercury are especially severe pollutants of land and water. Mercury (which bioaccumulates) and lead pose severe health hazards.

A science of restoration ecology in which fundamental ecological science is applied to minimize environmental pollution and restore a fauna and flora to polluted land, mines, shores, and estuaries is developing.

Review Questions

⏴ = Challenge Question

1. What are the features that distinguish human pollution of the environment from that by other social organisms?

2. ⏴ Explain why it may be impossible to achieve increasing agricultural production without creating unacceptable levels of nitrate in drinking water.

3. ⏴ Consider the toilet that you most frequently use. Find out where your sewage goes and how it is treated. What pollution problems are you contributing to as a result of your sewage disposal?

4. Describe the causes of acid rain and the way in which it damages terrestrial and aquatic communities.

5. What is the ozone layer? Why is its stability a matter of ecological concern?

6. Define the characteristics that make some pesticides particularly dangerous pollutants.

7. Describe the ways in which the use of metals by humans has created problems of environmental pollution.

8. Define the greenhouse effect and list the pollutants that contribute to it.

9. ⏴ Is it sensible to treat environmental pollution as more than just another density-dependent effect of population growth? Discuss why you consider it different—or do not.

10. ⏴ It is often argued that environmental pollution can be prevented only by making it costly to the polluter. Discuss the ways in which this is, or might be, done.

Natural ecosystems have been put at risk by a plethora of human influences, particularly in the face of a burgeoning human population. Conservation is the science concerned with increasing the probability that the Earth's species and communities (or, more generally, its biodiversity) will persist into the future. We need to appreciate the scale of the problem, understand the threats posed by human activities, and consider how our knowledge of ecology can be brought to bear to provide remedies.

Conservation

Chapter Contents

14.1 Introduction

14.2 Threats to species

14.3 Threats to communities

14.4 Conservation in practice

Key Concepts

In this chapter you will

- appreciate that in seeking to conserve the Earth's species and communities, we are often woefully ignorant of what there is to conserve.
- understand that endangered species are usually rare but not all rare species are endangered.
- recognize that a variety of human influences, including overexploitation, habitat disruption, and introduction of exotic species, have made species more rare and increased their probability of extinction.
- appreciate that populations that become very small may experience genetic problems.
- understand that some species are at risk for a single reason but often a combination of factors is at work.
- appreciate that conservation involves the development of species management plans but also often requires a broader, community perspective.

14.1 **Introduction**

what is biodiversity?

The term *biodiversity* makes frequent appearances in both the popular media and the scientific literature—but it often does so without an unambiguous definition. At its simplest, it is species richness, the number of species present in a defined geographical area (Chapter 10). Biodiversity, though, can also be viewed at scales smaller and larger than the species. For example, we may include genetic diversity within species, perhaps seeking to conserve genetically distinct subpopulations and subspecies. Above the species level, we may wish to ensure that species without close relatives are afforded special protection, so that the overall evolutionary variety of the world's biota is as broad as possible. At a larger scale still, we may include in biodiversity the variety of community types present in a region—swamps, deserts, early and late stages in a woodland succession, and so on. Thus, biodiversity may itself, quite reasonably, have a diversity of meanings. Yet it is necessary to be specific if the term is to be of any practical use. Ecologists and conservationists must define precisely what it is they mean to conserve in their particular circumstances, and how to measure whether this has been achieved.

estimates of the number of species on Earth range from 3 to 30 million or more

Most often the focus of concern of conservation biologists is the rate of extinction of species in the face of human influence. To judge the scale of this problem, we need to know the total number of species that occur in the world, the rate at which these are going extinct, and how this rate compares with that of prehuman times. Unfortunately, there are considerable uncertainties in our estimates of all these things. About 1.8 million species have so far been named, but the real number must be much larger. Estimates have been derived in a variety of ways. One approach, for example, uses information on the rate of discovery of new species to project forward, group by taxonomic group, to a total estimate of up to 6–7 million species in the world. However, the uncertainties in estimating global species richness are profound, and our best guesses range from 3 to 30 million or more (May, 1990). One set of estimates is shown in Figure 14.1.

modern extinction rates compared to historic extinction rates

An important lesson from the fossil record is that the vast majority of (probably all) species eventually become extinct—more than 99 percent of species that ever existed are now extinct (Simpson, 1952). However, given that individual species are believed, on average, to have lasted about 1 to 10 million years, and if we estimate conservatively that the total number of species on Earth is 10 million, we would predict that only an average of between 100 and 1000 species (0.001–0.01 percent) would become extinct each century. The current observed rate of extinction of birds and mammals of about 1 percent per century is 100 to 1000 times this "natural" background rate. Furthermore, the scale of the most powerful human influence, that of habitat destruction, continues to increase, and the list of endangered species in many taxa is alarmingly long (Table 14.1).

The evidence, then, although somewhat inconclusive because of the difficulty of making accurate estimates, suggests that our children and grandchildren may live through a period of species extinction comparable to "natural" mass extinctions evident in the geological record. But should we care? For many, the answer is a resounding and unhesitating yes. Whether the answer seems obvious or debatable,

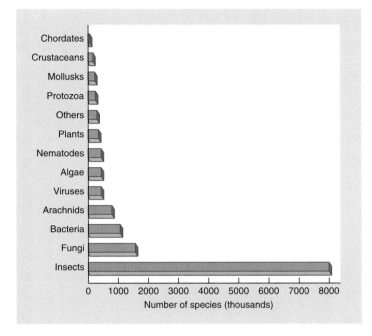

Figure 14.1
Conservative working estimates of current species richness of different groups of organisms. (From Gaston, 1998; based on Heywood, 1995.)

however, it is important to consider *why* we should care—*why* biodiversity is valuable (Box 14.1).

Conservation biology relies on an understanding of the threats faced by individual species (Section 14.2) and by whole communities (Section 14.3). After presenting this background, we consider in Section 14.4 the options open to conservation biologists to maintain or restore biodiversity.

14.2 > Threats to Species

The basic aim of biological conservation is to prevent individual species, or sometimes whole communities, from becoming extinct either regionally or globally. But how do we define the risk of extinction that a species faces? A species can be described as

the classification of threat to species

1. *vulnerable* if there is considered to be a 10 percent probability of extinction within 100 years.
2. *endangered* if the probability is 20 percent within 20 years or 10 generations, whichever is longer.
3. *critical* if within 5 years or 2 generations the risk of extinction is at least 50 percent (Figure 14.2).

In terms of these criteria, for example, 43 percent of vertebrate species have been classified as threatened (i.e., they fall into one of the categories) (Mace, 1994).

Table 14.1

The numbers and percentages of named animal and plant species in major taxa judged to be in danger of extinction. The higher values associated with plants, birds, and mammals may reflect our greater knowledge of these taxa

TAXONS	NUMBER OF THREATENED SPECIES	APPROXIMATE TOTAL SPECIES	PERCENTAGE ENDANGERED
Animals			
Molluscs	354	10^3	0.4
Crustaceans	126	4.0×10^3	3
Insects	873	1.2×10^6	0.07
Fishes	452	2.4×10^4	2
Amphibians	59	3.0×10^3	2
Reptiles	167	6.0×10^3	3
Birds	1029	9.5×10^3	11
Mammals	505	4.5×10^3	11
Total	3565	1.35×10^6	0.3
Plants			
Gymnosperms	242	758	32
Monocotyledons	4421	5.2×10^4	9
Monocotyledons: palms	925	2820	33
Dicotyledons	17,474	1.9×10^5	9
Total	22,137	2.4×10^5	9

(After Smith et al., 1993.)

Figure 14.2

Levels of threat as a function of time and probability of extinction. (After Akçakaya, 1992.)

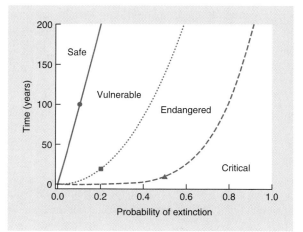

Species that are at high risk of extinction are almost always rare, but not all rare species are at risk. We need to ask what precisely we mean by rare. A species may be rare

the several ways of being rare

1. in the sense that its geographical range is small.
2. in the sense that its habitat range is narrow.
3. in the sense that local populations, even where they do occur, are small.

Species that are rare on all three counts, such as the giant panda (*Ailuropoda melanoleuca*) and the giant cave weta (*Gymnoplectron giganteum*, a primitive insect from New Zealand), are intrinsically vulnerable to extinction. However, species need only be rare in one sense in order to become endangered. For example, the peregrine falcon (*Falco peregrinus*) is broadly distributed across habitats and geographic regions, yet, because it exists always at low densities, local populations in the United States have become extinct and have had to be reestablished with individuals bred in captivity (refer to Box 13.3). The grand Otago skink (*Leiolopisma grande*), on the other hand, is found in only one part of New Zealand and is restricted to schist gullies containing native tussock grasses—but where it occurs, it does so at relatively high densities. It is at risk because changes to pastoral land use threaten its special habitats.

Nevertheless, rare species, by virtue of their rarity, are not necessarily at risk of extinction. In fact, it seems that many, probably most, species are naturally rare. We have already said (Section 2.1) that everything is always absent from almost everywhere. Succinctly: many species are born rare—but others have rarity thrust upon them. Other factors being equal, it will be easier to make a rare species extinct, simply because a localized effect may be sufficient to push it to the brink. Next, therefore, we deal with the various categories of human influence that increase the chances of species extinction.

some species have rarity thrust upon them

The giant panda (*Ailuropoda melanoleuca*) is rare on three counts—it has a small geographic range, requires a specialized habitat, and where it does occur local populations are small.

Box 14.1   *Topical ECOncerns*

What Is the Value of Biodiversity?

To most people, biological diversity is undeniably of value. But it does not always lend itself to the economic valuation on which political decisions are usually based. Standard economics has generally failed to assign value to ecological resources, so that the costs of environmental damage or depletion of living resources have usually been disregarded. A major challenge for the future is the development of a new *ecological economics* (Costanza, 1991) in which the worth of species, communities, and ecosystems can be assigned financial value to be set against the gains to be made in industrial and other human projects that may damage them.

Many species are recognized, now, as having actual *direct value* as living resources; many more species are likely to have a potential value, which as yet remains untapped (Miller, 1988). For example, wild meat, fish, and plants remain a vital resource in many parts of the world, although most of the world's food is derived from plants that were originally domesticated from wild plants in tropical and semiarid regions. In the future, wild strains of these species may be exploited for their genetic diversity, and quite different species of plants and animals that are appropriate for domestication may be found. Second, as we saw in Section 12.5, the potential benefits

of natural enemies if they could be used as biological control agents for pest species are enormous; most natural enemies of most pests remain unstudied and often unrecognized. Finally, about 40 percent of the prescription and nonprescription drugs used throughout the world have active ingredients extracted from plants and animals. Aspirin, probably the world's most widely used drug, was derived originally from the leaves of the tropical willow, *Salix alba*. The nine-banded armadillo (*Dasypus novemcinctus*) has been used to study leprosy and prepare a vaccine for the disease; the Florida manatee (*Trichechus manatus*), an endangered mammal, is being used to help understand hemophilia; the rose periwinkle (*Catharanthus roseus*), a marine snail from Madagascar, has yielded two potent drugs effective in treating blood cancer.

Other species have *indirect economic value*. For example, multitudes of wild insect species are responsible for pollinating crop plants. In a different context, the monetary value of ecotourism, which depends on biodiversity, is becoming ever more considerable. Each year, nearly 200 million adults and children in the United States take part in nature-oriented recreation and spend about $4 billion on fees, travel, lodging, food, and equipment. Moreover, ecotourists, who visit a country wholly or partly to

14.2.1 Overexploitation

large animals are prone to overexploitation

The essence of *overexploitation* is that populations are harvested at a rate that is unsustainable, given their natural rates of mortality and capacities for reproduction (Section 12.3). We have already discussed the idea that in prehistoric times humans were responsible for the extinction of many large animals, the so-called megaherbivores, by overexploiting them (Section 10.6). In more recent times, the history of the great whales has followed a similar pattern, and today we are still taking our toll of other vulnerable giants. Sharks provide an interesting example. Among the most feared of species (although attacks are much more rare than thought in the popular imagination), large numbers are taken for sport, many others to make shark fin soup, and a large proportion of the estimated annual 200 million shark kills are accidental by-catches of commercial fishing. Evidence is mounting that many species of shark have been declining in abundance, a trend that should come as no surprise given their late ages of maturity, slow reproductive cycles, and low fecundities (Manire

experience its biological diversity, spend approximately $12 billion a year worldwide on their enjoyment of the natural world (Primack, 1993). On a smaller scale, a multitude of natural history films, books, and educational programs are "consumed" annually without harming the wildlife on which they are based. More ingenuity is required to find ways to measure the indirect economic benefits of natural biodiversity; for example, biological communities can be of vital importance in maintaining the chemical quality of natural waters, in buffering ecosystems against floods and droughts, in protecting and maintaining soils, in regulating local and even global climate, and in breaking down or immobilizing organic and inorganic wastes.

The final category is *ethical value*. Many people believe that there are ethical grounds for conservation, arguing that every species is of value in its own right and would be of equal value even if people were not here to appreciate or exploit it. From this perspective even species with no conceivable economic value require protection.

It would be wrong, though, to see things only from the point of view of conservation—not that there are really arguments *against* conservation as such, but there are arguments in favor of the human activities that make

conservation a necessity: agriculture, the felling of trees, the harvesting of wild animal populations, the exploitation of minerals, the burning of fossil fuels, irrigation, the discharge of wastes, and so on. To be effective, it may be that the arguments of conservationists must ultimately be framed in cost–benefit terms because governments will always determine their policies against a background of the money they have to spend and the priorities accepted by their electorates.

A government conservation authority is considering a proposal to designate a marine reserve at a rocky promontory of great scenic beauty. The site is very diverse in species, including a few that are rare. Commercial and recreational fishers wish to continue fishing at this unusually productive site, local people have mixed feelings about an expected influx of tourists, while conservationists (who mostly live a long way from the site) believe that the conservation value is such that no fishing should be permitted and visitor numbers should be strictly controlled. Imagine that you are an arbitrator chairing a meeting of all interested parties. What arguments do you think they will put forward? What decision would you reach and why?

and Gruber, 1990). Sharks are among the most important predators in the marine environment, and their enforced rarity may have widespread repercussions in ocean communities.

A feature of animals that are collected for ornamentation, whether for their body parts or as exotic pets, is that their value to collectors increases as they become more scarce. Thus, instead of the normal safeguard of a density-dependent reduction in consumption rate at low density (Section 8.5) the very opposite occurs. The phenomenon is not restricted to animals. New Zealand's endemic mistletoe (*Trilepidia adamsii*), for example, parasitic on a few forest understory shrubs and small trees, was undoubtedly overcollected to provide herbarium specimens. Always a rare species, it became extinct (recorded from 1867 to 1954 but not seen since) as a result of overcollecting combined with forest clearance and perhaps an adverse effect on fruit dispersal caused by reductions in bird populations.

UNANSWERED QUESTION:

what are the consequences for ocean communities of the increased rarity of sharks as top predators?

the threat posed by collectors

14.2.2 Habitat disruption

Habitats may be adversely affected by human influence in three main ways. First, a proportion of the habitat available to a particular species may simply be destroyed, for urban and industrial development or for production of food and other natural resources such as timber. Second, habitat may be degraded by pollution (Chapter 13) to the extent that conditions become untenable for certain species. Third, habitat may be disturbed by human activities to the detriment of some of its occupants.

habitat may be destroyed

Forest clearance has been, and is still, the most pervasive of the forces of habitat destruction. Much of the native temperate forests in the developed world was destroyed long ago, and current rates of deforestation in the tropics are 1 percent or higher per annum. As a consequence, more than half of wildlife habitat has been destroyed in most of the world's tropical countries. The process of habitat destruction often causes the habitat available to a particular species to become more fragmented than it has been historically. This can have several repercussions for the populations concerned, a point we take up again in Section 14.2.6.

or degraded

Degradation by pollution can take many forms, from the application of pesticides that harm nontarget organisms; to acid rain with its adverse effects on organisms as diverse as forest trees, amphibians in ponds, and fish in lakes; to global climate change that may turn out to have the most pervasive influence of all. Aquatic environments are particularly vulnerable to pollution. Water, inorganic chemicals, and organic matter enter from drainage basins, with which streams, rivers, lakes, and continental shelves are intimately connected. Land-use changes, waste disposal, and water impoundment and abstraction can profoundly affect their patterns of water flow and quality (Allan & Flecker, 1993).

or disturbed

Habitat disturbance is not such a pervasive influence as destruction or degradation, but certain species are particularly vulnerable. For example, censuses of gray bats (*Myotis grisescens*) in caves in Alabama and Tennessee in 1968–1970 and again in 1976 indicated population declines that ranged from 0 to 100 percent. Disturbance by visitors to the caves seems to have been responsible; where visits were fewer than one per month only 0–20 percent of bats were lost, but caves with more than four visits per month suffered population declines of between 86 and 95 percent (Tuttle, 1979). Nature recreation, ecotourism, and even ecological research are not without risk of disturbance and the decline of the populations concerned.

14.2.3 Introduced species

Invasions of exotic species into new geographical areas sometimes occur naturally and without human agency. However, human actions have increased this trickle to a flood. Human-caused introductions may occur either accidentally as a consequence of human transport, or intentionally but illegally to serve some private purpose, or legitimately to procure some hoped-for public benefit by bringing a pest under control, producing new agricultural products, or providing novel recreational opportunities. Many introduced species are assimilated into communities without much obvious effect. However, some have been responsible for dramatic changes to native species and natural communities.

For example, the accidental introduction of the brown tree snake *Boiga irregularis* onto Guam, an island in the Pacific, has through nest predation reduced 10 endemic forest bird species to the point of extinction. The gradual spread of the snake from its bridgehead population in the center of the island has been paralleled by the timing of the loss of bird species to the north and south (Figure 14.3). Similarly, the introduction as a source of human food of the predaceous Nile perch (*Lates nilotica*) to the enormously species-rich Lake Victoria in East Africa has driven most of its 350 endemic species of fish to extinction or near-extinction (Kaufman, 1992).

introduced predators

Conservation biologists are particularly concerned about the effects of introduced species wherever there are communities of native organisms that are largely *endemic* (that is, live nowhere else in the world). Indeed, one of the major reasons for the world's huge diversity is the occurrence of centers of endemism so that similar habitats in different parts of the world are occupied by different groups of species that happen to have evolved there. If every species naturally had access to every place on the globe, we might expect a relatively small number of successful species to become dominant in each biome. The extent to which this homogenization can happen naturally is restricted by the limited powers of dispersal of most species in the face of the physical barriers to dispersal that exist. By virtue of the transport opportunities offered by humans, these barriers have been breached by an ever increasing number of exotic species. The effect of introductions has been to convert a hugely diverse range of local community compositions into something much more homogeneous.

introductions leading to homogenization

It would be wrong, however, to conclude that introducing species to a region will inevitably cause a decline in species richness there. For example, there are numerous species of plants, invertebrates, and vertebrates found in continental

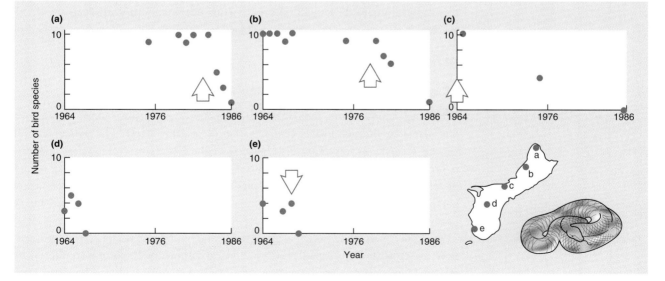

Figure 14.3

Decline in the number of forest bird species at five locations on the island of Guam. Large arrows indicate the first sightings of the brown tree snake *Boiga irregularis* at each location (in location d, the snake was first sighted in the early 1950s). (After Savidge, 1987.)

Box 14.2

What Determines Genetic Variation?

Genetic variation is determined primarily by the joint action of natural selection and *genetic drift* (whereby the frequency of genes in a population is determined by chance rather than evolutionary advantage). The relative importance of genetic drift is higher in small isolated populations, which as a consequence are expected to lose genetic variation. The rate at which this happens depends on the *effective population size* (N_e). This is the size of the "genetically idealized" population to which the actual population (N) is equivalent in genetic terms.

N_e is usually less, often much less, than N, for a number of reasons (detailed formulas can be found in Lande and Barrowclough, 1987):

1. If the sex ratio is not 1:1: for instance, with 100 breeding males and 400 breeding females, $N = 500$ but $N_e = 320$.

2. If the distribution of progeny from individual to individual is not random: for instance, if 500 individuals each produce 1 individual for the next generation on average ($N = 500$), but the variance in progeny production is 5 (with random variation this would be 1), then $N_e = 100$.

3. If population size varies from generation to generation, then N_e is disproportionately influenced by the smaller sizes: for instance, for the sequence 500, 100, 200, 900, 800, mean $N = 500$ but $N_e = 258$.

How many individuals are needed to maintain genetic variability? Franklin (1980) suggested that an effective population size of about 50 would be unlikely to suffer from inbreeding depression, and 500 might be needed to maintain longer-term evolutionary potential. Such rules of thumb should be applied cautiously, and, bearing in mind the relationship between N_e and N, the minimum population size N should probably be set at 5 to 10 times N_e, that is, 2500–5000 individuals (Nunney & Campbell, 1993).

Europe but absent from the British Isles (many because they have so far failed to recolonize after the last glaciation). Their introduction would be likely to augment British biodiversity. The significant detrimental effect noted previously arises where aggressive species provide a novel challenge to endemic biotas ill equipped to deal with them.

14.2.4 Possible genetic problems in small populations

loss of evolutionary potential

Theory tells conservation biologists to beware genetic problems in small populations that may arise through loss of genetic variation (see Box 14.2). The preservation of genetic diversity is important first because of the long-term evolutionary potential it provides. Rare forms of a gene (alleles), or combinations of alleles, may confer no immediate advantage but could turn out to be well suited to changed environmental conditions in the future. Small populations tend to have less variation and hence lower evolutionary potential.

the risk of inbreeding depression

A more immediate potential problem is inbreeding depression. When populations are small there is a tendency for individuals breeding with one another to be related. The offspring of inbreeding parents tend to have levels of heterozygosity far below that of the population as a whole. Individuals with low levels of heterozygosity often have reduced fitness. There are many examples of inbreeding depression—breeders of domesticated animals and plants, for example, have long been aware of reductions in fertility, survivorship, growth rates, and resistance to disease. Furthermore, as well as the problems of a generally low level of heterozygosity, all populations carry specific recessive alleles that are deleterious, or even lethal, when homozygous (human examples include sickle-cell anemia and cystic fibrosis). Close relatives are more likely to carry the same deleterious recessives. Hence, individuals that breed with close relatives are more likely to produce offspring in which the same harmful allele is derived from both parents and the deleterious effect is therefore expressed.

In their study of 23 local populations of the rare plant *Gentianella germanica* in grasslands in the Jura Mountains (Swiss–German border), Fischer and Matthies (1998) found a negative correlation between reproductive performance and population size (Figure 14.4a–c). Furthermore, population size decreased between 1993 and 1995 in most of the studied populations, but population size decreased more rapidly in the smaller populations (Figure 14.4d). These results are consistent with the hypothesis that genetic effects resulted in a reduction in fitness in the small populations. However, they may equally have been caused by differences in local habitat conditions (small populations may be small because they have low fecundity resulting from low-quality habitat) or by disruption of plant-pollinator interactions (small populations may have low fecundity because of low frequencies of visitation by pollinators). To determine whether genetic differences were, indeed, responsible, seeds from each population were grown under standard conditions in a common

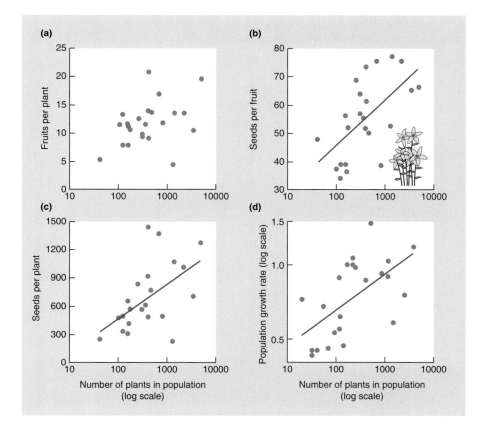

Figure 14.4

Relationships for 23 populations of *Gentianella germanica* between population size and (a) mean number of fruits per plant, (b) mean number of seeds per fruit, and (c) mean number of seeds per plant; (d) the relationship between population growth rate from 1993 to 1995 (ratio of population sizes) and population size (in 1994). (From Fischer & Matthies, 1998.)

garden experiment. We have already witnessed the power of such transplant experiments to reveal intrinsic differences among populations of a species within its geographic range (Section 2.3). After 17 months in the common garden, there were significantly more flowering plants and more flowers (per planted seed) for seeds from large populations than from small populations. We can conclude that genetic effects are of importance for population persistence in this rare species.

14.2.5 A review of risks

Some species are at risk for a single reason, but often, as in the case of the New Zealand mistletoe discussed earlier, a combination of factors is at work. A review of the factors responsible for recorded vertebrate extinctions showed that habitat loss, overexploitation, and species introductions were all of considerable significance, although habitat loss was less prominent in the case of reptiles and overexploitation hardly figured in the case of fishes (Table 14.2). As far as threatened extinctions are concerned, habitat loss is most commonly the major hazard, and risk of overexploitation remains very high, especially for mammals and reptiles.

Table 14.2

Review of the factors responsible for recorded extinctions of vertebrates and an assessment of risks currently facing species categorized globally as endangered, vulnerable, or rare by the International Union for the Conservation of Nature (IUCN)

| GROUP | PERCENTAGE DUE TO EACH CAUSE* | | | | | |
	HABITAT LOSS	OVER EXPLOITATION†	SPECIES INTRODUCTION	PREDATORS	OTHER	UNKNOWN
Extinctions						
Mammals	19	23	20	1	1	36
Birds	20	11	22	0	2	37
Reptiles	5	32	42	0	0	21
Fishes	35	4	30	0	4	48
Threatened extinctions						
Mammals	68	54	6	8	12	—
Birds	58	30	28	1	1	—
Reptiles	53	63	17	3	6	—
Amphibians	77	29	14	—	3	—
Fishes	78	12	28	—	2	—

* The values indicated represent the percentage of species that are influenced by the given factor. Some species may be influenced by more than one factor, thus, some rows may exceed 100%.

† Overexploitation includes commercial, sport, and subsistence hunting and live animal capture for any purpose.

(After Reid & Miller, 1989.)

It is interesting that no example of extinction due to genetic problems has so far come to light. Perhaps inbreeding depression has occurred, though undetected, as part of the "death rattle" of some dying populations (Caughley, 1994). Thus, a population may have been reduced to a very small size by one or more of the processes described, and this may have led to an increased frequency of matings among relatives and the expression of deleterious recessive alleles in offspring, leading to reduced survivorship and fecundity and causing the population to become smaller still—the so-called extinction vortex (Figure 14.5). Perhaps the small populations of *Gentianella germanica* (Figure 14.4) have entered an extinction vortex.

14.2.6 The dynamics of small populations

Much of conservation biology is a crisis discipline. Inevitably, conservationists are confronted with too many problems and too few resources. Should we focus attention on the various forces that lead species to extinction and attempt to persuade governments to act to reduce their prevalence? Or should we restrict our activities to identifying areas of high species richness where we can set up and protect reserves? Or should we identify species at most risk of extinction and work out ways of keeping them in existence? Ideally, we should do all these things. The greatest pressure, however, is often in the area of species preservation; the remaining population of giant pandas in China (or yellow-eyed penguins in New Zealand or spotted owls in North America) has become so small that if nothing is done the species may become extinct within a few years or decades. There is a pressing need to understand the dynamics of small populations.

These are governed by a high level of uncertainty—whereas large populations can be described as being governed by the law of averages (Caughley, 1994). Three kinds of uncertainty or variation that are of particular importance to the fate of small populations can be identified.

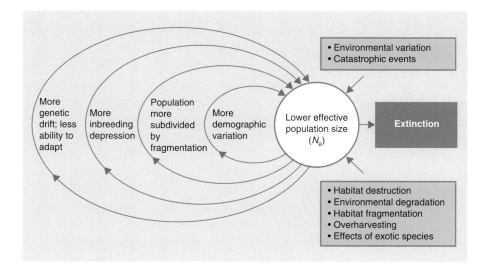

Figure 14.5
Extinction vortices may progressively lower population sizes, leading inexorably to extinction. (From Primack, 1993.)

1. Demographic uncertainty: random variations in the number of individuals that are born male or female, or in the number that happen to die or reproduce in a given year, or in the genetic "quality" of the individuals in terms of survival/ reproductive capacities can matter very much to the fate of small populations. Suppose a breeding pair produces a clutch consisting entirely of females—such an event would be unnoticed in a large population but would be the last straw for a species down to its last pair.

2. Environmental uncertainty: unpredictable changes in environmental factors, whether disasters (such as floods, storms or droughts of a magnitude that occurs very rarely) or more minor factors (year to year variation in average temperature or rainfall) can also seal the fate of a small population. A small population is more likely than a large one to be reduced by adverse conditions to zero or to numbers so low that recovery is impossible.

3. Spatial uncertainty: many species consist of an assemblage of subpopulations that occur in more or less discrete patches of habitat (habitat fragments). Since the subpopulations are likely to differ in terms of demographic uncertainty, and the patches they occupy in terms of environmental uncertainty, the dynamics of extinction and local recolonization can be expected to have a large influence on the chance of extinction of the overall metapopulation (see Section 9.3).

the case of the heath hen

To illustrate some of these ideas, take the demise in North America of the heath hen (*Tympanuchus cupido*). This bird was once extremely common from Maine to Virginia. Being highly edible and easy to shoot (and also susceptible to introduced cats and affected by conversion of its grassland habitat to farmland), it had by 1830 disappeared from the mainland and was only found on the island of Martha's Vineyard. In 1908 a reserve was established for the remaining 50 birds and by 1915 the population had increased to several thousand. However, 1916 was a bad year. Fire (a disaster) eliminated much of the breeding ground, there was a particularly hard winter coupled with an influx of goshawks (environmental uncertainty), and finally poultry disease arrived on the scene (another disaster). At this point the remnant population is likely to have become subject to demographic uncertainty; for example, of the 13 birds remaining in 1928 only 2 were females. A single bird was left in 1930, and the species was extinct in 1932.

the importance of habitat or island area

Of the high-risk factors associated with local extinctions of plant and animal species, habitat or island area is probably the most pervasive. Figure 14.6 shows the negative relationships for a variety of taxa between annual extinction rate and area. No doubt the main reason for the vulnerability of populations in small areas is the fact that the populations themselves are small. This is illustrated in Figure 14.7 for bird species on islands and for bighorn sheep in various desert areas in the southwestern United States.

habitat fragmentation

In fact, loss of habitat frequently results not only in a reduction in the absolute size of a population but also in the division of the original population into a metapopulation of semiisolated subpopulations. Further fragmentation can result in a decrease in the average size of fragments, an increase in the distance between them, and an increase in the proportion of edge habitat (Burgman et al., 1993). A question of fundamental importance, then, is whether a species is more at risk simply

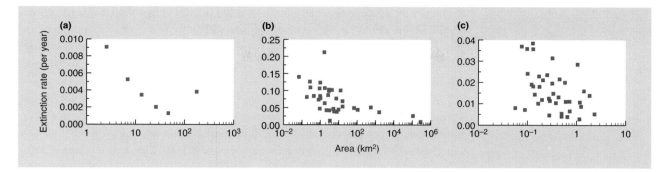

Figure 14.6
Percentage extinction rates as a function of habitat area for (a) zooplankton in lakes in the northeastern United States, (b) birds on northern European islands, (c) vascular plants in southern Sweden. (Data assembled by Pimm, 1991.)

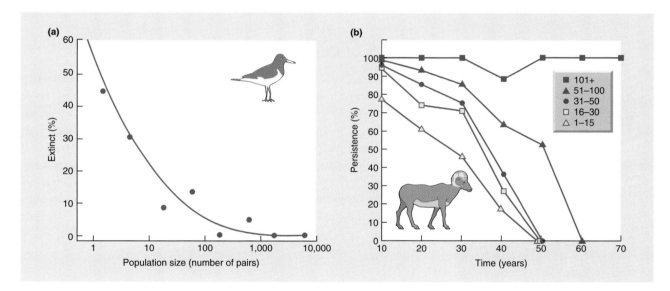

Figure 14.7
(a) Extinction rate of island birds is higher for small populations. (b) The percentage of populations of bighorn sheep in North America that persists over a 70-year period is lowest where the initial population size was small (open triangles; 1–15 individuals) and is highest where initial population size was large (solid squares: more than 101 individuals). (After Berger, 1990.)

because its population is subdivided. In other words, would a single population of a given size be less or more at risk than one divided into a number of subpopulations in habitat fragments? The general conclusion drawn from theoretical explorations is that the answer lies in the balance between the connectedness of different subpopulations, on the one hand, and the correlation between the dynamics of different subpopulations, on the other. Thus, where the probability of dispersal between fragments (that is, connectedness) is high, metapopulations will tend to persist for longer than unfragmented populations. The reason is that when individual subpopulations become extinct, there is a good chance that they will be restarted by a colonist from another subpopulation. However, where extinction events in different subpopulations are strongly correlated (because environmental variation acts identically in all fragments) metapopulations will be more at risk than unfragmented populations. The reason here is that individual subpopulations, being small, are vulnerable to extinction, and when one becomes extinct, they all tend to (Hanski, 1989).

A long-term study of the checkerspot butterfly (*Euphydryas editha*) at Jasper Ridge (western United States) illustrates some of these phenomena. Figure 14.8a shows that if we view the whole population (metapopulation) as a unit, there is a fair degree of consistency in overall abundance, despite a certain amount of year-to-year variation. However, when the population is examined as spatially distinct subpopulations, we see evidence, first, that environmental variation is not well correlated between subpopulations (patterns of increase and decrease are not very similar), and also that in Area G the subpopulation went extinct twice, recolonizing the first time but not the second (Figure 14.8b).

14.3 ▷ Threats to Communities

So far in this chapter, attention has been focused on individual species, treating them as though they were largely independent entities and applying what we know about population dynamics. However, it hardly needs to be pointed out that conservation of biodiversity also requires a broader perspective in which we apply our knowledge of communities and ecosystems. There are many reasons for this.

chains of extinctions—the case of the flying fox

If we ignore community interactions, a chain of extinctions may follow inexorably from the extinction of particular native species, which therefore deserves special attention. Flying foxes in the genus *Pteropus*, which occur on South Pacific islands, are the major, and sometimes the only, pollinators and seed dispersers for hundreds

Figure 14.8
Abundance of checkerspot butterflies (a) when viewed as a single population and (b) when subdivided into three spatially distinct units. (After Ehrlich & Murphy, 1987.)

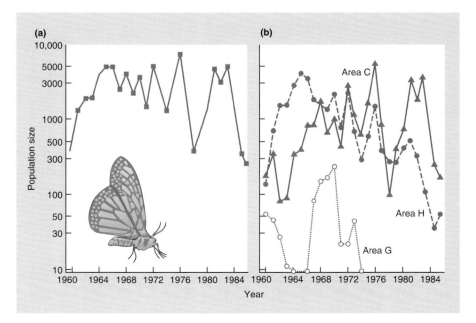

of native plants (many of which are of considerable economic importance, providing medicines, fiber, dyes, prized timber, and foods). Flying foxes are highly vulnerable to human hunters, and there is widespread concern about declining numbers. On Guam, for example, the two indigenous species are either extinct, or virtually so, and there are already indications of reductions in fruiting and dispersal (Cox et al., 1991).

A similarly unwanted outcome may occur if a particularly influential exotic species is introduced into a community without regard to food-web consequences. Our example is provided by the introduction into Flathead Lake and its tributaries in Montana, in the United States, of opossum shrimp (*Mysis relicta*) to provide food for kokanee salmon (*Oncorhynchus nerka*) (itself a previous introduction). Instead of benefiting the fishery as intended, the shrimps heavily exploited the zooplankton food of the salmon while avoiding predation themselves because of their behavioral adaptation of migrating to deep water during the day. The collapse of the kokanee population has apparently dramatically reduced numbers of bald eagles and grizzly bears that gather each autumn to feed on spawning salmon and their carcasses (Figure 14.9). Ecotourism has also suffered a serious decline because the charismatic megafauna are now less in evidence.

We have already noted that species richness declines with habitat area when previously continuous habitat becomes fragmented. But how does trophic organization change in the process of fragmentation? If we are not careful, we may devise a plan to conserve some endangered species but forget to allow for the ecological functional links required to keep the system going. Conservation managers need to

new species for old

functional integrity in relation to fragment size

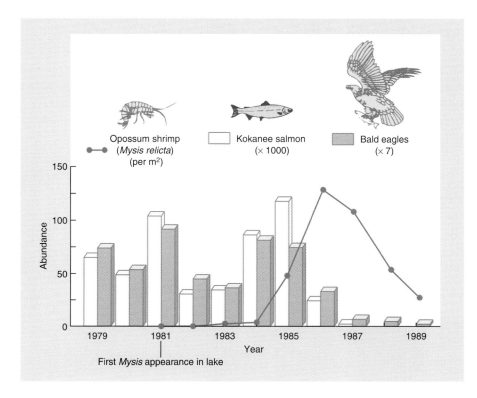

Figure 14.9
As the opossum shrimp population increased and exploited zooplankton in Flathead Lake, kokanee salmon declined, as did the eagle population that relied on the salmon. (After Spencer et al., 1991.)

Grizzly bears (*Ursus arctos*) and bald eagles (*Haliaeetus leucocephalus*) feed on spawning salmon when these are available. The collapse of the kokanee salmon (*Oncorhynchus nerka*) population in Flathead Lake and its tributaries means that these are much rarer sights than they used to be.

UNANSWERED QUESTION:

ecological redundancy—how much do species really matter?

be wary about the scale of their plans. Where the focus is on a particular species, for example the elephants discussed later in Section 14.4.1, the recommended minimum viable population may, by chance, define a minimum viable area that is big enough to contain all the necessary functional links. But ecologically defined boundaries may extend farther than required from just the perspective of any particular species. For example, there is little point in setting up a wetland reserve unless it extends sufficiently far into the surrounding landscape to provide the necessary habitat for terrestrial phases of aquatic insects and to ensure that terrestrial energy and nutrient fluxes provide appropriate inputs to the aquatic ecosystem.

In the community and ecosystem contexts, we can pose the question, Can we lose a few, several, or many species and maintain, for example, productivity, nutrient cycling, or community resilience in the face of disturbance? Three possible scenarios are illustrated in Figure 14.10. In each case, the community process operates at a "natural" level at the high (natural) level of species richness and the process reduces

Figure 14.10

Hypothetical relationships between a basic ecosystem process (such as productivity, nutrient cycling, stability in composition) and species richness.

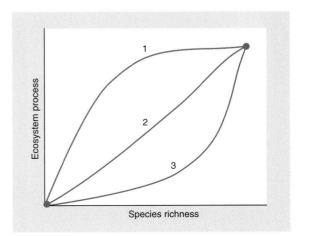

to zero when no species are left. But there are three curves between these two extremes. Curve 1 shows little change to the process until many species have been lost, corresponding to a situation in which there is high "ecological redundancy." Loss of biodiversity is of least significance in this scenario. Curve 2 has every species contributing to community functioning; if this corresponded to reality we should beware any loss of biodiversity, but the system may be able to cope with a certain level of loss. Curve 3 has the system collapsing with even the loss of one or a few (perhaps keystone) species. Which of these models is closest to reality? The simple answer is that we have almost no idea.

In one study, though, Mikkelson (1993) gathered together information from five contrasting situations in which fragments of different sizes were available—one set of mountaintops and two sets of islands (each of which comprised continuous habitat 6000–14,000 years ago) and two groups of nature reserves in areas of Australia where habitat fragmentation began at the turn of the century. The overall picture was that species richness declined as a result of fragmentation, but *trophic organization* (the proportional representation of plants, herbivores, and carnivores) varied very little between different sized fragments. This, and other studies, therefore, indicate that large fragments are indispensable if we wish to conserve high biodiversity. On the other hand, Mikkelson's conclusion provides some cheer since it implies that, up to a point at least, fragmentation need not lead to the dislocation of all fundamental ecological processes.

14.4 ▷ Conservation in Practice

14.4.1 Species management plans

Given the environmental circumstances and species characteristics of a particular rare species, what is the chance it will become extinct in a specified period? Alternatively, how big must its population be to reduce the chance of extinction to an acceptable level? These are frequently the crunch questions in conservation biology.

The ideal classical approach of experimentation, which might involve setting up and monitoring for several years a number of populations of various sizes, is unavailable to those concerned with species at risk because the situation is usually too urgent and there are inevitably too few individuals to work with. How then are we to decide what constitutes the minimum viable population? Three approaches will be discussed in turn: a search for patterns in evidence already gathered in long-term studies, subjective assessment based on expert knowledge, and development of population models. Each has its limitations, which we will explore by looking at particular examples.

Data sets such as that in Figure 14.7b are unusual because they depend on a long-term commitment to monitoring a number of populations—in this case, bighorn sheep in desert areas. If we set an arbitrary definition of the necessary *minimum viable population* (MVP) as one that will have at least a 95 percent probability of

trying to determine the minimum viable population

from available biogeographic data: a risky approach

persistence for 100 years, we can explore data like these to provide an approximate estimate of MVP. Populations of fewer than 50 individuals all went extinct within 50 years whereas only 50 percent of populations of 51–100 sheep lasted for 50 years. Evidently, for an MVP here we require more than 100 individuals. Indeed, for these sheep such populations demonstrated close to 100 percent persistence over the maximum period studied of 70 years. The value for conservation of studies like this, however, is limited because they deal with species that are generally not at risk. It is only safe to use them to produce recommendations for management of endangered species if the species of concern and the ones in the study are sufficiently similar in their vital statistics, and if the environmental regimes are similar.

by subjective expert assessment

Information that may be relevant to a conservation crisis exists not only in the scientific literature but also in the minds of experts. Bringing experts together in a conservation workshop allows well-informed decisions to be reached. The recommended actions may not be entirely correct, but when a decision about an endangered species has to be made without the possibility of further research, this approach has much to commend it. It makes use of available data, knowledge, and experience in a situation when a decision is needed and time for further research is unavailable. Moreover, it explores the various options in a systematic manner and does not duck the regrettable but inevitable truth that unlimited resources will not be available. However, it also runs a risk. In the absence of all necessary data, the recommended best option may simply be wrong (see Box 14.3).

by simulation modeling: the case of the African elephant

Simulation models provide another way of gauging viability. In such models, every individual is treated separately in terms of the probability (with an attached degree of uncertainty) that it will survive or produce a certain number of offspring within each of a succession of periods. The program is then run many times, each giving a different population trajectory because of the random elements involved. The outputs, for each set of model parameters used, are estimates of the probability of extinction (the proportion of simulated populations that become extinct) over a specified interval.

One particular application of a simulation model concerns African elephants (*Loxodonta africana*). Overall numbers are in decline, and few are expected to survive over the next few decades outside high security areas, mainly because of habitat loss and poaching for ivory. The elephant population was modeled in 12 five-year age classes through discrete five-year time steps. Values for age-specific survivorship and density dependent reproductive rates were derived from a thorough data set from Tsavo National Park in Kenya, because its semiarid nature has the general characteristics of land planned for game reserves now and in the future. Environmental stochasticity was incorporated as possible drought events affecting sex- and age-specific survivorship—again, realistic data from Tsavo were used, based on a mild drought cycle of approximately 10 years superimposed on a more severe 50-year drought and an even more severe 250-year drought. Table 14.3 indicates the survivorship of females under "normal" conditions and the three drought conditions. The relationship between habitat area and the probability of extinction was examined in 1000-year simulations; at least 1000 replicates were performed for each model. Extinctions were taken to have occurred when no individuals remained or when only a single sex was represented.

Table 14.3

Survivorship for 12 elephant age classes in normal years (occur in 0.47 of 5-year periods), and in years with 10-year droughts (0.41 of 5-year periods), 50-year, and 250-year droughts (0.1 and 0.02 of 5-year periods, respectively)

| AGE CLASS (YEARS) | FEMALE SURVIVORSHIP | | | |
	NORMAL YEARS	10-YEAR DROUGHTS	50-YEAR DROUGHTS	250-YEAR DROUGHTS
0–5	0.500	0.477	0.250	0.01
5–10	0.887	0.877	0.639	0.15
10–15	0.884	0.884	0.789	0.20
15–20	0.898	0.898	0.819	0.20
20–25	0.905	0.905	0.728	0.20
25–30	0.883	0.883	0.464	0.10
30–35	0.881	0.881	0.475	0.10
35–40	0.875	0.875	0.138	0.05
40–45	0.857	0.857	0.405	0.10
45–50	0.625	0.625	0.086	0.01
50–55	0.400	0.400	0.016	0.01
55–60	0.000	0.000	0.000	0.00

(After Armbruster & Lande, 1992).

The results imply that an area of 500 square miles is required to yield a 99 percent probability of persistence for 1000 years (Figure 14.11). In fact, the researchers recommend to managers an even more conservative minimum area of 1000 square miles for reserves, recognizing in particular the unreliability of their estimates for survivorship in the youngest age class. Of the parks and game reserves in central and southern Africa, only 35 percent are larger than 1000 square miles.

In an ideal world, a population viability analysis would enable us to produce a specific and reliable recommendation for an endangered species of the population size, or reserve area, that would permit persistence for a given period with a given level of probability. But this is rarely if ever achievable because the biological data are hardly ever good enough. The modelers know this. It is important that conservation managers appreciate it too. Nevertheless, we have seen how models allow us to make the very best use of available data and may well give us the confidence to make a choice among various possible management options and to identify the relative importance of different risk factors. The sorts of management interventions that may then be recommended include translocating of individuals to augment the numbers and/or genetic diversity of target populations, raising the carrying capacity by artificial

(continued on page 516)

Subjective Expert Assessment Concerning the Sumatran Rhino (*Dicerorhinus sumatrensis*)

The Sumatran rhino (*Dicerorhinus sumatrensis*) persists only in small, isolated subpopulations in an increasingly fragmented habitat in Sabah (East Malaysia), Indonesia, and West Malaysia, and perhaps also in Thailand and Burma. Timber harvest, human resettlement, and hydroelectric development threaten unprotected habitat. There are only a few designated reserves, which are themselves subject to exploitation, and only two individuals were held in captivity at the time of a workshop to discuss their fate.

The vulnerability of the Sumatran rhino, the way this

vulnerability varies with different management options, and the most appropriate management option given various criteria were assessed by a technique known as *decision analysis*. A decision tree based on the estimated probabilities of the species becoming extinct within a 30-year-period (equivalent to approximately two rhino generations), is shown in the figure below.

The tree was constructed in the following way. The two small squares are decision points: the first distinguishes between intervention on the rhino's behalf and noninter-

Decision tree for management of the Sumatran rhino.

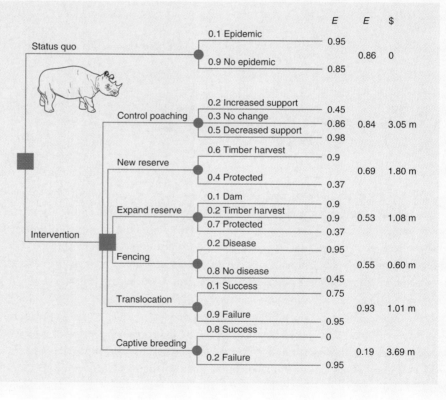

feeding, restricting dispersal by fencing, fostering young (or cross-fostering young by related species), reducing mortality rate by controlling predators or poachers or providing vaccination, and of course habitat preservation.

vention (status quo); the second distinguishes the various management options. For each option, the line branches at a small circle. The branches represent alternative scenarios that might occur, and the numbers on each branch indicate the probabilities estimated for the alternative scenarios. Thus, for the status quo option, there was estimated to be a probability of 0.1 that a disease epidemic will occur in the next 30 years, and hence a probability of 0.9 that no epidemic will occur.

If there is an epidemic, the probability of extinction, p, was estimated to be 0.95 (i.e., 95 percent probability of extinction in 30 years), whereas with no epidemic p was 0.85. The overall estimate of species extinction for an option, E, is then given by

E = (probability of first scenario $\times$ p for first scenario)
   + (probability of second scenario $\times$ p for second
      scenario)

which, in the case of the status quo option, is

$$E = (0.1 \times 0.95) + (0.9 \times 0.85) = 0.86$$

The values of p and E for the other options are calculated in a similar way. The final column in the figure then lists the estimated costs of the various options.

Consider two of the management options in a little more detail. The first is to fence an area in an existing or new reserve, managing the resulting high density of rhinos with supplemental feeding and veterinary care. Disease here is a major risk: the probability of an epidemic was estimated to be higher than in the status quo option (0.2 as opposed to 0.1) because the density would be higher, and p if there were an epidemic was again 0.95. On the other hand, if fencing were successful, p was expected to fall to 0.45, giving an overall E of 0.55. The fenced area would cost around U.S. $0.6 million over 30 years.

For the establishment of a captive breeding program, animals would have to be captured from the wild, increasing E if the program failed to an expected 0.95. However, E would clearly drop to zero if the program succeeded. The cost, though, would be high, since it would involve the development of facilities and techniques in Malaysia and Indonesia (around $2.06 million)and the extension of those that already exist in the United States and Great Britain ($1.63 million). The probability of success was estimated to be 0.8. The overall E was therefore 0.19.

Where do these various probability values come from? The answer is from a combination of data, educated use of data, educated guesswork, and experience with related species. Which would be the best management option? The answer depends on what criteria are used in the judgment of best. Suppose we wanted simply to minimize the chances of extinction, irrespective of cost. The proposal for best option would then be captive breeding. In practice, though, costs are most unlikely ever to be ignored. We would then need to identify an option with an acceptably low E but also an acceptable cost.

With the benefit of hindsight, we can now report that about $2.5 million has been spent catching Sumatran rhinos; 3 died during capture, 6 died post capture, and of the 21 rhinos now in captivity only 1 has given birth and she was pregnant when captured (data of N. Leader-Williams reported in Caughley, 1994). Leader-Williams suggests that the $2.5 million could have been used effectively to protect 700 square kilometers of prime rhino habitat for nearly two decades. This could in theory hold a population of 70 Sumatran rhinos, which, with a rate of increase of 0.06 per individual per year (shown by other rhino species given adequate protection), might give birth to 90 calves during that period.

14.4.2 Ex situ conservation

Captive propagation is attractive in its directness, its immediacy, and the opportunity it provides to preserve options. It can be used to provide demographic or genetic reservoirs for enhancing existing natural populations or establishing new ones, or

Examples of species that are well established in captivity but extinct or endangered in the wild—the Pere David deer (*Elaphuurs davidianus*), the Mongolian or Przewalski's wild horse (*Equus przewalski*), the scimitar-horned oryx (*Oryx leucoryx*), the giant tortoise (*Geochelone elephantopus*), and the blackbuck (*Antilope cervicapra*).

to furnish a final refuge for species with no immediate hope of survival in the wild. Thus the Pere David deer (*Elaphuurs davidianus*) and the Mongolian or Przewalski's wild horse (*Equus przewakski*), well established in captivity, are believed extinct in nature, whereas the endangered but still surviving wild populations of blackbuck

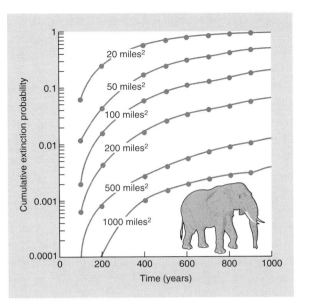

Figure 14.11
Cumulative probability of elephant population extinction over 1000 years for six habitat areas. (After Armbruster & Lande, 1992.**)**

(*Antilope cervicapra*), Arabian oryx (*Oryx leucoryx*), peregrine falcon (*Falco peregrinus*), giant tortoise (*Geochelone elephantopus*), and European otter (*Lutra lutra*), among others, have all been augmented from captive-bred populations.

Zoos have recently become the center of a spirited debate. At one extreme, some see zoological and botanical gardens (and germ banks and seed banks) as small arks that provide refuge for endangered species from a flood of species extinctions; others see them as living museums—once a species enters a zoo it is essentially dead (Ginsberg, 1993). The preceding cases demonstrate the merit of the Ark perspective. However, captive breeding for conservation should involve three stages— (1) planning of a conservation strategy, (2) captive breeding, and (3) reintroduction— and these need to be integrated with habitat protection and in situ conservation. Regrettably, many species are simply inappropriate for captive breeding and eventual reintroduction (e.g., breeding technology unavailable, appropriate habitat unavailable, causes of endangerment irreversible) and available resources are insufficient to build an Ark big enough for a significant proportion of the world's biota.

zoo or ark?

14.4.3 Protected areas

Producing individual species survival plans may be the best way to deal with species recognized to be in deep trouble and identified to be of special importance (e.g., keystone species, evolutionarily unique species, charismatic large animals that are easy to "sell" to the public). However, there is no possibility that all endangered species could be dealt with one at a time. For instance, the US Fish and Wildlife Service calculated it would need to spend about $4.6 billion over 10 years to recover

getting the most from the conservation dollar

fully all species recorded as endangered in the United States (U.S. Department of the Interior, 1990), whereas the annual budget for 1993 was $60 million (Losos, 1993). We can, though, expect to conserve the greatest biodiversity if we protect whole communities by setting aside protected areas.

Protected areas of various kinds (national parks, nature reserves, multiple-use management areas) have grown in number and area through the 20th century, with the greatest expansion occurring since 1970 (Table 14.4). However, the 4500 protected areas in existence in 1989 still only represented 3.2 percent of the world's land area. At best, and given the political will, perhaps 6 percent of land area may eventually be provided protection—the rest would be considered necessary to provide the natural resources needed by the human population (Primack, 1993).

Priorities for marine conservation, which have lagged behind terrestrial efforts, are now being urgently addressed. Most of the world's biota is found in the sea (32 of the 33 known phyla are marine, 15 exclusively so), and marine communities are subject to a number of potentially adverse influences, including overfishing, habitat disruption, and particularly pollution by land-based activities. Many of the new generation of marine protected areas are designed as multiple use reserves, accommodating many different users (environmentalists, cultural harvesters, recreational and commercial fishers, and so on) (Agardy, 1994); it is clear that conservation and sustainable use can often proceed hand in hand as long as planning has a scientific basis and the negotiated objectives are clear.

marine protected areas—conservation and sustainable use can go hand in hand

Table 14.4

The pattern of establishment of protected areas before and during the 20th century. Only protected areas greater that 1000 ha are included. The dates of establishment are unknown for 711 areas covering 194,395 km²

DATE ESTABLISHED	NUMBER OF AREAS	TOTAL AREA PROTECTED (km²)
Unknown	711	194,395
Pre-1900	37	51,455
1900–1909	52	131,385
1910–1919	68	76,983
1920–1929	92	172,474
1930–1939	251	275,381
1940–1949	119	97,107
1950–1959	319	229,025
1960–1969	573	537,924
1970–1979	1317	2,029,302
1980–1989	781	1,068,572

(After Reid & Miller, 1989.)

It is important to devise priorities so that the restricted number of new protected areas, in terrestrial and marine settings, can be evaluated systematically and chosen with care. We know that the biotas of different locations vary in species richness (centers of diversity), the extent to which the biota is unique (centers of endemism), and the extent to which the biota is endangered (hot spots of extinction, for example, because of imminent habitat destruction). One or more of these criteria could be used to prioritize potential areas for protection (Figure 14.12).

A perhaps rather surprising application of island biogeography theory (Section 10.5.1) is in nature conservation. This is because many conserved areas and nature reserves are surrounded by an "ocean" of habitat made unsuitable, and therefore hostile, by people. Can the study of islands in general provide us with design principles that can be used in the planning of nature reserves? The answer is a cautious yes; some general points can be made.

centers of diversity or of endemism or of extinction?

the design of nature reserves

1. One problem that conservation managers sometimes face is whether to construct one large reserve or several small ones adding up to the same total area. If each of the small reserves supported the same species, then it would be preferable to construct the larger reserve in the expectation of conserving more species (this recommendation derives from the species–area relationships discussed in Section 10.5.1).

2. On the other hand, if the region as a whole is heterogeneous, then each of the small reserves may support a different group of species and the total conserved might exceed that in a large reserve. In fact, collections of small islands tend to

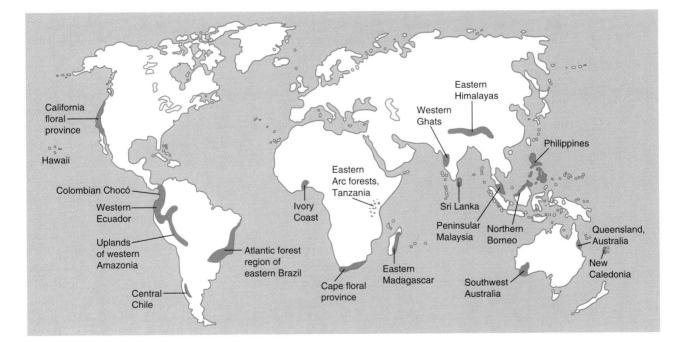

Figure 14.12
Hot spots of high endemism and significant threat of imminent extinction for tropical rain forest and other biomes. (After Myers, 1988, 1991.)

contain more species than a comparable area composed of one or a few large islands. The pattern is similar for habitat islands and, most significantly, for national parks. Thus, several small parks contained more species than larger ones of the same area in studies of mammals and birds in East African parks, of mammals and lizards in Australian reserves, and of large mammals in national parks in the United States. It seems likely that habitat heterogeneity is a general feature of considerable importance in determining species richness.

3. A point of particular significance is that local extinctions are common events, and so recolonization of habitat fragments is critical for the survival of fragmented populations. Thus, we need to pay particular attention to the spatial relationships among fragments, including the provision of dispersal corridors (Fahrig & Merriam, 1993). There are potential disadvantages—for example, corridors could increase the correlation among fragments of catastrophic effects such as the spread of fire or disease (Hess, 1994)—but the arguments in favor are persuasive. Indeed, high recolonization rates (even if this means conservation managers themselves move organisms around) may be indispensable to the success of conservation of endangered metapopulations. Note especially that human fragmentation of the landscape, producing subpopulations that are more and more isolated, is likely to have had the strongest effect on populations with naturally low rates of dispersal. Thus, the widespread declines of the world's amphibians may be due in significant measure to their poor potential for dispersal (Blaustein et al., 1994).

14.4.4 Restoration, rehabilitation, or replacement

Conservation is an appropriate aim when there is something to conserve. In many cases, however, matters have gone too far. Species—or, quite likely, whole communities—have disappeared, been destroyed, or have been altered beyond recognition. Restoration, rather than conservation, may then seem appropriate.

restoring a "natural" community

But what, precisely, is to be restored? First, there may be an attempt to restore the plants and animals that are considered to be the natural inhabitants of that area of land—"natural" in the sense that they were the species living there before the disturbance that ultimately led to the need for restoration. There are three main problems: (1) we may have no clear idea what the authentic, original species are; (2) their natural pattern of establishment may take years; (3) shortcutting these natural processes, even if possible in principle, would require detailed knowledge of the ecologies of many species, which we rarely if ever have.

removing unwanted species

A community may have been changed dramatically because of the effects of one or more introduced species. In such cases, restoration involves removing and not just adding species. New Zealand's offshore islands provide some significant success stories. Over the centuries, domestic stock was purposely introduced to many islands. Successful removals of cattle, goats, pigs, and cats occurred between 1916 and 1936. The much more daunting task of tackling the smaller, faster reproducing, and less conspicuous species, such as Norway rats, ship rats, and mice, has occupied teams of New Zealand ecologists for the past two decades, and with considerable success; over 120 pest eradication campaigns have been conducted to restore island habitats in a way that improves them for endangered species already present and that opens up opportunities to translocate individuals from more vulnerable settings on the mainland (Towns & Ballantine, 1993).

community rehabilitation or replacement

An alternative approach is necessary where abiotic conditions have been profoundly, perhaps irreversibly, altered. This occurs, for instance, when mine wastes of various sorts are deposited above ground (Section 13.7). Here, restoration as such may not be realistic. Bradshaw (1984) has suggested the terms *rehabilitation* for the establishment of a community that is similar to but not the same as the original and *replacement* for the establishment of a quite different community. The pragmatic approach of replacement will often be necessary, establishing whichever species can be established, in order to provide aesthetic or recreational facilities, or perhaps even a productive environment in terms, say, of forestry or grassland.

restoration ecology is an acid test of ecological understanding

Whether the aim is pragmatic or romantic (pristine restoration), the exercise is one in community rather than population ecology. There is a need, in other words, for an understanding of the principles and processes described throughout this book.

14.4.5 Finale—a healthy approach to conservation

the triage approach to setting priorities

In desperate times, painful decisions have to be made about priorities. Thus, wounded soldiers arriving at field hospitals in the First World War were subjected to a *triage* evaluation: Priority 1—those who were likely to survive but only with rapid intervention, Priority 2—those who were likely to survive without rapid intervention, Priority 3—those who were likely to die with or without intervention. Conservation managers are often faced with the same kind of choices and need to demonstrate some courage in giving up on hopeless cases, and prioritizing those species and habitats for which something can be done.

the biodiversity health care compendium

Some argue that the species approach is very much an emergency room strategy when what is needed is better preventative and primary care—something that can most appropriately be accomplished by focusing on communities and ecosystems. We have also seen that intensive care (captive breeding in zoos or botanical gardens) and reconstructive surgery (restoration ecology of communities) need to figure in the biodiversity health care compendium. All these approaches have their place in protecting and enhancing the health of the living world.

taking a balanced view

The spectrum of opinions on conservation is complete. It ranges from the environmental terrorist, who is prepared to destroy property and put human life at risk for what is seen as unacceptable exploitation of the environment, to the other extreme of the exploitational terrorist, who is prepared to destroy rare habitat just as it is about to achieve protected status. There are zealots on both sides of the spectrum too. On the one hand, there are the industrialists, fishers, farmers, and foresters who accept none of the conservationist case and are not prepared to look objectively at the scientific evidence; on the other, are the environmental zealots—preservationists who seem unwilling to accept any exploitation of the natural world, some even pronouncing that fishing or hunting or logging is intrinsically wrong. The middle ground is occupied by both exploiters and conservationists whose basic philosophy holds that natural resources can be used and should be used in a sustainable and balanced manner (Chapter 12). A thorough understanding of the principles and applications of ecological science should enable all to pay healthy regard to the scientific aspects of what, in its broader context, is very much an ethical and sociopolitical problem.

Summary

The scale of conservation

Conservation is the science concerned with increasing the probability that the Earth's species and communities (or, more generally, its biodiversity) will persist into the future. Biodiversity is, at its most basic, the number of species present, but it can also be viewed at smaller scales (e.g., genetic variation within populations) and larger scales (e.g., the variety of community types present in a region). About 1.8 million species have so far been named, but the real number is probably between 3 and 30 million. The current observed rate of extinction might be as much as 100 to 1000 times the background rate indicated by the fossil record.

Rarity and the risk of extinction

A species may be rare in the sense that its geographical and or habitat range is small or in the sense that local populations, even where they do occur, are small. Many species are naturally rare, but just by virtue of their rarity species are not necessarily at risk of extinction. However, other factors being equal, it will be easier to make a rare species extinct. Some species are born rare; others have rarity thrust upon them as a result of the actions of humans.

Human influences that increase extinction probability

The principal causes of decline are overexploitation, habitat degradation, and introduction of exotic species. Overexploitation occurs when people harvest a population (for food or trophies) at a rate that is unsustainable. Humans adversely affect habitat in three ways—a proportion of available habitat may simply be destroyed, or it may be degraded by pollution, or it may be disturbed by human activities to the detriment of some of its occupants. Human-caused introductions of exotic species, which may occur accidentally or intentionally, have sometimes been responsible for dramatic changes to native species and natural communities.

Small populations and genetic problems

Rare alleles of a gene may confer no immediate advantage but could turn out to be well suited to changed environmental conditions in the future—small populations that have lost rare alleles through genetic drift have less potential to adapt. A more immediate potential problem is inbreeding depression—when populations are small there is a tendency for individuals breeding with one another to be related, and this trend may lead to reductions in fertility, survivorship, growth rates and resistance to disease

Some species are at risk because of a combination of factors

A given population may have been reduced to a very small size by one or more of the processes described, and this may have led to an increased frequency of matings among relatives and expression of deleterious recessive alleles in offspring, leading to reduced survivorship and fecundity rates and causing the population to become smaller still—the so-called extinction vortex.

The dynamics of small and fragmented populations

Much of conservation biology is a crisis discipline concerned with small populations in immediate danger of extinction. A high level of uncertainty governs the dynamics of small populations, whereas large populations can be described as being governed by the law of averages. Three kinds of uncertainty or variation that are or particular importance to the fate of small populations can be identified: demographic uncertainty, environmental uncertainty, and spatial uncertainty. Moreover, loss of habitat frequently results not only in a reduction in the absolute size of a population but also in the division of the original population into a number of fragments.

Conservation often requires a broader, community perspective

A chain of extinctions may follow inexorably from the extinction of particular native species. A similarly unwanted outcome may occur if an exotic species is introduced into a community without regard to food-web consequences. Moreover, when habitats become fragmented we need to pay attention not only to the fate of particular species but also to the capacity of food webs to maintain their functional integrity.

Species management plans

Given the environmental circumstances and species characteristics of a particular rare species, what is the chance it will become extinct in a specified period? There are three approaches to determine the minimum viable population: a search for patterns in evidence already gathered in long-term studies, subjective assessment based on expert knowledge, and development of simulation population models.

Captive breeding

Captive propagation is attractive in its directness and the opportunity it provides to preserve options. However, many species are simply inappropriate for captive breeding and/or eventual reintroduction, and available resources are insufficient to build an Ark big enough for a significant proportion of the world's biota.

Protected areas

Given limited funds to purchase protected areas, it is important to devise priorities so that they can be evaluated systematically and chosen with care. We know that the biotas of different locations vary in species richness, the extent to which the biota is unique, and the extent to which the biota is endangered; one or more of these criteria could be used to prioritize potential areas for protection. The principles of island biogeography theory provide some clues about the most appropriate shape and disposition of protected areas. A balanced approach to conservation requires a compendium of ecological "healthcare" measures: preventative and primary care (setting aside protected areas), emergency care (for critically endangered species), intensive care (captive breeding in zoological and botanical gardens), and reconstructive surgery (restoration ecology).

Review Questions

▲ = Challenge Question

1. ▲ Of the estimated 3 to 30 million species on Earth, only about 1.8 million have so far been named. Provide a detailed justification for your answer to the following question: How important is it for the conservation of biodiversity that we can name the species involved?

2. Species may be rare on three counts; what are these? Provide examples, from your own experience if possible,

of three rare species and explain the nature of their rarity.

3. ▲ Researchers collected data on the relative abundance of 16 Peruvian mammal species in forest areas that contrasted in whether they were subject to light (Yavari Miri) or heavy (Tahuayo) hunting by local people. As an index of vulnerability to hunting they used the reduction in

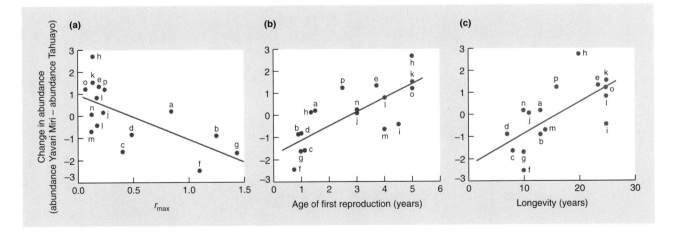

Figure 14.13

Relationships among (a) r_{max}, (b) age of first reproduction, (c) longevity, and the vulnerability of mammals to population declines measured as the change in abundance between lightly (Yavari miri) and heavily hunted (Tahuayo) areas of forest. Mammals: a, white-lipped peccary; b, collared peccary; c, red brocket deer; d, gray brocket deer; e, lowland tapir; f, black agouti; g, green acouchy; h, woolly monkey; i, howler monkey; j, red wakari monkey; k, brown capuchin; l, white-fronted capuchin; m, monk saki monkey; n, titti monkey; o, spider monkey; p, squirrel monkey. (After Bodmer et al., 1997.)

relative abundance in the heavily versus lightly hunted areas. This is plotted against intrinsic rate of population increase (r_{max}), age of first reproduction, and longevity in Figure 14.13. Provide explanations for the relationships shown in the figure. Would you expect the variables r_{max}, age of first reproduction, and longevity to be intercorrelated? If so, how? Many species of large animals have become extinct in the last 50,000 years. What light do the results of this study shed on the possible role of overexploitation by humans in historic extinctions? On the basis of these results, what advice would you give wildlife managers about conserving mammals in Peruvian forests?

4. Explain the various consequences, in conservation terms, of introducing exotic species into a community.

5. Explain what is meant by the environmental, demographic, and spatial uncertainty that afflicts small populations.

6. Explain, with examples, how the loss or introduction of a single species can have conservation consequences throughout a whole ecological community.

7. What is a minimum viable population? Outline different ways of determining the minimum viable population for a particular species.

8. Discuss the value of zoos and botanical gardens in nature conservation.

9. Why is restoration ecology said to be an acid test of ecological understanding?

10. The famous ecologist of the early 20th century A. G. Tansley, when asked what he meant by nature conservation, defined it as maintaining the world in the state he knew as a child. From your perspective, as we enter the new millennium, how would you define the aims of conservation biology?

Web Research Questions

1. Before global warming became number one on the environmental agenda, acid rain received frequent mention in the media. Have the problems associated with acid rain got better or worse, or are they now simply the least of our worries?

2. Biodiversity has become a buzzword in the media. How do ecologists define biodiversity and how well do those in the media understand its meaning?

3. Explore any lists of endangered species that you can find and evaluate the relative importance of different categories of risks of extinction to the group(s) you come up with.

4. Sharks. Most people fear them and might prefer to see them dead. Others see them as a valuable resource to exploit (for shark fin soup or big game hunting). Environmentalists see them as peculiarly susceptible to overexploitation and in need of conservation. Consider the diverse range of views and propose an overall strategy for their exploitation and/or conservation.

5. Discuss the pros and cons of traditional tertiary treatment (stripping of nutrients by chemical means) at sewage works as opposed to passing secondarily treated sewage through artificial wetlands.

References

Abbott, I. (1978) Factors determining the number of land bird species on islands around south-western Australia. *Oecologia,* 33, 221–223.

Abrahamson, W.G. (1975) Reproductive strategies in dewberries. *Ecology,* 56, 721–726.

Abramsky, Z. & Rosenzweig, M.L. (1983) Tilman's predicted productivity-diversity relationship shown by desert rodents. *Nature,* 309, 150–151.

Abramsky, Z. & Sellah, C. (1982) Competition and the role of habitat selection in *Gerbillus allenbyi* and *Meriones tistrami:* a removal experiment. *Ecology,* 63, 1242–1247.

Agardy, M.T. (1994) Advances in marine conservation: the role of marine protected areas. *Trends in Ecology and Evolution,* 9, 267–270.

Akçakaya, H.R. (1992) Population viability analysis and risk assessment. In: *Proceedings of Wildlife 2001: Populations* (D.R. McCullough, ed.). Elsevier, Amsterdam.

Al-Hiyaly, S.A., McNeilly, T. & Bradshaw, A.D. (1988) The effects of zinc contamination from electricity pylons—evolution in a replicated situation. *New Phytologist,* 110, 571–580.

Allan J.D. & Flecker, A.S. (1993) Biodiversity conservation in running waters. *Bioscience,* 43, 32–43.

Allen, K.R. (1972) Further notes on the assessment of Antarctic fin whale stocks. *Report of the International Whaling Commission,* 22, 43–53.

Alliende, M.C. & Harper, J.L. (1989) Demographic studies of a dioecious tree. I. Colonization, sex and age-structure of a population of *Sarex cinerea. Journal of Ecology,* 77, 1029–1047.

Anderson, R.M. (1982) Epidemiology. In: *Modern Parasitology* (F.E.G. Cox, ed.), pp. 205–251. Blackwell Scientific Publications, Oxford.

Anderson, R.M. & May, R.M. (1991) *Infectious Diseases of Humans: Dynamics and Control.* Oxford University Press, Oxford.

Anderson, R.M., Jackson, H.C., May, R.M. & Smith, A.D.M. (1981) Population dynamics of fox rabies in Europe. *Nature,* 289, 765–771.

Andrewartha, H. G. (1961) *Introduction to the Study of Animal Populations.* Methuen, London.

Angel, M.V. (1994) Spatial distribution of marine organisms: patterns and processes. In: *Large Scale Ecology and Conservation Biology* (P.J. Edwards, R.M. May & N.R. Webb, eds), pp. 59–109. Blackwell, Oxford.

Armbruster, P. & Lande, R. (1992) A population viability analysis for African elephant (*Loxodonta africana*): how big should reserves be? *Conservation Biology,* 7, 602–610.

Aston, J.L. & Bradshaw, A.D. (1966) Evolution in closely adjacent plant populations. II. *Agrostis stolonifera* in maritime habitats. *Heredity,* 21, 649–664.

Ayre, D.J. (1985) Localized adaptation of clones of the sea anemone *Actinia tenebrosa. Evolution,* 39, 1250–1260.

Ayre, D.J. (1995) Localized adaptation of sea anemone clones: evidence from transplantation over two spatial scales. *Journal of Animal Ecology,* 64, 186–196.

Bach, C.E. (1994) Effects of herbivory and genotype on growth and survivorship of sand-dune willow (*Salix cordata*). *Ecological Entomology,* 19, 303–309.

Baltensweiler, W., Benz, G., Bovey, P. & Delucchi, V. (1977) Dynamics of larch budmoth populations. *Annual Review of Ecology and Systematics,* 22, 79–100.

Batzli, G.O. (1983) Responses of arctic rodent populations to nutritional factors. *Oikos,* 40, 396–406.

Bazzaz, F.A. (1979) The physiological ecology of plant succession. *Annual Review of Ecology and Systematics,* 10, 351–371.

Bazzaz, F.A. (1996) *Plants in Changing Environments.* Cambridge University Press, Cambridge.

Beaver, R.A. (1979) Host specificity of temperate and tropical animals. *Nature,* 281, 139–141.

Becker, P. (1992) Colonization of islands by carnivorous and herbivorous Heteroptera and Coleoptera: effects of island area, plant species richness, and 'extinction' rates. *Journal of Biogeography,* 19, 163–171.

Begon, M., Sait, S.M. & Thompson, D.J. (1995a) Persistence of a predator–prey system: refuges and generation cycles? *Proceedings of the Royal Society of London, Series B,* 260, 131–137.

Bellows, T.S. Jr. (1981) The descriptive properties of some models for density dependence. *Journal of Animal Ecology,* 50, 139–156.

Bentley, S. & Whittaker, J.B. (1979) Effects of grazing by a chrysomelid beetle, *Gastrophysa viridula,* on competition between *Rumex obtusifolius* and *Rumex crispus. Journal of Ecology,* 69, 79–90.

Berger, J. (1990) Persistence of different-sized populations: an empirical assessment of rapid extinctions in bighorn sheep. *Conservation Biology,* 4, 91–98.

Berner, E.K. & Berner, R.A. (1987) *The Global Water Cycle: Geochemistry and Environment.* Prentice-Hall, New Jersey, USA.

Blaustein, A.R., Wake, D.B. & Sousa, W.P. (1994) Amphibian declines: judging stability, persistence, and susceptibility of populations to local and global extinctions. *Conservation Biology,* 8, 60–71.

Bodmer, R.E., Eisenberg, J.F. & Redford, K.H. (1997) Hunting and the likelihood of extinction of Amazonian mammals. *Conservation Biology,* 11, 460–466.

Bosch, R. van den, Leigh, T.F., Falcon, L.A., Stern, V.M., Gonzales, D. & Hagen, K.S. (1971) The developing program of integrated control of cotton pests in California. In: *Biological Control* (C.B. Huffaker, ed.), pp. 377–394. Plenum Press, New York.

Bradshaw, A.D. (1984) Ecological principles and land reclamation practice. *Landscape Planning*, 11, 35–48.

Breznak, J.A. (1975) Symbiotic relationships between termites and their intestinal biota. In: *Symbiosis* (D.H. Jennings & D.L. Lee, eds), pp. 559–580. Symposium 29, Society for Experimental Biology, Cambridge University Press, Cambridge.13.5.2

Briand, F. (1983) Environmental control of food web structure. *Ecology*, 64, 253–263.

Brookes, M. (1998) The species enigma. *New Scientist*, June 13, 1998.

Brown, J.H. & Davidson, D.W. (1977) Competition between seed-eating rodents and ants in desert ecosystems. *Science* 196, 880–882.

Brown, V.K. & Southwood, T.R.E. (1983) Trophic diversity, niche breadth, and generation times of exopterygote insects in a secondary succession. *Oecologia*, 56, 220–225.

Bryant, J.P. & Kuropat, P.J. (1980) Selection of winter forage by subarctic browsing vertebrates: the role of plant chemistry. *Annual Review of Ecology and Systematics*, 11, 261–285.

Brylinski, M. & Mann, K.H. (1973) An analysis of factors governing productivity in lakes and reservoirs. *Limnology and Oceanography*, 18, 1–14.

Burgman, M.A., Ferson, S. & AkÇakaya, H.R. (1993) *Risk Assessment in Conservation Biology*. Chapman & Hall, London.

Caughley, G. (1994) Directions in conservation biology. *Journal of Animal Ecology*, 63, 215–244.

Charnov, E.L. (1976) Optimal foraging: attack strategy of a mantid. *American Naturalist*, 110, 141–151.

Clark, C.W. (1981) Bioeconomics. In: *Theoretical Ecology: Principles and Applications*, 2nd ed. (R.M. May, ed.), pp. 387–418. Blackwell Scientific Publications, Oxford.

Clausen, J, Keck, D.D. & Hiesey, W.M. (1948) Experimental studies on the nature of species. III. Environmental responses of climatic races of *Achillea*. Carnegie Institute of Washington, 581, 1–129.

Clements, F.E. (1905) *Research Methods in Ecology*. University of Nevada Press, Lincoln.

Cody, M.L. (1975) Towards a theory of continental species diversities. In: *Ecology and Evolution of Communities* (M.L. Cody & J.M. Diamond, eds.), 214–257. Belknap, Cambridge, MA.

Coe, M.J., Cumming, D.H. & Phillipson, J. (1976) Biomass and production of large African herbivores in relation to rainfall and primary production. *Oecologia*, 22, 341–354.

Cole, J.J., Findlay, S. & Pace, M.L. (1988) Bacterial production in fresh and salt water ecosystems: a cross-system overview. *Marine Ecology Progress Series*, 4, 1–10.

Comins, H.N., Hassell, M.P. & May, R.M. (1992) The spatial dynamics of host-parasitoid systems. *Jouranl of Animal Ecology*, 61, 735–748.

Connell, J.H. (1961) The influence of interspecific competition and other factors on the distribution of the barnacle *Chthamalus stellatus*. *Ecology*, 42, 710–723.

Connell, J.H. (1978) Diversity in tropical rainforests and coral reefs. *Science*, 199, 1302–1310.

Connell, J.H. (1980) Diversity and the coevolution of competitors, or the ghost of competition past. *Oikos*, 35, 131–138.

Cook, J.A., Chubb, J.C. & Veltkamp, C.J. (1998) Epibionts of *Asellus aquaticus* (L.) (Crustacea, Isopoda): an SEM study. *Freshwater Biology*, 39, 423–438.

Cook, L.M., Dennis R.L.H. & Mani, G.S. (1999) Melanic morph frequency in the peppered moth in the Manchester area. *Proceedings of the Royal Society, Series B*, 266, 293–297.

Cooper, J.P. (ed.) (1975) *Photosynthesis and Productivity in Different Environments*. Cambridge University Press, Cambridge.

Corey S., Dall, D. & Milne, W. (1993) *Pest Control and Sustainable Agriculture*. CSIRO, East Melbourne.

Costanza, R. (ed.) (1991) *Ecological Economics: The Science and Management of Sustainability*. Columbia University Press, New York.

Cox, F.E.G. (ed.) (1982) *Modern Parasitology*. Blackwell, Oxford.

Cox, F.E.G. (ed.) (1982) Immunology. In: *Modern Parasitology*, pp. 173–203. Blackwell Scientific Publications, Oxford.

Cox P.A., Elmquist, T., Pierson, E.D. & Rainey, W.E. (1991) Flying foxes as strong interactors in South Pacific island ecosystems: a conservation hypothesis. *Conservation Biology*, 5, 448–454.

Crawley, M.J. (1983) *Herbivory: The Dynamics of Animal-Plant Interactions*. Blackwell Scientific Publications, Oxford.

Crawley, M.J. (1986) The structure of plant communities. In: *Plant Ecology* (M.J. Crawley, ed). Blackwell, Oxford.

Currie, D.J. (1991) Energy and large-scale patterns of animal and plant species richness. *American Naturalist*, 137, 27–49.

Currie, D.J. & Paquin, V. (1987) Large-scale biogeographical patterns of species richness in trees. *Nature*, 39, 326–327.

Davidson J. & Andrewartha, H.G. (1948a) Annual trends in a natural population of *Thrips imaginis* (Thysanapotera). *Journal of Animal Ecology*, 17, 193–199.

Davidson J. & Andrewartha, H.G. (1948b) The influence of rainfall, evaporation and atmospheric temperature on fluctuations in the size of a natural population of *Thrips imaginis* (Thysanoptera). *Journal of Animal Ecology*, 17, 200–222.

Davidson, D.W. (1977) Species diversity and community organization in desert seed-eating ants. *Ecology*, 58, 711–724.

Davis, M.B. (1976) Pleistocene biogeography of temperate deciduous forest. *Geoscience and Management*, 13, 13–26.

Davis, M.B., Brubaker, L.B. & Webb, T. III (1973) Calibration of absolute pollen inflex. In: *Quaternary Plant Ecology* (H.J.B. Birks & R.G. West, eds), pp. 9–25. Blackwell Scientific Publications, Oxford.

Deevey, E.S. (1947) Life tables for natural populations of animals. *Quarterly Review of Biology*, 22, 283–314.

Denno, R.F., McClure, M.S. & Ott, J.R. (1995) Interspecific interactions in phytophagous insects: competition reexamined and resurrected. *Annual Review of Entomology*, 40, 297–331.

Deshmukh, I. (1986) *Ecology and Tropical Biology*. Blackwell Scientific Publications, Oxford.

Detwiler, R.P. & Hall, C.A.S. (1988) Tropical forests and the global carbon cycle. *Science*, 239, 42–47.

Diamond, J.M. (1983) Taxonomy by nucleotides. *Nature*, 305, 17–18.

Dirzo, R. & Harper, J.L. (1982) Experimental studies of slug–plant interactions. IV. The performance of cyanogenic and acyanogenic morphs of *Trifolium repens* in the field. *Journal of Ecology*, 70, 119–138.

Dobben, W.H. van (1952) The food of the cormorants in The Netherlands. *Ardea*, 40, 1–63.

Dobson, A.P. & Carper, E.R. (1996) Infectious diseases and human population history. *Bioscience*, 46, 115–126.

Dobzhansky, T. (1950) Evolution in the tropics. *American Scientist*, 38, 209–221.

Dohi, H., Yamada, A. & Entsu, S. (1991) *Journal of Chemical Ecology*, 17, 1197–1203.

Doube, B.M., Macqueen, A., Ridsdill-Smith, T.J. & Weir, T.A. (1991) Native and introduced dung beetles in Australia. In: *Dung Beetle Ecology* (I.

Hanski & Y. Cambefort, eds), pp. 255–278. Princeton University Press, Princeton, USA.

Douglas, Angela E. (1994) *Symbiotic Interactions.* Oxford University Press, Oxford.

Duynisveld, W.H.M., Strebel, O. & Bottcher, J. (1988) Are nitrate leaching from arable and nitrate pollution of groundwater avoidable? *Ecological Bulletins*, 39, 116–125.

Edwards, P.B. & Aschenborn, H.H. (1987) Patterns of nesting and dung burial in *Onitis* dung beetles: implications for pasture productivity and fly control. *Journal of Applied Ecology* 24, 837–851.

Ehrlich, P.R. & Murphy, D.D. (1987) Conservation lessons from long-term studies of checkerspot butterflies. *Conservation Biology*, 1, 122–131.

Einarsen, A.S. (1945) Some factors affecting ring-necked pheasant population density. *Murrelet*, 26, 39–44.

Eis, S., Garman, E.H. & Ebel, L.F. (1965) Relation between cone production and diameter increment of douglas fir (*Pseudotsuga menziesii* (Mirb). Franco), grand fir (*Abies grandis* Dougl.) and western white pine (*Pinus monticola* Dougl.), *Canadian Journal of Botany*, 43, 1553–1559.

El Titi, A. (1989) Farming systems research at Lautenbach Oedhein, FRGermany. In: *Current Status of Integrated Farming Systems Research in Western Europe* (P. Vereijken & D.J. Royle, eds), pp. 21–35. IOBC WPRS Bulletin, 12(5).

Elton, C. (1927) *Animal Ecology.* Sidgwick & Jackson, London.

Elton, C. (1933) *The Ecology of Animals.* Methuen, London.

Elton, C.S. (1958) *The Ecology of Invasion by Animals and Plants.* Methuen, London.

Emiliani, C. (1966) Isotopic palaeotemperatures. *Science*, 154, 851–857.

Endler, J.A. (1980) Natural selection on color patterns in *Poecilia reticulata*. *Evolution*, 34, 76–91.

Erwin, T.L. (1982) Tropical forests: their richness in Coleoptera and other arthropod species. *Coleopterists Bulletin*, 36, 74–75.

Fahrig, L. & Merriam, G. (1994) Conservation of fragmented populations. *Conservation Biology*, 8, 50–59.

Fenner, F. (1983) Biological control, as exemplified by smallpox eradication and myxomatosis. *Proceedings of the Royal Society, Series B*, 218, 259–285.

Fenner, F. & Ratcliffe, R.N. (1965) *Myxomatosis.* Cambridge University Press, London.

Ferguson, R.G. (1933) The Indian tuberculosis problem and some preventative measures. *National Tuberculosis Association Transactions*, 29, 93–106.

Fischer, M. & Matthies, D. (1998) Effects of population size on performance in the rare plant *Gentianella germanica*. *Journal of Ecology*, 86, 195–204.

Fitter, A.H. (1991) The ecological significance of root system architecture. In: *Plant Root Growth: An Ecological Perspective* (D. Atkinson, ed). Blackwell, Oxford.

FitzGibbon, C.D. (1990) Anti-predator strategies of immature Thomson's gazelles: hiding and the prone response. *Animal Behaviour*, 40, 846–855.

FitzGibbon, C.D. & Fanshawe, J. (1989) The condition and age of Thomson's gazelles killed by cheetahs and wild dogs. *Journal of Zoology*, 218, 99–107.

Flecker, A.S. & Townsend, C.R. (1994) Community-wide consequences of trout introduction in New Zealand streams. *Ecological Applications*, 4, 798–807.

Flessa, K.W. & Jablonski, D. (1995) Biogeography of recent marine bivalve mollusks and its implications of paleobiogeography and the geography of extinction: a progress report. *Historical Biology*, 10, 25–47.

Flint, M.L. & van den Bosch, R. (1981) *Introduction to Integrated Pest Management.* Plenum Press, New York.

Flower, R.J. & Battarbee, R.W. (1983) Diatom evidence for recent acidification of two Scottish lochs. *Nature*, 305, 130–133.

Ford, E.B. (1975) *Ecological Genetics*, 4th ed. Chapman and Hall, London.

Ford, M.J. (1982) *The Changing Climate: Responses of the Natural Fauna and Flora.* George Allen and Unwin, London.

Frank, D.A. & McNaughton, S.J. (1991) Stability increases with diversity in plant communities: empirical evidence from the 1988 Yellowstone drought. *Oikos*, 62, 360–362.

Franklin, I.A. (1980) Evolutionary change in small populations. In: *Conservation Biology, An Evolutionary-Ecological Perspective* (M.E. Soulé & B.A. Wilcox, eds), pp. 135–149. Sinauer Associates, Sunderland, MA.

Fridriksson, S. (1975) *Surtsey: Evolution of Life on a Volcanic Island.* Butterworths, London.

Gadgil, M. (1971) Dispersal: population consequences and evolution. *Ecology*, 52, 253–261.

Gaston, K.J. (1998) *Biodiversity.* Blackwell Science, Oxford.

Gibbs, L. (1998) *Love Canal: The Story Continues.* New Society Publishers (in press).

Ginsberg, J.R. (1993) Can we build an Ark? *Trends in Ecology and Evolution*, 8, 4–6.

Godfray, H.C.J. & Crawley, M.J. (1998) Introductions. In: *Conservation Science and Action* (W.J. Sutherland, ed), pp. 39–65.

Gorman, M.L. (1979) *Island Ecology.* Chapman & Hall, London.

Gostick, K.G. (1982) Agricultural Development and Advisory Service (A.D.A.S.) recommendations to farmers on manure disposal and recycling. *Philosophical Transactions of the Royal Society of London, Series B*, 296, 329–332.

Gould, W.A. & Walker, M.D. (1997) Landscape-scale patterns in plant species richness along an arctic river. *Canadian Journal of Botany*, 75, 1748–1765.

Grant, P.R. (ed.) (1998) *Evolution on Islands.* Oxford University Press, Oxford.

Green, R.E. (1998) Long-term decline in the thickness of eggshells of thrushes, *Turdus* spp., in Britain. *Proceedings of the Royal Society of London, Series B*, 265, 679–684.

Green, T.R. & Ryan, C.A. (1972) Wound-induced proteinase inhibitor in plant leaves: a possible defense mechanism against insects. *Science*, 175, 776–777.

Grime, J.P. (1973) Control of species density in herbaceous vegetation. *Journal of Environmental Management*, 1, 151–167.

Groves, R.H. & Williams, J.D. (1975) Growth of skeleton weed (*Chondrilla juncea* L.) as affected by growth of subterranean clover (*Trifolium subterraneum* L.) and infection by *Puccinia chondrilla* Bubak and Syd. *Australian Journal of Agricultural Research*, 26, 975–983.

Haefner, P.A. (1970) The effect of low dissolved oxygen concentrations on temperature–salinity tolerance of the sand shrimp. *Crangon septemspinosa*. *Physiological Zoology*, 43, 30–37.

Haines, E. (1979) Interaction between Georgia salt marshes and coastal waters: a changing paradigm. In: *Ecological Processes in Coastal and Marine Systems* (R.J. Livingston, ed). Plenum Press, New York.

Hairston, N.G., Smith, F.E. & Slobodkin, L.B. (1960) Community structure, population control, and competition. *American Naturalist*, 44, 421–425.

Hall, D.C., Norgard, R.B. & True P.K. (1975) The performance of independent pest management consultants. *California Agriculture*, 29, 12–14.

Hall, S.J. & Raffaelli, D.G. (1993) Food webs: theory and reality. *Advances in Ecological Research*, 24, 187–239.

Hansen, J., Ruedy, R., Glascoe, J., and Sato, M. (in press). GISS analysis of surface temperature change. *Journal of Geophysical Research*

Hanski, I. (1989) Metapopulation dynamics: does it help to have more of the same? *Trends in Ecology and Evolution*, 4, 113–114.

Hanski, I. (1994) Metapopulation ecology. In: *Spatial and Temporal Aspects of Population Processes* (O.E. Rhodes Jr., ed.). University of Georgia, Georgia.

Hanski, I., Pakkala, T., Kuussaari, M. & Lei, G. (1995) Metapopulation persistence of an endangered butterfly in a fragmented landscape. *Oikos*, 72, 21–28.

Harcourt, D.G. (1971) Population dynamics of *Leptinotarsa decemlineata* (Say) in eastern Ontario. Ill. Major population processes. *Canadian Entomologist*, 103, 1049–1061.

Harper, J.L. (1955) The influence of the environment on seed and seedling mortality. VI. The effects of the interaction of soil moisture content and temperature on the mortality of maize grains. *Annals of Applied Biology*, 43, 696–708.

Harper, J.L. (1977) *The Population Biology of Plants*. Academic Press, London.

Harper, J.L. & Ogden, J. (1970) The reproductive strategy of higher plants: I. The concept of strategy with special reference to *Senecio vulgaris* L. *Journal of Ecology*, 58, 681–698.

Harper, J.L. & White, J. (1974) The demography of plants. *Annual Review of Ecology and Systematics*, 5, 419–463.

Hassell, M.P., Latto, J. & May, R.M. (1989) Seeing the wood for the trees: detecting density dependence from existing life-table studies. *Journal of Animal Ecology*, 58, 883–892.

Heal, O.W. & MacLean, S.F. (1975) Comparative productivity in ecosystems—secondary productivity. In: *Unifying Concepts in Ecology* (W.H. van Dobben & R.H. Lowe-McConnell, eds), pp. 89–108. Junk, The Hague.

Hendrix, S.D. (1979) Compensatory reduction in a biennial herb following insect defloration. *Oecologia*, 42, 107–118.

Hess, G.R. (1994) Conservation corridors and contagious disease: a cautionary note. *Conservation Biology*, 8, 256–262.

Heywood, V.H. (ed.) (1995) *Global Biodiversity Assessment*. Cambridge University Press, Cambridge.

Hilborn, R. & Walters, C.J. (1992) *Quantitative Fisheries Stock Assessment*. Chapman & Hall, New York.

Hildrew, A.G. & Townsend, C.R. (1980) Aggregation, interference and the foraging by larvae of *Plectrocnemia conspersa* (Trichoptera: Polycentropodidae). *Animal Behaviour*, 28, 553–560.

Holt, R.D. & Hassell, M.P. (1993) Environmental heterogeneity and the stability of host–parasitoid interactions. *Journal of Animal Ecology*, 62, 89–100.

Hoyer, M.V. & Canfield, D.E. (1994) Bird abundance and species richness on Florida lakes: influence of trophic status, lake morphology & aquatic macrophytes. *Hydrobiologia*, 297, 107–119.

Hudson, P.J., Dobson, A.P. & Newborn, D. (1992) Do parasites make prey vulnerable to predation? Red grouse and parasites. *Journal of Animal Ecology*, 61, 681–692.

Hudson, P.J., Dobson, A.P. & Newborn, D. (1998) Prevention of population cycles by parasite removal. *Science*, 282, 2256–2258.

Huffaker, C.B. (1958) Experimental studies on predation: dispersion factors and predator–prey oscillations. *Hilgardia*, 27, 343–383.

Hunter, M.L. & Yonzon, P. (1992) Altitudinal distributions of birds, mammals, people, forests, and parks in Nepal. *Conservation Biology*, 7, 420–423.

Huryn, A.D. (1998) Ecosystem-level evidence for top-down and bottom-up control of production in a grassland stream system. *Oecologia*, 115, 173–183.

Hutchings, M.R., Kyriazakis, I., Gordon I.J. & Jackson F. (1999). Trade-offs between nutrient intake and faecal avoidance in herbivore foraging decisions: the effect of animal parasitic status, level of feeding motivation and sward nitrogen content. *Journal of Animal Ecology*, 68, 310323.

Hutchinson, G.E. (1957) Concluding remarks. *Cold Spring Harbour Symposium on Quantitative Biology*, 22, 415–427.

Huxley, C.R. & Cutler, D.F. (1991) *Ant-Plant Interactions*. Oxford University Press, Oxford.

Inouye, D.W. (1978) Resource partitioning in bumblebees: experimental studies of foraging behaviour. *Ecology*, 59, 672–678.

Inouye, R.S. & Tilman, D. (1995) Convergence and divergence of old-field vegetation after 11 yr of nitrogen addition. *Ecology*, 76, 1872–1877.

Inouye, R.S., Huntly, N.J., Tilman, D., Tester, J.R., Stillwell, M. & Zinnel, K.C. (1987) Old-field succession on a Minnesota sand plain. *Ecology*, 68, 12–26.

International Organisation for Biological Control (1989) *Curent Status of Integrated Farming Systems Research in Western Europe* (P. Vereijken & D.J. Royle, eds). IOBC WPRS Bulletin 12(5).

IUCN/UNEP/WWF (1991) *Caring for the Earth. A Strategy for Sustainable Living*. Gland, Switzerland.

Ives, A.R. (1992) Continuous-time models of host-parasitoid interactions. *American Naturalist*, 140, 1–29.

Jain, S.K. & Bradshaw, A.D. (1966) Evolutionary divergence among adjacent plant populations. I. The evidence and its theoretical analysis. *Heredity*, 21, 407–411.

Janis, C.M. (1993). Tertiary mammal evolution in the context of changing climates, vegetation and tectonic events. *Annual Review of Ecology and Systematics*, 24, 467–500.

Janzen, D.H. (1966) Coevolution of mutualism between ants and acacias in Central America. *Evolution*, 20, 249–275.

Joern, A. & Lawlor, L.R. (1980) Food and microhabitat utilization by the grasshoppers from arid grasslands: comparisons with neutral models. *Ecology*, 61, 591–599.

Johnston, D.W. & Odum, E.P. (1956) Breeding bird populations in relation to plant succession on the piedmont of Georgia. *Ecology*, 37, 50–62.

Jones, C.G., Lawton, J.H. & Shaachak, M. (1997) Positive and negative effects of organisms as physical ecosystem engineers. *Ecology*, 78, 1946–1957.

Jones, M. & Harper, J.L. (1987) The influence of neighbours on the growth of trees. I. The demography of buds in *Betula pendula*. *Proceedings of the Royal Society of London, Series B*, 232, 1–18.

Kaufman, L. (1992) Catastrophic change in a species-rich freshwater ecosystem: Lessons from Lake Victoria. *Bioscience*, 42, 846–858.

Keddy, P.A. (1982) Experimental demography of the sand-dune annual. *Cakile edentula*, growing along an environmental gradient in Nova Scotia. *Journal of Ecology*, 69, 615–630.

Keeling, C.D., Whorf, T.P., Wahlen, M. & van der Plicht, J. (1995) Interannual extremes in the rate of rise of atmospheric carbon dioxide since 1980. *Nature*, 375, 666–670.

Keith, L.B. (1983) Role of food in hare population cycles. *Oikos*, 40, 385–395.

Keith, L.B., Cary, J.R., Ronstad, O.J. & Brittingham, M.C. (1984) Demography and ecology of a declining snowshoe hare population. *Wildlife Monographs*, 90, 1–43.

Kerbes, R.H., Kotanen, P.M. & Jefferies, R.L. (1990) Destruction of wetland habitats by lesser snow geese: a keystone species on the west coast of Hudson Bay. *Journal of Applied Ecology*, 27, 242–258.

Kettlewell, H.B.D. (1955) Selection experiments on industrial melanism in the Lepidoptera. *Heredity*, 9, 323–342.

Kigel, J. (1980) Analysis of regrowth patterns and carbohydrate levels in *Lolium multiforum* Lam. *Annals of Botany*, 45, 91–101.

Kingston, T.J. (1977) *Natural Manuring by Elephants in the Tsavo National Park, Kenya.* PhD thesis, University of Oxford.

Kingston, T.J. & Coe, M.J. (1977) The biology of a giant dung-beetle (*Heliocorpis dilloni*) (Coleoptera: Scarabaeidae). *Journal of Zoology*, 181, 243–263.

Koella, J.C., Sörensen, F.L. & Anderson, R.A. (1998) The malaria parasite, *Plasmodium falciparum*, increases the frequency of multiple feeding of its mosquito vector, *Anopheles gambiae. Proceedings of the Royal Society of London, Series B*, 265, 763–768.

Kratz, T.K., Webster, K.E., Bowser, C.J. et al. (1997) The influence of landscape position on lakes in Northern Wisconsin. *Freshwater Biology*, 37, 209–217.

Krebs, C.J. (1972) *Ecology.* Harper & Row, New York.

Krebs, C.J., Boonstra, R., Boutin, S. *et al.* (1992) What drives the snowshoe hare cycle in Canada's Yukon. In: *Wildlife 2001: Populations* (D.R. McCullough & R.H. Barrett, eds), pp. 886–896. Elsevier, New York.

Krebs, J.R. (1978) Optimal foraging: decision rules for predators. In: *Behavioural Ecology: An Evolutionary Approach* (J.R. Krebs & N.B. Davies, eds), pp. 23–63. Blackwell Scientific Publications, Oxford.

Krebs, J.R., Erichsen, J.T., Webber, M.I. & Charnov, E.L. (1977) Optimal prey selection in the great tit (*Parus major*). *Animal Behaviour*, 25, 30–38.

Kutschera, L. (1960) *Wurzelatlas mitteleuropäischer Ackerunkräuter und Kulturpflanzen.* DLG Verlag, Frankfurt-am-Main.

Lande, R. & Barrowclough, G.F. (1987) Effective population size, genetic variation, and their use in population management. In: *Viable Populations for Conservation* (M.E. Soulé, ed), pp. 87–123. Cambridge University Press, Cambridge.

Larcher, W. (1980) *Physiological Plant Ecology*, 2nd ed. Springer-Verlag, Berlin.

Lawlor, L.R. (1980) Structure and stability in natural and randomly constructed competitive communities. *American Naturalist*, 116, 394–408.

Lawrence, W.H. & Rediske, J.H. (1962) Fate of sown douglas-fir seed. Forest Science, 8, 211–218.

Lawton, J.H. & May, R.M. (1984) The birds of Selborne. *Nature*, 306, 732–733.

Lawton, J.H. & Woodroffe, G.L. (1991) Habitat and the distribution of water voles: why are there gaps in a species' range? *Journal of Animal Ecology.* 60, 79–91.

Le Cren, E.D. (1973) Some examples of the mechanisms that control the population dynamics of salmonid fish. In: *The Mathematical Theory of the Dynamics of Biological Populations* (M.S. Bartlett & R.W. Hiorns, eds), pp. 125–135. Academic Press, London.

Leverich, W.J. & Levin, D.A. (1979) Age-specific survivorship and reproduction in *Phlox drummondii. American Naturalist,* 113, 881–903.

Levins, R. (1969) Some demographic and genetic consequences of environmental heterogeneity for biological control. *Bulletin of the Entomological Society of America,* 15, 237–240.

Lewis, J.R. (1976) *The Ecology of Rocky Shores.* Hodder & Stoughton, London.

Likens, G.E. (1989) Some aspects of air pollutant effects on terrestrial ecosystems and prospects for the future. *Ambio*, 18,172–178.

Likens, G.E. & Bormann, F.H. (1995) *Biogeochemistry of a Forested Ecosystem* (2nd edition). Springer-Verlag, New York.

Likens, G.E., Driscoll, C.T. & Buso, D.C. (1996) Long-term effects of acid rain: response and recovery of a forest ecosystem. *Science*, 272, 244–245.

Likens, G.E. & Bormann, F.G. (1975) An experimental approach to New England landscapes. In: *Coupling of Land and Water Systems* (A.D. Hasler, ed), pp. 7–30. Springer-Verlag, New York.

Likens, G.E., Bormann, F.H., Pierce, R.S. & Fisher, D.W. (1971) Nutrient–hydrologic cycle interaction in small forested watershed ecosystems. In: *Productivity of Forest Ecosystems* (P. Duvogneaud, ed). UNESCO, Paris.

Lindeman, R.L. (1942) The trophic–dynamic aspect of ecology. *Ecology*, 23, 399–418.

Losos, E. (1993) The future of the US Endangered Species Act. *Trends in Ecology and Evolution*, 8, 332–336.

Lotka, A.J. (1932) The growth of mixed populations: two species competing for a common food supply. *Journal of the Washington Academy of Sciences*, 22, 461–469.

Louda, S.M. (1982) Distributional ecology: variation in plant recruitment over a gradient in relation to insect seed predation. *Ecological Monographs*, 52, 25–41.

Louda, S.M. (1983) Seed predation and seedling mortality in the recruitment of a shrub, *Haplopappus venetus* (Asteraceae), along a climatic gradient. *Ecology*, 64, 511–521.

Lubchenco, J. (1978) Plant species diversity in a marine intertidal community: importance of herbivore food preference and algal competitive abilities. *American Naturalist*, 112, 23–39.

Lubchenko, J., Olson, A.M. *et al.* (1991) The sustainable biosphere initiative: an ecological research agenda. *Ecology*, 72, 371–412.

Lukens, R.J. & Mullany, R. (1972) The influence of shade and wet on southern corn blight. *Plant Disease Reporter*, 56, 203–206.

Lussenhop, J. (1992) Mechanisms of microarthropod–microbial interactions in soil. *Advances in Ecological Research*, 23, 1–33.

MacArthur, J.W. (1975) Environmental fluctuations an species diversity. In: *Ecology and Evolution of Communities* (M.L. Cody & J.M. Diamond, eds), pp. 74–80. Belknap, Cambridge, MA.

MacArthur, R.H. (1955) Fluctuations of animal populations and a measure of community stability. *Ecology*, 36, 533–536.

MacArthur, R.H. (1972) *Geographical Ecology.* Harper & Row, New York.

MacArthur, R.H. & Pianka, E.R. (1966) On optimal use of a patchy environment. *American Naturalist,* 100, 603–609.

MacArthur, R.H. & Wilson, E.O. (1967) *The Theory of Island Biogeography.* Princeton University Press, Princeton, NJ.

Mace, G.M. (1994) An investigation into methods for categorizing the conservation status of species. In: *Large-Scale Ecology and Conservation Biology* (P.J. Edwards, R.M. May & N.R. Webb, eds), pp. 293–312. Blackwell, Oxford.

MacDonald, D.W. (1980) *Rabies and Wildlife.* Oxford University Press, Oxford.

Mackie, G.L., Qadri, S.U. & Reed, R.M. (1978) Significance of litter size in *Musculium securis* (Bivalia: Sphaeridae). *Ecology*, 59, 1069–1074.

MacLulick, D.A. (1937) Fluctuations in numbers of the varying hare (*Lepus americanus*). *University of Toronto Studies, Biology Series*, 43, 1–136.

Magurran, A.E. (1998) Population differentiation without speciation. *Philosophical Transactions of the Royal Society of London, Series B*, 353, 275–286.

Manilove, R.J. (1985) *On the population ecology of Avena fatua L.* Unpublished thesis, University of Liverpool.

Manire, C.A. & Gruber, S.H. (1990) Many sharks may be headed toward extinction. *Conservation Biology*, 4, 10–11.

Marshall, I.D. & Douglas, G.W. (1961) Studies in the epidemiology of infectious myxomatosis of rabbits. VIII. Further observations on changes in the innate resistance of Australian wild rabbits exposed to myxomatosis. *Journal of Hygiene*, 59, 117–122.

Martin, P.S. (1984) Prehistoric overkill: the global model. In: *Quaternary Extinctions: A Prehistoric Revolution* (P.S. Martin & R.G. Klein, eds). University of Arizona Press, Tuscon, Arizona.

Marzusch, K. (1952) Untersuchungen über di Temperaturabhängigkeit von Lebensprozessen bei Insekten unter besonderer Berücksichtigung winter-schlantender Kartoffelkäfer. *Zeitschrift für vergleicherde Physiologie,* 34, 75–92.

May, R.M. (1981c) Patterns in multi-species communitites. In: *Theoretical Ecology: Principles and Applications,* 2nd ed. (R.M. May, ed), pp. 197–227. Blackwell Scientific Publications, Oxford.

Mayr, E. (1942) *Systematics and the Origin of Species.* Columbia University Press, New York.

McIntosh, A.R. & Townsend, C.R. (1994) Interpopulation variation in mayfly antipredator tactics: differential effects of contrasting predatory fish. *Ecology,* 75, 2078–2090.

McIntosh, A.R. & Townsend, C.R. (1996) Interactions between fish, grazing invertebrates and algae in a New Zealand stream: a trophic cascade mediated by fish-induced changes to grazer behavior. *Oecologia,* 108, 174–181.

McNaughton, S.J. (1977) Diversity and stability of ecological communities: a comment on the role of empiricism in ecology. *American Naturalist,* 111, 515–525.

Metcalf, R.L. (1980) Changing role of insecticides in crop protection. *Annual Review of Entomology,* 25, 219–256.

Mikkelson, G.M. (1993) How do food webs fall apart? A study of changes in trophic structure during relaxation on habitat fragments. *Oikos,* 67, 539–547.

Miller, G.T. Jr. (1988) *Environmental Science,* 2nd ed. Wadsworth, Belmont CA.

Morgan, R.P.C. (1985) Assessment of soil erosion in England and Wales. *Soil Use and Management,* 1, 127–131.

Moss, G.D. (1971) The nature of the immune response of the mouse to the bile duct cestode, *Hymenolepis microstoma. Parasitology,* 62, 285–294.

Murdoch, W.W. & Stewart-Oaten, A. (1975) Predation and population stability. *Advances in Ecological Research,* 9, 1–131.

Murdoch, W.W., Briggs, C.J., Nisbet, R.M., Gurney, W.S.C. & Stewart-Oaten, A. (1992) Aggregation and stability in meta-population models. *American Naturalist,* 140, 41–58.

Murton, R.K., Westwood, N.J. & Isaacson, A.J. (1974) A study of woodpigeon shooting: the exploitation of a natural animal population. *Journal of Applied Ecology,* 11, 61–81.

Myers, J.H. (1988) Can a general hypothesis explain population cycles of forest Lepidoptera. *Advances in Ecological Research,* 18, 179–242.

Myers, N. (1991) The biodiversity challenge: expanded 'hot-spots' analysis. *Environmentalist,* 10, 243–256.

National Research Council (1990) *Alternative Agriculture.* National Academy of Sciences, Academy Press, Washington, D.C.

Nedergaard, J. & Cannon, B. (1990) Mammalian hibernation. *Philosophical Transactions of the Royal Society, Series B,* 326, 669–686; also in *Life at Low Temperatures* (R.M. Laws & F. Franks, eds), pp. 153–170. The Royal Society, London.

Neilson, R.P., Prentice, I.C., Smith, B., Kittel, T. & Viner, D. (1998) Simulated changes in vegetation distribution under global warming. Available as Annex C at http://www.epa.gov/globalwarming/reports/pubs/ipcc/annex/index.html.

NERC (1990) *Our Changing Environment.* Natural Environment Research Council, London. (NERC acknowledges the significant contribution of Fred Pearce to the document.)

Newsham, K.K., Fitter, A.H. & Watkinson, A.R. (1995) Arbuscular mycorrhiza protect an annual grass from root pathogenic fungi in the field. *Journal of Ecology,* 83, 991–1000.

Nielsen, B.O. & Ejlerson, A. (1977) The distribution of herbivory in a beech canopy. *Ecological Entomology,* 2, 293–299.

Niklas, K.J., Tiffney, B.H. & Knoll, A.H. (1983) Patterns in vascular land plant diversification. *Nature,* 303, 614–616.

Nowak, C.L., Nowak, R.S., Tausch, R.J. & Wigand, P.E. (1994) Tree and shrub dynamics in northwestern Great Basin woodland and shrub steppe during the Late Pleistocene and Holocene. *American Journal of Botany,* 8, 265–277.

Nunney, L. & Campbell, K.A. (1993) Assessing minimum viable population sizes: demography meets population genetics. *Trends in Ecology and Evolution,* 8, 234–239.

Ogden, J. (1968) *Studies on Reproductive Strategy with Particular Reference to Selected Composites.* Ph.D. thesis, University of Wales.

Owen-Smith, N. (1987) Pleistocene extinctions: the pivotal role of megaherbivores. *Paleobiology,* 13, 351–362.

Paine, R.T. (1966) Food web complexity and species diversity. *American Naturalist,* 100, 65–75.

Park, T. (1948) Experimental studies of interspecific competition. I. Competition between populations of the flour beetle *Tribolium confusum* Duval and *Tribolium castaneum* Herbst. *Ecological Monographs,* 18, 267–307.

Park T. (1954) Experimental studies of interspecific competition. II. Temperature, humidity and competition in two species of *Tribolium. Physiological Zoology,* 27, 177–238.

Pauly, D. & Christensen, V. (1995) Primary production required to sustain global fisheries. *Nature,* 374, 255–257.

Peake, A.J. & Quinn, G.P. (1993) Temporal variation in species–area curves for invertebrates in clumps of an intertidal mussel. *Ecography,* 16, 269–277.

Pearl, R. (1927) The growth of populations. *Quarterly Review of Biology,* 2, 532–548.

Pearl, R. (1928) *The Rate of Living.* Knopf, New York.

Perrins, C.M. (1965) Population fluctuations and clutch size in the great tit, *Parus major* L. *Journal of Animal Ecology,* 34, 601–647.

Peters, R.H. (1983) *The Ecological Implications of Body Size.* Cambridge University Press, Cambridge.

Petren, K., Grant B.R. & Grant, P.R. (1999) A phylogeny of Darwin's finches based on microsatellite DNA variation. *Proceeding of the Royal Society of London, Series B,* 266, 321–329.

Pianka, E.R. (1967) On lizard species diversity: North American flatland deserts. *Ecology,* 48, 333–351.

Pianka, E.R. (1983) *Evolutionary Ecology,* 3rd ed. Harper & Row, New York.

Pimentel, D. (1993) Cultural controls for insect pest management. In: *Pest Control and Sustainable Agriculture* (S. Corey, D. Dall & W. Milne, eds), pp. 35–38. CSIRO, East Melbourne.

Pimentel, D., Krummel, J., Gallahan, D., *et al.* (1978) Benefits and costs of pesticide use in U.S. food production. *Bioscience,* 28, 777–784.

Pimm, S.L. (1991) *The Balance of Nature: Ecological Issues in the Conservation of Species and Communities.* University of Chicago Press, Chicago and London.

Pisek, A., Larcher, W., Vegis, A. & Napp-Zin, K. (1973) The normal temperature range. In: *Temperature and Life* (H. Precht, J. Christopherson, H. Hense & W. Larcher, eds), pp. 102–194. Springer-Verlag, Berlin.

Pitcher, T.J. & Hart, P.J.B. (1982) *Fisheries Ecology.* Croom Helm, London.

Power, M.E. (1990) Effects of fish in river food webs. *Science,* 250, 411–415.

Prance, G.T. (1987). Biogeography of neotropical plants. In: *Biogeography and Quaternary History of Tropical America* (T.C. Whitmore & G.T. Prance, eds), pp. 46–65. Oxford Monographs on Biogeography, 3.

Price, P.W. (1980) *Evolutionary Biology of Parasites.* Princeton University Press, Princeton.

Primack, R.B. (1993) *Essentials of Conservation Biology*. Sinauer Associates, Sunderland, MA.

Pyke, G.H. (1982) Local geographic distributions of bumblebees near Crested Butte, Colorado: competition and community structure. *Ecology*, 63, 555–573.

Rätti, O., Dufva, R. & Alatalo, R.V. (1993) Blood parasites and male fitness in the pied flycatcher. *Oecologia*, 96, 410–414.

Ratcliffe, D.A. (1970) Changes attributable to pesticides in egg breakage frequence and eggshell thickness in some British birds. *Journal of Applied Ecology*, 7, 67–107.

Reichle, D.E. (1970) *Analysis of Temperate Forest Ecosystems*. Springer-Verlag, New York.

Reid, W.V. & Miller, K.R. (1989) *Keeping Options Alive: The Scientific Basis for Conserving Biodiversity*. World Resources Institute, Washington, DC.

Reznick, D.N., Shaw, F.H., Rodd, F.H. & Shaw, R.G. (1997) Evaluation of the rate of evolution in natural populations of guppies (*Poecilia reticulata*). *Science*, 275, 1934–1937.

Richards, O.W. & Waloff, N. (1954) Studies on the biology and population dynamics of British grasshoppers. *Anti-Locust Bulletin*, 17, 1–182.

Rickards, J., Kelleher, M.J. & Storey, K.B. (1987) Strategies of freeze avoidance in larvae of the goldenrod gall moth *Epiblema scudderiana*: winter profiles of a natural population. *Journal of Insect Physiology*, 33, 581–586.

Ricklefs, R.E. (1973) *Ecology*. Nelson, London.

Ridsdill-Smith, T.J. (1991) Competition in dung-breeding insects. In: *Reproductive Behaviour of Insects* (W.J. Bailey & T.J. Ridsdill-Smith, eds), pp. 264–294. Chapman & Hall, London.

Rieck, A.F., Belli, J.A. & Blaskovics, M.E. (1960) Oxygen consumption of whole animal tissues in temperature acclimated amphibians. *Proceedings of the Society of Experimental Biology and Medicine*, 103, 436–439.

Robinson, D. (1991) Roots and resource fluxes in plants and communities. In D. Atkinson, ed., *Plant Root Growth: An Ecological Perspective*. Blackwell, Oxford.

Rosenzweig, M.L. (1971) Paradox of enrichment: destabilization of exploitation ecosystems in ecological time. *Science*, 171, 385–387.

Roush, R.T. & McKenzie, J.A. (1987) Ecological genetics of insecticide and acaricide resistance. *Annual Review of Entomology*, 32, 361–380.

Sakai, A. & Otsuka, K. (1970) Freezing resistance of alpine plants. *Ecology*, 51, 665–671.

Sale, P.F. & Douglas, W.A. (1984) Temporal variability in the community structure of fish on coral patch reefs and the relation of community structure to reef structure. *Ecology*, 65, 409–422.

Salisbury, E.J. (1942) *The Reproductive Capacity of Plants*. Bell, London.

Sarukhán, J. & Harper, J.L. (1973) Studies on plant demography: *Ranunculus repens* L., *R. bulbosus* L. and *R. acris* L. I. Population flux and survivorship. *Journal of Ecology*, 61, 675–716.

Savidge, J.A. (1987) Extinction of an island forest avifauna by an introduced snake. *Ecology*, 68, 660–668.

Schall, J.J. (1992) Parasite-mediated competition in *Anolis* lizards. *Oecologia*, 92, 58–64.

Schindler, D.W. (1978) Factors regulating phytoplankton production and standing crop in the world's freshwaters. *Limnology and Oceanography*, 23, 478–486.

Schluter, D. & McPhail, J.D. (1993) Character displacement and replicate adaptive radiation. *Trends in Ecology and Evolution*, 8, 197–200.

Schoener, T.W. (1983) Field experiments on interspecific competition. *American Naturalist*, 122, 240–285.

Schoenly, K., Beaver, R.A. & Heumier, T.A. (1991) On the trophic relations of insects: a food-web approach. *American Naturalist*, 137, 597–638.

Schulze, E.D. (1970) Dre CO_2-Gaswechsel de Buche (*Fagus sylvatica* L.) in Abhäbgigkeit von den Klimafaktoren in Freiland. *Flora, Jena*, 159, 177–232.

Schulze, E.D. (1989) Air pollution and forest decline in a spruce (*Picea abies*) forest. *Science*, 244, 7765–7783.

Schulze, E.D., Fuchs, M.I. & Fuchs, M. (1977a) Spatial distribution of photosynthetic capacity and performance in a mountain spruce forest in northern Germany. I. Biomass distribution and daily CO_2 uptake in different crown layers. *Oecologia*, 29, 43–61.

Schulze, E.D., Fuchs, M.I. & Fuchs, M. (1977b) Spatial distribution of photosynthetic capacity and performance in a mountain spruce forest in northern Germany. III. The significance of the evergreen habit. *Oecologia*, 30, 239–249.

Silvertown, J.W. (1982) *Introduction to Plant Population Ecology*. Longman, London.

Simberloff, D.S. (1976) Experimental zoogeography of islands: effects of island size. *Ecology*, 57, 629–648.

Simpson, G.G. (1952) How many species? *Evolution*, 6, 342.

Sinclair, A.R.E. (1973) Regulation, and population models for a tropical ruminant. *East African Wildlife Journal*, 11, 307–316.

Sinclair, A.R.E. & Norton-Griffiths, M. (1982) Does competition or facilitation regulate migrant ungulate populations in the Serengeti? A test of hypothesis. *Oecologia*, 53, 354–369.

Slobodkin, L.B., Smith, F.E. & Hairston, N.G. (1967) Regulation in terrestrial ecosystems, and the implied balance of nature. *American Naturalist*, 101, 109–124.

Smith, F.D.M. May, R.M., Pellew, R., Johnson, T.H. & Walter, K.R. (1993). How much do we know about the current extinction rate? *Trends in Ecology and Evolution*, 8, 375–378.

Smith, F.E. (1961) Density dependence in the Australian thrips. *Ecology*, 42, 403–407.

Smith, J.N.M., Krebs, C.J., Sinclair, A.R.E. & Boonstra, R. (1988) Population biology of snowshoe hares. II. Interactions with winter food plants. *Journal of Animal Ecology*, 57, 269–286.

Soil Use and Management, 1, 127–131.

Sousa, M.E. (1979a) Experimental investigation of disturbance and ecological succession in a rocky intertidal algal community. *Ecological Monographs*, 49, 227–254.

Sousa, M.E. (1979b) Disturbance in marine intertidal boulder fields: the nonequilibrium maintenance of species diversity. *Ecology*, 60, 1225–1239.

Spencer, C.N., McClelland, B.R. & Stanford, J.A. (1991) Shrimp stocking, salmon collapse, and eagle displacement. *Bioscience*, 41, 14–21.

Spiller, D.A. & Schoener, T.W. (1994) Effects of a top and intermediate predators in a terrestrial food web. *Ecology*, 75, 182–196.

Spradbery, J.P. (1970) Host findings of *Rhyssa persuasoria* (L.), an ichneumonid parasite of siricid woodwasps. *Animal Behaviour*, 18, 103–114.

Sprent, J.I. & Sprent, P. (1990) *Nitrogen Fixing Organisms: Pure and Applied Aspects*. Chapman and Hall, London.

Stauffer, R.C. (1975) *Charles Darwin's Natural Selection: Being the second part of his big species book written from 1856 to 1858*. Cambridge University Press, London.

Steffen, W.L., Walker, B.H., Ingram, J.S.I. & Koch, G.W. (eds) (1992) *Global Change and Terrestrial Ecosystems: The Operational Plan*. IGBP, ICSU, Stockholm, Sweden.

Stenseth, N.C., Falck, W., Bjornstad, O.N. & Krebs, C.J. (1997) Population regulation in snowshoe hare and lynx populations: asymmetric food web configurations between the snowshoe hare and the lynx. *Proceedings of the National Academy of Science of the USA*, 94, 5147–5152.

Stevens, C.E. (1988) Comparative physiology of the vertebrate digestive system. Cambridge University Press, London.

Stone, G.N. & Cook J.M. (1988) The structure of cynipid oak galls: patterns in the evolution of an extended phenotype. *Proceedings of the Royal Society of London, Series B*, 265, 979–988.

Strong, D.R. (1992) Are trophic cascades all wet? Differentiation and donor-control in speciose ecosystems. *Ecology,* 73, 747–754.

Strong, D.R. Jr., Lawton, J.H. & Southwood, T.R.E. (1984) *Insects on Plants: Community Patterns and Mechanisms.* Blackwell Scientific Publications, Oxford.

Sutton, S.L. & Collins, N.M. (1991) Insects and tropical forest conservation. In: *The Conservation of Insects and Their Habitats* (N.M. Collins and J.A. Thomas, eds) pp. 405–424. Academic Press, London.

Symonides, E. (1979) The structure and population dynamics of psammo-phytes on inland dunes. II. Loose-sod populations. *Ekologia Polska,* 27, 191–234.

Symonides, E. (1983) Population size regulation as a result of intra-population interactions. I. The effect of density on the survival and development of individuals of Erophila verna (L.). *Ekologia Polska,* 31, 839–881.

Tansley, A.G. (1904) The problems of ecology. *New Phytologist,* 3, 191–200.

Thomas, C.D. & Harrison, S. (1992) Spatial dynamics of a patchily distrib-uted butterfly species. *Journal of Applied Ecology,* 61, 437–446.

Thomas, C.D. & Jones, T.M. (1993) Partial recovery of a skipper butterfly (*Hesperia comma*) from population refuges: lessons for conservation in a fragmented landscape. *Journal of Animal Ecology,* 62, 472–481.

Thomas, C.D., Thomas, J.A. & Warren, M.S. (1992) Distributions of occu-pied and vacant butterfly habitats in fragmented landscapes. *Oecologia,* 92, 563–567.

Thompson, J.N. (1994) *The Coevolutionary Process.* University of Chicago Press, Chicago.

Thrall, P.H. & Burdon, J.J. (1997) Host-pathogen dynamics in a metapopula-tion context: the ecological and evolutionary consequences of being spatial. *Journal of Ecology,* 85, 743–754.

Tilman, D. (1982) *Resource Competition and Community Structure.* Princeton University Press, Princeton, NJ.

Tilman, D. (1986) Resources, competition and the dynamics of plant commu-nities. In: *Plant Ecology* (M.J. Crawley, ed), pp. 51–74. Blackwell Scien-tific Publications, Oxford.

Tilman, D., Mattson, M. & Langer, S. (1981) Competition and nutrient kinetics along a temperature gradient: an experimental test of a mechanis-tic approach to niche theory. *Limnology and Oceanography,* 26, 1020–1033.

Tjallingii, W.F. & Hogen Esch, Th. (1993) Fine structure of aphid stylet routes in plant tissues in correlation with EPG signals. *Physiological Entomology,* 18, 317–328.

Tokeshi, M. (1993) Species abundance patterns and community structure. *Advances in Ecological Research,* 24, 112–186.

Tonn, W.M. & Magnuson, J.J. (1982) Patterns in the species composition and richness of fish assemblages in northern Wisconsin lakes. *Ecology,* 63, 137–154.

Towns, D.R. & Ballantine, W.J. (1993) Conservation and restoration of New Zealand island ecosystems. *Trends in Ecology and Evolution,* 8, 452–457.

Townsend C.R., Thompson R.M., McIntosh, A.R., et al. (1998) Disturbance, resource supply, and food-web architecture in streams. *Ecology Letters,* 1, 200–209.

Townsend, C.R. & Crowl, T.A. (1991) Fragmented population structure in a native New Zealand fish: an effect of introduced brown trout? *Oikos,* 61, 348–354.

Townsend, C.R., Dolédec, S. & Scarsbrook, M.R. (1997) Species traits in relation to temporal and spatial heterogeneity in streams: a test of habitat templet theory. *Freshwater Biology,* 37, 367–388.

Townsend, C.R., Scarsbrook, M.R. & Dolédec, S. (1997) The intermediate disturbance hypothesis, refugia and biodiversity in streams. *Limnology and Oceanography,* 42, 938–949.

Townsend, C.R. & Hildrew, A.G. (1980) Foraging in a patchy environment by a predatory net-spinning caddis larva: a test of optimal foraging theory. *Oecologia,* 47, 219–221.

Townsend, C.R., Hildrew, A.G. & Francis, J.E. (1983) Community structure in some southern English streams: the influence of physiochemical factors. *Freshwater Biology,* 13, 521–544.

Turkington, R. & Harper, J.L. (1979) The growth, distribution and neighbour relationships of *Trifolium repens* in a permanent pasture. IV. Fine scale biotic differentiation. *Journal of Ecology,* 67, 245–254.

Turner, J.R.G., Lennon, J.J. & Greenwood, J.J.D. (1996) Does climate cause the global biodiversity gradient? In: *Aspects of the Genesis and Mainte-nance of Biological Diversity* (M. Hochberg, J. Claubert & R. Barbault, eds). Oxford University Press, London, New York.

Tuttle, M.D. (1979) Status, causes of decline, and management of endangered gray bats. *Journal of Wildlife Management,* 43, 1–17.

UNEP (1991) *Environmental Data Report,* 3rd ed. Basil Blackwell, Oxford.

United Nations (1998) *Global change and sustainable development: critical trends.* Report of the Secretary General, The United Nations, New York. Also available on the World Wide Web at www.un.org/esa/sustdev/trends.html

US Department of the Interior (1990) *Audit Report: The Endangered Species Program.* US Fish and Wildlife Service Report 90–98.

Utida, S. (1957) Cyclic fluctuations of population density intrinsic to the host–parasite system. *Ecology,* 38, 442–449.

Valentine, J.W. (1970) How many marine invertebrate fossil species? A new approximatio. *Journal of Paleontology* 44, 410–415.

Volterra, V. (1926) Variations and fluctuations of the numbers of individuals in animal species living together. (Reprinted in 1931. In: R.N. Chapman, *Animal Ecology.* McGraw Hill, New York.)

Waage, J.K. & Greathead, D.J. (1988) Biological control: challenges and opportunities. *Philosophical Transactions of the Royal Society of London, Series B,* 318, 111–128.

Walsh, J.A. (1983) Selective primary health care: strategies for control of disease in the developing world. IV. Measles. *Reviews of Infectious Diseases,* 5, 330–340.

Waring, R.H. & Schlesinger, W.H. (1985) *Forest Ecosystems: Concepts and Management.* Academic Press, Orlando, FL.

Warren, P.H. (1989) Spatial and temporal variation in the structure of a freshwater food web. *Oikos,* 55, 299–311.

Watkins, C.V. & Harvey, L.A. (1942) On the parasites of silver foxes on some farms in the South West. *Parasitology,* 34, 155–179.

Watkinson, A.R. (1984) Yield-density relationships: the influence of resource availability on growth and self-thinning in populations of *Vulpia fascicu-lata. Annals of Botany,* 53, 469–482.

Watkinson, A.R. & Harper, J.L. (1978) The demography of a sand dune annual: *Vulpia fasciculata.* I. The natural regulation of populations. *Journal of Ecology,* 66, 15–33.

Webb, W.L., Lauenroth, W.K., Szarek, S.R. & Kinerson, R.S. (1983) Primary production and abiotic controls in forests, grasslands and desert ecosys-tems in the United States. *Ecology,* 64, 134–151.

Werner, H.H. & Hall, D.J. (1974) Optimal foraging and the size selection of prey by the bluegill sunfish *Lepomis macrochirus. Ecology,* 55, 1042–1052.

Werner, P.A. (1975) Predictions of fate from rosette size in teazel (*Dipsacus fullonum* L.). *Oecologia,* 20, 197–201.

Werner, P.A. & Platt, W.J. (1976) Ecological relationships of co-occurring golden rods (*Solidago:* Compositae). *American Naturalist,* 110, 959–971.

Westover, K.M., Kennedy, A.C. & Kelley, S.E. (1997) Patterns of rhizosphere community structure associated with co-occurring plant species. *Journal of Ecology,* 85, 863–874.

Whitehead, A.N. (1953) *Science and the Modern World.* Cambridge University Press, Cambridge.

Whittaker, R.H. (1975) *Communities and Ecosystems,* 2nd ed. Macmillan, London.

Whittaker, R.H. (1977) Evolution of species diversity in land communities, *Evolutionary Biology,* 10, 1–67.

Williams, C.B. (1944) Some applications of the logarithmic series and the index of diversity to ecological problems. *Journal of Ecology,* 32, 1–44.

Williams, W.D. (1988) Limnological imbalances: an antipodean viewpoint. *Freshwater Biology,* 20, 407–420.

Wills, C., Condit, R., Foster, R.B. & Hubbell, S.P. (1997) Strong density- and diversity-related effects help to maintain tree species diversity in a neotropical forest. *Proceedings of the National Academy of Sciences U.S.A.,* 94, 1252–1257.

Winemiller, K.O. (1990) Spatial and temporal variation in tropical fish trophic networks. *Ecological Monographs,* 60, 331–367.

Wit, C.T. de, Tow, P.G. & Ennik, G.C. (1966) Competition between legumes and grasses. *Verslagen van landbouwkundige onderzoekingen,* 112, 1017–1045.

Woiwod, I.P. & Hanski, I. (1992) Patterns of density dependence in moths and aphids. *Journal of Animal Ecology,* 61, 619–629.

Woodwell, G.M. (1970) Effects of pollution on the structure and physiology of ecosystems. *Science,* 168, 429–433.

World Meteorological Organization (1985) Trace gas effects on climate. *Atmospheric Ozone 1985.* Global Ozone Research and Monitoring Project, Report No. 16, Volume III, Chapter 15.

Worthington, E.B. (ed) (1975) *Evolution of I.B.P.* Cambridge University Press, Cambridge.

Wurtsbaugh, W.A. (1992) Good-web modification by an invertebrate predator in the Great Salt Lake (USA). *Oecologia,* 89, 168–175.

Yodzis, P. (1986) Competition, mortality and community structure. In: *Community Ecology* (J. Diamond & T.J. Case, eds), pp. 480–491. Harper & Row, New York.

Zeevalking, H.J. & Fresco, L.F.M. (1977) Rabbit grazing and diversity in a dune area. *Vegetatio,* 35, 193–196.

For the student interested in further readings, the Ecological Society of America sponsored the publication of a collection of 40 original scientific papers that were chosen as classics in the science of ecology. The collection is published under the title Foundations of Ecology, *edited by L. A. Real and J. H. Brown, University of Chicago Press, 1991. Most of the papers represent high points in the history of ecology and give a newcomer to the science a direct insight into the way that ideas and practices have developed.*

Index

Note: Page numbers with an *f* indicate figures; those with a *t* indicate tables.

Productivity
    lattitudinal gradients and, 376
    limitations to, 393–396, 394f, 395f
    primary, 391–402
        fate of, 396–402, 397f–401f
        geographic patterns in, 293f, 391–393, 392t
        resource richness and, 356–360, 356f–362f
    secondary, 396–398, 397f
    temperature and, 394f, 394–395
Proghorn antelope, 131f
Protected areas, 517–520, 518t, 519f
Protozoa
    for beetle control, 277, 278f
    as detritivores, 404f
Prunus laurocerasus, 114f
Pseudomyrmex ferruginea, 250, 251f
Pteropus sp., 508–509
Puccinia graminis, 263f, 263–264, 265
P-values, 13–15
p(t) variable, 329
Pyrodictium occultum, 82–83

Q

Quarantine, 33–35. *See also* Disease.
Quarrying pollution, 485–489, 487f
Quercus sp., 142
    galls on, 256–258, 257f
    Lyme disease and, 327
    as modular organism, 167, 169f
    niche of, 114f
    photosynthetic rate of, 335t
    reproduction of, 198f
Quotas, sustainable yields and, 433f, 433–434, 434f

R

Rabbits. *See also* Hares.
    defenses of, 110
    digestion of, 248f, 249
    embryos of, 171
    myxomatosis and, 166, 265–267, 306
    plant species richness and, 306, 307f
Rabies
    fox, 33–38, 34f–37f
    mathematical models for, 35–38, 36f, 37f
Radiation
    nuclear, 483t, 483–484, 485f, 486f
    solar, 120, 121f
        aquatic environments and, 126
        oceans and, 149–151
Radon, 483
Rain
    acid, 31–33, 468–469, 479–481
    hydrological cycle and, 410–412, 411f
    lake ecology and, 149, 150f
    nutrients in, 408–409
    pH of, 479
    in savanna, 140

solar radiation and, 121f
    temperatures and, 124f
    topography and, 122f
Rain forest(s), 135–139, 137f, 139f, 367. *See also* Forest(s).
    biodiversity of, 135–138
    as biome, 131f, 135–139, 137f–140f
    evolution of, 136
    global warming and, 156–157
    imperiled, 519f
    life-forms of, 133f
    Malaysian, 359–360, 362f
    productivity of, 392t
    soils of, 138–139
Rain shadow, 122
Rana pipiens, 86f
Random sampling, 18
Rangifer tarandus, 143
Rank-abundance diagram, 354–357
Ranunculus sp., 89f, 114f, 316f
Raspberry moth, 103
Ratites, 69, 70f
Rats, 298, 458
Raunkiaer's life-forms, 132–134, 133f
    in rain forests, 136
    in temperate forests, 142
RDZs (resource depletion zones), 98
Recruitment, net, 192, 193f
Red Rock Canyon (Nev.), 130f
Redundancy, ecological, 510f, 510–511
Refection, 249
Regression coefficient, 323
Rehabilitation, 520–521
Reindeer, 143
Reproduction. *See also* Fecundity.
    age and, 175f, 176, 177
    cost of, 193f, 195–196
    environmental conditions for, 82f
    growth vs., 172, 173f, 194
    life cycles and, 171–172, 172f, 173f
    life histories and, 175f
Reptiles. *See also specific types, e.g.,* Lizards.
    endangered, 496t, 504t
    evapotranspiration and, 359f
    species-area relationships for, 370f
Resource(s)
    animal, 102–110, 104f–106f, 108t
    in aquatic environments, 126, 144
    competition in, 204
    conditions vs., 80
    environmental, 80
    harvesting of, 430–436, 431f–436f
    interactions and, 85–86, 88f
    niches and, 112–115, 226
    overexploitation of, 498–500
    patterns in
        large-scale, 120–122, 121f–125g
        small-scale, 122–126
        temporal, 126–128, 127f, 128f, 226
    plant, 94–101, 95f, 98f–101f
    population stability and, 317

productivity of, 356–360, 356f–362f
    of rain forest, 135
    specialization and, 103, 104f
    water as, 442f, 442–444
Resource depletion zones (RDZs), 98
Response curves, 81–82, 82f
Restoration, 520–521
Restoration ecology, 489–490
Rhea (bird), 69f, 69–70, 70f
Rhinoceros, 514, 515
Rhizobium sp., 252–254, 256
    electron micrograph of, 255f
    life cycle of, 254f
    as symbionts, 268, 270f
Rhizomes, 168f
Rhizopertha dominica, 193f
Rhizosphere, 244–245
Rhyssa persuasoria, 300
Ricklefs, R.E., 4
Rigor, statistical, 12–15, 14f
Rinderpest, 193f
Riparian vegetation, 146–148
Rodents
    ants and, 209–210, 210f
    control of, 298
    disease from, 458
    in Israel, 217–218, 219f
    species richness and, 362f
Rodolia cardinalis, 448
Roots
    diseases of, 252
    fungi and, 250–251
    hairs of, 97–98, 98f
    mineral nutrients and, 100, 101f
    parasites of, 240f
    rhizosphere of, 244–245
Rothamsted experiment, 358, 361
Roundworms, 238t, 241f
r species, 196–197, 198f
Rumex sp., 280f
R units, 355, 356f
Rush moths, 85
Rust. *See also* Fungi.
    black stem, 263f
    flax, 265
Rye grass, 281f

S

Sagina procumbens, 99f
Sagittaria sagittifolia, 89f
Salazaria mexicana, 99f
Salinization, 441
Salix sp., 86f, 193f, 276f
Salmon, 175, 509f, 510f
Salmo trutta. See Trout, brown.
Salvinia molesta, 277–278, 278f, 279f
Salvis dorrill, 99f
Sampling
    accuracy with, 18, 19f
    for population count, 169–171